**Multi-Storey Buildings
in Steel**

Advice and co-operation:

Associazione fra i Costruttori in Acciaio Italiani (ACAI), Milan
Beratungsstelle für Stahlverwendung, Düsseldorf
British Constructional Steelwork Association (BCSA), London
Centre Belgo-Luxembourgeois d'Information de l'Acier (CBLIA), Brussels
Centro Italiano Sviluppo Impieghi Acciaio (CISIA), Milan
Constructional Steel Research and Development Organisation (CONSTRADO), London
Deutscher Stahlbau-Verband (DStV), Cologne
Office Technique pour l'Utilisation de l'Acier (OTUA), Paris
Stichting Staalcentrum Nederland, Amsterdam
Stichting Centrum Bouwen in Staal, Rotterdam
Syndicat de la Construction Métallique de France, Puteaux

Multi-Storey Buildings in Steel

F. Hart
W. Henn
H. Sontag

Editor of the English edition
G. Bernard Godfrey, MICE, FIStructE

CROSBY LOCKWOOD STAPLES
GRANADA PUBLISHING
London Toronto Sydney New York

Authors:
Dipl.-Ing. Franz Hart, Senior Professor at the Technical University of Munich
Dr.-Ing. Walter Henn, Senior Professor at the Technical University of Brunswick
Dr.-Ing. Hansjürgen Sontag, Berlin

Co-authors:
Dipl.-Ing. Joachim Borstel, Dipl.-Ing. Peter Cziffer, Dipl.-Ing. Ingo Grün,
Dipl.-Ing. Gunter Henn, Dipl.-Ing. Hans-Georg Köhler, Dipl.-Ing. Anneliese Zander

Advisers:
René Boll, Ingénieur de l'École Spéciale des Travaux Publics, Paris
Werner Bongard, Dr.-Ing., Cologne
Marcel Bourguignon, Ingénieur, Brussels
Michel Folliasson, Architecte DPLG, Architecte en Chef des Bâtiments Civils et Palais Nationaux, Paris
Louis Fruitet, Ingénieur ECP, Professeur d'Unité Pédagogique d'Architecture, Paris
Hans Gladischefski, Dipl.-Ing., Düsseldorf
Christian Trognon, Paris

Editorial control: Institut für Internationale Architektur-Dokumentation, Munich
Konrad Gatz: direction
Hans-Jürgen and Eva-Regina Meier-Menzel: diagrams
Fritz Hierl: illustrations
Dr. Margret Wanetschek-Gatz: assistance
Klaus Halmburger: production

This work stems from the initiative of the Deutscher Stahlbau-Verband with the financial support of the European Coal and Steel Community (EGKS) under Article 55 of the EGKS Agreement

Publishers of the German edition: Deutscher Stahlbau-Verband (DStV)
Publishers of the French edition: Syndicat de la Construction Métallique de France
Publishers of the Italian edition: Associazione fra i Costruttori in Acciaio Italiani (ACAI)
Publishers of the Dutch edition: Stichting Centrum Bouwen in Staal
Publishers of the English edition: Crosby Lockwood Staples

Granada Publishing Limited
First published in the Federal Republic of Germany 1974 by Deutscher Stahlbau-Verband
English language edition first published in Great Britain 1978 by Crosby Lockwood Staples, Frogmore, St Albans, Hertfordshire AL2 2NF and 3 Upper James Street, London W1R 4BP

English translation copyright © 1978 by Crosby Lockwood Staples

Translated from the German by C. V. Amerongen, MSc, MICE

All rights reserved. No part of this publication may be reproduced, stored in a retrieval system, or transmitted, in any form or by any means, electronic, mechanical, photocopying, recording or otherwise, without the prior permission of the publishers.

ISBN 0 258 96974 1

Printed in Great Britain by
William Clowes & Sons, Limited
London, Beccles and Colchester

Contents

Part One
Structural Steel and Architecture F. Hart

One hundred years of steel-framed structures – developments
and achievements 9
    Forerunners of multi-storey steel-framed buildings, 1790–1872
    The Chicago School, 1880–1910
    Evolution of framed construction in Europe: France, Belgium,
    West Switzerland, 1890–1930
    Efforts to develop a new architecture based on framed construction in Germany, 1910–1930
    Skyscraper architecture in the USA, 1890–1940
    'Gridiron' architecture, 1940–1955
    International architecture in steel. The curtain wall, 1945–1960
    Reactions against the strict glass-and-metal architecture
    The exposed steel frame
    Steel: the constructional medium for wide-span and complicated structures

Steel-framed structures today – applications and possibilities 41
    Steel in residential and school buildings
    Steel in other types of buildings
    Steelwork and internal services
    Exposed or concealed steelwork
    Steel and the traditional building materials – steel buildings
    in historical surroundings
    Steel and glass
    Steel and concrete
    Steel or reinforced concrete

Part Two
Examples of Multi-Storey Steel-Framed Buildings
W. Henn

Residential buildings 59
Communal buildings 73
Hotels 78
Hospitals 83
Schools 88
Universities and research centres 94
Exhibition buildings and passenger handling facilities 105
Shopping centres 114
Official buildings 120
Administrative buildings 132

Part Three
Principles of Design and Construction H. Sontag

Fundamentals of planning 165
    Characteristics of a steel structure
    Economy of a steel structure
    Dimensional co-ordination

Types of multi-storey steel-framed buildings 179
    Examples: Day nurseries · Schools · Universities · Office blocks ·
    Department stores · Residential buildings · Hospitals · Multi-storey car parks

Types of supporting structure 187
    Structural frameworks for multi-storey steel-framed buildings
    Arrangement of horizontal components
    Arrangement of vertical components
    Bracing for steel-framed buildings
    Joints in buildings

Structural steel frames 225
    Columns
    Beams and girders
    Stiffening elements

Floors and staircases 271
    Floors
    Steel decks
    Concrete floors
    Fire protection of floors
    Suspended ceilings
    Horizontal services
    Floor finishes
    Staircases

Roofs, external walls, internal walls 299

Fire protection 327
    Effects of fire
    Preventive measures against fire
    Stability of structures under fire conditions
    Fire behaviour and fire protection of steel components
    Twelve rules for fire safety in steel buildings

Execution of the scheme 337
    Preliminaries
    Fabrication
    Erection

Steel 353
    Steel
    Corrosion protection

Detailed contents lists are given at the beginning of Part Two, 'Examples of Multi-Storey Steel-Framed Buildings', and at the beginning of the individual sections of Part Three, 'Principles of Design and Construction'.

Preface to English Edition

In present-day building a bewildering array of planning, constructional and aesthetic opportunities is available. Notwithstanding, in most cases the following considerations are of great significance.
1. The quest for economy; this opens up the possibility for the widespread industrialisation of building methods and, hence, rationalisation of the design processes.
2. The quest for simple assembly methods; these demand the application and provision of constructional systems which will facilitate the erection, alteration and enlargement, or even the dismantling, of a building. These considerations explain the progressive trend in all kinds of multi-storey buildings towards prefabrication and towards structural steelwork.

Like every other constructional material, steel has special individual properties which must be understood if it is to be used to best advantage. These determine the design of the building, as well as its details, and affect not only the skeleton but also the carcase of the building. Many well-tried, functional and inexpensive solutions are available nowadays for multi-storey steel-framed buildings but, as the use of structural steelwork has not always kept pace with the possibilities available, it must be concluded that the advantages of structural steelwork have not been brought to the notice, in particular, of building owners and their architects.

It is therefore the object of this publication to fill these gaps in knowledge. Although originally prepared by German authors and published in Germany, with the financial support of the European Coal and Steel Community, it has been translated into a number of other European languages.

This present translation was prepared at the suggestion of the British Constructional Steelwork Association and with the support of the Constructional Steel Research and Development Organisation (CONSTRADO). Despite the fact that the contents of the publication relate to European and North American practice, the publishers expect that the material will appeal wherever the English language is understood.

This publication is in three parts. The first part gives a brief historical glimpse justifying the rather late appearance of structural steelwork in the development of multi-storey buildings, then describes the evolution some fifty years ago of a true architecture in steel and, finally, the many possible concepts of design offered to the architectural profession today.

The second part gives a representative international cross-section of steel-framed buildings which were erected during or since the 1960s. With the aid of 62 carefully selected structures, design concepts and structural forms are presented for all the most important groups of buildings, the adaptability of structural steelwork to all kinds of constructional problems being very clearly demonstrated.

The third part, which specifically constitutes a construction atlas, gives a systematic description of the structural possibilities which are available within the statutory regulations. After a general description of the decision-making criteria for structural steelwork and the present position of modular co-ordination and its operation, the various kinds of structural systems are treated from the point of view of planning considerations and architects' requirements. The development of details within the structural framework and the interrelationship of the structural floors and staircases are demonstrated in theory and in practice. Then the correlation of the loadbearing structure with the claddings, partitions, roofs and technical services is carefully explained. In view of its special importance, an individual chapter is devoted to fire protection. Then there are sections dealing with the preparation, fabrication and erection of steelwork and, finally, a chapter on steel as a material.

In considering future editions of the publication it is the intention of the original German authors and publishers, in collaboration with other interested national organisations, to review the contents and, where necessary, to bring the material up-to-date. This applies particularly to the 'Examples of Multi-Storey Steel-Framed Buildings' in the second part and, to a lesser extent, to the 'Principles of Design and Construction' in the last part.

January 1978   G B Godfrey

Structural Steel and Architecture

FRANZ HART

One Hundred Years of Steel-Framed Structures Developments and Achievements

Forerunners of multi-storey steel-framed buildings, 1790–1872

The greater part of this book was written in 1972, exactly one hundred years after the construction of Saulnier's factory building at Noisiel-sur-Marne, which can be regarded as the first steel-framed building in continental Europe. Viewed in the context of the overall development of iron technology and structural steelwork, however, 1872 is quite a late date. Let us briefly review some important earlier events. In 1720, at Coalbrookdale, Abraham Darby began to smelt iron successfully in a blast furnace using coke, instead of charcoal, and thus established the conditions for the mass-production of pig iron. The technical improvement of the puddling process made it possible in 1784 to use coke also for the conversion of pig iron into wrought iron, thus reducing the lead that cast iron had gained. With Henry Bessemer's invention of the converter in 1855 and the introduction of the Siemens-Martin or open-hearth process in 1864, the era of mild steel began.

In addition to the tremendous expansion in the production of iron there was progress in the further processing and shaping of the metal. As early as the eighteenth century, iron plates were being rolled in Britain and, by 1830, rails for railway lines were being manufactured by rolling, while 1854 saw the production, in France, of the first I-sections of wrought iron. The I-beam, the basic element of modern structural steelwork and also the first strictly standardised structural component, is the direct descendent of the old iron rail, the product which can be regarded as the hall-mark of the emergent industrial era and on which, as it were, all the driving forces of that age are focussed: commerce and transport, steam power and mechanical engineering, heavy industry, metallurgy and applied science.

The most significant iron structure is the bridge over the Severn at Coalbrookdale, completed in 1779. With its span of just over 100 ft, it is the structure in which iron first came into its own as a constructional material for long spans. The cast iron arch bridge, however, was soon outclassed by bridges designed to exploit the favourable properties of wrought iron in tension and flexure: suspension bridges, girder bridges and truss bridges.

Among the early British and American chain suspension bridges, the most outstanding is Thomas Telford's bridge over the Menai Straits in North Wales, opened to traffic in 1826, which has a main span of 173 m. Subsequently, when the chain had given way to the cable, the record span for such structures had reached 300 m by 1850 and almost 500 m by 1870 with the construction of the Brooklyn Bridge in New York. The daring Britannia Tubular Bridge over the Menai Straits, having twin wrought iron box girders, 9 m deep, with a maximum span of 140 m, was completed by Robert Stephenson in 1850. The truss bridge was particularly suited to the requirements of railway engineering and remained the dominant type from the middle to the end of the century. Representative examples of such bridges are the Cathedral Bridge in Cologne, with spans of 100 m, and the Royal Albert Bridge at Saltash, designed by I. K. Brunel, both completed in 1859. In fact, all the most significant structural systems which have shaped the evolution of steel bridges had already emerged by the middle of the nineteenth century. In 1851, the construction of large iron-framed single-storey buildings reached a spectacular level of achievement in the Crystal Palace, London. About the same time, with the construction of Kings Cross Station in London and the Gare de l'Est in Paris, the series of large railway stations characterised by the arched iron-ribbed roof began to evolve. In 1867, St Pancras Station, London, established a European record with its span of 78 m.

How can it be explained, in a period marked by such major achievements in bridge engineering and the construction of enormous single-storey buildings, that the application of iron to multi-storey buildings failed to proceed beyond rudimentary beginnings and that framed construction did not at that time manage to establish itself wholly and permanently? The reasons may be summarised under four headings and these are in principle the same inhibitions which have until now impeded, or still impede, the general acceptance and efficient application of structural steelwork by architects.

1. In comparison with other applications of manual skill and industrial technology, building construction has always been conservative. The architect's conservatism is in itself a critical factor. It is bound up with the fact that people expect their dwellings, their religious structures and their public buildings to provide more than mere protection against the weather and a functionally adequate interior. Man strives to perpetuate himself in his buildings, thus proclaiming his aspiration to culture and his sense of history.

2. 'Architecture' in the historical sense of the word, the grandiose permanence with which architects through the ages have transformed the Greek temple or the mediaeval cathedral from a simple timber structure or a humble assembly hall into works of art expressing tremendous power and highly developed craftsmanship, has ceased to be possible since the advent of technology. The building owners and architects of the nineteenth century developed a sort of artificial tradition with their historical eclecticism. Nowadays, we no longer look with disdain, as we used to do even twenty years ago, upon the architecture of the last century as on a sort of historical fancy dress pageant. But the fact is that the pursuit of historical styles has long obscured the truth (or, at any rate, the practical consequences of this truth) that with the introduction of iron as a constructional material the old concepts of architecture — support and load, anatomy and space — and its structural features — post and lintel, arch and abutment — were superseded or challenged.

3. The architects' historical prejudice increasingly widened the gulf between architect and structural engineer. The division that occurred in the traditional profession of architecture, resulting in the rise of the engineer as a professional man in his own right, and the emergence of modern structural theory were developments that began concurrently with the industrial revolution and constituted an important distinctive feature of the new age that had dawned in building construction. The rapidly progressing refinement of the methods of structural analysis

and strength of materials soon carried them beyond the range of most architects and so it is hardly surprising that they returned to the study of historical buildings as the supposed prerequisite to 'monumental architecture' and left 'industrial building construction' to the engineers. Big bridges and wide-span buildings, examples of which were mentioned above, had already become virtually the province of the civil and the structural engineer. Although the engineers were also influenced by the concepts of their time, as is exemplified by their persistence in using the arched girder as a feature of design, they possessed the necessary breadth of vision when the greatest demands were made on their technical judgement, namely, the development of new constructional forms suited to the use of iron.

4. In multi-storey buildings, which had thus become the exclusive field of the architects, the external pressures driving the designers of bridges and single-storey buildings to increasingly daring achievements were largely lacking. Neither the number of storeys nor the floor spans and loadings went beyond the familiar concepts traditionally established in the construction of major public and private buildings. Even when, in certain more ambitious buildings, the incombustible floor slab comprising iron joists — an innovation of the early days of industrial architecture — was installed in lieu of timber joists or the vault, this did not involve any significant change in the general structural pattern of multi-storey buildings. In their external appearance the traditional architectural features of elevational treatment remained unchanged. That is why, in the great cities such as Paris, Milan and Rome, the majority of buildings dating from the latter part of the nineteenth century harmonise so well with the older and more priceless architectural creations.

The relationship indicated here perhaps emerges even more clearly when one considers the celebrated exceptional cases where architects of those days tried to design wide-span iron roof structures. An example is provided by H. Labrouste's meticulously detailed two-bay cast iron arched structure for the Ste Geneviève Library in Paris, yet this ironwork is kept concealed, so that it does not appear on the exterior of the building. Actually, the enclosing wall with its unusually shaped features is even more impressive; evidently the architect felt himself to be on safer ground with this. Even in the big railway stations built towards the close of the nineteenth century, for example at Frankfurt-am-Main, this contrast between the airy iron arched roof spanning the platforms and the massive monumental entrance building rising in front of it is striking.

The first steps towards the development of the structural steel frame for multi-storey buildings had been taken at quite an early period. In order to gain more space and increase the loadbearing capacity of the floors for the installation of machinery in the British cotton mills, the timber posts were replaced by cast iron columns and subsequently floor beams of the same material were substituted for timber beams. The best known of these early multi-storey industrial buildings was erected for the firm of Philip and Lee at Salford in Lancashire in 1801. The designers were Boulton and Watt of steam engine fame. Similar textile mills with internal cast iron frames had in fact already been built in the 1780s, but the one at Salford surpasses them in its daring dimensions and its mature structural design, which made it a prototype for the further development of this form of construction. The building is 42 m in length, 14 m in width and exceptional for its time in having no fewer than seven storeys. Cast iron girders spaced at 2.7 m centres span in the transverse direction, supported at the third-points by a double row of cast iron columns. Here for the first time the floor girders are a kind of I-section with a flat brick infilling.

It was not until 1845, when William Fairbairn used wrought iron I-beams in lieu of cast iron in a refinery building, that this type of structure underwent a significant change. In the same year A. Zorès brought the rolled I-section into the construction of residential buildings. From then onwards the wrought iron beam and, later, the mild steel beam came to be more widely used in other types of building, in addition to industrial structures. A number of factors had conspired to help Zorès in achieving this innovation: a strike of carpenters in the building trade, the rise in the cost of timber beams, more stringent fire regulations and the need for floors of greater span.

As developed, the steel beam floor did indeed offer the advantage of better fire resistance and substantially higher loadbearing capacity than the timber beam floor. Also, it was superior to the vault, not only because of the reduced amount of labour involved in constructing floors with steel beams, but especially because of the substantial reduction in storey height and wall thickness. The fact that with the arrival of the iron beam the vault, as a structural system and as the most important design feature of monumental architecture, was rendered obsolete justifies referring to this development as the dawn of a new epoch. The fact that the jack arch held its own for a long time in combination with iron and steel beams in floor construction is a characteristic demonstration of the persistence of the historical approach whereby nineteenth-century architects sought to assimilate or to neutralise technical innovations.

The division between historically orientated romanticism and the realities of technical progress is even more strikingly shown in the structural component which could be said to epitomise the architecture of the Victorian era: the cast iron column, that unusually slender loadbearing member embellished with the historical decorative features of capital, base and fluting. Early cast iron columns of highly original and expressive design were installed in the kitchen of the Royal Pavilion that John Nash built at Brighton in 1821. These are unusually tall and slender tubular members whose capitals are shaped as clusters of palm leaves made of beaten iron plate. Later, the cast iron column was to degenerate into a mass-produced item which could be ordered from a catalogue in any desired height and style: Doric, Tuscan, Corinthian, Gothic or Saracen.

The 'beam construction method' developed by Boulton and Watt dominated industrial architecture throughout the nineteenth century. It cannot, however, truly claim to be called a structural frame in the sense that the iron or steel components formed a skeleton serving as the actual loadbearing system. In these early multi-storey industrial buildings the external walls had to carry a very substantial proportion of the floor loads and also to perform the important bracing or stiffening function, that is, resisting and transmitting the wind loads. This is clearly reflected in the American and European building codes or by-laws of the day. These required the external walls of such buildings to be, on average, half a brick thicker than those of residential and office buildings which were provided with internal loadbearing and stiffening walls. This requirement, which was retained, for example, in the German by-laws up to the Second World War, is astonishingly optimistic. For the permissible loads and stresses on masonry quoted in modern regulations the walls of such buildings would have to be made considerably thicker.

Liverpool, Albert Dock 1845

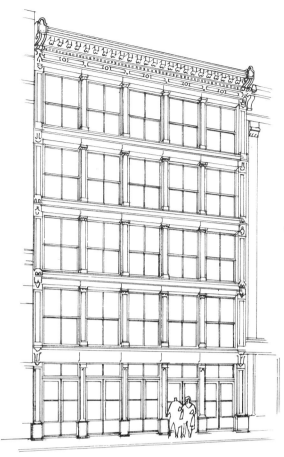

River front, St Louis, Gantt Building 1877

The next step on the road to the steel skeleton, the transformation of the external wall into a loadbearing metal frame, was long in coming, not least because it was inevitably to have a profound effect on the architectural treatment of the façade. Not surprisingly, the first attempts in this direction were again concentrated on that favourite product of the cast iron era, the column. They traced their origins from the arcade, that venerable architectural feature which had played so prominent a part in house construction and in urban architecture generally since ancient times. The wide-spanning arcades in the stark brick fronts of such British dockside warehouses as St Katharine's Dock, London (1828), and Albert Dock, Liverpool (1845), with their thick classical tapered cast iron column shafts of splendid rust colour, surmounted by slim Doric capitals, are indeed most impressive.

Another precursor of the framed external wall which came close to modern construction comprised wrought iron frameworks which were used as far back as the beginning of the nineteenth century to span the increasingly wide shop windows which were coming into vogue more particularly in Paris. These components were a kind of flat arched truss or parallel chord lattice girder enclosed within the masonry. In the United States in the period between 1850 and 1880 a large number of warehouses, department stores and office buildings were constructed whose façades were entirely supported by metal loadbearing members. It would appear that the impetus to this development was given by James Bogardus, a versatile inventor, designer and manufacturer. Among the most important of his creations is the building for the publishing firm of Harper and Brothers (1854). The front of this five-storey building is composed entirely of prefabricated cast iron components; internally it has an iron framework, as used in the British industrial buildings of the period, but applied here in the form of rolled wrought iron beams for the first time in the USA. The architecture of the façade apparently derived its inspiration from Venetian buildings of the High Renaissance and is characteristic of the trend towards heavy, opulent forms, which were subsequently to give way to the eclecticism of the second half of the century.

Of greater appeal to us at the present time are the timeless cast iron façades of the buildings on the river front in St Louis, USA. Here, the traditional architectural features that are used to enliven the façades of historic buildings have been reduced to the barest minimum. The austere cornice mouldings, the elegant plain columns, capitals and bases, really serve only to emphasise the elementary contrast of strong horizontal and graceful vertical features, of prismatic and cylindrical loadbearing members. In Europe, too, some notable iron façades were built around this time, for example, Gardner's Iron Building, Glasgow (1856). Especially attractive in Oriel Chambers, Liverpool (1864), are the slender sandstone pillars alternating with iron-framed bay windows. Windows of this kind were later to play a major part in the architecture of the Chicago School.

The big fires which occurred in Chicago and especially also in Boston in the 1870s shattered the illusion of the indestructibility of iron structures. The sobering discovery that an incombustible material is not necessarily fire-resistant resulted, in Europe even earlier than in America, in more stringent fire protection regulations and thus effectively killed this first emergence of a purely metal-based architecture.

The first multi-storey building in continental Europe entirely designed as a steel-framed structure is the Menier chocolate factory at Noisiel-sur-Marne, near Paris (1871–1872), by Jules Saulnier. Here again, as in the earlier British cotton mills, it was the operational demands of industrialised building that compelled the engineer to take full advantage of the structural possibilities and the strength of iron as a constructional material.

The factory is built out over the Marne, being supported on four massive piers shaped as cutwaters. The external framework stands on a peripheral sill beam, designed as a deep box section which transmits the entire load of the building and the wind forces to the eight points of support. There are no internal stiffening walls, nor are the end walls, because of the cantilevered construction of the building, able to transmit horizontal forces. Therefore, a rational solution for bracing the structure had to be devised. The exposed framework is stiffened in the longitudinal direction of the building by a network of diagonal members inclined at 60°, forming a system of diamond-shaped meshes which reduce in size in a narrow strip at each corner to form a kind of wind-bracing. To stiffen the building transversely, the floor beams are rigidly connected by means of lattice brackets to the main stanchions of the external frame.

These stanchions, located in pairs over the supporting piers, stand out somewhat because of their heavier section, but otherwise the façade with its thin infilling of brick is entirely smooth. The window dimensions are exactly defined by the diamond network of diagonal members. A subtle differentiation in the height of the windows has been obtained by slightly shifting the position of the joints in the first and the third storeys.

The Menier factory anticipates modern structural steelwork in several respects: the cantilevered corner and especially the diagonal network of bracing, which nowadays still plays a major part in the wind-bracing of skyscrapers. On the other hand, its framework is unmistakably inspired by mediaeval timber-framed buildings, a splendid confirmation of the doctrines of Viollet le Duc, which will be mentioned later. Saulnier's system had no direct imitators of any consequence. The principle of the metal frame was able to make a more lasting impact only when high-rise commercial building became an urgent constructional problem demanding new structural solutions. This development started in Chicago about 1880.

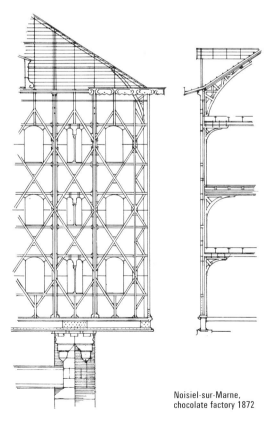

Noisiel-sur-Marne, chocolate factory 1872

The Chicago School, 1880–1910

The modest pioneering settlement at the mouth of the Chicago River where it flows into Lake Michigan attained township status in 1830. By 1871, its population had risen to 30 000. In those days Chicago consisted almost entirely of timber houses of the so-called balloon-frame construction, which is still used in the USA. In that year, however, the town was almost entirely destroyed by fire. Initially, the work of rebuilding the town proceeded in fits and starts. Nevertheless, about the year 1880 there was an unparalleled burst of building activity.

The opening-up of the Middle West, the expansion of the railway system and waterways and the exploitation of the country's mineral resources, had made Chicago the transfer point and production centre of a vast hinterland, the world's largest grain market, the focal point of the timber trade and the food industry, of machinery manufacture and machine tool production. The building industry was scarcely able to cope with the heavy demand for commercial and office buildings, warehouses and shops. The cost of land rose sharply, the density of building within the pre-established street network increased and soon multi-storey structures evolved into skyscrapers. Only with the aid of structural steelwork was it possible to achieve maximum utilisation of the sites and the floor space inside the buildings and also to increase the speed of erection. As early as 1895 the new constructional method had become firmly established in all the major American cities, but at that period Chicago had more high-rise steel-framed structures than all the other cities put together.

There were also often special requirements and circumstances which made structural steelwork possible, influenced it and enforced it. First, there was the topographical situation which, together with the traffic problems, long prevented any extension of the Loop, Chicago's central commercial area. From the outset, importance was attached to open plan buildings with scope for subsequent modification of the internal layout. Several of the early framed buildings alternated in function between warehouses and office premises. Even at that time, subsequent upward extension by the erection of additional storeys was envisaged at the planning stage and often carried out.

The high-rise commercial building would not have been a practical proposition, however, if the mechanical engineers had not managed to produce the necessary service installations at the right time. The first and most important prerequisite was the passenger lift. The safety lift had been invented by Elisha Graves Otis, who demonstrated it at the Crystal Palace Exhibition in New York in 1853. In 1857, he installed the first lift for public use in a department store on Broadway. From then onwards, New York gained a certain lead in the construction of tall buildings and indeed claimed the distinction of having produced the first skyscrapers, nine- or ten-storey commercial buildings of masonry construction, which were erected in the mid-1870s. While the lift was evolving from a steam-powered via hydraulic operation to an electrically operated piece of machinery, the other technical services were also being developed: telephones, pneumatic despatch system, central heating and air conditioning. Apart from the technical prerequisites, one must not lose sight of the imagination and courage which created the first modern steel-framed buildings in Chicago. The architects were inspired by the still active pioneering spirit, which gave their buildings their distinctive power and freshness and also an unmistakable family likeness.

The founder and leader of the Chicago School was William le Baron Jenney. In 1868, he opened his architectural design office in Chicago and with the construction of the Leiter Building I in 1879 his first gamble proved successful. The determination with which this structure was realised within the confines of its formal expression has hardly ever been surpassed since that time. In its sheer strength and power the building is reminiscent of ancient Roman architecture. Structurally, this five-storey building, to which two storeys were added later, was still something of a hybrid: timber beams on wrought iron girders, carried by cast iron columns in the interior and by masonry piers along the perimeter. The daring slenderness of these piers and the great width of the window openings, necessitating a wrought iron girder to act as lintel and edge beam, were novel features. That the masonry is supported and stiffened by the internal iron framework is clearly shown in the prominently featured bearing plates for anchoring the main beams to the top of the piers. Even more advanced is the layout adopted for the Leiter Building I: a clearly conceived planning grid, an open layout and a reduction in the dimensions of the structural components. In these respects it remains unsurpassed, for example, by any reinforced concrete building. This becomes even more strikingly evident when we compare it with the ground plan of the Monadnock Building, the last high-rise building of conventional masonry construction.

In Jenney's next major building, constructed for the Home Insurance Company (1883–1885), each external pier encases a loadbearing steel column. In its elevational treatment, this building offers nothing like the unity and structural clarity of the Leiter Building I: the prominently featured ground floor, the enclosing round arches in the top storey, the balustrade surmounting the

Chicago, Leiter Building I 1879

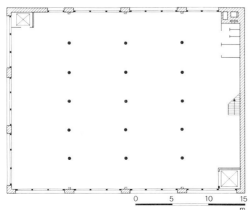

Chicago, Reliance Building 1894

cornice, these are eclectic architectural motifs which are at variance with the character of steel-framed structures.

Jenney's two main achievements, the Leiter Building II (1889) and the Fair Building (1891), strike a more modern note. Here the historic reminders of the architectural features of bygone days have been reduced to a minimum: only very delicately moulded bases and capitals raise the piers to the status of pilasters; indeed, they could be quite readily omitted, and one cannot help feeling that Jenney himself would have preferred to have omitted them. These two buildings are complete steel-framed structures. It is evident that the masonry pier has no load-bearing function to perform, not even that of supporting its own weight, merely forming a casing around the structural steel column within.

The natural transition from the external wall to the loadbearing structural steel frame had first been achieved by Holabird and Roche in the 14-storey Tacoma Building, erected in 1884 and, unfortunately, subsequently demolished. In this building the bay windows extending all the way up the façade were particularly distinctive, an architectural feature which was to be continued in Chicago until beyond the end of the century. In their most important work, the Marquette Building (1894), Holabird and Roche achieved something very akin to the clear-cut horizontal and vertical arrangement exemplified in the Fair Building and Leiter Building II; indeed, they even surpass them in the reduction of traditional ornamentation.

However, one should avoid rating the work of the architects of the Chicago School the higher the more it conforms to present-day ideas about framed construction. They were not out to develop a new architecture for its own sake: their assignment was to erect tall buildings of stable and fire-resisting construction. To what extent they were able to utilise or adapt the architectural repertoire of their own time depended largely on the client's requirements. Indeed, those architects would have had no time to indulge in much theoretical speculation. It is particularly the spirit of ebullient unconcern with which they took advantage of favourable circumstances and overcame difficulties that characterises their work and links them to one another.

This is very clearly revealed when one examines more closely the two structures in which the spirit of the Chicago School appears to display itself most powerfully: the Reliance Building and the Monadnock Building. The former does indeed deserve the prize for streamlined slenderness of its façade elements and the extremely high proportion of glazed surfaces. Here, framed construction is immediately recognisable. Its bay windows are twice as wide but much flatter than usual. They do not create the impression of mere embellishment, but are instead an integral part of the façade structure, so that we have in effect a folded surface. The actual loadbearing system is largely concealed: each bay window element comprises three external columns of the steel frame, the middle one being hidden behind the protruding window front, whereas the two outer ones are half covered by the mullions. The structural columns are exposed only at the corners of the building and in the bottom storeys. All these refinements would not have been at all effective, however, if an additional ten storeys had not been added to the Reliance Building beyond the five storeys originally intended. Thus, the progressive character of this architecture, the open form, the succession of completely identical units (the principle which had, for large single-storey structures, already been given form and substance in Paxton's Crystal Palace), was highlighted more by a fortunate coincidence than by deliberate design.

The Monadnock Building, which is the most original of Chicago's multi-storey buildings and the one which has since gained the most credit for its monumental value, is not a steel-framed structure at all. In its time, it was in fact the world's tallest structure having loadbearing brick walls. The conservative building owner had brusquely turned down the first designs submitted by the architects Burnham and Root (a steel-framed structure with terra-cotta cladding) and insisted on an all-brick structure with no adornment. Root, at first dismayed by his client's reaction, gradually worked up a genuine interest in this brick creation and devoted special effort to it. The bold concave batter with which the external wall rises from the bottom storey, which protrudes to form a kind of plinth, the gently curved overhang in lieu of a cornice at the top, and especially the rounded corners flaring out at the upper extremity, are features that give this building its incredible dynamism and enhance the impression of tremendous strength suggested by the deeply recessed window reveals.

By means of the bay window, arranged at every third grid line on the plan, Root did in fact manage to introduce some steel framework into the structure, bring relief to its massiveness and also enhance the effect. More particularly, he harmonised his building with the adjacent contemporary buildings and achieved this so effectively that the basic structural difference is not perceived at first sight.

The buildings that have survived from those days do not in themselves tell us very much about the development of structural steelwork. Among the technical publications of the period, only *Architectural Engineering* by J. K. Freitag (New York, 1901) gives detailed information, its aims being approximately the same as those of this present book. That the engineer, as distinct from the architect, has played an important part in American high-rise building construction is evident even from the title of that book, which could be amplified as indicating 'the application of engineering methods to the structural design and construction of buildings'.

With reference to 'Chicago construction', Freitag distinguishes two types embodying two stages of development. First, there is 'skeleton construction', established by Jenney and given mature application in the Tacoma Building: a completely interconnected structural framework which receives all floor, roof and wall loads and transmits them to the footings of the columns, but which is still dependent on the stiffening effect of the masonry walls for the transmission of wind loads. This 'skeleton' was, however, soon superseded by the second type of structure: the 'cage'. Here the structural frame is a self-contained rigid system in which the wind-bracing has become an integral part of the supporting framework of the building. Since, in this case, the frame is virtually independent of the enclosing walls, it is possible to form large window areas, use interchangeable partitioning for internal layout and reduce the wall thicknesses and loads. More particularly, there is a considerable speeding-up of construction: the erection of the infilling or cladding for the external walls can now be carried out in a number of storeys simultaneously.

Chicago, Monadnock Building 1891

Typical floor plans, Monadnock Building

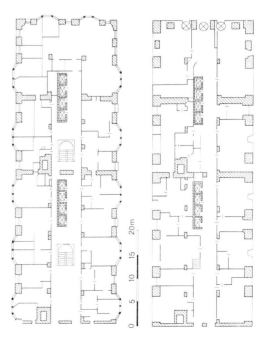

The present-day structural engineer may be surprised to learn that fire protection figures very prominently in the above-mentioned book. The Chicago engineers and architects were disposed to regard 'fireproofing' not merely as a necessary evil, but as something deserving particular attention. The consternation caused by the disastrous fire of 1871 must have made a lasting impression. Freitag gives statistics of fire damage; he reports experience gained with fires which attacked steel-framed buildings in the 1890s and describes in detail the measures for providing fire protection around all load-bearing components.

The foundation is the third problem for which steel-framed construction demanded new solutions or, to be more precise, modifications of the known types: strip footings, piles, caissons. That Chicago played a leading part in developing the 'floating foundation' was due to the soil conditions encountered there: plastic clay of great depth, but offering the advantages of uniform settlement for buildings with correctly designed foundations. The type of foundation usually provided under conventional masonry buildings, wide and not very deep masonry strip foundations, could not be adopted for the individual footings under steel-framed multi-storey buildings. The hugh massive pyramids of masonry which would have been required for such foundations would either have filled up the urgently needed space in the basement or, alternatively, if they had been installed at greater depth, would have involved additional expense and increased risk, as the settlement in the plastic clay would have been greater and more irregular the deeper the foundation penetrated into it. For this reason designers adopted a concrete foundation slab on which, instead of stepped masonry footings, railway rails were laid in alternate directions in several layers which were successively embedded in concrete. In this way the depth of the foundation was halved. Later, rolled steel I-beams were used instead of rails. With the increasing number of storeys a situation was soon reached where groups of columns were installed on a common grillage or where sometimes the entire base had to be designed as one large grillage foundation.

In the early days of steel-framed construction the piled foundation made relatively little headway. It was, however, the preferred form of construction in New York and Boston with their rocky subsoil. With timber piles, the lowering of the level of the ground-water was a source of anxiety, there being several instances of piled foundations coming to grief as a result. Park Row Building in New York, which with its 36 storeys was the world's tallest building around 1900, is still founded on wooden piles. The advanced form of the pneumatic caisson, constructed of timber or steel, was employed to cope with particularly difficult soil conditions and ensure structural safety. These ambitious techniques for foundation construction, which had already been tested in railway bridge building, were further developed to meet the demands of steel-framed high-rise buildings in New York.

Reverting to the technology of structural steelwork, it should be noted that in 1885 the rolled wrought iron beam was superseded by the rolled mild steel beam, which was first produced by the Carnegie Steel Company in the USA. After this innovation, the cast iron column soon receded into obsolescence, as did also the complex shapes for columns composed of curved or chamfered wrought iron sections, these being superseded by standardised rolled steel sections or box sections. In addition, at an early period, the rivet had replaced the bolt for structural connections. Thus, about the turn of the century, all the essential elements of construction had been evolved which were to carry structural steelwork through the next forty years of its evolution.

Perhaps of greater interest to the architect than the above-mentioned technical achievements are the problems of façade detailing presented by steel-framed construction. For one thing, it became necessary to build windows of unusually large width. At first, the architects contented themselves with a simple arrangement

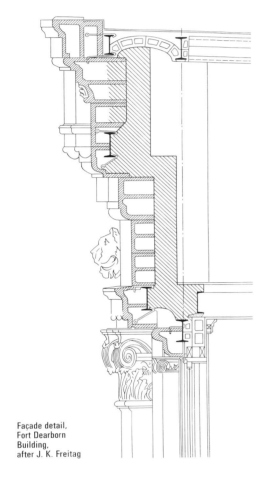

Façade detail, Fort Dearborn Building, after J. K. Freitag

Wind-bracing, the other requirement for an efficient steel-framed structure, was already provided in the 1890s in the three ways which are still in current use:

1. Diagonal bracing, usually in the form of intersecting round bars.
2. Portal frames, which were used where large openings in walls were required and when diagonal bracing could not be accommodated.
3. Lattice girders or trusses of the greatest possible construction depth, rigidly connected to the loadbearing columns.

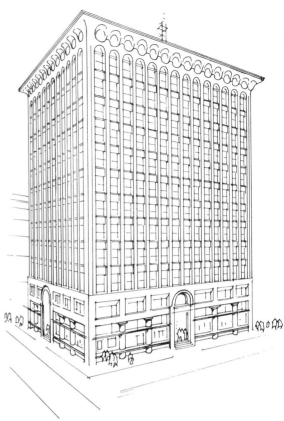

Buffalo, Guaranty Building 1895

comprising three or four vertically sliding windows per bay. From this was evolved the typical 'Chicago window': undivided fixed glazing in the middle, flanked by two narrower horizontally divided side lights. Mullions, were already being utilised as a means of obtaining a differentiated internal layout with movable partitions.

A particularly important and interesting detail is the spandrel, that is, the area between the windows of two consecutive storeys. With the transition to the cage-type structural frame the window lintel, the floor edge girder and the cill of the window above had to be designed and detailed as a single entity. Some of the spandrel sections illustrated in Freitag's book, with their mixture of steel and historical façade details, now strike one as rather odd. On closer inspection, however, it appears that the structural and physical problems — temperature movement of the steel columns, relief of the window construction, compensation for settlement of the cladding and the masonry backing — have been carefully thought out.

Having proved highly successful as a lightweight fireproof infilling for floors and as an internal lining, terra-cotta became very popular as an external cladding material. The ancient technique of facing a building with ceramic slabs offered the architects numerous possibilities for enrichment and colour. The delicate relief treatment that burnt clay demands can provide a finely graded dimensional scale and convey an impression of lightness which is appropriate to framed construction. Terra-cotta ornamental features confined to cornices, window surrounds and cills harmonise very well with brick facing on a high-rise building such as, for instance, the Marquette Building.

Finally, for prestige buildings a whole range of materials was available: natural stone for the bottom storeys and brick facing for the main upper part of the building, with terra-cotta features for lintels, mouldings and cornices. The fact that natural stone facing, despite its disadvantages, inadequate fire resistance, difficulties of fixing to steelwork and backing, had always managed to hold its own on commercial high-rise buildings and was especially popular towards the close of the last century is due to that century's fundamental sense of history, which although thrust into the background by the Chicago School was not entirely extinguished.

As steel-framed construction was perfected and spread throughout the United States, a change in architects' attitudes took place with a movement towards the academic historical approach. The first historians of modern architecture, Giedion for example, blame the 1893 Chicago international exhibition for this, more particularly the ornate decorative style which was imported from the Ecole des Beaux Arts, France, and which reigned supreme at the exhibition. It was considered to have nipped in the bud the upsurge of new architecture and to have held up progress for the next fifty years. At the present time, this interpretation no longer appears acceptable. It has since been recognised that the exhibition of 1893 was not the origin of the neo-classicist movement, but a symptom of it. The Chicago School's achievement and the architectural forms it stood for were too spontaneous to provide the basis for a new convention in design which would be accepted by the public and the architectural profession throughout the country, to meet the increased and more exacting demands in terms of size and prestigious character of buildings. The time was not yet ripe for functional steel-framed architecture without historical antecedents.

The problems that skyscraper architecture had met in the 1890s are forcefully demonstrated in the work of Louis Sullivan. After completing his studies in Paris he had joined D. Adler's architectural practice and worked with him until 1895.

The first high-rise commercial building designed as a fully developed steel-framed structure by this firm is the Wainwright Building at St Louis (1890–1891). Here Sullivan attempted to solve the architectural problem in the sense of a self-contained visually well-balanced composition. In conformity with the classical arrangement, substructure surmounted by walls terminating in a cornice at the top, he subdivided the building into three zones: three plain bottom storeys of masonry construction, followed by the office storeys in brickwork, with closely spaced protruding piers and recessed ornamental spandrel walls, the whole being crowned by a powerful richly-ornamented frieze with a widely-projecting top slab. The piers are much wider at the corners than at the intermediate points; the actual loadbearing system is revealed in the wide spacing of the columns on the ground floor.

In the Guaranty Building at Buffalo, built in 1894–1895, Sullivan not only refined this principle of architectural articulation, but also enhanced the scale effect of the vertical treatment of the façade: here, the marked divisions between the three zones and the solidity of the corner piers and cornice are much less obvious, all the elements being combined into one organically harmonised entity. At the base of the building, however, the loadbearing columns are more distinctly expressed; on the ground floor some of them are round columns, early forerunners of the pilotis which Le Corbusier was later to postulate as a component of modern framed structures.

The ornate architectural treatment of these two skyscrapers failed to catch on – it was too personal and too ambitious. Sullivan must have felt this himself; at any rate, in his last famous building, the Carson, Pirie and Scott store in Chicago (1899–1901), he reverted to a simple elevational treatment closely following the pattern of the structural framework. The ornamental features are confined to the two bottom storeys and to delicately moulded window surrounds. Here, the classical principles of architectural design have been superseded. For Sullivan himself, the building proved to be a bitter disappointment; his contemporaries failed to understand him and saw in it a relapse into the primitive beginnings of framed construction. In the last 24 years of his life he received no more major assignments.

Finally, attention must be drawn to some projects in which the pioneering spirit of Chicago continued to display itself until the First World War. These were more modest buildings, attracting less notice from the public and the critics, in which, with less architectural pretension and with simpler means, the designers managed to accomplish what Sullivan had, in the Carson, Pirie and Scott building, distilled as the quintessence of the Chicago School. With their plain brick facing, their fine differentiation of vertical features and horizontal bands, they have an almost timeless quality. A characteristic example of these early genuine steel-framed buildings is the Liberty Mutual Insurance Building (1908).

Chicago, Liberty Mutual Insurance Building 1908

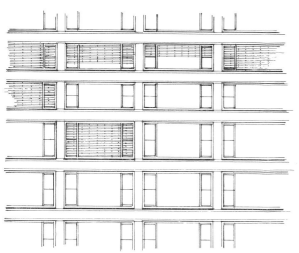

Chicago, detail from front elevation of Carson, Pirie and Scott store 1901

Evolution of framed construction in Europe: France, Belgium, West Switzerland, 1890–1930

It is no coincidence that it was in France and Belgium that the first steel-framed multi-storey buildings were developed, because the material and intellectual conditions were particularly favourable there. Even in the early days of ironwork France had, at various times, challenged Britain's claims to leadership in this field. The first wrought iron roof frames were constructed in France before cast iron bridges were built in Britain. With their glass and iron vaulted roofs at the Galerie d'Orléans and the Jardin des Plantes in Paris, the architects Fontaine and Rouhault introduced an innovation of fundamental significance for nineteenth-century architecture.

From about 1860 onwards, France gained a clear lead in building iron bridges and, especially, in the construction of iron-framed single-storey buildings. Nor could the achievements of the great French bridge engineer G. Eiffel, in the boldness of their conception, be rivalled by the work of the contemporary German designers of truss bridges. Following the achievements at the international exhibitions in Paris in 1855, 1867 and 1878, structural steelwork culminated in the 110 m span three-pinned arch structure for the Galerie des Machines at the 1889 exhibition. This building, possibly the most interesting and impressive structure of its kind ever built, would perhaps not have had the misfortune to be demolished in 1910 had it not been overshadowed by the Eiffel Tower. The latter, initially unpopular, had by then become firmly established as a symbol of Paris and it must certainly have prompted not only artists but also architects to acquire a new sense of space and structure.

In architecture, the French had always been distinguished among the Latin nations for their rational approach, their clear thinking as designers. French architectural theoreticians of the eighteenth century were the first to draw attention to the rational qualities, the constructional and formal logic of the mediaeval Gothic cathedrals, and applied these strict criteria to a critical assessment of modern buildings. About 1850, when the Gothic Revival in Britain reigned supreme, the French architect Viollet le Duc produced his ten-volume *Dictionnaire raisonné*, a comprehensive compendium of the architecture and building technology of the Middle Ages, which made him the leading exponent of what was termed 'mediaeval rationalism'.

He attributes all developments of architectural form to structural necessities and achievements. For him the groined vault, the system of piers and buttresses of the Gothic cathedrals represent the end of a long evolution which moved towards a progressively clearer realisation of the skeleton principle in the structure of the vaulted basilica, the great achievement of the French engineering spirit. 'La construction gothique est ingènieuse', asserts Viollet le Duc in his controversial article, 'Construction'. The passionate polemics with which he championed the mediaeval builders' achievements were directed against his contemporary colleagues of the academic school of thought, who clung to the architectural ideals of the Ancients and of the Renaissance and who rejected the innovations of iron construction.

'The Romans construct in the way that the bee builds its cell; this is marvellous, but it is not progress: the honeycombs at the time of the Romans look exactly like those at the time of Noah. Give a Roman architect cast iron, iron plate and glass and he will not know what to do with them. The modern spirit is of a different kind...'. In this context he uses the words 'modern' and 'Gothic' almost as synonyms.

Aided by the movement led by Ruskin and Morris in Britain, the rational doctrines advocated by Viollet le Duc constituted an important precondition for that movement in international architecture which preceded the 'modern' school of thought and which for the first time made a complete break with the 'historical' outlook.

'Art Nouveau', when it arrived, was hailed as something really novel. It provided an artistic impetus which was as intense as it was short-lived, a necessary transition stage. In order to topple the deep-rooted concepts regarding traditional architectural forms, the leading architects evidently had to start with ornamentation. They had to offer a completely new, versatile, self-contained repertoire of ornamental shapes and patterns, derived directly from nature, as already demonstrated in contemporary painting, graphic arts and interior decoration by, among other artists, Edward Munch, Aubrey Beardsley and the British designers of the Morris School. The structural material that was to provide the realisation of this language of architectural form in the actual structure was iron. The techniques which were employed in making the brittle metal conform to the curving, swaying shapes of plants and which involved, to our way of thinking, a peculiar combination of cast iron, curved sections and sheet metal cut to various shapes, may have been inspired by Viollet le Duc's proposed designs for the embellishment of iron structures. By exploiting and emphasising the slenderness and the malleability of the iron loadbearing component, the Art Nouveau architects reverted to a course of evolution which had started far back in the early days of cast iron construction in the interior of the Royal Pavilion at Brighton, designed by Nash, and which had temporarily come to a halt in the iron façades at St Louis around 1860. Although they did not achieve the true synthesis of the contemporary form of expression with the current loadbearing system, in their best work they did achieve an astonishing harmony of structure and decor. What is more, they kept alive the idea of the externally exposed metal frame.

It is particularly in this respect that the first creations of Art Nouveau architecture, the buildings erected in Brussels to the designs of Victor Horta, are also the most important. With the Tassel Building (1892–1893) he rose to sudden fame. The layout on plan, the flowing sequences of rooms and internal spaces at different levels, was not something fundamentally novel for an urban building in the Belgian capital. What was new was the intense spirit of spatial movement, the flourish and the decorative power of the visibly exposed ironwork in the staircase, the lattice girders for the landings and stair strings growing out from a cast iron central column with the curving tendrils of the infilling members.

Brussels, Maison du Peuple 1899

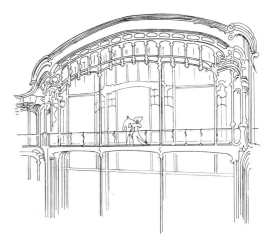

Brussels, Magasin à l'Innovation 1901

Much richer and denser is the ornamental ironwork, even more convincing is its association with the exposed structural ironwork, in the lecture hall of the Maison du Peuple, a trade union building (1899). The profile of the elegant ornamental window framework, an inscribed trapezium, with multiple curved and rounded lines, shows a remarkable resemblance to the outline of the ground plan of the building as a whole, which is situated in a circus, a circular open space, with two streets radiating from it. This double concave–convex curved front is resolved into a transparent iron framework; only the staircase with the main entrance at one corner and two narrow strips at the transition with the neighbouring premises are of solid masonry. In the upper storeys, the bays between the steel columns are halved by secondary columns; as a result, the windows are so narrow that they adapt themselves accurately to the curvature of the front, while preserving the 'small town' scale of the elevational treatment.

For its façade alone, a masterpiece of imaginative planning in an urban setting, the Maison du Peuple would have deserved preservation. It is therefore incomprehensible that this building was pulled down in 1967, despite repeated protests. The adornment of its external framework is comparatively restrained for Art Nouveau: the flat arched profiles of the horizontal members in the lower part of the building, the curved balcony brackets in the top storey. The difficult connections and joints in the elevational arrangement of the building are detailed with incredible sureness and thoroughness. Five different materials, the glass and timber frames of the windows, the structural ironwork, the latticework and railings, the light-coloured brickwork of the piers interspersed with granite and the granite portals, are combined into a whole of such elegance that one is at first apt to overlook the tremendous amount of craftsman's effort that went into its construction.

It may have been this difficulty that prompted Horta to simplify the combination of materials in his next major glass-and-iron façade, the Magasin à l'Innovation, a store in Brussels (1901). In this building the masonry is confined to a narrow granite border. The façade itself is of double-leaf construction. The large areas of glazing are fixed in the internal structural framework; decorative features embodying foliage motifs sprout from the elegant external ironwork. In the series of big European department stores which began with Bon Marché, built by Boileau and Eiffel in Paris (1876), the Magasin à l'Innovation represents a notable advance in elevational treatment and its relation to the internal structural frame. The only aesthetically objectionable features are the baroque-like curves and mouldings of the top terminating arches. This building was unfortunately destroyed by fire in 1967. Even more excessive is the ornamentation on the Samaritaine store which F. Jourdain built in Paris (1905) and which represents the mature, or indeed overripe, expression of Art Nouveau architecture in that city.

On the other hand, the commercial building that the architect G. Chedanne erected in the Rue Réaumur, Paris (1903–1905), presents a surprisingly austere and clean-cut appearance. Here, harmony between elevational treatment and supporting structure is completely achieved; the Art Nouveau style has been vanquished. This is not 'iron' architecture, but perhaps the very first example of a true multi-storey façade in structural steelwork. Here, fifty years in advance of Mies van der Rohe, the characteristic structural component of the mild steel era, the plate girder with its prominent flange and stiffened web, has had full justice done to its visual plastic qualities and has been utilised as an architectural feature.

This building in the Rue Réaumur gives the impression that France was already on the right road to achieving the true constructional steel architecture of our century and thus taking the lead in international developments. That this did not in fact occur is due to the emergence of reinforced concrete, which, starting from France itself, impeded the progress of structural steelwork in Europe for the next half century.

The famous building at 25 Rue Franklin, the first reinforced concrete structure of architectural merit, was built by A. Perret in 1903. In this instance, the structure was still faced with glazed tiles; in his later buildings Perret always exposed the concrete. The supporting system is clearly emphasised in the contrast between the light-coloured structural members and the coloured infilling; the large flower pattern is still entirely Art Nouveau. Here, the efficiency of the reinforced concrete frame is demonstrated by an additional artifice. In conformity with the sophisticated internal layout, the front of the building is deeply recessed in the middle and projects at the sides, where it cantilevers over the generously glazed ground floor. The latter contains offices, the upper floors being used as residential flats.

Nevertheless, on comparing this building with the best work of the Chicago School, the impression is given that here the structural has been made subservient to the architectural. Perret the engineer has been subordinated to Perret the architect. The horizontal and vertical features in the façades do not produce the impression of having a structural function to perform, that is of being loadbearing members, but merely appear as dividing elements in the areas between the windows. This impression emerges even more strongly in Perret's later buildings, such as the block of flats in the Place de la Porte de Passy (1930) and the vast complexes erected in the rebuilding of Le Havre (1950). Even such a product of his early period as the garage in the Rue Ponthieu, a purely functional structure, is not free from pretensions to architectural merit.

The same classically conventional attitude is manifest in the work of Perret's contemporary Tony Garnier, who shares with him the credit of having originated the international concrete architecture of this century. Though less popular and, as an artistically creative architect, not quite in Perret's class, he was nevertheless widely influential. In the planning of his industrial city, he was the first to conceive the complete organism of a modern urban community with residential districts, schools, hospitals,

Paris, commercial building in the Rue Réaumur 1905

Paris, Perret building in the Rue Franklin 1903

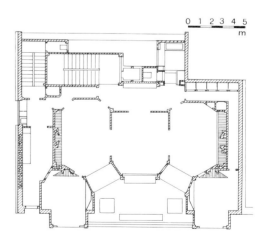

railway stations, etc., entirely in reinforced concrete. From 1908 onwards, Garnier was able to put many of his ideas into effect in Lyons, his home town. Since that time, town planners and architects, in so far as they think in terms of a building material at all, have tended to think in reinforced concrete. Garnier did, however, also possess an excellent sense of what could be achieved with structural steelwork, as is evident from the 80 m span cattle shed of the abattoir in Lyons. Here he modified the structural system and the spatial concept of the great machinery exhibition hall erected in Paris in 1889; this he did in a manner entirely appropriate to the different requirements of the time and function of his building, transforming the three-pinned arches into more austerely shaped three-pinned portal frames, in so doing reducing the glazed areas but improving the natural lighting. That he considered it necessary to provide this impressive steel structure with a solid gable-end wall of stepped construction and pierced by tall segmental-arched windows is, however, just one more indication of the basic conservative tendency of architecture in general, as has already been mentioned.

Thus, when building construction recommenced after the First World War, structural steelwork had receded into the background, both quantitatively and in the minds of the architects and their clients. However, the competition between steel and reinforced concrete, which has since grown keener, has in its overall effect turned out to be very fruitful as a stimulating force in the evolution of building technology and the engineering sciences.

Even as far back as the 1920s, the structural steelwork engineers must have felt the need to perfect their technical and scientific methods if they were to achieve ascendancy over reinforced concrete, at least in the construction of tall and wide-span buildings. The steel frame of the multi-storey building had, at that period, scarcely progressed beyond the state of development reached in America by 1900. More particularly, the structural analysis of steel-framed buildings was relatively undeveloped: each floor girder, joist and column was considered as an independent member and simple connections were assumed for the joints.

Because of the monolithic character of the material, reinforced concrete designers had been compelled at an earlier stage to investigate the spatial relationship, the effects of continuity and restraint, and to pass from the analysis of statically determinate systems to the analysis of continuous systems. The discovery and development of the T-beam was an important step in this direction, without which reinforced concrete would not have been able to compete successfully with structural steelwork at all. The transition from the T-beam to the ribbed floor, the two-way reinforced slab, the flat-slab floor, the portal frame and the multi-storey frame are further stages in achieving structural continuity in reinforced concrete buildings.

Initially, the steelwork designer had difficulty in following the example of the reinforced concrete industry with the constructional devices then at his disposal (gusset plates, angle cleats and riveted connections), as is apparent from a typical structural joint in the Torengebouw, a tower block built in Antwerp in 1931. On comparing this with a joint of similar function in a steel-framed building erected in Zürich in the 1950s, it will be evident, even to the layman, what efforts were needed to arrive at a really elegant solution of the three-dimensional design problem and what tremendous progress has been made in improving the methods of making structural connections as a result of the transition from the rivet to the weld and to the high-strength friction-grip bolt.

With its 26 storeys the Torengebouw was, at the same time, the tallest building in Europe. It shows that the development of structural steelwork in Belgium and France had been by no means at a standstill and that a start had been made in catching up with the Americans in high-rise buildings.

The most important and epoch-making progress in the use of steel in architecture within the French-speaking sphere in those years was made by Le Corbusier. Among the leaders of modern architecture, he must certainly rank as the most versatile creative force. His pioneering ideas embrace all sectors of the art of building; function, design and shape. He began his professional career as an architect in the design office of Perret, whom he always acknowledged as his teacher.

Thus Le Corbusier was, from the outset, particularly committed to concrete as a building material. However, with his concrete buildings he went far beyond the limits that had halted his teacher: he really evolved the reinforced concrete frame as a means of architectural expression; moreover, in the course of his life, he discovered and developed all the potentialities of plastic expression that concrete could offer. Without him, the direction in which international architecture has moved in the last twenty years would be inconceivable.

Not so widely known is that Le Corbusier also repeatedly occupied himself with structural steelwork and made important contributions to it. In his design for the Pavillon Suisse in the Cité Universitaire in Paris (1930–1932), he applied his idea of pilotis, an important item in the famous five-point programme, whereby the building is raised clear of the ground to keep the area at ground level unobstructed. This he achieved by means of a combination of reinforced concrete and steel construction, thus establishing an architectural feature which has since been imitated and modified countless times and which more particularly in recent years has enjoyed renewed popularity.

Nowhere is this structural concept literally so well founded as in the above-mentioned example. The deeply recessed base of the building comprises six widely spaced pairs of reinforced concrete columns which rise directly from 20 m deep bored piles and which carry twin longitudinal girders from which the powerful slab cantilevers out. With this form of substructure it was virtually a necessity to build the four upper storeys as a structural steel frame, the slim members of which are clearly featured in the main façade. The infilling is entirely of glass, with a subtle rhythmic spacing of the mullions and transoms, the direct precursor of the curtain wall.

In 1927, on the Weissenhof housing estate at Stuttgart, Le Corbusier built his Steel House, a two-storey residential building for two families raised on pairs of steel columns, thus conspicuously displaying the metal frame. Of greater future potential as a design solution for a modern residential building was the Maison

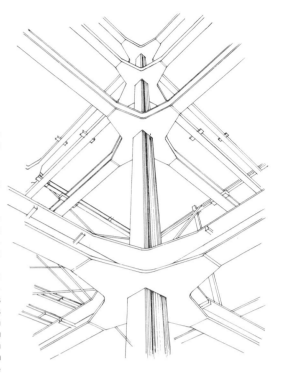

Oerlikon, near Zürich, rigid steel frame for department store 1955

Antwerp, structural joint in the Torengebouw 1931

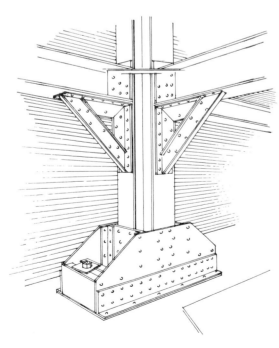

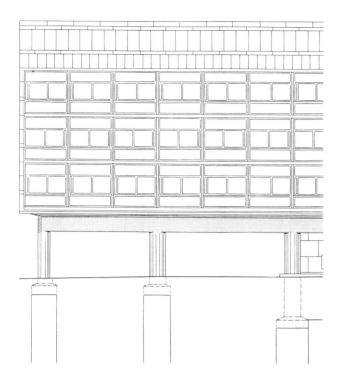
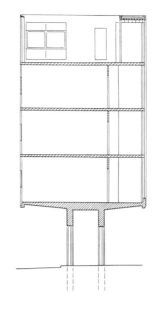

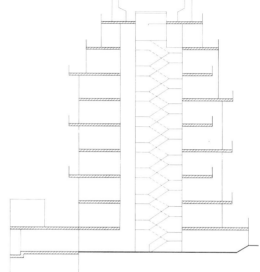

brief was to build a house entirely of glass and steel, including even the stair treads, the movable internal partitions and the built-in furnishings. With sporting singlemindedness and impeccable taste, the designer produced a luxury version of the 'machine for living', as Le Corbusier had demanded in his book *Vers une architecture* as early as 1922.

Paris, front elevation and section of the Pavillon Suisse, Cité Universitaire

Clarté at Geneva (1930–1932), the forerunner of the Unité d'Habitation at Marseilles, for which Le Corbusier had also planned a steel frame, before eventually deciding to build it in 'béton brut'. The building at Geneva, of very great depth on plan, comprises 45 residential flats, each two storeys in height and so disposed around two staircases as to provide a series of dwellings of varying size and offering attractive internal layouts. An additional refinement of this building, which rises from a wide first storey, is that its upper storeys are twice stepped back some distance from the main façade.

This highly sophisticated spatial configuration is accommodated within a welded steel frame composed of standard rolled sections and based on a rigidly applied planning grid. The main front consisting entirely of glass and steel is thereby given its basic rhythm and clear-cut orderly arrangement. Here, this rhythm has been achieved by a very simple means: the continuous balconies with elegant metal balustrading, which are provided only on every alternate storey, the alternation of steel and glass, the subtle variations in the spacing of the mullions, the free arrangement and alternate positioning of the awning blinds. One really gets the impression of 'clarté' – brightness. It is indeed apparent that the building has been designed from within its interior, that the architect has grappled with the financial and functional problems of his assignment, with building regulations that were inadequate for a conception of such magnitude, and with every technical detail, finally to emerge with an optimum of carefully planned up-to-date residential amenity for the occupants. As a result of the detailed studies which he undertook when the industrialist Wanner had commissioned him to design steel-framed apartment buildings, Le Corbusier derived an important stimulus from a small somewhat neglected structure which nevertheless deserves to be mentioned in the history of steel building construction: Dr Dalsace's house in the Rue Guillaume, Paris, built by P. Charreau (1929–1931), who till then had practised as an interior designer in the conventional style. Here, his

Geneva, front elevation and section of the Maison Clarté 1931

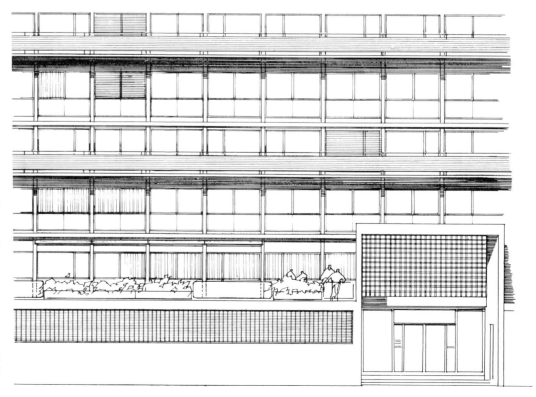

Efforts to develop a new architecture based on framed construction in Germany, 1910–1930

In Germany the application of framed construction to multi-storey buildings did not make much headway until after the First World War. As in other Western countries, the steel beam floor and, from about 1905 onwards, the reinforced concrete floor were used for the sake of their loadbearing and fire-resisting properties within enclosing walls of conventional masonry. Even after 1920, architects still preferred a composite form of construction, internal structural frame of reinforced concrete in combination with external walls in masonry, and among the true framed buildings of the 1920s in Germany those with a frame in structural steelwork are in the minority.

Nevertheless, even in the development of structural steelwork German architecture from about 1910 to 1930 has a tale to tell, perhaps not so much through the actual buildings erected as on account of the new concepts of form and shape which, first evolved by the leaders of the Bauhaus group, Gropius and Mies van der Rohe, were to sweep across the world from America after the Second World War and to bring about the rise of a true steel-based architecture. The breakthrough of this 'modern' style was heralded as long ago as 1910 by three important achievements in the sphere of industrial building. Each of the three architects concerned had attempted, with the aid of a steel frame, to design a factory as a functional concept. With respect to architectural character, however, the buildings in question are very different from one another, corresponding to the divergent artistic trends of that transition period. The turbine house of AEG in Berlin, designed by P. Behrens, with its prominent features, the robust stanchions of the portal frames, the pronounced sculptured effect of its roof structure and its sophisticated symmetrical gable end, is of well-nigh timeless monumentality. The classical character of this industrial building has often been criticised, but to this day it has lost nothing of its fascination.

The chemical works at Luban, near Posen, is the masterpiece of H. Poelzig, the exponent of 'expressionism' in German architecture. This generously dimensioned structure, which conveys an impression of lively movement, was developed entirely from the requirements of the industrial processes that it accommodates. What gives the building its distinctively expressive power, the leading idea of the whole design, is the consistent alternation and strong contrast of the architecture. All parts of the building that serve the manufacturing operations have a ribbed structure: a loadbearing network of flat steel bars in which the infilling bricks are set on edge. The staircases, amenity blocks, stores, etc., are of solid brickwork with deeply recessed arched windows.

The Fagus factory at Ahlfeld, designed by Walter Gropius, ranks as the most progressive architectural creation of those years, a forerunner of 'constructivism', or 'functionalism', although the loadbearing structure is less clearly expressed than in the AEG turbine house and in the factory at Luban. To contemporaries, the all-glazed elevations of the three-storey factory block with the projecting transparent corner must have appeared as something sensationally daring. This façade has been called the first curtain wall. It would be more correct to refer to it as one of the earliest and most important of the many forerunners of the curtain wall, for the realisation of a true glass-and-metal curtain wall, in the present-day sense, was then still beyond the range of what construction technology could offer.

Directly after the First World War the pent-up intellectual energy could not at first find an outlet in the construction of big buildings. Instead, the main emphasis was on theoretical studies and ideal designs. Prominent among these was Mies van der Rohe's fantastic project for a glass high-rise building in Berlin (1919) – a dreamlike apotheosis of a fully glazed steel-framed structure. In this project, Mies was fifty years ahead of his time and twenty years ahead of what he himself was later able to accomplish in Chicago.

With the economic expansion that got under way from 1923 onwards, the construction of tall commercial and administrative buildings in Germany, too, became a top priority. In looking back upon the German building activities of the 1920s, its consistently high level of quality is a notable feature. Quite unmistakable is the passionate striving to achieve a modern, independent form of expression for framed buildings; but another distinctive characteristic is the truly German thoroughness with which the extreme solutions were raised to the status of fundamental problems, to alternative philosophies, thus splitting the army of architects into two warring camps on the issue of whether the emphasis should be on vertical or on horizontal features in the elevational treatment of their buildings.

The right-wing conservative members of the German architectural profession favoured emphasis on verticality, shown, for example, in a particularly pronounced manner in the Stumm Building in Düsseldorf. The Chile Building in Hamburg immediately captured the public imagination and postcards depicting it were produced as soon as it had been completed. The closely spaced columns soaring up to the arches under the cornice, the ceramic ornamentation and the loadbearing structure revealed in the widely spaced arches on the ground floor are all strongly reminiscent of Sullivan's Guaranty Building in Buffalo.

Those architects who clung to tradition were evidently influenced by historical forms in their approach to framed construction – the disposition of the posts in timber structures and the stone ribs of Gothic cathedrals. In those days, architectural theory was insistent on the concept of linear construction, expressed as vertically arranged elements, in contrast to the mass construction appropriate to storage buildings and the like. In the 'perpendicular' façades of high-rise buildings there is thus nothing to show whether they enclose a steel or a concrete frame. Indeed, provided that they were suitably stiffened by crosswalls, many of them could equally well be of loadbearing masonry.

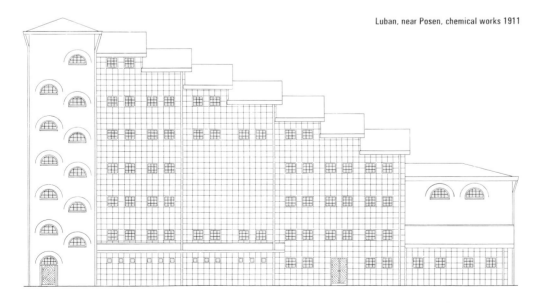

Berlin, reconstruction of the market hall 1923

Luban, near Posen, chemical works 1911

With even greater decisiveness and a certain arrogance, the progressive left wing of the profession emphasised the horizontal features in the elevational treatment of framed buildings. As with a shrill trumpet blast, this attitude proclaimed itself in a building which was destroyed in the last war and is now almost forgotten: the conversion of the Mosse publishing firm's premises in Berlin, designed by E. Mendelsohn and R. Neutra, 1921–1923. At the corner the new extension penetrates into the somewhat unimpressive sandstone pier-and-arch architecture of the old building; it rises three storeys above the latter and extends backwards from the corner like great wings, looking for all the

Berlin, Shell House 1931

world like some fearsome monster baring its fangs. In addition to the structural difficulties involved in constructing additional storeys, the spandrels at the rounded corner, with their long span and torsional loading conditions, compelled the designers to use structural steelwork. Nonetheless, in all probability, this was traditional ironwork rather than a true structural steel frame.

The horizontality of the German avant-garde made its appearance in a structurally acceptable and mature form in a design by Mies van der Rohe for an office building in Berlin, 1922. Here, bands of windows extend round the building without intermediate piers and afford an unobstructed view of the interior, showing the structural frame with set-back columns and beams cantilevering out from them. The glass façade is suspended from the massive spandrels, yet another forerunner of the curtain wall. This multi-storey building has been specifically designed as a framed structure which could not possibly be mistaken for a masonry building. However, it is not a steel frame but a typical early concrete frame based on T-beam construction. The closest approach to Mies's design in buildings actually constructed is to be found in the Schocken department stores designed by E. Mendelsohn, especially the one at Chemnitz, with its curved street frontage and strikingly effective continuous bands of glazing.

Although the German efforts in the 1920s to achieve a new form of expression for framed construction were, on the whole, very fruitful, they did not directly promote the idea of building with steel. All the same, there are three major buildings in Berlin among the best examples of work achieved in the 1920s which can justifiably claim their place in the history of steel in architecture. In the multi-storey switchgear building of the Siemens-Schuckert works at Siemensstadt, by Hans Hertlein, 1926–1928, the steel frame spans the continuous production areas without intermediate support; only the upper storeys are divided into offices. On the longitudinal façades the brick facing is subdivided by sturdy piers, but the vertical continuity is interrupted by the flush surfaces of the stepped-back office floors and the prominent solid staircase cores. Thus, the external appearance of the building reveals in which part of the structure the steel frame has to cope with long spans and heavy imposed loads. There is nothing romantically awe-inspiring about the vertical emphasis of the elevational treatment; it has an agreeable down-to-earth quality combined with a sense of dignity and status.

The most brilliant framed building in the Berlin of this period, Shell House, is the Rhenania-Ossag administrative building, built by E. Fahrenkamp in 1931. On account of the shape of the site this building, which encloses a courtyard, is trapezoidal on plan. The oblique front facing the Landwehr Canal presents a stepped configuration. In addition, the architect has effectively enhanced this motif by stepping the building in section so that it rises in stages to the dominating high-rise block, an architectural idea which is once again much in vogue at the present time. The staggered arrangement of the various parts of the building is particularly effective on account of the horizontal treatment of the façade, in the restrained form of detailing popular at the time: set-back mullions, continuous spandrel bands given emphasis by projecting mouldings. It is very successfully combined with another often used – often misused – style, disparagingly called 'battleship architecture' in the Germany of those days: radiused re-entrant and projecting angles, with travertine cladding carried continuously round them and with corresponding rounded glazing at the outer and rounded piers at the inner corners. As a result of these features the building gives such an impression of flexibility and resilience that among framed buildings of that period in Germany it comes nearest to suggesting the presence of the steel frame within. It can still hold its own very well in comparison with the New National Gallery erected a few years ago, that monument to structural steelwork and to the master of the medium, Mies van der Rohe.

Another progressive steel-framed building, in advance of its time and for many years the last such structure of international status erected in Germany, is Columbus House designed by E. Mendelsohn, 1931. For the multi-purpose urban commercial building it represents as perfect a solution as that achieved by Le Corbusier for residential building construction. The exacting planning requirements – shops and arcades on the ground floor, flexibility in the use of the upper floors as offices and restaurants – have been met on the irregularly shaped site by the structural resources of the steel frame. Among the distinguishing features are the widely spaced set-back columns, the projecting lintel girders with their (for that period) tremendously long spans over the ground floor and the closely-spaced loadbearing columns in the upper storeys. All these are features which have, in recent years, reappeared in a number of steel-framed buildings erected in Berlin and which have shown themselves to be particularly advantageous both structurally and economically.

Berlin-Siemensstadt, multi-storey switchgear building 1928

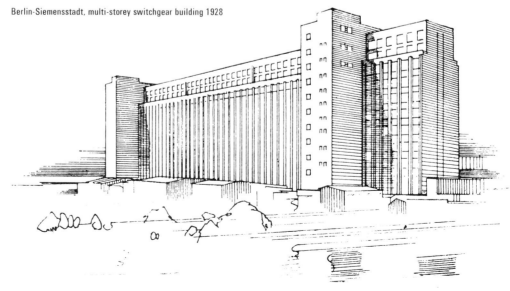

Skyscraper architecture in the USA 1890–1940

The desire for historical character was demonstrated in the later work of the leading architectural firms of the Chicago School, at any rate in their major buildings, which their owners wanted to be impressive or dignified.

It is instructive to compare Burnham's masterpiece, the Reliance Building, dating from 1894, with the Fisher Building which the same architect built two years later. At first sight the two structures show considerable similarity: the incorporation of the bay windows into the façade system, the extensive glazing, the decorative features concentrated on the horizontal bands. Compared with the flat quatrefoil ornaments on the Reliance façade, the tracery on the spandrel areas of the Fisher Building appears distinctly Gothic. This impression is due more particularly to the fact that here the vertical members, the main columns as well as the window mullions, are similar to the clustered piers of the perpendicular period. This façade is noticeably more academic than that of the earlier building, but it possesses a grace appropriate to the structural steel frame and is sufficiently resourceful to gloss over the inconsistencies above. Under the cornice the main columns are linked by round arches and a frieze of narrow arched windows. The bay windows, which do not fit in with this cornice architecture, have been terminated a storey lower down.

The impartiality which gives this anachronistic solution a certain charm is lost in the later Chicago buildings, which subscribe and succumb to the classical influence in architecture emanating from the east coast of the United States. The 'New York fashion' appealed more particularly to the architectural firm of Jenney and Mundie, probably under their clients' influence. This first found expression in the New York Life Assurance Building and the Fort Dearborn Buildings, structures which can in no way stand comparison with, say, the two Leiter Buildings. In the late 1890s Chicago's leading position in high-rise building construction was taken over by New York, not only in the number of buildings erected and the record heights attained, but also in architectural merit. Over a period of some thirty years the spectacle of historical eclecticism was repeated there, with all the styles of the past reflected in skyscraper architecture. At first, classical treatment predominated, adopting its examples from all periods, from Brunelleschi to Palladio.

A characteristic 'Renaissance' building, successful of its kind, is the Broadway Chambers Building in New York. The classical threefold subdivision is especially pronounced, being given added emphasis by the alternation of the colours of the materials used: the two bottom storeys and the transitional mezzanine floor are distinguished by grey granite piers; the next eleven storeys are in dark brickwork, with narrow windows grouped in pairs, but otherwise entirely unadorned; while the top part, which is executed in brightly coloured terracotta, again introduced by a transitional storey, terminates in an assembly of piers and arches, à la Sansovino, extending through two storeys, and a low attic storey serving as a frieze under the cornice.

When critically considering this and similar examples of skyscraper construction, it should be clearly realised what sense of shape, what mastery over classical forms, was needed to overcome or conceal the difficulty of the whole architectural project. The classical subdivision of façades by means of plinths, pilasters, columns, arches, string-courses and cornices presupposes a building with well-balanced proportions of length, width and height, the whole of which can be taken in at a single glance. On the other hand, the composition of the Broadway Chambers Building, with its careful arrangement of perspective and scale, could hardly be viewed as a whole by the pedestrian far below in the confined space of the street. It is therefore not surprising that the early photographs of these Manhattan skyscrapers were always taken from adjacent tall buildings.

The proud isolation, the self-sufficiency of true Renaissance architecture based on symmetry and harmony, was still perceptible and still reasonably credible as long as the buildings were not more than 20 storeys high and so long as they stood out as individual giants among lower structures. But they became grotesquely distorted, architecturally speaking, when their heights increased to 40 and more storeys and they were erected in ever closer proximity to one another. The best architects were well aware of these problems and sought solutions for them. Even at an early period attempts were made to depart from the closed prismatic form of tower block by adopting a stepped arrangement. Thus, a central projecting feature rose high into the sky, to justify this aspiration to great height. It could, for example, be shaped in outline like the Campanile at St Mark's in Venice, whereas the side bays or wings were only about half as high. As a result of such an arrangement, however, the proportions became more extreme and the architecture even more pretentious.

It was thus really not at all surprising that the classical fashion in skyscraper eclecticism was followed by a wave of romantic mediaevalism. A stronger impulse in that direction emanated from the 52-storey Woolworth Building, 1913, designed by Cass Gilbert, the architect of the Broadway Chambers Building. Here, too, in the main façade a central tower block soars to twice the height of the relatively compact block from which it rises. Visually, however, the whole structure does not split up into three distinct parts, but is held together as a unit by continuous vertical composition.

The Gothic style is, if anything, better suited to skyscraper architecture than is the classical style. For one thing, the character of ribbed construction and the compulsive vertical development are inherent in Gothic. Secondly, the compound or clustered piers as features of the elevation of the building offers the possibility of interconnecting and harmonising structural members differing in height but of equal cross-section, or of equal height but different cross-section, without disrupting the scale and the structural logic of the building. This is exemplified by the aisle and nave piers in five-bay cathedrals, or by the stouter columns at the crossing and under the towers. The pier profiles from which the quatrefoil pier and the clustered pier were eventually derived are significant structural elements of mediaeval churches. That they were, in the USA, transformed into elevational features on skyscrapers must not be

Chicago, detail from front of Fisher Building 1896

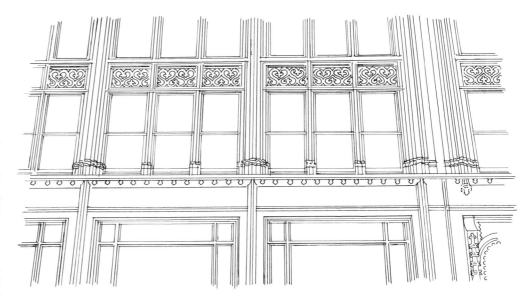

regarded as sacrilege on the part of the architects concerned. After all, the ancient Romans took the columns and beams featured on the exterior of the Greek temple and utilised these motifs to adorn the internal walls of their vaulted buildings, where they were, strictly speaking, even less appropriately applied.

In any case, the pier forms he adopted on the Woolworth Building enabled Gilbert successfully to express the different load conditions of the structural members in the building and thus obtain a rhythmic articulation to enliven the façades. Though starting from a reactionary attitude, he thereby succeeded in arriving at what was basically a more functional solution than that achieved by Sullivan with his avant-garde approach. Indeed, the column shafts with their restrained profiles do not produce an historical effect, as contrasted with, say, the vertical features on the Fisher Building in Chicago. The Gothic decorative elements — baldacchinos, arches, pinnacles, finials, etc. — are concentrated at the top of the building and on the delicate horizontal string-courses, more particularly on the spire which constitutes a much more satisfying terminal feature to the structure as a whole than a horizontal cornice would have done at that great height.

The Woolworth Building displays the first signs of a crisis in eclectic or, as it is more frequently called in the present century, traditional skyscraper architecture. The modern structural system is so clearly revealed, the building has evidently to so great an extent been designed from within, that the horizontal trimmings appear as no more than a thin veneer, a threadbare or transparent garment. A similar evolution can also be observed in the best traditional European buildings in the years immediately before and after the First World War, such as the Town Hall in Stockholm or the dwelling houses built in Britain by Shaw and Voysey. The first phase of American high-rise commercial architecture, which was determined or dominated by historical traditionalism, can be said to have ended with the Chicago Tribune Tower, a building which would not on its own merits have caused such a sensation, had it not emerged as the result of an international competition for which there were no fewer than 263 entrants, a competition which was one of the most exciting events in the architectural history of this century.

The design which was awarded the first prize, and was subsequently carried out, was that of the New York architects J. M. Howells and R. M. Hood. It made use of devices and features very similar to those adopted in the Woolworth Building. As an integrated architectural composition it did indeed deserve a prize — at any rate, among the American entries, all of which were stylistically at the same level of moderate or obsolescent eclecticism. With their Tribune Tower scheme Howells and Hood had an easier task to tackle than had Gilbert in designing the Woolworth Building: it is only about half as high as the latter and, apart from the rear connecting block, it is square on plan. Here, too, there is marked differentiation in the width of the piers, the main piers have been kept away from the corners, and the whole centre part of the building projects forward a little, while the corners are chamfered, thus preparing for the octagonal shape of the stepped-back tower superstructure. The main piers rise above the building, and at their upper ends they extend inwards as mighty arched buttresses to the corners of the tower, thus appearing to grip and secure the whole structure. The shaft of the skyscraper is entirely unadorned and rises with its vertical articulation from a two-storey smooth base. The decorative treatment is confined to the transition zone, the tower and the arches, which closer inspection reveals to be, not pointed arches, but segmental arches approximating to a semicircular shape.

The second prize in this competition was won by Eliel Saarinen, whose design is definitely superior to that which was awarded the first prize, both in architectural quality and in the progressiveness of its conception. The vertical division of the façades by piers or pilaster strips, the boldly articulated middle part, are features which it shares in common with the Howells and Hood design actually executed. The horizontal division into three distinct parts is, however, absent or at any rate much less pronounced here; the top part of the tower rises organically from the building like a plant stem from within an enclosing sheath. The ornamental treatment is not reminiscent of Gothic; it is more suggestive of some mystically nordic Art Nouveau. If the building had been built to this design, it would have appeared nearly twice as high on account of its meticulously scaled detailing and would have become one of the major sights of Chicago, a city so rich in monumental modern buildings.

The well-known and often published design which Walter Gropius submitted must have had a startling if not shocking effect on the panel of judges for the Chicago Tribune competition. In this scheme an austere prismatic high-rise block is surmounted by a narrower rectangular prism soaring tower-like above it; the tower is, however, not brought forward and located centrally, but is instead stepped back and to one side. The whole building is consistently gripped within an enclosing framework, infilled almost entirely with typical Chicago windows. Only on the flanks of the tower is there a minor anomaly: two narrower bays presenting a closed surface and serving to enhance the verticality of the tower and to root it more effectively into its substructure. The whole building is a decisive reversion to the strictest and, by present-day standards, most advanced solutions of the Chicago School. No credence can be given to Giedion's assertion that Gropius had not known the early steel-framed buildings of Chicago. On the other hand, this design very clearly foreshadows the type of modern framed building which came into vogue, more particularly in Germany, in the 1950s and was characterised by the grid-like elevational features, but which soon turned out to be an architectural blind alley. Thus, it happens that Gropius's design for the Tribune Tower, in spite of all its daring, strikes us as somewhat antiquated, just as it did his contemporaries and, surprisingly, almost more so than do the prize-winning designs. One feature adds considerably to this impression, something which even twenty years ago would have been looked upon as rather too modish and too likely to date the building: the deeply recessed loggias and boldly projecting thick balcony slabs which are continued at varying levels round each corner as though they function as gigantic clasps. Their object was evidently to impart a sense of strength and energy to this building which, despite its carefully balanced proportions, might have seemed to lack stability. Incidentally, they also introduced into a Chicago building a whiff of avant-garde Berlin horizontalism.

The overall impression that one gains from a study of the Tribune Tower competition is that the design of the steel-framed high-rise building was not something that could be solved by one architect of genius in a single tour-de-force of creative achievement, but that, instead, it needed the combined experience and efforts of several decades. Thus, on the whole, the American projects reveal a clearer insight into the nature of

Design for the Chicago Tribune competition, Eliel Saarinen 1922

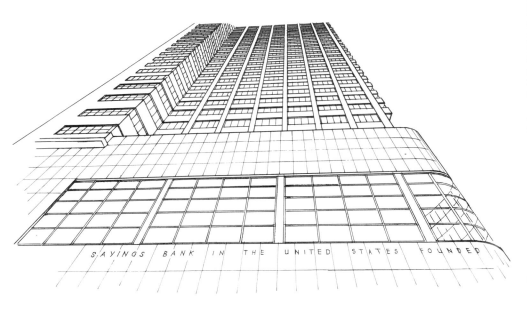

Philadelphia Savings Fund Building 1932

the architectural problems and greater sureness in tackling them; among the unacceptable designs the oddest are to be found among those submitted by European entrants. The well-known design by Adolf Loos, who conceived the skyscraper as an enormous Doric column with the windows of its thirty storeys rising one above the other in its flutings, must surely be regarded as a caricature, a malicious throwback to moribund historicism. It is worth recalling that in that same year Le Corbusier published his controversial *Vers une architecture* with the object of delivering the death-blow to eclecticism.

The second period of the American skyscraper lasted approximately from 1920 to the Second World War. It is characterised especially by the boom and the tremendous upsurge in building activity that continued until the world crisis of 1929, by the setting up of new height records (Chrysler Building, 320 m; Empire State Building, 380 m), and, above all, by a definite change in the architectural conception of framed building construction. The historically inspired detailing, the ornamental treatment of the terminal or transitional features on the façades, are reduced more and more and finally abandoned. In lieu of the traditional elements of external architectural treatment comes a system of recessed and projecting surfaces, of vertical and horizontal bands, whereby the structural system, that is, the steel frame, is given external expression, thus projecting a sense of power and strength. The ideal image of the high-rise commercial building is no longer that of the isolated self-sufficient tower, the pretentious front or axially symmetrical stepped configuration. Instead, rectangular shapes on plan, free asymmetric groupings and less rigidly disposed stepped features are preferred. In fact, this last mentioned trend was promoted – and imposed – by the new regulations governing the stepping-back of the upper storeys of tall buildings, such as, for instance, the regulations introduced in New York in 1916.

A characteristic solution is that adopted for the Daily News Building in New York, erected in 1931. The vertical bands, emphasised by lighter colouring, shoot upwards without interruption, without starting features at the base or terminal features at the top. The stepped-back configuration which, as it were, goes spiralling up around the tall structure, in no way diminishes the sense of verticality; on the contrary, it enhances it and creates pleasing perspective effects. It is almost unbelievable that this building was designed by the same team of architects that produced the Chicago Tribune Tower. However, the vertical elevational treatment of the Daily News Building loses some of its persuasiveness and indeed appears rather sketchy and schematic in comparison with the McGraw-Hill Building, also designed by Howells and Hood and built only a year later. Here, it is the horizontal features that boldly stand out in relief, emphasised by the light colour of the ceramic facing, while the vertical features comprising the main columns, secondary columns and mullions are all combined by a darker colour and are recessed. On the Daily News Building all the verticals are similar. This is a blemish, as only every third column is structural.

An outsider which oversteps the conventions of the second period in the history of skyscraper architecture is the Philadelphia Savings Fund Building (1932). The architects were G. Howe and W. Lescaze. With its boldly pronounced verticals on the longitudinal face and the spandrel panels continuing round to the cantilevered end face, this building gives very definite external expression to the structural frame. It resembles the Inland Steel Building in Chicago, dating from 1957, which is characteristic of the latest trends in steel-framed structures. The only difference is that at Philadelphia the facing is of stone, not steel. However, this building, too, has shortcomings in architectural design which date it. The narrow protruding bay extending the height of the end face engages with the prominently projecting substructure with its rounded corners, which are characterised by having the glazing of the mezzanine floor continued round them, all very reminiscent of the horizontalism of the Berlin School around 1930.

A building which is representative of the creative power of that period but also rises above it to acquire a timeless quality of greatness is the Rockefeller Center in New York. Designed by a team under the leadership of R. M. Hood, it was begun in 1931 and its construction continued until the Second World War. As a lavishly planned recreational, commercial and press centre, an island of orderly building development promoted by private enterprise, the Rockefeller Center has long remained unique in the United States and has never been remotely equalled by other attempts in this direction, such as the Lincoln Center or the UNO complex in New York. The group as a whole comprises fifteen buildings. Above a row of low buildings along Fifth Avenue rise a number of tall slab blocks, so disposed around the dominating RCA Building in a radial pattern, like the sails of a windmill, that the shadows cast by the structures are as unobtrusive as possible and a great many attractive intersections and vistas are obtained.

The towering 70-storey RCA Radio Building has one end adjacent to a plaza which is laid out as a garden and which serves also as a skating-rink in winter. The elongated stepped shape of the building on plan is the result of careful planning: utilisation of space and efficient arrangements for natural lighting have been harmonised with a rationally conceived structural system. In the upward direction, too, the stepped configuration of this building is functionally determined: the projecting parts on plan are stepped back at the respective floors on which the various liftshafts terminate. The loadbearing columns are spaced at 27 ft centres, a distance corresponding to the normal room depth within the building. Externally they are clearly revealed in the regular alternation of the two narrower with one wider vertical member. These verticals are all located in one plane and project only a little in relief from the plane of the spandrel panels.

This basic rhythm and its modular dimension recur in all the buildings of the Rockefeller Center, but are not rigidly adhered to: instead, they emerge here and there in a simplified or more compact form. In the lower-rise buildings and in certain parts of buildings the piers are all of equal width and are flush with the spandrel panels. Especially pleasing is the contrast with the closed external surface of the big Music Hall joined to the rear of the RCA block. Here in particular is demonstrated how much the austere unity of the material, polished greyish-green sandstone, contributes to the harmony of the whole. The vertical treatment is devoid of monumental pathos and shows merely a trace of romanticism; it appears here as the festive and yet dignified manifestation of the controlled power of well-conceived modern framed construction. For the first time, the unity of sophisticated functional building, modern structural steelwork and spectacular shape has been achieved in the context of major urban development.

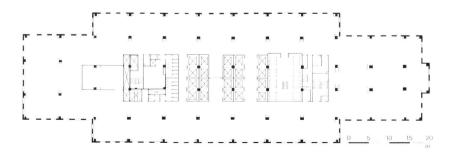

New York, Rockefeller Center, RCA Building 1931/32

'Gridiron' architecture, 1940–1955

After the Second World War, framed construction based on a supporting skeleton spread rapidly into use throughout Europe. From the high-rise commercial building and from industrial construction it spread to schools, hospitals, council offices, etc. Masonry or brick construction was pushed into the domain of housing and even there it did not hold undisputed sway. This trend was due not so much to practical as to psychological needs. It is understandable that more particularly the German architects, after the collapse of national socialist dictatorship and the catastrophes of war, were fired by the urge to adopt and give wide publicity to the achievements of the Bauhaus, Gropius's influential school of architecture, which the Nazis had suppressed. These achievements appeared to them as the embodiment of freedom and progress, but they were inexperienced in the new constructional methods and were, moreover, soon too overburdened with work. Thus quite often, by violating the principles of building technology and the rules for obtaining a congenial interior environment, they discredited the 'new architecture' in the public esteem. Even in the formal aspect, in their misconceived handling of modern structural features, there was a widespread superficiality which was apparent even to the layman. This criticism is applicable more especially to the 'gridiron' architecture which, in Germany, was practised with amazing persistence for a number of years.

In the narrower sense this description would be applicable to framed buildings whose façades comprise a uniform grid pattern of protruding vertical and horizontal members all equally thick or thin. The recessed windows and spandrel panels play only a subsidiary part in determining the external appearance. As a rule, the verticals are spaced at 1.50 m to 1.80 m centres, room widths being a multiple of this basic dimension or module. Whether the vertical members are main and secondary columns or whether all the columns share equally in supporting the floor loads is not immediately obvious, though it is sometimes possible to draw conclusions from the fact that, on the ground floor, only every second or third vertical member is continued down to ground level. The grid is either continuous around the building or — especially in the case of slab blocks — terminates at the blank end walls.

The prototype for this sort of framed construction was created by H. Salvisberg with the Bleicherhof Building in Zürich (1938). Throughout the 1940s this example was followed by a number of commercial buildings in its immediate vicinity, all so closely resembling one another in their external patterns that this style has sometimes been called the 'Zürich School'. When eventually, around 1950, construction began to boom in Germany and interest was focussed on the urban office and administrative building, German architects enthusiastically seized upon this Swiss example as the one nearest and most conveniently to hand. The present author vividly remembers his first post-war visit to Zürich with a party of German architects from Munich. Their Swiss colleagues felt both flattered and embarrassed by their visitors' compliments on the new business quarter of the city and asserted that the grid system of façade design was not a Swiss invention. There were, they said, finer examples in Finland and elsewhere.

It is quite true that in Scandinavia and Italy

several of these grid façades had been constructed about the same time or even a little earlier. The influences that led to this solution of the framed building problem are clearly shown in the Bleicherhof: it is, in a manner of speaking, a hybrid of the Columbus House in Berlin, the Zeppelin House at Stuttgart and the Casa del Fascio at Como. In the 1920s, Salvisberg had practised successfully as an architect in Berlin; he was familiar with the aspirations both of the progressive and of the conservative German architects with regard to the elevational treatment of framed buildings. He certainly also knew the other type of façade, in which the supporting frame is concealed behind the smooth skin of a stone facing, as in the Hoffmann-Laroche administrative building at Basel. His verticals do not act as loadbearing columns nor do the horizontals act as beams or lintels; instead, the whole arrangement is an assembly of window surrounds which emulate the noble proportions of Italian palace windows. The character of modern framed construction is completely belied, however. The entire configuration becomes even more pretentious and tedious when this external framework is in addition enclosed within lateral and top and bottom wall strips forming a border around the façade as a whole.

Significantly, the Bleicherhof type of elevational treatment had scarcely any imitators in France — although there, too, as in all countries in the 1930s, a pronounced neo-classical trend was apparent. The French were immune against this kind of compromise architecture; after all, in the work of Perret they had the example of a cultured, structurally plausible combination of modern building construction with traditional elevational treatment.

The German architects of the 1950s were not slow to realise the shortcomings of the grid façade, its limited impact, the lack of plastic and graphic quality, and they endeavoured to avoid the danger of sterility and tediousness by varying the vertical and the horizontal façade elements, by alternating the main and the secondary columns and by giving rhythm to the spacing of the windows or the columns. It would, however, be unprofitable to examine these experiments in detail here because, for one thing, they scarcely produced any result which, in terms of architectural achievement or structural idea, could stand comparison with the epoch-making productions of the leading French and German architects of the 1920s or with the best examples of American high-rise buildings erected about 1930. Furthermore, the period of evolution represented by the grid façade was a retrograde development rather than an advance for structural steelwork.

Designers in concrete deliberately took advantage of the upsurge in the construction industry not only to secure the great majority of framed buildings for concrete as the structural medium, but also to persuade architects and engineers that the reinforced concrete frame, which was simpler and more convenient to design and detail, could compete successfully — in terms of both economy and slenderness of columns — with the structural steel frame, even for buildings of thirty or more storeys. This claim was based, inter alia, on the tremendously increased compressive strength of concrete.

This situation is exemplified by the construction of the BASF Building in Ludwigshafen (1954). Here, for the first time, the reinforced concrete frame exceeded a height of 100 m and broke the European record previously held by the Torengebouw. As an exercise in public relations this achievement received great prominence, but architecturally the building was of no great significance — for instance, it cannot stand comparison with the Phoenix-Rheinrohr Building in Düsseldorf (now known as the Thyssen Building), which was designed by the same architect and with which, in 1960, Germany regained its link with international architecture and with the mainstream of development in structural steelwork.

Among the more important German buildings designed with grid façades in the 1950s there are, however, a few in which steel was able to hold its own and indeed showed itself to be superior to reinforced concrete in terms of economy and rational construction procedure. The new building designed by Zinssen for Continental Gummiwerke AG, Hanover (1952) was one of the first German post-war buildings to receive international acclaim. It is remarkable because the architect, by diversification of the top storey, has skilfully managed to avoid the pitfalls of monotony and drabness that are often associated with the grid scheme, more particularly when it is maintained without relief within a large architectural complex comprising buildings of different heights. It must be admitted, however, that it is not obvious, from an inspection of the exterior of this building, that it has a structural steel rather than a reinforced concrete frame.

As early as the mid-1950s a sense of disenchantment and even of revulsion against the grid principle began to display itself. Architects sought a way out in two directions. The first consisted in adopting a structurally emphasised 'plastic' configuration of the façade or in livelier modelling of the building as a whole — as was done, for example, by G. Weber in his administrative building for Hoechst Farbwerke, boomerang-shaped on plan. Some quite extraordinary ground plans became internationally fashionable at the time: spindle-shaped, streamlined, flattened hexagonal, etc., as exemplified by Gropius's design for the Backbay Center in Boston (which was never built), the Pirelli Building in Milan, and the competition-winning design for the Phoenix-Rheinrohr Building in Düsseldorf. The other way out of the impasse was to return to the flat continuous façade, generally of brick or masonry construction, as applied, for example, with pleasing success to

Hanover, head office of Continental Gummiwerke AG 1952

Bleicherhof not only strikes a perfect balance between horizontality and verticality, but also retains the dignity of the stone-faced frontage, a compromise solution which could succeed so perfectly and with such sound workmanship only in neutral Switzerland.

At the time, this architecture would not have received so generally favourable a reception, had it not been for the fact that basically it was quite conventional and classical in character, though on a much reduced 'bourgeois' scale compared with its example, Terragni's Casa del Fascio (1936). The latter had, in its day, been highly praised as a tour-de-force of progressive architecture under a totalitarian government, though by present-day standards it seems hardly more fascist than the work of the traditionalists Piacentini, Muzio, etc. In the grid façade the

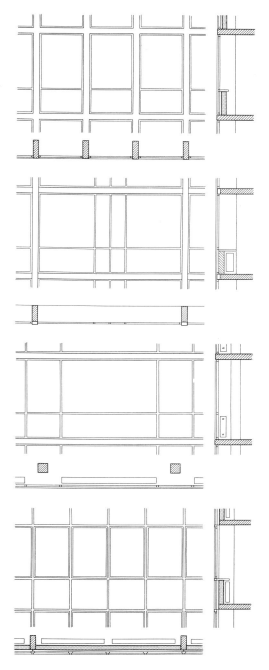

Development stages from grid construction to curtain wall

Nevertheless, the curtain wall is not a product entirely imported from the USA. There is also a European course of evolution (several of its stages have already been mentioned) leading from the conventionally infilled structural frame, and more particularly from the grid-type frame, step by step to the curtain wall. To begin with, the grid pattern of the façade was opened up, the vertical members being confined to the actual structural columns. The wider spaces between these members were then filled, not with masonry spandrels and windows, but with storey-height external wall units made of metal and glass, wood and glass, or other combinations of materials as, for example, in the Maison de Clarté, which was built as long ago as 1931. With his Cité de Refuge in Paris (1932), Le Corbusier also anticipated this first step. The vertical members recede behind the façade; of the structural frame only the floor strips remain visible on the outside; the cladding units join together to form continuous horizontal bands. From this conception, it needs only one final step to the fully fledged curtain wall, in which the structural frame is entirely within the cladding suspended in front of it. With the curtain wall begins a new stage in the development of modern structural steelwork. Of course, it did not at once become the only acceptable system. It represents no more than an extension, a prevailing fashion, within the architect's design repertoire, just as every other stage in the evolution of the steel frame has given rise to structural and architectural principles of design that are still valid.

The curtain wall does indeed constitute an important innovation in that, under its domination, structural steelwork merges a series of regional, separate, secondary lines of evolution into one major international trend. From this point of view, it is worth mentioning that the first truly suspended and fully prefabricated metal curtain walling on a multi-storey building (actually a reinforced concrete structure, the headquarters of the French Building Federation in Paris) was used as far back as 1949 by Jean Prouvé, the brilliant designer to whom structural steelwork owes so many important developments and stimulating new ideas.

Paris, headquarters of the Building Federation 1949

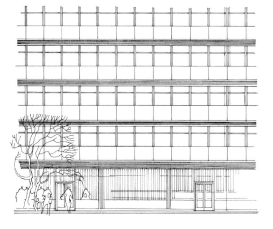

the steel-framed building for Kalichemie in Hanover (1951). The smooth brick skin of such buildings betrays the presence of the steelwork only in the slenderness of the vertical members, but in this way it often produces a more powerful and more restful effect than the agitated narrow-meshed grid façade.

However, these efforts were not destined to have any lasting success. Towards the end of the 1950s there was a rapid increase in the trend towards the curtain wall, the metal-and-glass cladding suspended in front of the structure which had in the meantime gained acceptance in America and was advancing internationally.

International architecture in steel
The curtain wall, 1945–1960

The breakthrough of an international style of architecture in steel was initiated by the work of Mies van der Rohe in Chicago. In 1938, he was appointed professor of architecture at the Illinois Institute of Technology – IIT. During his 20-year tenure he designed first a new campus for the Institute and then many of the buildings that this scheme comprised. It was based on a square planning grid. His fundamental grid dimension of 24 ft is most effectively shown in, and indeed derives its justification from, the almost invariable principle of leaving the structural steel frame exposed, its members painted black to stand out boldly from the glazed areas and the light-coloured brick infilling.

Nonetheless, the structural system is by no means reduced to a regimented pattern. Instead, for each building the framework has been separately worked out in accordance with its own particular functional requirements and the varying arrangements of the rooms, two- and three-storey teaching blocks, combinations of such blocks and wide single-storey halls, and elaborated in every detail: the solutions adopted for the corners, the disposition of the windows and the connections. If Mies van der Rohe had built nothing else but the IIT complex, that achievement alone would have been an important milestone in the architecture and town planning of our century and in the history of structural steelwork.

In his international influence he reigned supreme. For a number of years he dominated design, succeeding also in developing a new system of architectural treatment, a great structural style for the high-rise building. For the skyscraper it was very much more difficult to develop architecture purely in steel than it had been for the low-rise buildings of the IIT. The handicap of having to encase the structural steel members for fire protection could not be avoided in these tall buildings. Perhaps this was one reason why Mies adopted reinforced concrete for his first multi-storey building, the Promontory Apartments in Chicago (1948). In this project, the exposed concrete structure is reduced to slender but prominent columns and recessed spandrel bands. The stepped reduction in the width of the columns with height reflects the diminishing compressive loading and highlights the coherence of the deeply indented construction joints.

Even more influential was Mies van der Rohe's second high-rise building project: the Lake Shore Drive Apartments in Chicago (1949–1950). On plan these two blocks of residential flats have the classic proportions of 3:5, the load-bearing columns of the structural steel frame being at 6.40 m centres on a square planning grid. The principle of allowing the column spacing to provide the fundamental rhythm in the elevational treatment and emphasise the steelwork in a metal-clad façade was achieved by an expedient: the broad-flanged columns are protected by concrete which in turn is enclosed within permanent formwork made of galvanised steel plates welded together. These not only provide a steel casing, but also act compositely with the concrete core so that the steel frame is additionally stiffened and the deflection of the building under wind-loading is reduced. The edges of the floors are similarly faced with steel plates which are situated in the same plane. On this basic structure is superimposed a second system of verticals which

Chicago, Lake Shore Drive Apartments 1949/50

extends over all the upper storeys and comprises steel mullions of joist section simply supported at top and bottom. The mullions are spaced at 1.60 m centres, this dimension being the module which determines the width of the windows and the position of the internal partitions. The mullions need no fire protection as they are non-structural and are located outside the actual building; they do, however, resist the wind pressure and support the aluminium sections in which the fixed glazing is set.

The structural logic of the system is obvious to the observer standing at street level. On the ground floor the glazing of the entrance halls is recessed, the columns with their steel plate casing stand freely in front of it, and the whole structural concept is explicit. At the bottom edge of the façade proper the shape of the mullions is clearly shown. Above, the lines of these vertical members converging in perspective make the building appear even more impressive, there being no need to divide it into separate zones or to provide any cornice to terminate it. In accordance with its internal function, the building is composed of units which are all identical. The sizes of the windows and the spacing of the mullions ensure a human scale. Nevertheless, the viewer cannot escape the impression that these buildings could be extended upwards by any desired amount. At the same time, as in all Mies's creations, the relationship of length and width and height is carefully balanced, as is also the location of the two blocks at right angles to each other. It is this arrangement that gives the ultimate expression to the façade details: from whatever angle one looks at the building, one lateral face appears greatly foreshortened, so that the protruding mullion sections cover the windows and appear to merge into a dark surface. Thus, despite all the transparency and linear decomposition of the real structural frame, the effect of physical mass in the building is preserved.

What intellectual power, how many long years of intensive work were needed to develop steel-framed structures into something really deserving the name of architecture? These are matters which can perhaps best be judged by briefly considering how Mies attempted to perfect his solution for the elevational treatment in the skyscrapers subsequently built at the edge of Lake Michigan. These efforts arose from two visual defects, one of which did not reveal itself until after completion of the Lake Shore Drive Apartments: the geometric imperfection of the I-section mullions. In point of fact, it was simply impracticable to achieve accurate alignment of these vertical members over a height of more than 20 storeys, it being beyond the capacity of rolling mill technology as well as of erection precision.

Mies was able to overcome these defects by interrupting the mullions at the level of each floor by a joint approximately 3 cm deep and combining them with the similarly interrupted spandrel panels to form storey-height units. The visual cut at each floor creates an optical illusion in that the slight variations in alignment are hardly noticeable.

A feature duly allowed for in the design was the varying width of the windows. In order to reveal behind every fourth mullion the sheet-metal casing of the steel column and yet give the mullions equal spacing, the pair of windows on each side of the loadbearing column had to be made narrower than the two centre windows in each bay of the façade. It is possible that Mies came round to the view that in his attempt to combine the modular façade articulation with the major structural forms of the steel frame he had overtaxed the viewer's capacity. At all events, in his design of the Esplanade Apartments and the Commonwealth Promenade Apartments (1955–1956) he preserved the rigorously equal spacing of the I-section mullions and also the perfect corner solution, but made all the windows equal in width, so that he had to abandon the idea of making the structural frame visible on the outside of the building. He set the loadbearing frame members about one foot back from the actual façade and obtained the advantage of being able to accommodate the heating pipes in the space thus gained. He thereby followed, and also promoted, the trend that had emerged in the meantime in international framed building construction with the widespread introduction of the curtain wall. There is irony in the fact that in these blocks of flats the Miesian steel façade system was used to cover up purely reinforced concrete systems.

The curtain wall in a more specific sense has been defined as a lightweight cladding suspended in front of the structural frame and composed of prefabricated storey-height (or even larger) units which perform all the functions of the external wall except the loadbearing function. What this form of cladding aims to achieve and what its advantages are can readily be appreciated: gain of internal space, saving in weight, speeding-up and rationalisation of construction, and, last but not least, the prestige and glamour radiating from glass-and-metal architecture as the expression of the technological perfection of our modern times. It is in this latter idealised sphere that Mies's high-rise buildings have made their real impact. The regularity and architectural quality of his steel façades have served, if not as the example, then certainly as the criterion for the best international solutions. In terms of design, the line of development leading from Lake Shore Drive via the Commonwealth Promenade Apartments requires only one small final step to arrive at the full-fledged curtain wall. In the United States, as in Europe, curtain walling evolved step by step from the infilled wall panel. Here, too, various transitional stages are discernible, as in the well-proportioned integral aluminium-and-glass façade of the Equitable Building at Portland, Oregon, designed by P. Belluschi (1948).

The American high-rise buildings with which the actual era of the curtain wall begins are the Alcoa Building in Pittsburgh, designed by Harrison and Abramowitz, and Lever House in New York, by Skidmore, Owings and Merrill, both built at about the same time, shortly after the Lake Shore Drive buildings. The steelwork of the Alcoa Building comprises a notable novel feature as the cellular steel Q-floor system was extensively used here for the first time, thus realising an all-metal framed structure.

The Alcoa façade is the prototype of the panel construction method: storey-height aluminium sheet panels, stiffened by profiled edges which serve also to interlock the panels and to secure them to the structural frame. The face of each panel is divided into two square areas, one of which is additionally stiffened by having been

Chicago, Lake Shore Drive Apartments, façade detail

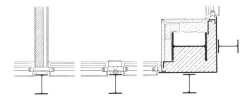

Commonwealth Promenade Apartments, façade detail

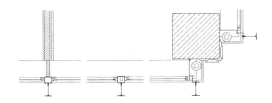

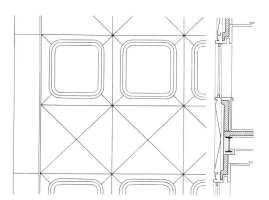

Pittsburgh, Alcoa Building, façade detail

dished, that is, press-formed into a shallow pyramidal shape, concave on the exterior surface of the wall. In the upper half of that panel this dished shape merges into the profiled surround of the vertically pivoted window which is opened pneumatically for cleaning only.

This is not yet a perfectly developed curtain wall, however. It falls short of this criterion in two respects: for one thing, the standard cladding panels on this building do not fill the entire grid area of the structural frame — a narrow strip is erected in front of each column. Secondly, this wall still requires a backing to perform the actual space-enclosing function and to provide a fire-resistant barrier. In this case, the backing was constructed by guniting: lightweight concrete pneumatically applied to wire netting. Although the Alcoa façade can be regarded as a curtain wall, the criterion which differentiates it from the vented claddings subsequently developed is that the window is, in the Alcoa arrangement, incorporated within the wall cladding unit itself.

From the technical point of view the aluminium panel cladding method is highly advantageous: it is relatively simple and economical, it involves no awkward sealing problems, the resilient profiled edge equalises temperature movements and dimensional inaccuracies, and the visual effect of the dished sheet-metal cladding units makes the façade immune to blemishes due to weathering stains or deposits.

The main problem associated with these press-formed aluminium units produced in the same way as motor car bodies is, as soon became evident in subsequent projects of this type in the USA, that this convenient method of shaping the units too easily led to formalistic trivialities which tended to reduce the façades of skyscrapers to the status of fancy wrappings.

The Alcoa Building, a structure of undoubted architectural quality, marked the beginning of a reaction against the strictly structural conception of high-rise building design and initiated a fashionable wave of decorative external treatment that the followers of Mies van der Rohe called 'playboy architecture'.

Lever House in New York embodies the first application of another type of curtain walling: the grid frame curtain wall. Here, the exposed metal components are reduced to an elegant network of horizontal and vertical members: the transoms and the only slightly wider mullions or, to be more precise, not the actual members, but the facing strips which sheath them. The rest of this cladding, which envelops the whole building, consists entirely of glass. This extremely graceful construction was made possible by the use of a backing composed of slag blocks and foam glass to fulfil the requirements of thermal insulation and fire protection. The structural columns are set back so far as to be scarcely visible by day. At night, when the finely articulated, soaring glass structure has been transformed as if by magic into tier upon tier of brightly lighted strips, the whole structural system is revealed — the columns along the flanks and the cantilevered end walls. In its intensity of architectural detailing, in its structural discipline, Lever House is hardly inferior to Mies van der Rohe's blocks of flats in Chicago. As for architectural quality, Mies himself virtually invites comparison with his Seagram Building (1955–1957) situated almost opposite, on Park Avenue.

This last-mentioned skyscraper is regarded by many as his most important work. It is in fact an enhancement or, indeed, an over-enhancement of the Lake Shore Drive architecture in every respect: the height has been almost doubled and the verticality is strongly emphasised by the close spacing of the I-section mullions. The building derives a special aura of grandeur from the fact that the topmost three storeys (accommodating the technical services) have no glazing and that here the mullions continue upwards in front of a closed metal surface which creates the impression of a diadem. Seagram's distinction is tremendously enhanced by the symmetrically disposed plaza in front of it, an unheard-of waste of space by New York standards, and, of course, not least by the famous bronze cladding panels.

Lever House can confidently face the challenge of this comparison. Here, too, a noble material has been employed — indeed, even more noble than bronze, namely, stainless steel. As befits such a material, it has been used more sparingly, the entire system of mullions and transoms being sheathed with it. Again, this building has a spacious forecourt, but now not forming a widening of the street in front. Instead, it is enclosed in the manner of an atrium within the low podium block, where it is screened from the street and is yet in open communication with it by virtue of the wide spacing of the ground-floor columns. The low-rise building inserts itself beneath the skyscraper, which appears to float above it. Lever House has no exalted pretensions; its modest unpretentiousness is indeed better suited to a building which, in terms of financing and depreciation, was carefully costed for a service life of only 25 years.

Lever House, like the Seagram Building, owed much of its initial visual impact and novelty to the contrast with the older and lower adjoining buildings. The original observer, on returning to the scene after five years, must have been rather disappointed. He might even have had difficulty in finding the two buildings, for in the interim a number of other metal-and-glass façades of lesser note had risen along Park Avenue. What is more, it could be perceived that the curtain wall is a very delicate and fickle means of architectural expression, that it demands a high degree of refinement in its proportions, its relief features and its materials, and that it offers only a relatively narrow range of possibilities of expression. The technological character of such a façade requires that it must be cultivated with even great subtlety and refinement in response to the keen stimuli of industrial design.

American architects have endeavoured, not unsuccessfully, to modify the elements of the grid frame curtain wall, the panel cladding system

New York, Lever House 1951

and the Miesian system of mullions, and to combine and refine them. Thus, apparently, the Lever Building's type of cladding was discovered to suffer from a certain weakness of aesthetic treatment in the pronounced differentiation of vertical and horizontal profiling. The experienced architect must, when watching rainwater stream down the façades of this building, have had misgivings about the protruding transoms. In the very elegant curtain wall of the Pepsi-Cola headquarters in New York, the Miesian mullions appear in combination with smooth aluminium spandrel panels.

teething troubles. Its design was further improved here, and it was enriched by the variety of material and profile combinations employed. A typical example was the panel construction method perfected by Jean Prouvé in collaboration with the firms of Studal and CIMT, Paris, comprising storey-height wall units with integral fixed glazing, or with vertically or horizontally sliding windows, and consisting of an inner and an outer skin of aluminium sheet or of aluminium combined with plywood, stiffened by an intermediate frame of steel sections serving also to interconnect the panels, the latter being provided with a heat-insulating backing or filling. Then there was the system used on the Pepsi Cola headquarters, vertical members and smooth spandrel panels, which comprised concealed mullions and transoms and a masonry backing. This reappeared in a more advanced form, a true curtain wall, on the Horten store in Düsseldorf and also, with very subtle profiling and technical refinement, on the two office buildings for the firms of Deckel and Osram, respectively, in Munich (pp. 140 and 313).

At first, structural steelwork received no direct boost from the propagation of the curtain wall. Indeed, it was immediately almost completely associated with reinforced concrete construction, which at the time was still predominant in Europe. This combination was favoured by architects which itself in turn accelerated the trend to set back the structural columns some distance from the curtain wall in order to give the latter an enhanced appearance of transparency and independence from the structural frame of the building. Thus, the extremely graceful curtain wall of A. Jacobsen's town hall at Roedovre is secured to the shallow outer edges of the reinforced concrete floors which cantilever the full depth of the offices from massive longitudinal girders on both sides of the central corridor. The structural system of the Commerzbank in Düsseldorf is of similar design, although here the reinforced concrete cantilevers are left exposed over the open ground floor. In the reinforced concrete buildings of that period the designers preferred to use so-called rocker members (pin-jointed at top and bottom) for the outer columns located directly behind the cladding and to let the wind forces be resisted by rigidly constructed service cores. The increased demands which the transparency and precision of glass-and-metal curtain walling, with its framework of transoms and mullions, imposed upon the designer, led many of the best architects to give more careful consideration to weighing the alternatives, steel or reinforced concrete, and to associate the method of construction selected (or a combination of the two) more actively with the design of the cladding. It is interesting as well as instructive to compare two major buildings completely free from any restrictive ties with the structural frame (in order to achieve an entirely flexible internal layout) and yet to reveal that frame.

In the Pirelli Building, Milan (1955–1957), by Gio Ponti in collaboration with P. L. Nervi, where the plan is somewhat spindle-shaped with pointed ends, the floors span longitudinally and are unusually long (20 m), being of prestressed concrete in order to keep the deflections to a medium. The two massive transverse shear walls, which taper upwards and are each divided into two piers, are visible through the curtain wall – yet just how this unusual structural system performs its function is not immediately apparent even to the expert who examines the building from the outside.

The administrative building of the Caisse Centrale d'Allocations Familiales, Paris, was built in 1955–1958 by the architects Lopez and Reby and the engineer Pascaud. It has a structural steel frame with robust girders spanning a distance of 9 m and cantilevering 5.5 m at both ends. The first and last of these transverse girders, together with similarly cantilevered longitudinal girders, show through the end faces of the building. The curtain walling on the flanks is suspended from the top row of cantilevers and continues past the floors below (p. 204). The alternation of spandrel strips with translucent polyester infilling and narrow bands of window which extend all round the building, as if to embrace it, and the syncopated arrangement of the staggered mullions all combine to give the façade a particular rhythmic charm and an amazing effect of contrast by day and night. France and Belgium produced a number of other excellent solutions to achieve clear-cut articulation of the façades. Thus, external columns either show through the curtain wall, as in the administrative building of the Caisse Centrale de Réassurances, Paris, built in 1958 by the architects Balladur and Lebeigle and very skilfully fitted into the restricted space that its narrow site provided; or, alternatively, the outer flanges of the external columns actually emerge on the outside of the curtain wall, as in the 14-storey building of Prévoyance Sociale, Brussels, by the architect H. van Kuyck, 1956. However, in order to give these loadbearing members the necessary slenderness in conjunction with the desired close spacing of the windows in the latter building, the main structural system, comprising robust multi-storey frames stiffened in both directions, had to be located entirely in the interior of the building.

To conclude this section, two buildings in Düsseldorf are worthy of note as outstanding achievements of the curtain wall era: the Mannesmann Building, designed by Schneider-Esleben and Knothe, and the Phoenix-Rheinrohr Building, by Hentrich and Petschnigg, both of which were begun in 1957. Which deserved to be rated architecturally superior to the other was long a matter of controversy among German architects. On the whole, opinion appeared to favour the Mannesmann Building.

The intensity and perfection of detailing, the harmony of cladding pattern, structural system and internal arrangement, the sweeping effect of the narrow soaring tower directly overlooking the Rhine but alongside the squat mass of the old administrative building – all this is indeed alluring. If the Mannesmann Building, despite these advantages, has meanwhile lost ground to its rival, it is mainly due to the re-emergence of the time-honoured principle of triple subdivision – particularly emphasised by the hinged bearings for the tubular external columns on the ground floor. The impression of grandeur is further enhanced by the severe closed configuration of the ground-plan and by the visual isolation of the free-standing building rising solidly from its narrow base at its prescribed distance from the old building. The elegant curtain wall comprising aluminium mullions and transoms, glass, and enamelled steel sheet continues all round the building. The tubular columns behind the façade are pin-jointed; the wind forces are transmitted to a concrete core which is located close to one of the longitudinal faces and further restricts the available space within the rather cramped area on plan.

The Phoenix-Rheinrohr Building benefits by its more spacious setting at the edge of Düsseldorf's public park. This skyscraper comprises

Paris, Caisse Centrale d'Allocations Familiales 1955/58

Equally attractive is the graceful façade of the Federal Standard Bank in San Francisco where the grid frame curtain walling has prominent vertical members faced with stainless steel and a rhythmic spacing of the mullions (a b a b); in alternate storeys the sequence is reversed (b a b a). In this way, a subtle diagonal structuring is obtained which is, as it were, superimposed upon the pattern of vertical members of the façade.

When the curtain wall, from about 1955 onwards, began its advance into Europe and made rapid progress, it had already overcome its three slabs, varying in height, with which the architects produced a brilliant example of the open form of design. The building is visually enlivened in a restrained manner by its stepped configuration in elevation and on plan. This design conception, which had been successfully exploited in the RCA Building, New York, twenty-five years earlier, has in this case been carried through to its logical conclusion. As in the Mannesmann Building, the tubular columns are placed behind the façades, but they are more widely spaced and are always located at the centres of the windows. The variable internal layout is simplified by this arrangement and there are no corner problems. The wind forces are not transmitted to a concrete core, but are resisted by bracings in the end faces of the three blocks (p. 214). The structural function of these blank faces is further emphasised by a change of material. Instead of aluminium, as on the longitudinal faces, the ends of the building have stainless steel cladding.

All the deliberations that must have preceded this design achieved in close collaboration between architects and engineers — structural calculations, access arrangements, internal layout, natural lighting, technical equipment, etc. — here appear to be combined, geometrically crystallised in an impressive centrally symmetrical figure: the 'beautiful' ground plan. This plan is a showpiece by means of which the layman, and indeed some experts, too, can clearly be shown what real architecture is all about and that there is more to it than just suspending a curtain wall in front of a structural frame.

Reactions against the strict glass-and-metal architecture

The reaction against functionalism in international architecture, the revolt against the severely constructive discipline that Mies van der Rohe imposed in Chicago, started at quite an early stage, as we have seen. This opposing trend was a setback for structural steelwork, initially severe but fortunately not permanent.
More serious and of greater significance than the superficial wave of fashion initiated by the aluminium cladding for the Alcoa Building is a modified architectural concept, a new expressive spirit, which reveals itself in the work of Eero Saarinen. His first major project, the extensive complex of the General Motors Technical Center at Warren, near Detroit, built 1946–1955, still shows the strict disciplinary approach of the Miesian school. The overall plan is a counterpart to the IIT in Chicago, though emotionally enhanced by the large artificial lake with fountains, the stainless steel tower rising out of the water and the domed exhibition hall.
The groups of buildings arranged on a strictly right-angled pattern and comprising offices and workshops clearly reveal their structural steel frames. In comparison with that used on skyscrapers, the cladding has been greatly simplified: the columns spaced at 1.60 m centres — very slender despite their casings and sheet-metal facings — are set somewhat forward; they enclose windows of green-tinted glass and spandrel panels of enamelled steel sheet. The whole taut graceful structure of the long principal façades is gripped between end walls entirely of glazed brickwork with gaily variegated colours reminiscent of Latin American traditional ceramic facings which architects, more particularly in Mexico, have tried to revive.
An even sharper contrast in materials was employed by Saarinen in the IBM Research Center at Yorktown Heights, NY. This two-storey building, approximately 1000 ft in length, is arranged in the shape of a circular segment around a hillock, the impression of vast length being enhanced by the vertical articulation of the façade. The concourse which extends along the entire front of the building is used for recreation and to admire the view. The actual laboratories and research modules located in the interior of the building are screened from the outside world and artificially lighted. The end walls and the sparsely glazed rear wall consist of rustic, almost massive masonry. Closer inspection reveals, however, that it is not composed of large blocks carefully jointed to produce a cyclopean effect, but of stone slabs providing permanent formwork for a concrete wall to which they are attached as facing. This somewhat dubious technique makes a particularly disagreeable impression when combined with the tapered features which support the curved canopy projecting out of the glass wall at the main entrance to the building.

Düsseldorf, Thyssen Building (formerly Phoenix-Rheinrohr Building) 1957/60

The urge to achieve livelier, not to say coarser, effects and the desire to include conventional building materials were no longer confined just to the surrounds and trimmings, but spread to the design of the curtain walling itself. This was brought about not only by a change in aesthetic taste, but also by practical considerations. The classic curtain wall composed of metal mullions and transoms with sheets of glass was, as soon became apparent, not only very awkward and incapable of much practical modification, but also, when properly constructed, rather expensive and in many instances simply not compatible with functional requirements and building regulations. On the other hand, the principle of the curtain wall, the subdivision of the façade into storey-height prefabricated units, had come to stay. As a result of architects' efforts to modify the design and use different materials to make the curtain wall less expensive, more impressive and more interesting, three types of cladding emerged:

1. The 'screen', a network of metallic components, pierced ceramic units, etc., suspended in front of, and concealing, the featureless external wall which has only a few windows or none at all.

2. Artificial stone slabs fixed in front of the wall and providing richer and bolder articulation of the façade, besides allowing the fenestration to be reduced.

3. Natural stone slabs fixed in metal frames or freely suspended in front of the wall.

prisingly transparent but produces pleasing shadow effects.

The screen is closely related to the brise-soleil, the baffle arrangement which, following a suggestion by Le Corbusier, was used on South American steel-framed buildings as far back as the 1940s. In the administrative building of the Pan-American Life Insurance Company (Skidmore, Owings and Merrill, 1950–1951), almost a contemporary of Lever House, vertical aluminium baffles spaced about 1 m apart are fixed between the continuous balcony slabs cantilevering from the floors. As a result, not only is sun protection provided, but also the possibility of cleaning the windows from the outside.

Progressive European architects were at first inclined to despise the screen as a symptom of decadence, but soon they were eagerly using it, for example, on the façades of department stores. A very successful feature from the commercial point of view was that a pierced metal screen could be used to give a facelift to some dilapidated stone façade, thus dressing it up in

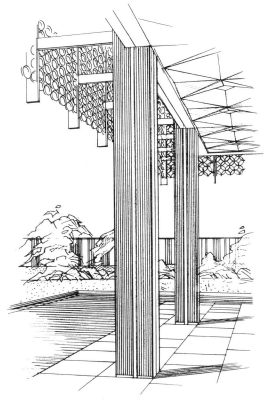

Detroit, sales office of Reynolds Metals Company

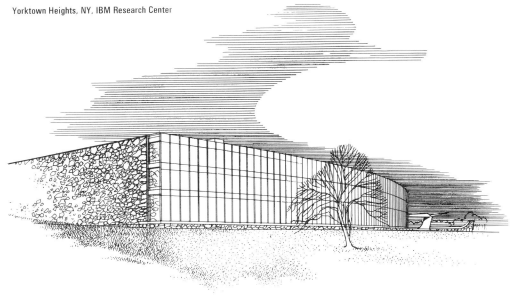

Yorktown Heights, NY, IBM Research Center

The principle of the screen as a time-honoured cosmetic feature can often be observed in historical architecture. Credit for its rediscovery for modern framed structures is due to the American aluminium industry, which obviously had a commercial interest in this development. The sales office of the Reynolds Metals Company, Detroit, embodies this form of cladding with the evident intention of publicising it. The architect, M. Yamasaki, has concealed the steel frame behind a sort of chain-mail which blends very well with the exposed steel columns on the ground floor and which, when viewed from within the building, is not only sur-

a fashionable new garment, as it were. Particularly, this application of the screen could, if correctly understood, serve as a salutary lesson to the architect not to indulge in false illusions as to the length of life of his creations but to grasp the truly positive aspect of this form of construction, namely, that the screen can be dismantled as easily as it is installed.

To fix precast concrete or artificial stone slabs in lieu of metal panels on the outside of the external walls of a building was an obvious development. An early attempt in this direction was made by the architects Harrison and Abramovitz in their design for the Wachovia

Bank at Charlotte, NC, unmistakably following the example set by the Alcoa aluminium panels in Pittsburgh. In this concrete cladding, the basic thin-walled elements are 'plastically' enlivened and also stiffened by folding and by profiling their edges. In this case, auxiliary columns had to be provided between the edges of the floor because the height of the elements is only half the storey height and they are staggered chequerboard fashion between window zone and spandrel zone. The heat-insulating vermiculite plaster was applied directly by pneumatic gun from the inside. An additional casing around the auxiliary columns, and also enclosing the services, creates the impression, from within the building, that the external wall consists of thick loadbearing piers. From the outside, too, there is a pronounced effect of massiveness, the glazing being even less than in the Alcoa façade.

More elegant and entirely in keeping with the essence of modern framed construction is the ingenious cladding of the 28-storey building for the Michigan Gasworks, Detroit, also by M. Yamasaki, 1961. Here, the window openings are extremely closely spaced — at about 50 cm centres. They extend from floor to ceiling and are bounded at top and bottom by a broken line, so that the pane is shaped like an elongated hexagon. These features produce a curious overall decorative effect, giving the impression almost of a screen, a sort of oriental harem grille, blending well in scale with the eclectic architecture of the adjacent skyscrapers. What is more, the material is strikingly attractive: the cladding units have been made with glistening white quartz chippings which closely resemble the marble slabs with which the external columns are faced and which give prominence to the loadbearing structure.

This cladding is essentially a mullion-type concept, the two-storey high units fixed on the outside of the wall each comprising a narrow rib out of which the shallow floor strips project as from a quill. Therefore, the vertical joint is not located in or on the mullion, as is normally the case, but in the middle of the spandrel. In addition, since the joint is situated in an obtuse angle, the whole arrangement is largely insensitive to dimensional inaccuracies.

Among the American architects of the anti-Mies trend, Yamasaki has developed a special talent to shake the complacency of the purists. He first tried out his artificial stone cladding on a four-storey university building, on which it still makes something of an eclectic Victorian-Gothic or Indian impression, because the superimposed framework is absent. At the time, the majority of architects thought it shocking that he then proceeded to apply this exotic decorative filigree work to a skyscraper. Yet this procedure, as an application of the additive principle, is perfectly legitimate. After all, architects did not shrink from applying the same type of bronze cladding as that on the Seagram Building to much lower buildings in places far away from that prototype, such as the Parliament Building in Stuttgart. The Yamasaki cladding has found similar widespread application; it has been copied not only in the USA, but also in Britain and Germany, sometimes transmuted into metal and not infrequently coarsened or watered down. In order not to lose sight of the development of structural steelwork, it should also be noted that the high-rise building of the Michigan Gasworks was, at the time, the world's largest all-welded steel-framed structure.

The Pan-American Building in New York, completed 1963, which was designed by the TAC architects' team with W. Gropius, E. Roth and P. Belluschi, shows a stricter approach to elevational treatment in its precast concrete cladding. Here, there is a panel construction system, Yamasaki's concrete being embellished by the use of quartz aggregate. The primary units, stiffened at the vertical edges by protruding ribs, are staggered in chequerboard fashion from storey to storey and enclose the spandrel panels and windows. The façade is characterised by the regular alternation of window and spandrel areas with boldly protruding mullions. Unmistakably, the second version of the Miesian cladding system served as the example for this. With the bolder and coarser relief necessitated by concrete, it was logical to emphasise the horizontal division between the vertical ribs and thus clearly reveal that they do not perform a loadbearing function, but serve merely to articulate the façade. Here, the structural columns do not appear in the curtain wall itself, but they are visible in the deeply recessed glazing of the two floors for mechanical services, which effectively emphasise and enhance the vertical treatment of the building.

The combination of concrete cladding units with a steel frame is not in itself any more inappropriate than that of metal cladding units on a reinforced concrete frame, even though it conflicts with the time-honoured principle that the lighter component should always be carried by the heavier. The designers of the Pan-American Building must certainly have weighed the disadvantage of the heavier loading placed upon the steel frame against the advantage of the lower cost of the concrete cladding. The fact that the decision went in favour of the latter can be regarded as further evidence of the efficiency of structural steelwork, which in this instance takes on an adventurous character: the 246 m high tower block straddles the tracks of the Grand Central Station and its foundations had to be so constructed as not to interrupt or interfere with railway operation.

In the general aspiration to achieve richer, representative effects in architecture, it was unavoidable that designers went back to natural stone, the well-tried building material of historical monumental architecture, and used it in curtain walling. Characteristically, it was banks and insurance companies that prompted the first attempts in this direction. Again it was Harrison and Abramovitz who started this fashion, having already demonstrated their fondness for natural stone in the marble facing of the huge end walls of the UNO Building in New York. In the case of the CIT Financial Corporation's administrative building in the same city, 1955, the masonry piers and spandrels are conventionally faced with polished granite slabs bedded in mortar: the stainless steel cover strips over the joints create the impression of a mullion-and-transom curtain wall.

Some years later a natural stone façade in the grand manner was realised by the architect Albert Kahn in his new building for the National Bank in Detroit. Here, slabs of polished crystalline marble are enclosed within stainless steel frames alternating with the storey-height stainless steel frames of the windows and spandrel panels. In terms of expressive power, however, this façade does not come up to the high pretensions of the expensive cladding material employed.

To the expert and connoisseur it appears to be something of a comedown when natural stone is enclosed in a framework of metal; but without this expedient it would not be possible to make storey-height independent panel units of natural stone. If stone slabs within the size limits imposed by the material are fixed as pier and spandrel facings directly to the structural frame or to the infilling, the resulting construction is not really a curtain wall, but a vented facing, the

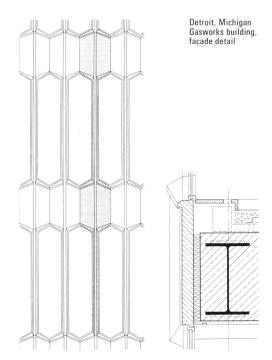

Detroit, Michigan Gasworks building, façade detail

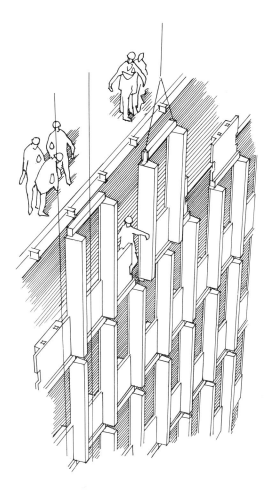

New York, erection of the cladding, Pan-Am Building

windows having to be fitted into the masonry backing. Claddings of this type are not specifically associated with framed building techniques, at least not with steel-framed construction, but they could perhaps be regarded as an improved version of the masonry-faced grid-type façade.

The revolt against functionalism, against the austere and cool Miesian ideal, had reached an acute stage by the end of the 1950s. In lieu of an objective architecture with emphasis on the structural features and characterised by smooth surfaces and simple colour contrasts, there now came design aiming at subjectively expressive, plastically animated picturesque effects. This architecture did not shrink from using historical or folklore motifs and embellishments — not to create some eclectic new style, but as requisites for achieving enhanced expressiveness. The swing was accomplished with such vehemence that people were inclined to suspect that it reflected some underlying law governing architectural evolution, a change in phase in the great alternation of classic and romantic trends as manifested at various stages in the history of architecture in modern times. The material best suited, and preferred, for this kind of forceful 'plastic' or 'sculptured' architecture is concrete, and thus this big switch was of major importance in the evolution of framed construction and greatly to the disadvantage in particular of structural steelwork.

In 1961 the 20-storey administrative building for the Hartford Insurance Company was built by Skidmore, Owings and Merril — a steel-framed structure with a concrete core and mushroom floors supported by columns arranged on a square grid. With this building, in Chicago of all places, the invasion of concrete into the domain of American structural steelwork, and more especially into that of high-rise commercial buildings, became an established fact. Here fully revealed is a principle of architectural design, initiating a new stage in the evolution of modern framed construction, which, as foreshadowed in the international steel-framed architecture of the 1950s, was also subsequently to find its most brilliant expression in structural steelwork: the prominently exposed structural skeleton with an obviously recessed façade.

The unusual sculptured quality of the structural system produced an almost shocking effect in the Hartford Building — more particularly in juxtaposition with the metal-clad façades of neighbouring buildings — where the mushroom column caps intersected the external wall surface. These features formed the transition from the shallow floor slabs to the outer columns and created a visual impression of vault-like curvature. Involuntarily, people were reminded (and the granite facing contributes to this impression) of the rather bombastic fascist monumental architecture of the Palazzo della Civiltà Italiana, which was erected for the international exhibition that was to have been held in Rome in 1942. That building, because of the numerous deep holes in its façade, was popularly nicknamed 'Gruyère'. Nowadays, the Hartford Building probably no longer produces that impression, as one has become rather accustomed to the startling sculptured effects achieved by international architectural concrete in the years that have since elapsed.

In 1964 the first of the two circular Marina City tower blocks was erected in Chicago. Here, the motif of slabs arching out from supporting columns, as in the Hartford Building, has been singled out for further elaboration and been developed to its full plastic potential. Jutting out from radial walls in a space-curve arrangement are semicircular balconies to which the sector-shaped residential flats have individual access and which give the structure its striking corn-cob configuration. The Marina City towers are particularly indicative of the new status that concrete construction had achieved at the time. They are built of structural lightweight concrete made with special lightweight aggregates, whereby a saving in weight was achieved without loss of compressive strength, besides providing a pleasing surface finish. Though such concrete does not attain the favourable strength:weight ratio of steel, it does move closer to it.

If we are to name one building that can be said to sum up the architectural and structural trends of this new stage of development and to point in new directions, it is the CBS Building, New York, a reinforced concrete structure by Eero Saarinen, 1961–1964. In comparison with some of the extravagances of other specimens of 'Miesless architecture', this building gives an impression of great discipline and simplicity. Actually, it is the sharpest challenge to Mies and, in particular, to his Seagram Building.

If it were ever possible to enhance the expressive power of a free-standing skyscraper, then it has indeed been achieved here. The tower is completely free, the façade articulation and the loadbearing structure are identical, both piers and window strips are 5 ft wide: there is only one basic dimension and virtually only one material. The entire building, together with the sunken plaza from which it rises, is faced and paved with granite slabs. No trace of any hierarchic three-stage elevational subdivision or stepped configuration is to be found; no increase in the basic planning dimension has been permitted, not even for the main entrance doors; everything is now unreservedly concentrated on verticality. Despite the extreme reduction in architectural design features, the CBS Building does not appear in the least austere or formal. Quite the contrary, the consummate mastery of shape and accomplished design displayed in this structure transcends the spirited elegance of the Reliance and the structural power of the Seagram Building, and enters the domain of magical fantasy, a domain in which hitherto only the architect's father, Eliel Saarinen, had operated, though admittedly with greater circumspection, with his design for the Chicago Tribune Tower, which was not actually built.

That the CBS Building appears so remarkably disembodied, almost phantom-like, is due primarily to the careful choice and treatment of the stone with which it is clad: dark grey Canadian granite which, as a result of thermal action, takes on the texture and lustre of newly quarried metamorphic rock. At the corners the piers merge together, forming a diagonal area of double width, one of the most striking corner solutions in modern framed construction. This chamfered corner treatment relieves the prismatic structure of a sense of hardness and imparts a refined precision to it.

But the really ingenious thing about Saarinen's CBS design is the very high degree of unity of structure and shape it achieves, the integration of the loadbearing system, sophisticated technical equipment and architectural design. In all the upper storeys the piers are of the same external section, but the internal cavity increases in size in the upward direction as the return flow ducts of the air-conditioning system (which are accommodated in these piers) increase in diameter. In comparison with this solution, the stepped-back piers of the Miesian steel-framed structural concept, which were evolved purely from considerations of steelwork design, appear almost archaic.

Yet the CBS Building, too, has an archaic quality which is shown in its shape on plan. The piers are set so close together that this can scarcely be called framed construction: it is more like a continuous outer wall pierced by window slits and comprising solid corners. Between this outer wall and the inner wall enclosing the core of the building are ribbed floors which help to stiffen these walls which are anchored like two enormous square tubes to the foundation. In this way, a structure of tremendous rigidity is obtained, thus making it possible to construct a skyscraper in reinforced concrete, previously regarded by experts as a utopian idea. The tube construction principle, here for the first time elevated to a dominant architectural feature, also began to be used for steel-framed buildings. A few years later it provided the means to set up new height records for American skyscrapers in which steel conclusively and impressively re-established its supremacy as a structural material.

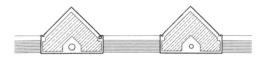

New York, CBS Building,
cross-section of piers and plan of floor grid

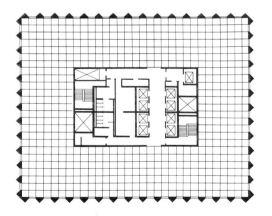

The exposed steel frame

All in all, international architecture since 1950 presents a confusingly complex picture: tremendous expansion of architectural and technical potentialities, abrupt changes, fierce controversy between contrary views and design trends. To form a well-founded opinion of one's own, and indeed even to keep oneself informed of what is happening in this struggle over new ideas, new conventions and new criteria, is made harder rather than easier for the practising architect by the flow of information issuing from a lively and active architectural press and by the growing influence exercised by theoretical speculation and polemic discussion.

Characteristic of the hectic conditions in present-day architecture is the fact that the most daring projects conceived by leading architectural firms, for example, Saarinen's dragon-winged shell structure for the TWA airport terminal building or the complex fully-sculptured prefabricated façade for the Lambert Bank in Brussels, designed by Skidmore, Owings and Merrill, are internationally publicised when still only in the model stage, deceptively realistic though these models admittedly are. It is almost as if the authors of such schemes fear that some other architect may forestall their ideas or otherwise that these might lose their novelty and interest by the time the buildings are actually constructed. Architectural theory has got itself into contortions over the concept of 'brutalism'. In his book bearing that name, R. Banham has described the circumstances with commendable accuracy: according to him, this frequently misunderstood catchword which was being used by critical architects and architectural critics had become a kind of legend, even before there actually was anything like a brutalistic trend in architecture.

It is not part of our present purpose to define or to criticise concepts of historical art or of architectural aesthetics, or to forecast future architecture. To confine ourselves to our actual subject, the steel-framed building, there are undeniably some positive developments and achievements to be recorded in the last fifteen or twenty years: despite having to face keener competition, structural steelwork as a whole is steadily moving ahead. An increasing trend to identify and publicise the prominently exposed external frame both functionally and structurally naturally leads to new forms in architectural steelwork. These endeavours presuppose, and necessitate as a consequence, the devising and development of fundamentally different solutions for fire and corrosion protection, new physical concepts and new structural systems.

Let us first turn our attention to an older structure which provides a positive starting point for considering current and future structural steelwork, the more so as it was, or perhaps despite the fact that it was, hailed as a landmark in the history of architecture: the prototype of 'brutalism'. This was the secondary modern school at Hunstanton, England, designed by the architects Alison and Peter Smithson in a competition in 1949, not long after the two most influential buildings of the immediate post-war period: Le Corbusier's Unité d'Habitation at Marseilles, 1948–1954, and the first institutional buildings produced by Mies van der Rohe in Chicago. The Hunstanton school reconciles, as it were, the concrete solidity of Marseilles with the architectural steelwork of the Illinois Institute of Technology. The symmetry of the building, its rectangularity and compactness, the decisiveness with which the steel frame stands out from the glazed surfaces and the brickwork — all these features derive unmistakably from Mies van der Rohe. But the detailing of the steelwork is devoid of pretentious geometric perfection, and the unconcerned manner in which the ancillary structural features and secondary materials, ducts and grilles of the mechanical services are allowed to show, is much freer, if not more careless, but basically more functionally appropriate.

With the Smithsons' school Britain regained a trace of the leading position in international architecture that she had enjoyed in the nineteenth century. The British had found a way out of the dilemma of pure sterile formality, on the one hand, and expressive indiscipline, or lack of restraint, on the other — thanks to their specific national temperament, their aversion to doctrinaire systems, their tendency to understatement and predilection for witty and eccentric solutions.

Visibly exposed steelwork as a principle of co-ordination and as a means of architectural design — and as brought to life in the IIT complex, although not without difficulties with the building authorities — had initially been passed over for multi-storey buildings. Protagonists of steel must have felt this consequence of the ascendancy of the curtain wall as a retrograde step, an impoverishment, as there had been no lack of attempts to leave the structural frame exposed in the cladding or to make it visible through the curtain wall. Around the mid-1950s, however, the first multi-storey buildings with exposed structural frames were erected. The initiative in this direction probably came again from Mies van der Rohe with his design for Crown Hall, the headquarters of the architectural faculty in the IIT, 1952–1956. Here, the external steel portal frame, which in itself is nothing fundamentally new in single-storey architecture, has been carried to its inevitable conclusion and raised to its highest geometric austerity.

With the Inland Steel office block, 1954–1957, Chicago regained its leading position in the development of the commercial multi-storey building. If the separation of functions is taken to be the basic principle of framed construction, then this building designed by Skidmore, Owings and Merrill must be rated as the ideal framed structure. The 960 m^2 area of each floor of the actual office block, 19 storeys in height, is completely free from any obstructing internal feature: within the modular grid any internal layout can be achieved. Lifts, stairs, lavatory blocks and other services are combined in a square tower connected to one end of the longitudinal façades. The legs of the 18 m span welded multi-storey portal frames are positioned on the outside of these façades. The cantilevered ends of the block and the recessed glazing in the two bottom storeys do their utmost to give full visual prominence to the uniaxial structural system. The legs of the frames are encased in concrete and are, like the rest of the main structure and the windowless service tower, sheathed with stainless steel.

In 1955, in Germany, a more modest building was erected which represented an important new development: it was probably one of the first multi-storey office buildings having externally exposed steel columns without fire protection. For this administrative building, which the firm of MAN designed and built for

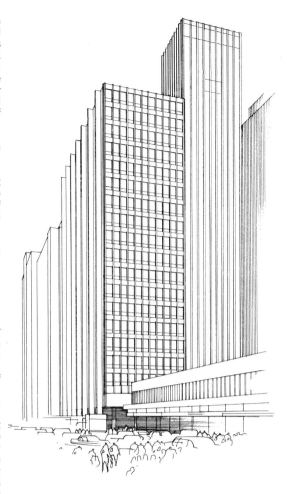

Chicago, Inland Steel Building 1954/57

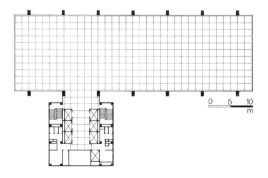

itself at Gustavsburg, the structural treatment was very carefully worked out to obtain a high-quality steel-framed building. Besides considerations of economy and short construction time, the designers' aim had been, from the outset, to leave the steelwork visible. The external columns were erected 15 cm in front of the façades. This arrangement was subjected to exhaustive tests with regard to the behaviour of

35

Gustavsburg, office building for MAN 1955

these members in the event of an outbreak of fire and with such success that the building authorities even allowed the use of wood for the cladding units.

From about 1960 onwards, the exposed structural framework became a generally accepted feature. Again it was Eero Saarinen who provided the main impulse in steel-framed construction with the administrative building for John Deere and Company at Moline (design begun in 1956, but not built until 1961–1964). What the architect as well as his clients, an old-established American firm manufacturing agricultural and earthmoving machinery, wanted was to display the steel structure in the fullest sense of the word. The frame has been externally exposed with the aid of a system of sun-protection galleries encircling the building on all floors. The columns and main girders are exposed on the longitudinal flanks and the projecting secondary girders on the end faces, while the encircling ancillary girders serve for the attachment of the vertical and horizontal sun-baffles and grilles. In this whole framework, no member is butt-welded to any other; at all the connections the main, secondary and auxiliary girders either bear one upon another or interpenetrate, so that all the sections are visible.

Although the large quantities of rolled sections and plate girders for sun protection are justifiable on account of their advantage in relieving the air-conditioning system of some of its functional load and although the system does not constitute a visual obstruction from the inside of the building, it would have added a considerable maintenance burden if this whole arrangement had not from the outset been planned as a demonstration of an important new technological development: weathering steel, in this case Cor-Ten. This is a special high-yield stress steel which, on exposure to the weather, forms a protective oxidised coating which prevents further rusting. This was the first structural use of weathering steel, but such steels had been originally developed by the American Steel Industry to resist abrasion and corrosion in railway wagons for the transportation of ore and coke. The purple-brown patina which forms on Cor-Ten steel, together with the play of shadows and reflections on the recessed glazing, gives the structural frame a particular distinction. In the manner in which it appears to grow out of the landscape it is a monument to the taming and cultivation of the American continent.

The concept of enlivening the exposed steel frame by means of sun-protection galleries appears in elegant form in the German Embassy in Washington, a building designed by Egon Eiermann in 1958 and completed in 1964. Here, only the outer columns of the structural frame are exposed unencased. The balconies which extend round the building are erected outside the actual frame: graceful cantilever brackets fabricated from perforated steel plate set on edge and interconnected by channel sections; slender tubular steel posts to which the railings and the vertical sun-baffles (made of timber boards) are attached. The balconies with their lattice-screen floors give additional sun protection and provide access for window cleaning; what is more, they serve as means of escape in the event of fire. It was probably this circumstance that made it easier to secure the approval of the building authorities for the exposed structural columns in combination with Oregon pine window frames.

A counterpart to Saarinen's Deere Building is Stirling's project for an administrative and research centre for Dorman Long in Middlesbrough, England. The 14-storey elongated building gains tremendous dynamism from the fact that the lower half of the fully glazed front slopes; this powerful effect is counterbalanced at the rear by the external service tower. The broken outline of the structure is followed by the exposed columns of the multi-storey frames, the beams of which span the full depth of the building. Wind-bracings and longitudinal stiffening girders complete the impression of vast power. The whole external steel frame is at a distance of 2 ft from the glass cladding, as required by the building authorities. Indeed, grappling with the official requirements appears to have been a major consideration in this design: the external stubs of the beams in each structural frame carry a fire-protective casing like a thick sleeve. With this design for the Dorman Long building, Stirling has shown that brutalist architecture is possible in steel, too, and that a building can thus be given a very distinctive image.

A physical consequence of the exposed structural frame is that the temperature variations of the external air are directly communicated to the structural members of the building, which is something that could be — and generally was — prevented in the conventional curtain wall and with the earlier infilling systems by means of suitable insulation. As a result of the detailed examination of the ever-increasing number of steel-framed and concrete-framed buildings of this new type erected since 1960, it has been possible to confirm that the transmission of temperature stresses into the supporting frame — which must, of course, be duly taken into account in the structural calculations as well as in the design of the air-conditioning plant — is not necessarily disadvantageous. On the contrary, the thermal storage capacity of the building facilitates the equalisation of temperature variations and stresses, whereas these phenomena are liable to cause difficulties in structures enclosed within curtain walls.

Thus, a new possibility emerges as the third stage of development of modern framed construction, after the infilled skeleton wall and the curtain wall, namely, the set-back or recessed façade. Apart from the bolder and more expressive elevational treatment that it provides, this form of construction offers various structural advantages. The solution of problems of sealing,

Moline, Illinois, John Deere Building 1962/64

of forming the connections, of dimensional accuracy of the cladding units, is substantially facilitated. A certain amount of weather and sun protection is obtained and it also becomes possible to install and maintain additional external sun-protection devices and to provide more convenient arrangements for cleaning the façade.

The transition from insulated to non-insulated structural frames, from closed claddings to externally exposed steelwork, is very compellingly demonstrated in low-rise buildings, a type of structure which emerged more particularly in the USA in the 1950s. These are two-storey or three-storey buildings, of essentially lateral development, sited in open country, with internal courtyards to admit daylight to the offices or, alternatively, comprising combined utilisation of all available internal space for offices, laboratories, workshops, drawing offices, store rooms, etc., occupying the full depth of the building. Besides the advantage of being more agreeably and peacefully situated in the country, the low-rise building has the merit of providing better internal communications and, if the structural system and the technical services are suitably planned, a high degree of flexibility in its multi-functional utilisation.

Typical early examples of low-rise buildings dating from the latter part of the 1950s are the Connecticut General Life Insurance Building at Hartford — which attracted notice as the first building to have systematically designed large open-plan offices — and the administrative building for the Reynolds Metals Company at Richmond, Virginia, both designed by Skidmore, Owings and Merrill. They have the smooth glass cladding and the refined dimensional scale of the curtain wall era. The closely spaced aluminium-sheathed outer structural columns in the upper storeys of the Reynolds Building are so slender that they might well be regarded as mullions, just as in Saarinen's institutional buildings for the General Motors Center, which also makes an interesting contribution to the evolution of low-rise architecture.

In marked contrast to the grace of these façades is the force of the prominent steel frames in the latest low-rise buildings, such as the Engineering Building of the Armstrong Cork Company at Lancaster, Pa., by Skidmore, Owings and Merrill (1965). The low storey at ground floor level, accommodating offices and social facilities, has columns spaced at intervals of 5.00 m and 8.75 m. Only the external columns of the longitudinal flanks rise into the 4 m high upper storey, the interior of which is entirely unobstructed by columns. Inside the continuous open-plan drawing offices, based on a 1.75 m modular grid, separate conference rooms which can be varied in size and location are formed by means of room-height glass panels. The glazing, which is continuous around the building and comprises mullions at 2.50 m centres, is set back 2.50 m behind the structural steelwork. The way in which the columns and fascia girders in the two storeys have been differentiated in respect to their loads and spans and yet have been brought into perfect geometric harmony, the way in which the connections have been detailed, all this derives unmistakably from the strict school of Mies van der Rohe, only being even harsher, even more direct in its architectural expression.

A tremendous revival in American high-rise construction started in 1963. Chicago retained its leading position: of the world's five tallest buildings there are now no fewer than three in that city, which saw the origin of the first multi-storey office buildings some 90 years ago. For this youngest generation of American skyscrapers the externally exposed structural frame is particularly important and characteristic — not only as a means of architectural design, but also as a basis for novel, highly efficient structural systems and design methods in steelwork which were the prerequisite conditions to enable the height of commercial and residential buildings to soar to 40, 60 and eventually to 100 and more storeys, without involving an excessive increase in construction cost per unit area of usable space.

The higher a building becomes, the more the transmission of horizontal forces and the provision of rigidity to resist wind forces become the dominant structural design problem, the deciding factor with regard to the efficiency and economy of structural steelwork. The various types of stiffening or bracing systems which have been evolved in the USA in the last ten years are milestones in a notable sequence of development in modern structural engineering. But they also represent the hard-fought rounds in a contest between steel and reinforced concrete as rival materials. From the outset of this evolution, concrete was in competition with steel and indeed, thanks to its monolithic character and the achievements in the technology of structural lightweight concrete, it gained advantages over steel and succeeded in extending its range of application, until steel-framed skyscrapers rose to heights so great that concrete was unable to keep up with its rival.

The most important structures of this period were the result of close collaboration between architects and engineers. In such projects the engineer is intimately associated with the design from the very outset — from preliminary investigations and the first draft schemes onwards. The more exacting functional requirements and economic conditions, the more intensive elaboration, the higher demands made by design in terms of intellectual effort and scientific supervision — these are factors that impart to such buildings a sense of force and power unprecedented in the architecture of modern times. Among the engineers who invented new structural and stiffening systems and who established new design methods for high-rise buildings only two will be mentioned here: Fazlur Khan and Myron Goldsmith, the latter an architect as well as an engineer; both are partners in the firm of Skidmore, Owings and Merrill and both teach at the IIT.

As early as 1959, with its design for the insurance building at Hartford, this firm had initiated the rivalry between steel and reinforced concrete in high-rise architecture. In that project the transmission of wind forces did not present a problem. Thanks to the great depth of the building, the concrete core was able to cope quite adequately with them. The prominently exposed horizontal and vertical components of the mushroom floor system performed no function other than transmitting the vertical loads; they embody the conventional architectural principle of beam and column, load and support.

With their design for the BMA Building at Kansas City, completed in 1964, Skidmore, Owings and Merrill managed to enhance the sculptured or plastic effect of the liberated multi-storey structural frame even further. The span of the square floor bays in this steel-framed building is substantially larger than in the Hartford building (36 ft as against 22 ft) and there are fewer of them (3 × 5). The windows are more deeply recessed, too, and this effect is enhanced by the dark tint of the glass and the aluminium framing. The tower, of approximately equal height, is slimmer and stands exposed on high ground. With these spans and with this length-to-width ratio, the reinforced concrete frame was unable to compete successfully, especially as the layout on plan did not permit a rigid core. The structural steel frame of the BMA Building is of high-tensile steel. The girders are rigidly welded to the columns in both directions; thus this building offers a typical example of wind-bracing by means of multi-storey rigid frames. The same principle is adopted in the external steelwork where the beams and columns are rigidly connected together. However, the steelwork as such is not directly recognisable, since the façade is clad entirely in marble.

On the other hand, the Civic Center in Chicago (1963–1966), designed by C. F. Murphy in collaboration with Skidmore, Owings and Merrill and another team of architects, is a thoroughbred steel structure, unsurpassed in the daring of its design and in the clarity of its architectural expression, a culminating point in the work of the Second Chicago School. Its forerunners and closest relatives, more particularly the BMA Building at Kansas City, and the Equitable Building (Skidmore, Owings and Merrill) and the Continental Center (C. F. Murphy Associates) both in Chicago, are surpassed by the Civic Center, not on account of its height (195 m, 31 storeys), but on account of the then unprecedented spans of the floor bays (26.50× 14.70 m).

The wide spacing of the columns was necessitated by the difficulties presented by the foundations. The columns had to be founded on

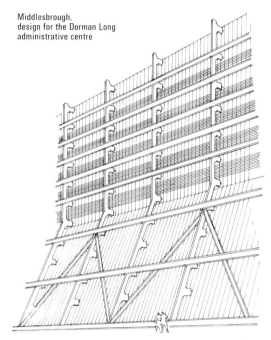

Middlesbrough, design for the Dorman Long administrative centre

caissons resting on bedrock at a depth of 30 m. A further reason was that the requirements as to flexibility of layout and variable utilisation were particularly exacting: offices, conference rooms, large and small courtrooms. This flexibility even extends to the third dimension: the large courtrooms extend through two storeys, but the headroom can, if desired, be reduced to normal storey height by subsequent insertion of floor bays. In both directions the floors are supported by 1.60 m deep welded lattice girders, enabling services to pass through the openings in their webs. The columns, of cruciform cross-section, are of high-tensile steel – the first instance of such columns being used on a large scale and in a really effective structural manner. Depending on their location within the planning grid, the columns are rigidly welded on two, three or four sides to the floor girders, no special connection being necessary (p. 315).

As a result of the differentiation between the verticals and horizontals and the tapering of the columns in the upward direction, the building as a whole achieves a lucidity of design which transcends the restrained external indication of the structural system on the first Lake Shore Drive façade. In addition, the metal cladding of the frame members, adopted from that earlier project, has been visually and structurally enlivened: the edge girders and the columns are encased in concrete, the latter in turn being enclosed in Cor-Ten steel plates welded together and provided with stiffening ribs which project the grid dimensions of the ground-plan outwards, as it were. The steel plates as well as the edge girders are anchored into the concrete by means of studs; this ensures better composite action and increases the rigidity of the whole frame so that the sideway under wind load is greatly reduced. The horizontal forces could not be adequately resisted by the multi-storey rigid frames alone and for this reason a hybrid stiffening system has been provided, as was previously done in the Seagram Building in New York (p. 220). In the upper half of the building there is portal frame action alone, but in the lower half the frames are braced internally with K-type lattice girders.

The Brunswick Building, completed 1962, another design by Skidmore, Owings and Merrill, stands opposite the Civic Center and gives emphatic expression to the advance of reinforced concrete as the rival material. To resist wind loads, the framework of the external walls has been utilised to augment the action of the concrete core. The stability of these walls is strongly accentuated: the piers rise in a convex curving sweep from the huge podium, clearly inspired by the Monadnock Building of 1894. Unfortunately, the podium block does not stand on the ground, but is elevated on widely spaced piers because of the caisson foundations and also because of the need for traffic to be routed underneath.

In their design for the 143 m high Chestnut De Witt Apartments, built in 1963, Skidmore, Owings and Merrill developed the so-called 'framed tube' type of reinforced concrete structural system. The elongated shape on plan and the flexibility required in the layout of the residential apartments are factors which ruled out the possibility of incorporating shear walls in the core of the structure. For this reason the horizontal forces are transmitted entirely to the periphery: the monolithic framed structure of the external walls functions as a huge tube rigidly fixed to the foundations.

The most outstanding building so far constructed on the framed tube principle is the 52-storey One Shell Plaza Building (Skidmore Owings and Merrill, 1968) at Houston, Texas. In sheer height, 218 m, and in the power of its novel, organically modelled architecture it represents a culmination of the 'framed tube', here raised to the stature of a 'tube within a tube'. The peripheral walls combine structurally with the internal tube of the concrete core in the same way as in the CBS Building, New York, and the Brunswick Building, Chicago.

Chicago, Civic Center 1963/66

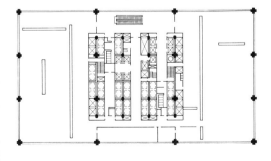

Plan of typical floor, Chicago Civic Center

An entirely analogous advance to progressively more effective methods of wind-bracing involving the more intensive utilisation of the rigidity and great width of the external wall panels was accomplished in steel, with the difference that the upper limit of the economically acceptable application of the various systems is 20, 40, and 60 storeys higher respectively than with reinforced concrete.

In the case of the 256 m high US Steel Building in Pittsburgh, the enclosing walls of the triangular core are designed as an assembly of vast vertical lattice girders with shear connections at the corners of the triangle, so that they function as a lattice tube rigidly fixed to the foundations and capable of resisting all the horizontal forces (pp. 160, 220). In the top storey this tubular structure is connected by rigid cantilevers and a framed crown to the external columns of the building. Under wind load, when the tubular core deflects, these columns absorb tensile and compressive forces and thus prevent twisting deformation of the roof surface; they thereby also reduce sideway of the structure as a whole. The unusually widely spaced, uncased main columns of the US Steel Building perform, besides their structural function, another function which is more particularly important in a representative building of the steel industry: they demonstrate the double success that steel construction has achieved in recent years in resisting its two traditional enemies, corrosion and fire. The box-section columns are of weathering steel, like the set-back cladding, and are filled with water, thereby forming the outer part of a closed-circuit cooling system in which the circulation of water starts automatically in the event of an outbreak of fire (p. 240).

With the aid of the 'framed tube', buildings 70 to 80 storeys in height can be constructed economically in steel. Theoretically this height could be doubled by using a 'lattice tube' form of construction for the external walls, that is, by adding diagonal members to give the outer framework extra rigidity. Thus the façades can be resolved into a close-meshed network of diagonals, as in the IBM Building at Pittsburgh (p. 222) or, alternatively, the main columns may be incorporated into the lattice system, as in the Alcoa Building at San Francisco, where the peripheral latticework, acting in combination with multi-storey frames, serves to resist horizontal wind and seismic forces (pp. 156, 216).

In the case of the 100-storey John Hancock Center in Chicago (architects B. J. Graham and Skidmore, Owings and Merrill, 1968), not only are the mighty diagonals connected to the vertical members by rigid joints, but the horizontal edge girders are also incorporated into the façade latticework. In this way, without any appreciable detriment to the fenestration, maximum rigidity for the tubular hull and optimum economy are achieved, the quantity of steel per m^2 gross floor area being no greater than in 50-storey buildings (p. 221). The pronounced upward tapering of the tower moreover greatly increases stability; at the same time, with this striking shape resembling that of an oil rig or a lattice pylon, and with the enormous scale of the cross-bracings, an extreme limit in sheer power of architectural expression has been reached or indeed exceeded. Here, the monumental nature of the structure assumes something of a dark and menacing character, not least because of the black anodised aluminium panels used for cladding the structural members. On account of the great height of the building, this sheathing, with fireproofing and thermal insulation concealed under it, was an inescapable requirement.

Of course, the floor areas and room depths, progressively diminishing the higher up the building they are located, are also functionally determined and planned and, indeed, they are distinguished by extraordinary versatility of utilisation. The Hancock Center is a town in itself. It comprises car parks, shops, offices, communal facilities, service undertakings of various kinds, residential flats in a range of sizes (from the 46th storey upwards) and finally, at the top, a panoramic restaurant and a television station. The internal structural frame is designed merely for the transmission of vertical loads; the columns and floor girders in the interior of the building are connected by bolted pin joints; floors can be removed and reinstated as desired.

The World Trade Center, New York, the construction of which started in 1966 (p. 162), with its 411 m high 110-storey twin towers, takes up again the urban planning idea that achieved realisation in the first Lake Shore Drive blocks in Chicago. The rigorous vertical articulation is reminiscent of Saarinen's CBS Building, but the architecture of the World Trade Center has none of the austerity of the Chicago School, none of the surging power of the CBS Building. This huge façade system produces the effect of a texture rather than a structure. The extremely closely spaced external columns have a wall thickness of 11 cm at the base; but even the substantially thicker ground-floor columns, each formed by combining three façade columns, do not produce the impression of piers which have to carry a high wall but, instead, look somewhat like warp threads tied together in groups to form the fringe at the end of a carpet.

Structurally this system is also a cantilever tube fixed at its base and serving to resist the wind forces. As in the Hancock Center, the internal columns have to transmit vertical loads only (pp. 212, 221). As a result of the rigid connection of the spandrel panels to the tubular columns, the external wall becomes a vast multi-bay Vierendeel girder: the whole framed tube is composed, as it were, of gigantic pressformed sheet-steel panels perforated by narrow window slots and stiffened at fairly close intervals by hollow-section ribs attached to them. The prefabricated units of which the huge external steel network is composed are interconnected by bolting; each of these units comprises three spandrel panels and three tubular columns. In principle, they have the same shape, deriving from the same underlying idea, as the concrete units that Yamasaki employed on the Michigan Gasworks Building in Detroit, except that now they have been enormously upgraded in terms of structural function and cost. A relatively high price has had to be paid for this visual refinement of the skyscraper façade, the steel consumption being substantially higher than in the John Hancock Center.

Even before the World Trade Center in New York had been fully occupied and put into service, the third of the super-skyscrapers was already nearing completion in Chicago. It is the 109-storey, 445 m high Sears Building, designed by B. Graham: the biggest office building and the biggest department store and mail order concern in the world. Here, the principle of the cantilevered tube appears as a bundle of nine gigantic square hollow sections, each with a length of side of 22.5 m comprising five bays of columns. In the storeys containing office accommodation the intersecting internal rows of columns were not a very objectionable feature from the viewpoint of internal layout; indeed, by skilful arrangement of the groups of lifts, it even proved possible to attain a considerable degree of flexibility in this respect for the offices intended for the firm's own use and for letting; but it was not possible to install diagonal bracings for structural stiffening. Thus Graham was led to adopt a Vierendeel frame system similar to the one that Yamasaki used for his great twin towers in New York. Although the column arrangement and the fenestration in the Sears Building are much more generously conceived, the high steel consumption is offset by the amazingly short construction time: the main structural work took only 15 months to complete. Here, too, the façade was assembled from three-storey high prefabricated units.

Of the three record-breaking skyscrapers the Sears Building is not only the highest, but probably also the most enduring in terms of architectural value: it is the one most firmly rooted in American high-rise building tradition. With its external steel frame, it represents the Second Chicago School. The stepped arrangement is reminiscent of the RCA Building and the Woolworth Building in New York and their predecessors right back to Sullivan: to him we owe a project for a high-rise building, dated from 1891, which comprises a very similar configuration of nine stepped square prisms.

In ending our survey of the first hundred years of steel-framed structures with the Sears Building, we do not mean to suggest that we regard the skyscraper as the most important item when considering the future of structural steelwork or of town planning. It can be stated, however, that modern framed construction, like all ambitious methods of building that have merged in the history of architecture, compels the architect to adopt simple and impressive geometric shapes in those circumstances where the most exacting demands have to be met. The groundplan of the Sears Building, the methodical regularity with which its nine squares are stepped back one by one as they soar skyward, possesses something of the character of the magic square with which the artists of bygone centuries were preoccupied. Geometry as the origin and basic principle of the art of building, the intellectual bond between architecture and engineering; structural steelwork as the medium particularly suited to educate architects in clear geometric thinking – this is the lesson to be drawn from the evolution of the American high-rise building.

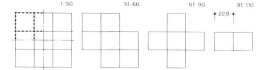

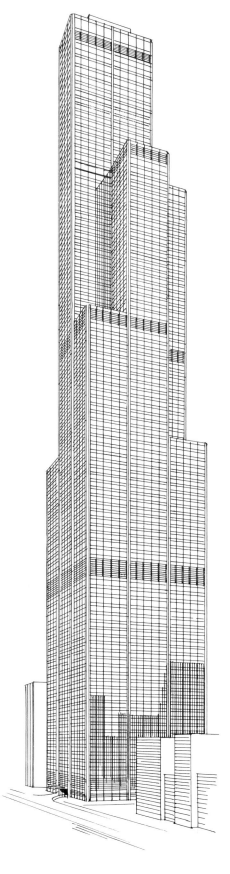

Chicago, Sears Building 1972/74

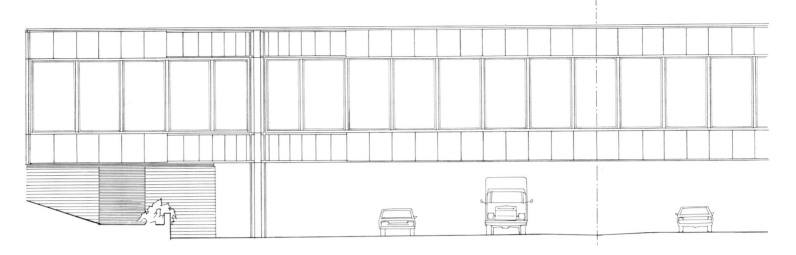

Abraham Lincoln Oasis, Illinois 1968

Steel – the constructional medium for wide-span and complicated structures

If we attempt to outline the many and varied possibilities that are now available to the architect for building in steel, we may, after the foregoing descriptions of the achievements of the steel frame in high-rise buildings, begin by considering a set of problems characterised by extremely exacting structural requirements or by great difficulty in execution – conditions which steel is particularly well suited to satisfy or indeed alone can satisfy.

First, there are those special cases of multi-storey structures in which the spans, the loads, the bending moments and the other conditions are so large or so severe that they are normally encountered only in bridge construction.

A representative example of a wide-span structural engineering project in the sphere of building construction is provided by the raising of the Parliament Building in Prague (p. 207). Here, a two-storey low-rise structure, measuring 60 × 80 m on plan and comprising offices, halls and an internal courtyard admitting daylight to an old assembly hall situated under it, has been raised to a height of about 25 m on four columns.

Within the concept of 'bridge buildings', a new type of structure has emerged: the motorway restaurant which straddles the dual carriageway. The extent of the range of design possibilities offered by this type of structure, straightforward in principle though it may be, is evident from a comparison between, say, the motorway restaurant at Montepulciano (p. 108), characterised by its dynamically powerful rigid frame construction, and the Abraham Lincoln Oasis restaurant on the Illinois motorway (architect D. Haid), where a similar spatial layout has been accommodated within the symmetry of a rigid bent with cantilever arms. Here, the austere geometric shape of plate girder construction is effectively animated by the arrangement of the web stiffeners and the compound flange plates detailed to meet the distribution of bending moments and shear forces.

Amazing, almost terrifying, is the impression created by steel-framed structures which transmit their entire load to the foundations through a few closely spaced supports at the base. This is exemplified by the Hotel du Lac, Tunis (p. 78), where the slab block cantilevers out longitudinally in both directions from its base like the wings of some monstrous bird of prey. In practical terms this shape emerges from the technical necessity to concentrate the whole load of the building on two groups of centrally arranged piles. At the same time, however, it is a sensational piece of architecture, offering splendid views from its upper storeys. Something similar is intended in the design of the Panorama Hotel near Brünn in Czechoslovakia. Here, the building corbels out on one side from storey to storey in the transverse direction, whereas it steps back in the longitudinal direction, so that it leans out menacingly over the tree-tops of an adjacent pine forest.

Less eccentric, but none the less notable as a major structural achievement in multi-storey architecture, is the system adopted for the Federal Reserve Bank at Minneapolis (p. 207). The 12-storey slab block spans a distance of 84 m between two massive supporting and stiffening piers with the aid of a tied arch girder, a sort of inverted suspension bridge. In the proposed subsequent extension of this building by the addition of further storeys, a second similar structural system, but now the right way up, is to be built on top of it. A structure embodying the same general principle was built by P. L. Nervi for a paper-mill near Mantua in 1962. This is an elongated building which is supported from two piers or towers by means of suspension cables. In the case of the Federal Reserve Bank the span is shorter, but the load is considerably larger; here the suspension and supporting structure are accommodated within the building itself, of which the outline remains unspoiled, but without detriment to its spatial arrangement. It was evidently important to the building owner as well as to the architect to allow the amply-dimensioned forecourt to extend freely and unobstructed beneath the building; for this reason, too, the car parking and other ancillary facilities have been accommodated underground.

The idea of explicitly expressing the lightness and transparency of a modern framed structure by providing a ground floor largely free from columns is something that crops up again and again: Le Corbusier was the first to exemplify it in his Pavillon Suisse. Since then, this idea has become accepted as self-evident, almost a conditio sine qua non for the more ambitious type of framed building. The structural solutions and the functional motivations have been considerably extended in recent years. Steel, reinforced and prestressed concrete now compete with one another in these 'buildings on stilts' or 'tree structures'. The cantilever and portal frame systems in which the loads from the building are concentrated on to a narrow base, or on to widely spaced peripheral points of support, are either distributed over all the storeys or, alternatively, are concentrated into a set of powerful girders or transverse portal frames – located directly over the ground floor or otherwise at the very top of the building – on which the whole structural frame of the upper storeys stands or from which it is suspended.

A high-rise building in which the floor steelwork for all the storeys cantilevers out from the core and which also very clearly demonstrates this structural principle externally, through the glazing bands extending round the building, is the Tour du Midi in Brussels (p. 158).

Cantilever construction can achieve enhanced dynamism when the cantilever is located rather higher above ground level and carries the entire weight of the external walls as exemplified by the office building at Puteaux (p. 152), which comprises a slab block projecting from its substructure in both directions. The controlled power embodied in this substructure harmonises well with the prominently exposed structural steelwork of the upper storeys.

In the case of the Royale Belge Building in Brussels (p. 154), there is a certain emotional appeal in the cantilevering substructure and in the contrast between this massive supporting system, with its twin reinforced concrete girders intersected by shallower steel girders, and the glass and steel architecture of the splendidly proportioned superstructure.

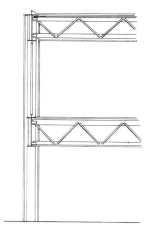

Steel-Framed Structures Today
Applications and Possibilities

The constant endeavour of architects to exploit the possibilities of span and the potential strength of steel to enrich their architectural creations is expressed in the relatively large number of suspended buildings emerging in recent years (p. 205). Initially, the cantilevering framework at the top of the core was of rigid construction comprising plate or lattice girders, as in the office block for Philips in Eindhoven or for Alpine Montan in Leoben. More recently, however, designers have tended to employ diagonal ties between the upper edges of the façade and the top of the core, as in case of the Siemens Building in Saint Denis (p. 144).

More discreet, but actually more effective, is the individual elegance of the suspended structure for an office building for an insurance company in London (p. 146). The twin supporting structures, the edge girders extending around the building and the cantilever construction are concealed behind closed panels in the façade.

Notable rigid frame buildings that call for mention are, inter alia, the Radisson South Hotel, Minneapolis, where the function of the building made it possible to base the design on the staggered-truss system; the Iranian hall of residence in the Cité Universitaire, Paris, comprising two blocks of four storeys suspended one above the other from box-section portal frames (pp. 76, 206); and the Television Centre at Berlin-Charlottenburg (p. 128).

A design concept which, although not quite new, is associated with the idea of giving a steel-framed building an elevated position, is gaining in importance because of the increasing complexity of functional requirements. Typical examples are the medical treatment block under the block accommodating the wards of a hospital, or the hall with counters for serving the customers under the offices of a bank, etc. Because of the relatively small cross-section of its columns, the steel-framed structure is particularly suitable for forms of construction involving wide spans (pp. 130, 154). A classic example of a low-rise block inserted under a tower block is presented by Lever House in New York (1951).

The Sports School at Magglingen comprises two blocks of approximately equal height, slid one over the other and separated from each other by an open entrance hall in which the slender continuous columns are exposed. The building is enlivened by its double elevation and staggered arrangements and harmonises well with the undulating landscape.

Structural steelwork is essential for buildings which have to be erected under particularly difficult topographical or climatic conditions or on very poor soil, such as in mountainous country or in the sea. Thus, a hostel for workmen on the construction site of the Grande Dixence dam in Valais, Switzerland (architect A. Perraudi), at an altitude of 2140 m and exposed to temperatures as low as −30°C, was built as a steel portal frame structure in dry construction, that is, without using plaster, mortar or in-situ concrete. The Alpine refuge in the Stelvio Pass had to be constructed under similar extremely adverse conditions of weather and inaccessibility (p. 73).

Even more compelling than in the case of the Hotel du Lac at Tunis was the need to concentrate the loads on a group of closely spaced piles in constructing the complex of exhibition pavilions in Lake Ontario at Toronto (p. 106). Each of the six square pavilions is elevated about 10 m above water level on four steel tubular struts, which project the same distance above the structure and at their upper ends support the cable stays sloping down to the corners of the building.

In this context, it is not inappropriate to mention a modern form of construction in which a bright future for steel is forecast: space structures — three-dimensional lattice systems which can be adapted to any size and lend themselves to multi-purpose use — which were first developed by French and German architects. Although such systems in the urban environment still appear somewhat utopian and we cannot imagine their solving the problem of high density building, their use has the merit of helping to generate or develop new ideas in planning. The first steps towards the realisation of such space structures have emerged in a number of designs for buildings to straddle railway stations or even entire railway sidings.

Minneapolis, Federal Reserve Bank, cross-section

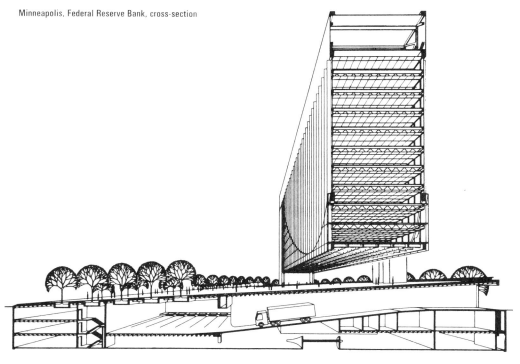

Steel in residential and school buildings

No less remarkable and fundamentally even more amazing than the success achieved by structural steelwork in large buildings is the fact that during the last twenty years or so it has penetrated — gradually at first and then at an increasing rate — into building sectors with no special span and loading requirements, in which other materials had been more competitive or in which prejudices and psychological barriers had to be broken down. This is particularly true of residential buildings, for in this sector the conservative attitude, the tendency to cling to the familiar and the customary, is especially strong.

Santa Monica, California, studio house

To dispel the basic conception of the house as a place of refuge, a sheltering cave, and to replace it with a new and less restrictive ideal of a home to meet the needs of modern man — this is the pioneering achievement of progressive architects who, chiefly through the example of homes designed for their own use, prepared the way for steel to be used in the construction of detached houses.

Here, too, the first step was taken by Mies van der Rohe with the famous Farnsworth house. The strictly modular layout, the flowing transition from one room to another, the technical services installed internally, all these were powerful stimuli. But however carefully this perfect white-painted steel framework has been blended into the landscape, with the platforms of the forecourt and the house with its terrace built out on stilts over low-lying swampy ground in the front, it is nevertheless a piece of monumental architecture, the manifestation of a new concept of the use of space, comparable with the Barcelona pavilion of 1929, rather than a home. In another architect's house, that of P. Johnson, a disciple of Mies, at New Canaan, which was designed not long after and which comes closest to the ideal, the self-assertive effect has been toned down and living rooms are incorporated in a lower storey concealed in the hillside.

In comparison with this architecture in steel, the house that Jean Prouvé built for himself at Nancy in 1929 appears startlingly casual and unpretentious and, to the present-day observer, much further ahead of its time, not a classical steel frame, but a skin structure of steel sheet, aluminium and timber panels.

The first coherent group, a kind of nursery garden of individual homes built in steel, was evolved in California in the 1950s, inspired by competitions promoted by a leading architectural journal. Best known in this group are the experimental houses designed by C. Ellwood. Here, the steelwork is especially graceful; instead of I-section columns, slim square tubular members are framed in timber so that the structural frame does not appear obtrusively, but instead gives the impression of modular articulation of the surfaces. Particularly attractive is the group of quadruple block houses in Hollywood embodying an approximately cruciform symmetrical arrangement of four terraced blocks, a composite form of construction that comprises longitudinal loadbearing brick walls and transverse steel frames.

Distinctive purity of form and its harmony with an all-metal assembly of steel frames, cellular steel floors and enamelled sheet steel spandrel panels characterise Pierre Koenig's experimental houses in Hollywood.

In all these buildings the steel structure has been made acceptable, as it were, and its effect enhanced by careful integration of the house and its appurtenances — verandahs, terraces, gardens — into the landscape. Evidently, architects find these hillside positions especially attractive and challenging, since they become a practicable proposition thanks to the possibilities offered by the structural steel frame with its slender members and capacity for wide spans.

The best Californian steel houses are distinguished by the superb layout of their rooms and a degree of gracious living. This architecture unmistakably takes up the tradition of the American house, which had reached successive peaks in the 'shingle style' in timber construction around 1880 and again in the early work of Frank Lloyd Wright and was to be given a fresh impetus subsequently by R. Neutra.

The architect's steel house is still in process of evolution in every country, two-storey and three-storey houses, sometimes with studios, being increasingly in evidence. A unique position within the Californian examples of these structures is occupied by the two-storey house and studio of Ch. Eames, the well known furniture designer, at Santa Monica. It is characterised by its emphasis on flatness, its Japanese-style sloping window panes, and the lively combination of materials — steel, sheet metal, timber and asbestos cement.

A steel house does not necessarily have to be set in the luxuriant vegetation of a subtropical garden to bring out its spatial qualities. This is evident from a comparison of a stylish hillside house at Buenos Aires (architect A. Bonet), and the hillside house with a roof garden at Liège (architect J. Mozin) — both reported in the architectural press some ten years ago. In the latter the steel frame makes a much more natural impression, the transparency of the spatial development up through three storeys to the roof garden being very effectively demonstrated, and on the relatively cramped site the integration of the building with its suburban environment has been very successfully achieved. The sureness of treatment of structural steelwork that characterises the work of the best French architects is clearly shown in a modest holiday home which the architect M. Lods built on a rocky slope at Le Tignet, a simple framework with an infilling of sandwich panels. In the houses built in recent years (pp. 59–62) certain extensions of the architectural possibilities emerge — towards the adoption of larger closed surfaces and discreet combinations of materials, as in the house at Landskrona (p. 60), more powerful expression in the externally exposed structural system of the house at Stallikon (p. 61), and an increased assertiveness in the flat treatment of weathering steel cladding, as in the architect's house in London (p. 62).

Nevertheless, the individually designed steel family dwelling remains a fairly exclusive architectural affair. So far, system building techniques have not made much headway in this sector, and in those cases where system building has been adopted, as in American 'frontier style' houses, it belies the character of structural steelwork. Of greater significance for the future of steel houses is the use of this material in housing estates intended for the general public. In this sector of development the steel frame has gained a foothold and has, especially in recent years, made considerable progress; it is competing successfully with large-panel precast concrete structures with which attempts have been made, more particularly in France, to meet the demands of social housing programmes (p. 184).

As long ago as the mid-1950s, a major move towards steel system building was made with the erection of the Porte-des-Lilas block of flats, Paris, which gave its name to the prefabrication system which was used on that project and which has since been developed further: multi-storey portal frames which are assembled by welding on the ground, lifted into position and provided with precast concrete floor slabs. The latter are also prefabricated on the site, stacked at ground level and subsequently hoisted up storey by storey. A typical industrialised system of more recent origin is

Le Tignet, holiday house in the Pas-de-Pique

Liège, hillside house with roofgarden

that used for the Grand Mare housing estate (p. 64). It is notable for its use of large components, shear walls extending the full height of the building, and floors, supported on lattice girders, spanning the whole width of each flat. Three other major building projects give some idea of the scope and architectural expertise of French residential building construction in steel: the Maison des Jeunes in the new town of Sarcelles, near Paris (architects Boileau and Labourdette), with its rhythmic arrangement of pier and spandrel units set between the flanges of the external columns; the high-rise block of flats in the Rue Croulebarbe, Paris (architects Albert, Boileau and Labourdette), which derives its special appeal from the exposed tubular columns and the tubular lattice wind-bracings on the narrow faces (p. 216); finally, the Boieldieu residential complex at Puteaux (architects Rabaund and Gilbert), with 479 units. The last-mentioned project is in composite construction, the steel floor beams being prestressed by propping prior to pouring the concrete. The external columns and the bracing in the corner cores are not visible on the outside. The Bourgmestre Machtens high-rise block of flats at Brussels and the Balornock housing estate (pp. 68, 70), the residential developments at Piombino and Taranto (pp. 63, 66), can be regarded as proof that structural steelwork is gaining ground internationally in the domain of social housing. They also show how numerous are the design possibilities, what variants in the rhythm of a more or less reduced fenestration are available in a type of building in which every constructional feature is determined by the strict precepts of economy.

In Japan, where the problems of housing shortage and lack of building land are particularly acute, the many and varied experiments with new structural steel systems are of considerable interest. In comparison with the highly complex systems which aim to separate the structural frame and open up the actual residential units, conventional steel-framed building systems, with wind-bracings placed in the party walls, lightweight concrete fire protection to the columns, continuous balconies and recessed loggias, may appear dull and humble, but also much more human.

Quantitatively and from the industrialised building point of view, the progress made by steelwork in school construction is even more significant. In the first primary and secondary school buildings constructed in steel in the 1950s the use of this material was not dictated by functional and economic considerations but, just as in the early steel residential buildings, it was the efforts of progressive-minded architects to bring the transparency and lightness of steel-framed construction into school building that showed the way. In this context, mention must be made of the primary school at Birmingham, Michigan (architect E. M. Smith), comprising a bold array of classroom pavilions and garden courtyards, and the secondary school at Joliet, Illinois (architects Skidmore, Owings and Merrill), also a single-storey complex surrounding a courtyard, while in Britain, the famous secondary school at Hunstanton evidently started a trend. Primary and secondary schools of similar advanced architectural character, comprising loosely arranged or more strictly composed groupings of buildings, one to three storeys in height, were also erected elsewhere. In the 1960s came a considerable expansion in school building activity to cope with the rising birth rate, the raised school-leaving age, the more extensive educational requirements, and more exacting demands for technical equipment. The conventional types of school — primary, secondary and grammar schools — and educational ideas and methods are changing. The hierarchical pattern of teaching is being broken down, and the traditional 'closed' character of the classroom is giving way to multifunctional flexible layouts designed to meet increasingly stringent requirements with regard to communication, general participation in activities, changes in the number of pupils and in teaching arrangements, etc. On comparing the pavilion-type school at Birmingham, already mentioned, with typical school building complexes of very recent origin, such as those at Isenburg and at Osterburken (p. 180), the fundamental difference becomes evident, just as does the superiority of the slender wide-span steel frame for the construction of such large structures. The required adaptability, together with pressure of costs and completion times, provided an effective impetus for the development of prefabricated building systems having uniform basic dimensions — the international modular dimensions 30 cm, 60 cm, 120 cm having been widely adopted.

In Britain about 50% of all primary and secondary schools are system-built, 75% of these prefabricated schools being in structural steelwork. In 1970, about 90% of all new schools put up in France were system-built, and half of these were in steel. In Italy, Germany and Austria, too, there has been a marked increase in system building and in steel construction. In the external appearance of these structures, the standardised basic dimensions produce a certain degree of uniformity, but also an agreeable co-ordination of scale. The external columns are usually set back, but the structural frame is clearly shown in the free and generously dimensioned fenestration, as well as in the vertical joints of the cladding units which are often backed by a vented cavity. Yet within these prescribed limits there are still latitude and scope for the architect to invent differentiated groupings and a sensitive articulation of façades. This can be clearly seen in such fine buildings as the German School in Brussels (architect K. Otto).

In recent years, structural steelwork has also made progress in a type of building intermediate between schools and residential buildings, namely, day nurseries and children's holiday homes (pp. 74, 75, 179). A structure notable for the generous spatial development made possible by a steel frame and because of its integration into splendid mountain scenery is the holiday home that Italsider has built for the children of its employees at Montechiaro (architect R. Severino).

Montechiaro, Italsider children's holiday home

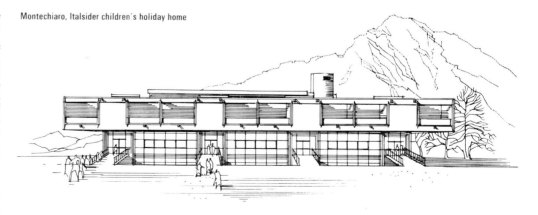

Steel in other types of buildings

In spatial arrangement, size of building and sophistication of technical facilities, a transition from secondary and grammar schools to universities is provided by such establishments as trade schools, technical colleges and polytechnics. Educational institutions of this category which have been designed with special care — the older in reinforced concrete, some of the more recent in steel — are to be found in Switzerland.

As a result of the increasing numbers of students, the widening ranges of disciplines and the more exacting demands upon technical equipment in the 1960s, the construction of buildings for universities underwent a great increase, though in this sector, in comparison with other educational fields, the proportion claimed by structural steelwork was relatively modest.

Great Britain gained a certain lead in university planning and building, thanks to its particular educational system, with its universities spread throughout the country, each with a relatively small number of students, their buildings designed on a human scale and sited to harmonise with the landscape. Thus healthy competition was able to develop between various traditional and industrialised construction methods. For steel it was something of an advantage that in this sector the same structural systems could, to some extent, be used as for school buildings. A good example of lightweight system building in steel is afforded by York University. The orthodox adherent of structural steelwork might be somewhat astonished by the sight of these two- and three-storey loosely grouped buildings under construction, with their unconstrained combinations of slender tubular columns and lightweight steel beams and with cladding partly of wood and partly of artificial stone. Yet the final result is very pleasing.

In France, too, structural steelwork at first made only slow progress, but towards the end of the 1960s a number of new university projects were undertaken, not only in the provinces and on the outskirts of Paris, but in the centre of Paris where the vast complex of the Faculty of Sciences (p. 100) was built with 400 000 m² of usable floor space.

Uniform centralised planning of facilities also characterised the schemes for the extension of existing or the construction of new universities in Germany, for example at Karlsruhe, Marburg and Bochum. It is instructive to observe how, after a promising start in the first buildings for the Ruhr University at Bochum, the structural steel frame was reproduced step by step in reinforced concrete or in prestressed concrete components. A culminating point of the concrete ideology in university architecture was reached at Marburg, where the principle of absolute flexibility of spatial layout and disposition of technical services was achieved at the expense of economy, although admittedly the result was of considerable consequence and architectural quality.

Even more rigorous requirements for flexibility and three-dimensional expansion were met by the planners of one uniform type of building for five new universities in the German province of North Rhine-Westphalia, designed in reinforced concrete. In the construction of the sixth of these universities, at Bielefeld, a proposal to redesign these buildings in steel was favourably received: steel would probably have undercut reinforced concrete even more in price had the planning and design been kept more open, and thus more appropriate to the use of steel, from the outset.

A special case is the Arts Faculty of the Free University at Berlin-Dahlem, the design for which emerged as the result of an international competition (p. 181). In this scheme, too, extreme flexibility of use, together with the possibility of subsequent structural conversion and upward extension, were required. This scheme is noteworthy for both the composite construction adopted for the structural system, precast concrete slabs connected to the supporting beams by high strength friction grip bolts, and the unusual modular planning. The structural frame, the cladding units, the suspended ceilings, etc., have their own different planning grids; the structural members of the frame do not intersect over the columns, but cross one another at different levels and pass by them. In this way, maximum freedom is obtained with the internal layout and in erecting the structure and installing the services.

In the massing of its blocks this much discussed building also breaks new ground: the constantly varying shape of the intimate little inner courtyards, the stark regularity and colourful character of the weathering steel cladding and the differences in level produced by sloping the floors.

The transition from university architecture to hospital architecture is formed by the category of scientific and technical research institutes — as regards both utilisation and the more exacting demands made upon the technical equipment for such buildings. The Jadwin Physics Laboratory at Princeton University shows how, with the aid of framed construction, complex functional and technological processes can be accommodated in a well-balanced and well-proportioned group of buildings. The closed, brick-faced blocks present a very compact appearance; but in the recessed ground floor and in the narrow glazed strip that gashes the superstructure the latent power of the steel frame is obvious. A counterpart to this building is the medical research laboratory at the University of California (p. 99). These two buildings, together with the complex at the Free University at Berlin-Dahlem, already mentioned, show a range of architectural expression which modern structural steelwork can hardly equal in any other class of building.

Another important field into which the steel frame has recently penetrated and in which it is establishing its claim is that of hospital architecture. Here, the inhibitions were initially stronger than in university architecture. Since then, however, some significant buildings have demonstrated that steel is fully competitive, as witness the major hospital centre at Gonesse in France (p. 185) or the prefabricated construction system first used in Oxford and subsequently applied successfully in building some thirty British hospitals. In Germany the breakthrough for steel in this domain came with the regional hospital at Detmold, which was duly completed for use within the specified cost and time limits.

Consider a typical example: the Lorenz Böhler casualty hospital in Vienna (p. 86), where exacting requirements for technical equipment of the buildings favoured and indeed necessitated the use of steelwork. It had already been laid down as a condition in the design competition that all the services accommodated in the ceilings and walls should at all times be accessible and capable of modification without

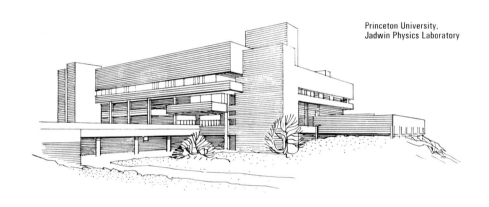

Princeton University, Jadwin Physics Laboratory

serious disruption of the normal functioning of the hospital. This condition could only be satisfied by a framed structure which was designed to provide as much cavity space as possible for accommodating these special services. The loadbearing external columns consist of conventional rolled steel sections which alternate with pairs of piers in which the air-conditioning ducts are housed. The main and secondary girders are of light Vierendeel construction, there being a space between the twin battened chords to provide access to the cavity in the double-leaf partitions. The floors comprise galvanised sheet steel planks which are only 4 cm thick.

The traditional problems associated with steelwork — fire protection and corrosion protection — played no part in the cost of this hospital. The hygienic and acoustic requirements applicable to the wall and floor panels led to a form of construction in which fire protection is, as it were, thrown in anyway; the lightweight structural frame is galvanised throughout. The cost of the purely structural work in this case was only 10% of the total cost of the building. The extra expense involved in providing this open lightweight steel system, with its greater headroom in comparison with conventional steel joist construction or with a reinforced concrete frame, is more than offset by the shorter and simpler arrangement of the services, the greater ease of erection, and the greater dimensional accuracy of the all-steel structural frame. In a project of this kind the external appearance of the building is not closely determined by the steel frame, but emerges as a consequence of the functional requirements, as do also the methods of fire protection.

Even more stringent requirements for flexibility of internal layout, constant adaptability to technical progress and accessibility of services without disruption of the hospital's normal activities were imposed in connection with the design of the Boston City Hospital (architects H. Stubbins and R. Allen). The centrepiece of the scheme, a 15-storey slab block with 1300 beds, which rises from a four-storey podium block, is a pure steel-framed structure. The floor girders are alternately plate girders and storey-height lattice girders whose top and bottom chords enclose a continuous space housing services and technical installations so that each ward and each sick-bay can be connected above or below to any desired service. There is something almost uncanny about the highly perfected interdependence of the structure and the sophisticated services built into it: care for the sick by advanced technology.

The extreme concentration that was required and achieved here is certainly not the only approach for future hospital development. Up-to-date care for the sick is also possible in smaller units, on a more human scale and under more informal conditions. Here again structural steelwork not only provides favourable technological conditions, but also creates a beneficial environment. All this is evident when, for comparison, we consider the maternity ward in the hospital in Kent (p. 83).

Other types of buildings — hotels, public transport buildings, department stores, shopping centres, government offices, etc. — need not be discussed here. In these sectors of modern architecture, also, structural steelwork owes its success to the same advantages of adaptability, lightness, strength and ease and efficiency of erection. Attention must, however, be drawn to a new field of application which is attractive to the steelwork designer in that it is the only type of multi-storey building in which the structural steel frame can be displayed almost without any space-enclosing surfaces to obscure it, namely car parks (p. 110).

Finally, some fairly recent phenomena in the domain of office and administrative buildings should be noted. The development of modern structural systems in the high-rise commercial building and the recently evolved external designs have already been discussed in this book. In these cases, the architecture was determined from the outset by the requirements for use with which the building had to comply. Although adaptability of internal layout was something that was considered desirable even at a very early stage in the history of office building, the planning of such buildings up to about 15 years ago nevertheless consisted basically in providing a number of large and small individual offices. The grid-type façades of the 1940s and 1950s clearly express this concept — and to that extent they fulfil a functional purpose that cannot be dismissed.

Open-plan offices, which originated in the USA, were widely adopted throughout Europe from about 1960 onwards and were further developed, particularly in Germany, where they became known as 'landscaped offices' (pp. 140, 182). The concept of the large undivided internal space undoubtedly promoted the trend towards making the façade and structure independent of one another, as was characteristic of the curtain wall era. In addition, it made possible a development which is approaching its culmination or has indeed already reached it: abandonment of the prismatic office box in favour of livelier, freer shapes on plan, without doubt inspired also by concepts of organic building which have in fact long played a part in modern architecture.

On the other hand, the principle of the planning grid is of such fundamental importance in arriving at a rational ground-plan and structural system that it is not possible just to dispense with the square or the rectangle as the basis of dimensional co-ordination and replace it by less constrained shapes. A way out of the difficulty is provided by the triangular or hexagonal grid, both of which are highly popular nowadays for reinforced concrete buildings. That steel can compete here, too,

Neuilly-sur-Seine, Havas-Conseil agency

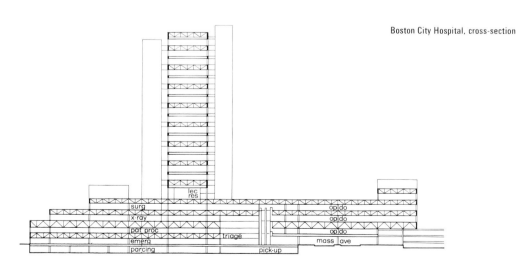

Boston City Hospital, cross-section

and that it can also provide scope for the architect, is demonstrated by the different designs for the administrative building for the Hamburg-Mannheimer Versicherungs AG in Hamburg (p. 197). Incidentally, one of the first office buildings that was successfully planned on a hexagonal grid is the administrative headquarters of ENI at S. Donato, near Milan (architects M. Nizzoli and M. Oliveri). This exciting building is steel-framed.

Another notable attempt to reduce the severity of orthogonal structural frames and to enliven buildings generally is to design them around a cylindrical core, as has been done with the headquarters of the Havas-Conseil publicity agency at Neuilly-sur-Seine (architects M. Andrault, P. Parat and J. P. Sarazin).

The animated structural organism which has been given form and substance here is a sharp protest against the stiff spatial configuration of the American skyscraper, against the axially symmetrical entrance that loses itself in the banks of lifts and against the inefficient means of communication between the various storeys. Here, on the other hand, we have a generous and extensive circulation system over the whole irregular area on plan; the ground floor of the main block is largely opened up and made readily accessible; visual and physical access to the various customer service departments is provided and distributed over the entire area in a series of glass-enclosed cylindrical bodies with entrance halls, reception offices and interviewing rooms. There are no fewer than five circular staircases, a ramp leading down to the underground garage, and a 'crater' extending rearwards into the courtyard from the large staircase tower of the main building. This crater is a funnel-shaped cavity serving to admit daylight to the three bottom storeys and giving the visitor the impression that the greater part of the building is buried in the ground. Behind the crater two four-storey suspended buildings, also cylindrical in shape, rise from the closed cylindrical shafts, these buildings, too, being interlinked by a circular staircase tower. The unusual richness, the flexibility of the spatial development, the interdependence of the vertical and horizontal circulation, all serve the interests of internal communication, make for better contacts, suggest adaptability and receptiveness. Thus, this architecture is truly functional, showing that it can promote the interests of publicity with all the means at its disposal, advertising both itself and the publicity agency that it serves.

The trend towards streamlined curved groundplans, which originated prior to the 1950s, is still continuing, probably revived by the design of the Town Hall in Toronto (architect V. Revel), with its tall slab blocks which enclose the circular council chamber like the two halves of a shell — a reinforced concrete structure which caused something of a sensation in its time and which arose in 1957 as the result of an international competition.

By contrast, the 210 m high Maine-Montparnasse tower in Paris (architect E. Beaudouin), the highest multi-storey building in Europe, has a spindle-shaped plan. Here, however, the closed concrete shear walls are arranged internally, acting as a core for the surrounding steel frame and glazing.

Steelwork and internal services

In the evolution of modern framed buildings the technical services, comprising such items as heating, ventilation, sanitary installations, sound insulation, acoustics, lighting, high and low voltage electrical systems, etc., take up an ever-increasing proportion of the volume of the building, of the construction cost, and of the design effort involved. The structural engineering itself no longer presents the most difficult design problem. The framed structure, especially in steel, in its efficiency and versatility has now developed to such an extent that it can cope even with the most exacting demands arising from the function of the building.

The structural system, which in the early stages of development was the driving force of the architecture, is now no longer the all-important thing; only in the case of structures of extremely great height or long span does structural engineering still provide decisive new stimuli. Also bound up with this is the fact that, despite the current predilection for the externally exposed structural frame, there are still far too many good steel-framed buildings in which the structural members remain concealed from view. To meet the exacting requirements of technical services, the design of the loadbearing structure is concentrated entirely on providing cavities for accommodating the various services, as in the big hospitals already described.

The increasing importance of technical facilities in buildings was recognised at a very early stage in the work of some leading architects. Thus, in British domestic architecture of the late nineteenth century, for example, R. N. Shaw designed his prominent groups of chimneys as a system of ducts within the structure. For similar reasons, Scandinavian and British architects have for years treated the exposed services as a particular functional means of expression, sometimes even in religious buildings. The British 'brutalists' turned exposed pipes, ducts and cables into official architectural features in their own right. Yet many an architect, his enthusiasm fired by all this honest-to-goodness informality and freedom, is subsequently disappointed and crestfallen when the heating, ventilating and other experts, who always take special pride in concealing their handiwork, point out that this approach offers neither economic nor technical advantages.

Nevertheless, the increasing demands upon the flexibility of the equipment and upon its ability to cope with rapid technical developments, as in university buildings, have led to the view — at any rate, among architects and building authorities, though perhaps not always among the users of the buildings in question — that it is no longer essential to conceal the services behind a suspended ceiling. Openly displayed pipework, ducts, etc., are more particularly justified, or indeed indispensable, in circumstances where their subsequent installation, with the various connections and branches that this involves, is not precisely known at the time the building is designed. The capacity of the steel supporting members of the floor to be penetrated and traversed by services is thus revealed. Visually, this can form a pleasing contrast or it can impose a certain ordered pattern on the network of services. Steel structures which are well designed to accommodate the services installed in them, such as the trade school at Berne (architect W. Pfister), testify to the aesthetic satisfaction that the architect can derive from exposing castellated beams or lightweight lattice girders.

Cadarache administrative building

If 'technical services' can be assumed to denote all the constructional measures that help to ensure the physical and mental comfort of the occupants of a building, this concept must include properly designed natural lighting and sun protection, that is, the design of the windows and their use for natural ventilation, the arrangements for cleaning the windows and especially the protective measures against adverse effects of sunshine. These oldest and most important aspects of 'environmental design' have indeed from time immemorial determined the external appearance of buildings and whole residential developments, no matter whether they are modest unobtrusive dwellings or ambitious architectural creations. That the revolutionary change brought about in the structural character of our buildings by modern framed construction calls for new solutions is something that Le Corbusier was among the first to realise, and it was he who established the brise-soleil as an up-to-date version of the window shutter. In his Maison Clarté he produced a particularly attractive result from the metal-framed windows, their pattern of glazing bars, and the sun-blind housing, although technically and in refinement of design they are surpassed by J. Prouvé's metal-framed windows in the house in the Place Mozart in Paris.

This tradition appears to have been taken up and continued, together with a reminder of the window shutters of apartment houses that have contributed so significantly to the fascinating impression of unity and historical continuity in the Parisian urban scene, in the library building

(p. 122) with its folding shutters of sheet aluminium, actuated by cranking from the inside, for sun protection. In contrast with the adjustable vertical fins which, starting with the Olivetti Building in Milan, were in fashion for a time in the 1950s, the charm of the fluctuating façade is not achieved here at the expense of the clearcut modular spacing of the windows.

Horizontal sun-blinds have also become a permanent feature and a considerable number of designs have been evolved for them, ranging from the rather sombre mass of the Cor-Ten steel louvres on Saarinen's Deere Building, through the impressive glistening splendour of the stainless steel strips on the Cadarache administrative building (architects Badani, Roux-Dorlut), to the functionalism of the protruding lattice grilles on the Unesco Building V in Paris (architects Zehrfuss; engineers Prouvé, Fruitet). The combination of sun-protection and window-cleaning facilities, of balcony grilles and vertical fins, such as Eiermann built as a kind of second skin around his external wall framework, an arrangement especially popular in Germany, has already been mentioned. (See also p. 130.)

The aim to raise the planning of the technical services to a design feature in its own right, equivalent in status to the loadbearing structure itself, is also apparent in the early attempts to indicate the air-conditioning units, heating elements or other technical appliances in the façade articulation. The design by W. Gropius for an office building for McCormick and Company in Chicago 1953 is significant in this context, the air-conditioning convectors appearing externally as closed panels in the spandrels.

Evolution does not stand still; it is pushing towards 'integrated environmental design' — integration of the structural system, technical services, spatial layout and architecture. So far, structural steelwork has not in this respect yielded such tangible and striking results as has reinforced concrete in some recent administrative buildings in the USA such as the office building for the American Republic Insurance Company at Des Moines (architects Skidmore, Owings and Merrill), where the concave profiled precast concrete floor girders spanning 20 m accommodate the ventilation ducts as well as lighting units. This combined ceiling and lighting system extends some distance outwards from the recessed glazing, thus emphasising the guiding principle of integrated design and also creating a physiologically satisfactory transition from natural to artificial lighting in the window zone of the large open-plan interior space.

With structural steelwork the problems associated with integrated design may be more complex as it is sometimes necessary to cater for corrosion and fire protection. The exposed water-filled box columns of Cor-Ten which support the US Steel Building in Pittsburgh (p. 160) are an instance in point. A similar method of fire protection has been used for the BFI Research Institute in Düsseldorf (p. 98). Even more advanced is the cladding system developed by the firm of Gartner, which combines water-filled columns with heating and ventilation facilities.

The Electrical Engineering block at the University of Delft (p. 104) is another example of integrated design. On all floors the longitudinal sides of the building have continuous externally glazed balconies, elaborate services being installed in front of the spandrel walls, so that these services are readily accessible and yet protected from the weather, while the balconies provide a more gradual transition from external to internal climatic conditions.

A fair idea of what the integrated design of steel-framed buildings can achieve is provided by the technical college at Brugg (p. 90). In this project, the choice of structural system, the design and detailing of the loadbearing members, the accessible and adaptable arrangement of the services, the rational construction, the fire protection system and the architectural treatment of the façades have all been very carefully co-ordinated. The intensity of design gives this steel-framed building a sense of purpose and a certain unobtrusive functionalism, but also the sureness of touch and relaxed ease that characterise the best achievements in architecture, as in everything else.

Chicago, design for an office building for McCormick and Company

Exposed or concealed steelwork

Many architects hold the view that in a true steel structure the loadbearing steel frame should be visible, structures which do not fulfil this requirement being considered unsatisfactory. This attitude of mind is understandable when we consider the evolution of international architecture in steel and its rivalry with concrete in recent years. It would, however, be rendering a disservice to steel to give unquestioning support to the prejudice in favour of the exposed steel frame.

A series of examples of actual buildings presented in the second part of this book comprises a surprisingly high proportion of true, carefully designed steel buildings in which the structural frame is not exposed and in which, indeed, in some instances there is no apparent indication of the frame at all. To supplement our examples we may merely consider a few recent buildings to obtain a fairly complete spectrum ranging from full concealment to complete exposure of the structural steelwork. This will show that the degree of exposure of the steel frame is not necessarily a valid criterion for architectural and functional quality and that, instead, the decision in this matter, involving careful consideration of the technical, operational and economic interests, most certainly belongs to the creative thinking that steelwork demands from the architect and the engineer.

Let us begin with a structure in which the designers, in their efforts to achieve the integrated design to which we have referred, arrived at an entirely smooth exterior in which the positions of the recessed structural columns are not immediately apparent. It is an administrative building at Lausanne-Ecublens (architect J. P. Cahen), for which an industrialised system previously employed in school construction was adapted. The double-skin cladding units were prefabricated from aluminium sheet, louvre blinds and air-conditioning units being fitted in the openings or in the inner skin respectively in front of the windows. The grid corresponding to the column positions can be identified by means of the narrow strips inserted after every fourth window.

The power that can be expressed by completely smooth cladding fixed in front of a structural steel frame is very convincingly demonstrated by the Bankrashof office building at Amstelveen (p. 148). The pattern of dark and light vertical and horizontal strips on the closed surface, which gives the building a kind of 'plastic' power and action, is reminiscent of the basic means of expression that Mondrian, Rietveldt and other Dutch avant-garde artists contributed to the new architectural movement fifty years ago. The structural columns are provided with fire encasement on the inside and sheathing on the outside which appears as dark lines down the shallow bands of glazing.

Nonetheless, the novel idea which gives this building its particular constructional charm is that of the asymmetrically arranged closed vertical strips behind which the wind-bracings are concealed and which give the tower block the dynamic quality of a controlled rotary motion. When one compares this simple version of the modern architectural idea of an externally exposed bracing system with the conventional solutions used by steelwork designers in the 1930s, the great change that has taken place in constructive thinking, in the structural conception of architecture, becomes immediately and strikingly clear.

Large housing schemes, whether terraced or

high-rise, in which the steel frames and the space-enclosing features have been very carefully designed for constructional economy, present us with a number of solutions which do not reveal to the casual observer whether they are framed in steel or concrete. Typical are the high-rise blocks at Balornock (p. 70) and the low-rise houses at Stora Tuna (p. 72), where the steel frame is concealed by a lightweight concrete infilling.

For buildings whose function requires closed external walls it may be entirely appropriate to place the steel frame completely inside, as in the Ahlens store in Stockholm (p. 116), where the frame is visible only through the display windows on the ground floor.

The transition from this first general type of external wall construction to a second type, in which the columns, clad or unclad, are revealed in the façade, is apparent from a comparison of two buildings: the Alpha Hotel in Amsterdam (p. 80), with the columns of its multi-storey portal frames concealed, and the Bourgmestre Machtens residential block in Brussels, where the structural columns, or their cladding, determine the rhythm of the façade (p. 68). In the institutional building at Lille-Annapes (p. 100) the mullions provide clear-cut vertical articulation, but there is a structural column behind only every other mullion.

On the other hand, the High School in Chicago (p. 92), a recent project from the Mies van der Rohe design office, exposes the widely spaced columns and edge girders which are encased in concrete to comply with the fire regulations and then sheathed with steel sheet – a strict solution in the spirit of the Chicago School.

A cultured and differentiated façade structure characterises the building of the Compagnie Saint-Gobain at Neuilly, dating from the early 1960s (architects Aubert and Bonin). In keeping with the activities of this company, the largest glass manufacturing concern in France, the entire façade, the band of high windows with the delicate rhythmic spacing of the glazing bars and the narrow dark strips of the floor zone, are all made of glass. The graceful arrangement of glazing bars is supported by and connected to the structural columns by slim vertical mullions. To locate external columns directly in front of the façade, and thus interconnect the façade articulation and the structural system, is particularly appropriate in buildings which are not of very great height or span and in which the most economical solution is provided by close spacing of the columns, corresponding to the planning module. This is exemplified, inter alia, by the new building for the Veterinary Faculty of the Free University of Berlin (pp. 214, 312). In such simply articulated façades the conventional I-section is being increasingly superseded by tubes, more particularly by RHS, which also offer certain advantages with regard to corrosion prevention and fire protection. A typical example of an elegant tubular column façade is the airport terminal at Irwine, California (architect C. Ellwood).

In the Vallourec research centre at Aulnoye (p. 134), the square hollow sections have been conspicuously placed in front of the façade. We thus arrive at the third group of façade systems. Although the motif has perhaps been rather overstated here and although the slender tubes project like pikes above the edge of the roof, there is justification for this – the building belongs to a firm of tube manufacturers.

In recent years the circular hollow section has also regained popularity as a structural member.

A notable advance in this direction is embodied in the elevational treatment adopted for the Faculty of Sciences buildings for the University of Paris (p. 102). The disposition of the tubular columns, varying in spacing and diameter between the ground floor and the upper storeys, particularly their mingling with the edge girders to which the long-span fish-belly girders over the ground-floor area are attached, is of incredible structural and architectural sensitivity. In addition, this novel architecture in steel has historical depth. In the elementary contrast of prismatic and cylindrical members it strikingly recalls the cast-iron façades of the Riverfront at St Louis; it also has its roots in the native Franco-Belgian structural steel tradition which reaches back through Horta to the cast iron period and to which Prouvé brought the impact of modern mechanical engineering and structural engineering in steel. Perhaps design tendencies are emerging here which make even more decisive and greater use of the expressive possibilities and the qualities of steel as a constructional material than do the austere rectangular shapes and profiles of the latest American steel structures.

Naturally, the selection of the profile for structural columns and its relationship with the surface of the outside walls must also be based on design considerations. The cruciform section is usually chosen not only for its practical merits and efficiency, but also because it provides a geometrically perfect corner solution and combines well with connecting wall components. It is very effectively employed, for example, in the children's home in Müllheim (p. 74).

In recent years, an increasingly large number of attempts have been made to enhance or dramatise the external steel frame by the insertion of diagonal bracing; the influence of the American super-skyscrapers, such as the Hancock Centre, is unmistakable.

When such diagonals are not members of long-span lattice girders in bridge-like buildings or suspended structures – such as the administrative building for the Burlington Corporation at Greensboro, NC, where the structural system is supported at only four points – but merely serve as bracing to the cladding panels or as a means of relieving short-span fascia beams, then perhaps they create a bogus impression.

When the structural columns are located even more obtrusively away from the outer faces of the building, the connections between the floor beams or rigid-frame girders and these columns are also exploited as architectural features, as in the administrative building of Fabrizia SA at Bellinzona (architects Snozzi, Vacchini). In the European Court of Justice in Luxembourg (p. 126), the external steel frame is placed 1.30 m in front of the façades, so that the connections to the steel beams are prominently displayed. What further enhances the effect is that in the fourth storey the steelwork cantilevers out another 1.80 m and is emphasised by the horizontal sun-baffles provided there. This heavy overhang of the upper storey has its counterpart in the massive protruding plinth. The building as a whole is reminiscent of Saarinen's Deere Building, the more so as weathering steel has also been used here.

Thus we come to the question of surface treatment for exposed steelwork which protrudes far in front of the external wall. The most pleasing, technically most advanced and, in terms of pure steelwork, nearest to ideal solution is obtained by the use of weathering steel, which is nowadays commercially available in many

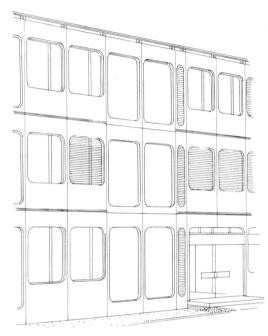

Lausanne-Eclubens, office building

countries and is already in fairly widespread use, although of course this material cannot in itself guarantee high architectural quality. On the contrary, it calls for sensitive treatment and considerable experience in handling it architecturally and also in the proper control of its subsequent behaviour.

The colour of the patina formed on this kind of steel is very attractive indeed, being considered to lend dignity and maturity to the façade it adorns, just as do copper and certain types of wood in the process of ageing.

Nonetheless, the architectural features must be carefully detailed. Otherwise, rain flowing down the face of the building will create uneven weathering and stain those parts of the building with which it comes into contact farther down. Alternatively, these blemishes may be avoided by the use of plastic coatings, but these leave further complications in their wake.

Notwithstanding, weathering steel has been used in many types of buildings and for a variety of architectural effects, ranging from the architect's own house in London (p. 62) through the Swiss National Transport Museum at Lucerne, the Sports School at Magglingen (p. 89), the Olympic Sports Centre at Munich (p. 96) to the Arts Faculty in the Free University of Berlin (p. 311). The two last-mentioned applications are particularly instructive. The architects for the Sports Centre attempted to keep the stains away from the window glazing by an ingeniously contrived system of horizontal gutter sections discharging down the mullions. However, wind-driven rain does not always obediently follow the path intended for it; in addition to which the large number of steel sections used in the finely articulated façade somewhat detract from the colourful effect of the steel patina. By contrast, the arrangement of steel sheet on the laminated thin-walled cladding units covering extensive areas of the façade – based on an idea conceived by J. Prouvé – for the Free University of Berlin is much more appropriate to this new material.

The colour of the patina, ranging from an early red to a weathered brown or purple, is more effectively displayed when it is in juxtaposition to light-coloured surfaces and used sparingly, as has been done, for example, on the exterior of the sports hall for the Munich Sports Academy. A notable and precisely detailed experiment in this direction was made in the design for the façade of the administrative building for the Caisse Patronale in Brussels (architect L. J. Baucher), where narrow vertical members of weathering steel are combined with filler elements made of chemically resistant and washable white plastic.

There is no objection to cladding steel columns or girders, or their fireproofing or other encasement, with sheet metal – light or dark aluminium, stainless steel – where this is necessary and it serves a real purpose in association with the rest of the façade construction. Indeed, in the evolution of steel cladding and curtain walling all possible combinations of this kind have been successfully applied at one time or another.

Finally, the protection of steel by painting, which has been the practice for so long and which has been brought to a fine art by the development of better rust-inhibiting primers, must also be considered a legitimate technical aspect and expression of architectural design. It offers the advantage that the designer can select the colour, specify it accurately and match it with the surroundings in which his structure will stand. The initial cost of such protective systems is low and, in so far as maintenance costs are concerned, any comparison with the more ambitious and expensive techniques should – as with all aspects of detailing in structural steelwork – also take due account of the intended or anticipated life of the structure. For example, when a building will be converted or demolished in less than 15 years, the extra expense on weathering steel, even if cost is not a major consideration, could hardly be justified. This is because the patina, even if it is formed naturally, is to some extent an artificial product (made possible by modern chemical and metallurgical progress) which produces an impression of great durability, of immortality, and thus aspires to a monumental quality which is not necessarily appropriate to a steel structure.

Steel and the traditional building materials
Steel buildings in historical surroundings

Not only are a great variety of external shapes possible with a structural steel frame, but also an unlimited number of material combinations. Thus, steel can be used in partnership with nearly all natural and artificial building materials, both old and new, for infilling or cladding or in prefabricated wall elements. Let us first consider the most important conventional materials: natural stone, brick and timber. Correctly used, they can display their specific advantage in steel-framed structures just as much as in traditional buildings.

In the early period of modern steel-framed construction, especially in American high-rise buildings, natural stone did not perform any loadbearing function; it served merely as an eclectic trimming to the façade, giving it historical character or decorative features. It was only after about 1925, when the use of solid ashlar masonry was being rapidly superseded by facing slabs, that in the USA and Europe a number of representative multi-storey buildings were erected in which the expensive natural stone skin enclosed the structural frame and yet revealed its presence – whether in the ratio of pier width to opening or in the delicate relief pattern of vertical and horizontal bands associated with the technique of stone facing slabs. The RCA Building in the Rockefeller Center, New York, is a magnificent example.

With the advance of pure glass-and-steel construction, the use of stone was at first restricted to flooring or wall covering in main reception areas. In the long run, however, architects as well as their clients were unwilling to forego the use of this high-grade natural material for the external adornment of their buildings, and thus there emerged the various techniques of the stone curtain wall, some examples of which are presented here. The rediscovery of natural stone in a new desire for architectural expression has undoubtedly also had a stimulating effect on the concrete industry to produce artificial stone with the appearance of marble or granite by the use of high-grade aggregates.

The combination of the steel frame and the natural stone façade is very convincingly displayed in the block of residential flats at Rueil-Malmaison (architects P. Somel and J. Duthilleul). This building, which is already twenty years old but is easily holding its own, compares favourably with many residential buildings of large-panel precast concrete construction, because of the weathered appearance that natural stone acquires and because of its link with the French urban architectural tradition, which has adhered to natural stone for a very long time. A particularly attractive feature of this building is the alternation of concealed frame components with externally exposed steel columns revealed in the wide recessed loggias. An even more remarkable structure is the 43-storey tower blocks for the Canadian Imperial Bank of Commerce in Montreal, completed in 1962. This is clad over its full height with slate imported from England.

An interesting experiment was carried out with the façade of the new building for General Motors in New York where the two-storey height lightweight concrete units were provided with an outer skin of polished crystalline marble which also served as permanent formwork for the concrete. The designers of the Standard Oil Company's skyscraper in Chicago have gone a step further. Here, the marble slabs are fixed

Bellinzona, offices of Fabrizia SA

directly to the triangular-section steel plate ribs of the loadbearing external wall.

A material which harmonises with every type of construction, including structural steelwork, and which has for centuries proved its worth in the historical development of timber-framed buildings, is brickwork. In addition to the colour and the dimensional qualities offered by bricks and the various types of bond, this material

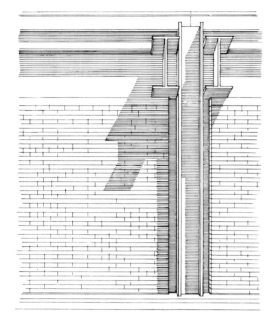

Moline, Illinois, John Deere Company, lecture hall detail

possesses superior resistance to deterioration with age, has advantageous physical properties and is fireproof. Like the iron beam, the brick was one of the first building components to be mass-produced in standard sizes.

In early industrial architecture, brickwork played an important part, as it did also in the American high-rise buildings in the early period of the Chicago School. The Beckman Tower, New York, with its unostentatious piers constructed of small red bricks, embodies the best design features that characterised New York skyscraper architecture about 1925 — a building which still compares favourably with the adjacent aluminium curtain walls and with the RCA Building.

In modern structural steelwork the brick has regained a position — in every type of building. An important stimulus in this direction came from Eero Saarinen, who was one of the first architects to use ceramic material for enlivening and enriching architecture in steel. The lecture hall in the John Deere complex at Moline provides a famous example: the brick infilling of the external walls, in combination with the conventional welded joints of the outer columns and the main girders, gives the building a distinctive character. The contemporary counterpart of the Deere Building, namely, Eiermann's German Embassy in Washington, also uses unrendered brickwork in the closed external surfaces of the bottom storeys.

The brick-infilled structural steel frame has gained particular significance in residential buildings. The Miesian steel-and-glass enclosure did not really become generally practicable for housing until it was used in combination with walls composed of brickwork and timber, as was done by C. Ellwood in the houses he built in California. Among recent residential buildings, it is more especially the artisans' dwellings at Piombino and Taranto (pp. 63, 66) in which the combination of the exposed steel frame and the double-leaf brickwork has shown itself to be truly domestic, constructionally sensible and economical. Moreover, at the same time, considerable architectural quality has been achieved.

Of all the natural building materials the one that comes closest to steel in character is timber. It is certainly no coincidence that it was in Chicago, of all places, that the structural steel frame found such a favourable soil on which to develop and flourish — in a city where, for a good many years, houses had been built by a rationalised method of timber construction, using mechanically sawn and standardised cross-sections in which the concept of modern framed construction was already present.

Timber construction has made great headway in the last ten years or so. Thanks to the progress achieved in structural techniques, especially the use of glued components, timber is increasingly emerging as a rival to steel, even in the construction of wide-span single-storey industrial buildings. What has hitherto hindered the combined use of wood and steel in multi-storey buildings is the fact that neither material is, or was, considered to qualify as being fire-resistant within the meaning of the building regulations. Where timber competes with steel — for example, in single-storey industrial buildings or in staircase construction — the timber experts repeatedly put forward the argument that a low-flammability or suitably impregnated type of timber is superior to steel which, though incombustible, does not retain its stability indefinitely in a fire. Whereas the combination of steel and timber used to be regarded with particular suspicion by building authorities, this prejudice has now appreciably diminished: the experts have come round to the view that there is no such thing as an absolutely incombustible building material and that one cannot justifiably generalise by subdividing the materials into admissible and inadmissible ones; instead, they should be considered in relation to the building as a whole, its purpose, its equipment and its accessibility, especially now that improved methods have been developed for the impregnation of timber to render it non-flammable.

In any case, the use of timber as an adjunct to modern structural steelwork is well established: timber as falsework or as studding for external walls, or boarding, or as a weather-resistant, attractively coloured, dimensionally and physically acceptable external or internal wall cladding material.

It is eminently suitable as a framing material for cladding elements because of its low weight, its ease of working, adaptation and fixing, and its anti-condensation properties, even when the material itself is not visibly exposed — as exemplified by the external cladding elements of the 30-storey blocks of flats at Balornock (p. 70), where the studding is clad on the outside with asbestos cement panels and is lined or infilled on the inside with damp-proof insulating slabs — in combination, incidentally, with wood floors and timber-framed windows. Of even simpler construction are the closed wall elements used for the family dwelling house at Landskrona (p. 60): a kind of timber framework infilled with insulating material, clad internally with wood and externally with impregnated plywood and corrugated asbestos cement sheets. In terms of the unity of the material, the timber framework is of course most convincing in situations where blank external wall surfaces are also faced with this material. Nowadays, it is no longer considered peculiar when timber frames and cladding appear in still more ambitious steel-framed buildings, for example, the administrative building for students' affairs at the Technological University of Brunswick (p. 94), here even in combination with portal frame columns set in front of the façade, or the law courts at Karlsruhe (p. 124). As already mentioned, Eiermann conceived the combination of a timber frame for the storey-height glazing and wooden boards as sun-break fins, together with exposed structural steelwork, in his design for the German Embassy in Washington. The woodwork in Oregon pine gives the strict glass-and-steel architecture a particular flavour of well-bred elegance and freedom which is very pleasant in an official building, but is also very suitable for public transport buildings, especially airport terminals. Thus, the architects have taken

Hollywood, experimental housing development

Gothenburg, Town Hall extension

advantage of this effect in the terminal buildings at the airports for Edinburgh and Copenhagen. Besides its merits of design and formal aesthetic quality, the combination of the steel frame with timber both internally and externally also has a psychological aspect. With the progressively increasing mechanisation of human life and environmental conditions, people are increasingly feeling the need to have at least some natural material in their homes and in their places of work. As a result of the incorporation of timber walls and partitions within the framework of steel and glass, the residential building, in particular, was made more congenial and humanised into a home. Here, the tubular steel frame harmonises quite naturally and easily with the glazed surfaces, the wood and the brick walls. A typical example of up-to-date functional design for a framed structure is provided by a medical research laboratory built on a wooded slope near San Francisco (p. 99). In this instance, the architects managed to make the very elaborate and highly sophisticated technical services accessible and visible in the framework of the floor structure and yet to create an atmosphere congenial to human beings. The beams and the floors are of timber, while the whole exterior of the building is covered with wooden shingles. Sprinkler equipment installed in the corridors and staircases was considered to provide adequate fire protection.

Therefore, in situations where the function or character of a building demands or suggests it, a combination of structural steelwork with the traditional materials can be achieved without difficulty, and sometimes quite advantageously. Also, it is possible to make a modern steel-framed building blend graciously, and with due regard to present-day concepts of care for ancient monuments, into historical architectural surroundings, among buildings constructed of stone or brick in the traditional way. This difficult and challenging problem is gaining in importance in view of the ever-increasing volume of building activity and the dwindling heritage of historical architecture, the increasing interest being shown by a wide section of the public in the preservation of irreplaceable cultural assets, and the more and more stringent requirements imposed for the protection of historic buildings.

Of the various possibilities of harmonising with a naturally developed urban environment — ranging from deliberate contrast, through dimensional and structural integration with modern resources, to a more or less faithful copy of the façade — the last-mentioned possibility is the least attractive to the true protagonist of steelwork. Yet it is here, in particular, that steel can prove very helpful, especially where a valuable old building has to be completely gutted and redeveloped to serve a different purpose, as has been done successfully on a number of occasions in Italy — for example, the conversion of a palazzo in Como into a municipal library.

An impressive example of the harmonious integration of the design resources of modern construction technology with surroundings characterised by protected historic buildings is afforded by the Town Hall extension at Gothenburg, Sweden, by Gunnar Asplund (1934–1937). From 1920 onwards, this architect had prepared a whole series of schemes, embodying traditional design and construction methods, for the extension to this classical style building. These schemes, displaying great expertise and the skilful control of shapes, were of a kind such as no present-day architect would have the single-minded dedication to produce. The solution he eventually evolved and which was actually built is a masterpiece, a culminating point of early functionalism, giving visual proof that in structural steelwork architecture as an art in the true sense of the word is certainly possible. It embodies a steel frame which reveals itself in delicate relief on the stuccoed façade, having asymmetrical flexible infilling with window units of varying size and composition, depending on the utilisation of the rooms as offices or as courtrooms. Admittedly, this Town Hall extension makes rather exacting demands upon the observer and, indeed, long after it was built it remained incomprehensible to many architects. If the design were submitted today, it would probably be turned down by the planning authorities.

It is instructive to compare the various degrees of contrast effect, of adaptation, or the transposition of historical construction features to three more recent Italian steel buildings: the Rinascente department store in Rome, with its resolutely exposed steel frame and windowless external wall, clad with reddish artificial stone units or pronounced sculptured configuration; the Jolly Hotel in Rome, likewise with an emphatically exposed steel framework and sun-baffles (p. 82); and finally the Chase Manhattan Bank in Milan (p. 138), whose shape resembles that of the adjacent domed Baroque church and in which the ground-floor zone attempts to transpose the traditional arcades into structural steelwork.

In all three projects the architects have succeeded in harmonising the buildings dimensionally with the local urban scene. The steel portal frame arcades are indeed a special case which cannot be directly adopted. If we consider the many variants that have been evolved from the architectural idea of the portico, the arcaded street, in places such as Bologna and Turin from early mediaeval times onwards — timber posts and beams, brick piers and arches, marble columns and vaults, reinforced concrete structures faced with natural stone — we must also consider it perfectly feasible for a structural steel feature to fit agreeably into such an urban situation. In any case, the exposed steelwork of the Jolly Hotel in Rome, though perhaps somewhat overstated in terms of present-day design trends, blends very well with the late-classical architecture of its surroundings.

This sympathy between the exposed steel frame and the classical façade articulation of the neighbouring houses was also successfully achieved in raising the height of a building, by the addition of several storeys, in the Rue Jouffroy in Paris some years ago (architect E. Albert). A typical old Parisian house, together with its forecourt, was covered in with the aid of a finely articulated tubular steel frame along the frontage of the street — incidentally, the first true application of circular hollow sections in recent construction. Here, even more clearly demonstrated than in the Faculty of Sciences building in the University of Paris, is a reminder of the cast iron era, the link with the French tradition of architecture in iron.

Paris, additional storeys in the Rue Jouffroy

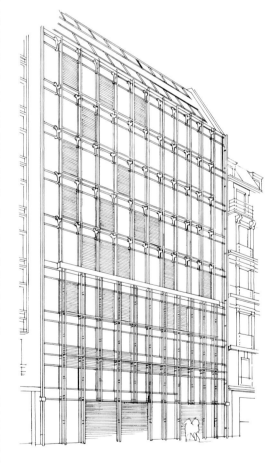

Steel and glass

The ideal concept of building entirely in glass and steel is undoubtedly one of the most typical and important driving forces in modern architecture. The first advances in this direction were made as far back as the early years of the nineteenth century, immediately after architects had begun to use iron instead of timber and masonry for structural purposes. France had an initial advantage in this evolution, for in that country the production of plate glass had made great progress as early as the seventeenth century with the invention of the casting process. The first domed granary roof, originally a timber structure, but destroyed by fire in 1799, was replaced by an iron framework with extensive glass-infilled areas. Later the Galerie d'Orléans emerged as the first true glass-and-iron vaulted structure, with an entirely novel sense of space and mass, a fundamentally changed relationship of internal and external spatial treatment — probably the most successful achievement of nineteenth-century architecture. The glass-and-iron roof or vault was to become a characteristic feature of urban life in the last century: covered arcades and galleries, market halls, station roofs and exhibition buildings, the most famous of which was the Crystal Palace in London. This last-mentioned structure, in particular, shows very clearly that the technology of glass manufacture was still lagging behind that of iron. The whole design of this building, its strictly modular co-ordination and its carefully timed erection plan, had to be based on the limited sizes in which sheets of glass were available in those days (not more than about 1.25 m in length).

In multi-storey building construction it was of course much more difficult than in single-storey shed-type structures to achieve, or at least to approach, the idea of the completely transparent house. This is not, however, the place to describe the whole evolution of modern framed construction from the point of view of the part that glass played in it. It is sufficient to note that all the difficulties and setbacks that had to be overcome in this domain nearly always resulted in reduced fenestration. Conversely, the smartest and most elegant solutions in steel-framed construction were generally closely associated with extensive glazing of the external walls.

The fact that it was so long before full-scale glass-and-steel architecture was able to gain a permanent foothold in multi-storey construction — about a hundred years after the Crystal Palace, which of course was a single-storey structure — is closely related to the lack of knowledge of the technology of glass manufacture. Thus the first attempts to construct extensively glazed façades on multi-storey buildings, such as the famous curtain wall that Gropius designed for the Fagus factory building, were obliged to remain modest efforts on account of the limitations imposed by the glass. It was not until the early 1920s that mass production of glass in large sheets became really practicable.

It so happened, however, that around that time in Europe, where significant new ideas for framed construction were emerging, reinforced concrete had thrust structural steelwork into the background. That is how the fully-glazed multi-storey building was first achieved, not with a structural steel frame, but with one of reinforced concrete. The earliest multi-storey buildings with overall glazing, such as the Van Nelle tobacco factory near Rotterdam, 1927, and the Boots drugs despatch centre at Beeston, England, are industrial or storage buildings, in which the extensive glazing of the external walls could more readily be achieved than in a residential or a commercial building. Here the mushroom floor, which R. Maillart had used in Zurich even before the First World War, has been exploited in an architecturally convincing manner, and the whole loadbearing structure in the external wall is reduced to a narrow strip.

The first fully steel-framed building in which glazing with large glass panels was consistently applied and made into an integral part of the structural fabric — an amazingly advanced design for the year 1939 — is the Maison du Peuple, a multi-purpose single-storey wide-span building at Clichy (architects Beaudouin and Lods; engineers Bodiansky and Prouvé).

The glass-and-steel façades of Mies van der Rohe's blocks of flats, and also the first curtain walls, were — by the standards of our present-day ideas and requirements — technically imperfect in that they were glazed with single thick panes. For a long time, designers blandly chose to ignore the heat losses, the coldness and the other physical and physiological drawbacks associated with single glazing and indeed, in America, where people had already been long accustomed to full air-conditioning and its

Pittsburgh, Heinz vinegar factory

shortcomings and where energy was plentiful and cheap, these drawbacks probably did not weigh as heavily as in European countries. Indeed, the enthusiastic residents of the Lake Shore Drive Apartments found it rather good fun to breathe 'peepholes' in the thick coating of frost that formed on their windows in winter, especially when a number of people were gathered in a room and the internal air humidity was therefore high.

By giving so much prominence to glass cladding, Mies van der Rohe did, however, provide a stimulus to the technological progress of glass manufacture. The inventions and new developments whereby the glass industry overcame in quick succession the inherent weaknesses of its product — deficient thermal insulation, brittleness, 'hothouse effect' due to the sun's rays — would not have been so speedily and so fully accomplished if there had not already existed a great demand for glass in large sizes. The glass chosen for the cladding of Lever House in New York was green-tinted Katacolor, a type of glass then extensively used for motor car rear windows. In selecting this material the architects, Skidmore, Owings and Merrill, were certainly not concerned exclusively with sun-protection — for in this respect tinted glass does not provide the complete answer — but were quite as much interested in ensuring a uniform colour for the whole glass cladding surface, matching the green wired-glass spandrel panels, and especially in giving the building as a whole a sense of mass. This idea must have fulfilled a generally felt need, for it immediately caught on and was widely adopted by architects. The most striking effect was achieved by Mies van der Rohe with the golden-brown glass, matching the bronze mullions, on the Seagram Building. This was exactly the colour of Seagram whisky and resulted in a marked increase in the sale of that product. Among the early curtain walls on American skyscrapers, that of the Corning Glass Building in New York (architects Harrison and Abramovitz, 1958) represents the extreme case of the glass façade: exceptionally slender vertical glazing bars, barely perceptible horizontal ones; the green-tinted sun-protection glass used all over the façades reveals scarcely any difference between the windows and the spandrel areas, where it is provided with a solid backing. A striking and, by European standards, unusual feature of this building was the pronounced waviness and irregularity of the drawn glass panels with their distorted, bizarre reflections of the adjacent buildings, for in Europe it was often considered essential that the curtain should consist of absolutely plane polished plate glass. Comparable to the Corning Building in having an almost entirely glass-covered façade, though with glass of a much darker tint, is the front of Saarinen's IBM Center. Yet to the viewer inside the building, looking out, the dark colour of the glass — like that of good quality sun-glasses — is hardly noticeable. The darker the tint of the window glazing, the more it acts outwardly as a mirror, and it is significant that international architecture, in turning to romantically expressive and symbolically oriented effects, has rediscovered and utilised this magical effect of glass as a construction material. Basically, this is reversion to its early origins now rather lost in the mists of antiquity. Glass as a transparent wraith-like medium, which functional architecture at first appeared to want and which was actually provided by modern mass-produced plate glass, is now no longer considered essential. Architects realised that this material could also be used as a coloured translucent, opaque or reflective surface; indeed, it had already been used in this way in Gothic cathedrals and Baroque palaces.

In their design for the Rare Books Library at New Haven the architects, Skidmore, Owings and Merrill, went so far as to incorporate slabs of onyx into the sculptured network of prefabricated cruciform cladding units and thereby conjured up memories of ancient palaces and ecclesiastical architecture, thus looking back to a time when glass was even more costly than a polished semi-precious stone.

In this context must also be viewed the project in which Mies's dream of the glass skyscraper in Berlin, conceived in 1919, was posthumously realised — more specifically in Lake Point Tower, Chicago, with its clover-leaf plan and its curved glass surfaces. Here, reinforced concrete once more stole a march on steel. In this 65-storey building, for a time the tallest reinforced concrete structure in the world, the special shape helped to resist wind pressure and to transmit wind forces to the foundations, the triangular concrete core providing adequate structural rigidity.

The vinegar factory at Pittsburgh (Skidmore, Owings and Merrill) calls for mention as a pure example of glass-and-steel architecture of the 1950s — a multi-storey industrial building of straightforward austerity with a finely proportioned rhythmical division of its mullions. As a possible European counterpart to this building, the car park and showroom at Barcelona may be mentioned (architect Echagüe).

From about 1950 onwards, glass technology made further important advances. Of all the various systems of insulating glass, the principle of double glazing, which makes the cost of heating a building with all-glazed external walls an economic proposition and prevents condensation as well as providing sound insulation, is the one that has been most widely used. Since double glazing units are, on the whole, not much heavier and thicker than thick single glass panes, they did not significantly change the development of façade construction as such. For example, on comparing the Miesian detailing on the outside of the Seagram Building with the technically improved cladding, giving better thermal insulation, on the Federal Center in Chicago, the difference in profile of the glazing units is so slight as to be hardly noticeable.

Simultaneously with the improvement of insulating glass, various types of safety glass were also developed: laminated multi-layer and toughened single-layer glass. With 'all-glass construction' for walls and wall openings, new and undreamt-of possibilities for glass architecture opened up. The advances in knowledge of the transmitting capacity of coloured glass in relation to various ranges of the spectrum have yielded fresh possibilities for mitigating the hothouse effect, besides offering enriched architectural expression.

In addition, in recent years various methods of applying coatings and thin metallic films to glass have been developed, whereby a large proportion of the heat is rejected by the glazing and the façade is given the character of a vast mirror. This effect was most strikingly utilised by Saarinen in his multi-purpose steel-framed Bell Telephone Building, the immense surfaces of which are clad with solar heat resistant glass.

An interesting thing about these new glazing techniques is that they can nearly all be used in combination with one another, that is, tinting or coating can be applied to double glazing or to safety glass, and so on. Thus, there is now a very wide choice, indeed almost a confusing variety, of possible uses and configuration — ranging from the plain to the patterned, from colourless transparent to coloured opaque, from the matt to the mirror-smooth glass surface. And there is another important new aspect too: the old controversial issue of sheet glass versus plate glass may soon be rendered meaningless by the introduction of float glass — produced by a relatively new process — if it becomes widely adopted and comes down in cost. This new type of glass combines the livelier sparkle of sheet glass with the accurate flatness of polished plate glass.

The whole charm and richness of lustre, colour and reflections that present-day glass technology can offer the architect will of course be most effectively displayed in combination with the slender proportions of an all-steel structural frame. Among the internationally famous steel structures of recent years, certain high-rise commercial and office buildings with beautiful storey-height glazing call for special mention: the office building at Stäfa, Switzerland (p. 132), the insurance company offices in London (p. 146) and the pension fund administration building in Luxemburg (p. 150). Glass-and-steel construction has attained a particularly effective charm and functional force in those buildings on which glass sunbreak fins are suspended as a kind of second skin in front of the glass façade proper. The first ambitious efforts in this direction were made by the architects Zehrfuss and Breuer on the Unesco Building in Paris and on the administrative building for the Van Leer Company in Amsterdam in the early 1960s. A strictly functional but carefully designed special-purpose building, such as the multi-storey steel-framed Heberlein engineering works at Wattwil, Switzerland (architect Custer), acquires, by the combination of two different kinds of glass for the outer skin and for the sunbreak fins, a curious charm and an amazing contrast of day and night effects which appears to soften the austere structural mass.

In the IBM head office building, a low-rise structure at Cosham, England (architect Foster), a record of nearly one hundred per cent glazing with large units has been established. Here, indeed, the impression is created of an all-glass prism in which the trees and clouds are reflected.

This complete suppression or concealment of the loadbearing and mullion system must not, however, be presented as the ultimate goal of evolution, the optimum in glass-and-steel construction. A graceful framework for assorted well-proportioned panes of glass in combination with elegantly detailed exposed steel stiffening ribs is, if anything, much more aesthetically pleasing, as is exemplified by the Museum of National Art in Paris (architect Dubuisson). A further example to show that architects are once again turning to coloured glass and its magic effects is provided by the big reception hall in the Blue Sky Hotel in Japan: a steel frame, completely infilled with large coloured panes in the walls and ceiling.

An outstanding achievement of modern glass-and-steel construction, in which, despite much more difficult functional requirements, the structure attains the fantastic effect produced by the Crystal Palace a hundred years earlier, is the Royal Belge administrative building at Boitsfort, Brussels. The 7.5 m high panels of plate glass, with their metal-and-glass stiffening mullions in the low-rise podium block, and the gold-

Paris, Musée des Arts et Traditions Populaires

coated insulating and solar heat resistant glass on the building itself, are the features that brilliantly bring out the daring quality of the exposed steelwork which appears to float in space.

Steel and concrete

The most important and most frequently encountered combination of constructional materials is that of steel with concrete. It is a remarkable coincidence that these two so essentially different materials are so completely compatible and complementary to each other — in their thermal expansion, in their composite action, in the corrosion protection that the concrete gives the steel — almost as if they had been made for each other. Since the invention of reinforced concrete, steel and concrete have, as it were, been destined for each other. Just as there can be no reinforced concrete without steel, so can there be, at least in multi-storey buildings, hardly any structural steelwork without concrete. Every steel-framed high-rise building stands on concrete foundations. In addition, as a rule, there are basements or underground car parks constructed in reinforced concrete. Plain or reinforced concrete is almost invariably used in floor slabs for steel-framed buildings, ranging from straightforward reinforced concrete floor through the various possibilities of composite construction to sheet-steel decking with lightweight concrete topping. In modern steel-framed buildings, reinforced concrete is used as a means of imparting structural rigidity: in shear walls (as cross-walls or end walls) and, even more extensively, in the service and circulation cores. There are various possible degrees of structural co-operation between the concrete core and the steelwork, ranging from the structurally independent reinforced concrete core, which is constructed in advance and to which the steel frame is then attached, to the internal structural steel frame which is subsequently embedded in concrete to form the core enclosure.

Besides being utilised to perform structural functions within the loadbearing system, reinforced concrete can also be used as an encasing or facing material: it offers one possible method of fire protection for steel and it is also frequently employed in wall or cladding components.

To what extent such components or parts of structures should preferably embody all-steel construction, be constructed entirely in reinforced concrete or be of composite construction depends on circumstances which vary from one instance to another, as will appear in subsequent sections of this book. It is not possible to lay down general rules for this. However, it may be possible to generalise to the extent of stating that it appears appropriate to use concrete within the context of steel-framed construction if it is thereby possible to fulfil several structural and functional requirements simultaneously, such as shear wall action, sound insulation and fire resistance of floor slabs, withstanding earth pressure and excluding moisture in combination with structural load distribution in the basement, etc. Besides considerations of economical construction, aesthetic design aspects may also play a part. To construct the internal core of the building in concrete is an obvious decision if, in the interests of flexible layout and smooth circulation, it is given a curved instead of a rectangular shape on plan, as has been done, for example, in the Chase Manhattan Bank in Milan (p. 138). More recently, completely cylindrical cores centrally arranged within buildings which are square or cylindrical on plan have been adopted (pp. 134, 136).

Just as in a steel framework certain individual structural members may be constructed in reinforced concrete, so the reverse may also be encountered. Thus, in the grid-type façades built in the 1950s, more particularly in Germany, the closely spaced external wall columns of a concrete building had to be intercepted by lintel girders spanning large openings on the ground floor. In many instances this could be achieved only by means of steel plate girders. Quite often, too, in a reinforced concrete structure the external columns may consist of steel sections, so that they may be accommodated unobtrusively within or behind the glazing. A particularly ambitious example of glazing in a multi-storey building dating from the 1950s, the Haus der Glasindustrie in Düsseldorf (architect B. Pfau), is constructed on this principle — a structure which certainly helped to spread the idea of pure metal-and-glass architecture in Germany.

Viewing all the possible combinations of steel and reinforced concrete, ranging from all-steel to entirely reinforced concrete construction, it is possible to observe continuous transition from one to the other, it being hard to draw a demarcation line between them. From time to time, it has been asked whether and to what extent such hybrid intermediate constructional forms are compatible with an honest architectural attitude of mind. Many architects felt, and indeed still feel, suspicious or ill at ease about the mixing of steel and reinforced concrete in structural features. On the other hand, as will be evident from an inspection of the examples of buildings presented in the next section of this book, the combination of a concrete stiffening and loadbearing core with a surrounding steel frame is so frequently encountered and is so logically acceptable within the functional layout and in the whole system of load transmission that it cannot be classed in any sense as a second-rate solution.

A classic example of how a combination of reinforced concrete and steel in a wide-span structure emerged from the technical and economic condition, and how the alternation of the materials became the actual design motif and indeed the hall-mark of the design, is provided by the Palazzo del Lavoro, a large exhibition building at Turin designed by P. L. Nervi who, until then, had always designed in reinforced concrete. The building was intended to have a reinforced concrete mushroom floor, daringly conceived to provide complete freedom and flexibility of layout, a superb example of an open-plan concept. In view of extreme pressure of time, a composite form of construction had to be adopted instead. Thus only the robust columns were built of in-situ reinforced concrete; the enlarged mushroom heads and the radiating floor ribs were, however, constructed as steel plate girders which offered the time-saving advantage that they could be fabricated off the site while the columns were being concreted. The contrast between the compact, upward-tapering columns and the cantilevering steel ribs of the floor structure they support gives this building its visually impressive character, the true mushroom floor effect, never attained previously with such telling intensity in pure reinforced concrete construction. This exhibition hall undoubtedly provided a strong stimulus to the wide adoption of the mushroom floor principle and, derived from it, the umbrella roof, a type of structure first applied to single-storey buildings, but latterly also to multi-storey ones (p. 114).

This example shows that a composite steel-and-concrete structure can be entirely relevant and that, in addition, a combination of two different media can achieve architectural expression in its own right if it becomes possible thereby to create unity of function, structure and shape. In the collected examples of steel-framed buildings presented in this book there are several in which this unity has indeed been achieved and in which the external appearance of the building, too, expresses the composite structural principle: for example, the suspended and bridge-like buildings already referred to. It is not, however, essential that the loadbearing and stiffening strength of the concrete core should be dramatically revealed on the outside to justify the presence of that structural feature. In some buildings a glance at the ground-plan will show the inherent need or justification for a compact internal core as, for example, in the Unilever Building in Hamburg (architects Hentrich and Petschnigg). This plan may be compared with a

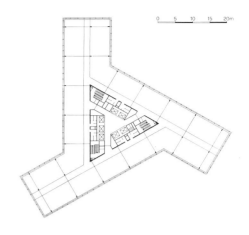

Hamburg, Unilever Building, plan

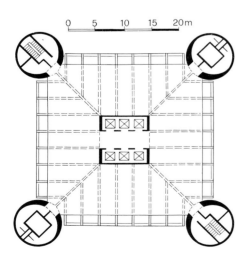

New Haven, Connecticut,
Knights of Columbus Insurance Company, plan

very recent French building, the Annexe to the Foreign Office at Nantes, in which a similar windmill-like arrangement of the constituent blocks characterises the layout and opens it up, but where a more striking effect has been obtained by locating the vertical circulation cores, not in the centre, but at the ends of these wings (architect J. Dumont).

The design concept of moving the vertical circulation cores outwards and thus elevating the contrast between the 'closed' concrete feature and the 'transparent' steel framework to an aesthetic motif is even more strikingly demonstrated in the highly individual building of the Knights of Columbus Insurance Company at New Haven (architects Roche and Dinkelo). This 23-storey building, which is square on plan, has prominent brick-faced reinforced concrete cylindrical corner towers which share with an internal core the function of transmitting the floor and wind loads. In this building, the structural steelwork is confined to the floors. In combination with the recessed glazing, the exposed outermost main and edge girders, in weathering steel, form pronounced horizontal elevational features linking the corner towers visually as well as structurally.

An ingenious and visually very effective architectural solution embodying such a combination of loadbearing concrete and associated steelwork has been provided by the architect E. Eiermann with the two towers in the Olivetti administrative centre at Frankfurt-am-Main. In this instance the multi-storey block as a whole is raised high above ground level on the pilotis principle, this arrangement being well justified by the relatively narrow site and by the need to provide extensive low-rise blocks as well. The powerful reinforced concrete shaft supporting the multi-storey building rises free to a height of 16 m, its upper third widening out in the shape of the inverted frustum of a pyramid. As a result of its completely closed white-painted surfaces, the structure acquires an enhanced clear-cut spatial quality and provides a striking contrast with the steel framework and the finely articulated sun-shading galleries.

What makes these Olivetti towers particularly attractive is the fact that the loadbearing concrete shafts project far above the top of the multi-storey block, farther than needed by the strictly technical requirements of accommodating the lift machinery and services. In one of his preliminary designs Eiermann had omitted the enlargement of the shaft under one of the towers and had, instead, envisaged this feature in the inverted position on top of the building, which would thus, in a sense, have been placed upside down. With regard to the present position of structural steelwork and architecture it is significant that this odd idea was conceived by a man who, only a few years previously at an international steel congress, had uttered the famous words that, to him, concrete was just an unappetising mushy mass, whereas the steel frame was the embodiment of the aristocratic principle in architecture.

The battle between steel and reinforced concrete, which is still being fought tenaciously in architectural and structural design offices, in the technical press, at congresses, in research establishments and in the universities and technical colleges — though in a manner not always apparent to the uninitiated outsider — is in fact, as already pointed out, a highly stimulating force in the evolution of modern building technology and structural engineering. Both methods of construction have always learned from each other and benefited by it. When one of the two had gained an advantage, the other had to make a special effort to catch up and, in so doing, often made use of theoretical and technical methods and progress already achieved by its rival. Thus, the progressive realisation of continuity, of spatial relationships within the structure, which the designer in concrete inescapably had to tackle, has also been of great benefit to structural steelwork. Conversely, reinforced concrete designers, prompted by the necessity to rationalise construction procedures, have more and more been compelled — more particularly with major building projects — to give careful consideration to the alternatives of in-situ or precast concrete or an efficient combination of the two. In this respect, they have adopted some useful ideas from the design and construction of structural steelwork, not to mention the fact that for exposed structural features and particularly severely loaded components they sometimes use rolled steel sections anyway.

In terms of architectural design steel and reinforced concrete have also stimulated each other and helped to achieve progress, as is indeed very clearly demonstrated in the latest trends in international architecture. Within the domain of modern framed construction there is in fact hardly any design principle or motif or form of expression which, having emerged and been evolved from structural steelwork or from reinforced concrete, was not sooner or later adopted by the other side, modified and sometimes even misunderstood or misused. Thus the mullions of the classic Miesian steel façade had been copied or adapted by designers in concrete, in Germany more particularly, as a means of architecturally enhancing the grid features of the façade, even before they had, with the advent of the curtain wall, become a commonplace feature in international architecture. The idea of arranging the loadbearing columns outside the building, free from the actual façade, and of raising the horizontal members of portal frames above roof level, thus providing a striking visual effect, did not escape the attention of the architects in concrete either: there are numerous characteristic examples to show this, especially in Italy.

The question whether the protruding structural frame and the recessed façade emerged first in steel or in reinforced concrete cannot so easily be decided. In American multi-storey construction, this new design concept for the framed structure emerged more or less simultaneously in both materials and its possibilities and architectural advantages were displayed in both methods of building. In any case, it is also revealed as clearly as in the latest American low-rise buildings in some European examples,

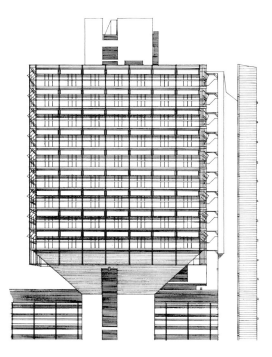

Frankfurt-am-Main, Olivetti administrative centre

for example, in the precast concrete constructional system for the University of Marburg, although in this case the structural concept has been coarsened, while the flexibility of the façade arrangement has been increased and the design pretension, if anything, enhanced.

Conversely, there are characteristic shapes and design ideas in reinforced concrete construction which have been adopted more or less felicitously and more or less faithfully in structural steelwork. In the case of an exhibition pavilion which demonstrates the multiple design possibilities offered to the architect by stainless steel cladding and which serves to publicise those possibilities, as in the Thyssen Pavilion, it is hardly appropriate to consider it a lapse in the architect's aesthetic judgment if he uses the

Nantes, Annexe to the Foreign Office, plan

0 5 10 15 20m

elegant sculptured effects readily attainable in concrete (p. 105). More questionable would appear the principle of reproducing in weathering steel, as in the case of the Transport Museum at Lucerne, a shape characteristic of concrete construction, such as the triangular piers of Saarinen's CBS Building. A particularly good example of the adoption of constructional features from the rival medium is afforded by the façade units used on the steel-framed building of the Standard Oil Company in Chicago, which at present ranks as the fifth-highest skyscraper in the world. In this case, triangular sheet-steel ribs were welded to the horizontal spandrel elements of the loadbearing and stiffening steel cladding to form units similar to those used in building the World Trade Center in New York, which in turn were foreshadowed in principle by the precast concrete units on the glass factory at Detroit, which Yamasaki had designed.

Steel or reinforced concrete

To many architects and engineers, including many who are in no sense professionally or commercially committed to either side of the construction industry, the alternative choice between steel and reinforced concrete has long been, and still is, a fundamental issue, almost a matter of conscience. Even two such competent and unbiased experts as the great engineer and architect, E. Torroja, and the distinguished historian of the art of structural engineering, H. Straub, agree that steel is a structural material, but not a building material in the true sense, because in their opinion it is lacking in body or mass. This view, which can now be regarded as obsolete, is typical of the deep-rooted attitudes of the architects, including some otherwise very open-minded ones, referred to at the beginning of this historical account in connection with prejudice against iron and steel and the tendency to cling to traditional concepts. Reinforced concrete was better suited to conform to those concepts, and for this reason it succeeded much more rapidly in evolving an independent and generally acceptable architectural language of its own. For the same reason, the establishment of pure architecture in steel was long-delayed.

The great impulses and trends in the architecture of our century, and the ripples of fashion that follow them after an interval of time, can in very general and greatly simplified terms be reduced to a common denominator: steel and concrete alternately in the ascendant. The temptation to view the situation in such simple terms is even greater in that the expressive merits of the two materials are embodied in the principal creations of their great protagonists Mies van der Rohe and Le Corbusier.

Architectural historians have engaged in speculation suggesting that in these two masters of their art are manifested the two roots and fundamental forces of functional architecture and of modern architectural trends: concrete, the embodiment of the Mediterranean, the Romanesque, evolving from masonry and solid stone construction; steel, the Anglo-Saxon Germanic structural medium, the successor to timber-framed building as practised in the northern countries.

But it is not really possible to mark down Mies and Le Corbusier as neatly representing steel and concrete respectively. Thus Mies's design for an office building dating from 1922 and his Promontory Apartments of 1948 are true reinforced concrete designs pointing out new directions in architecture. Conversely, Le Corbusier's Maison de Clarté is one of the most important of the earlier European steel-framed buildings. Indeed, it is a particularly significant, and within the context of this book, a telling point that the information centre which was built at Zürich to commemorate Le Corbusier, who himself was much involved with the preliminary designs for this building in the closing years of his life, is a steel-framed structure, though admittedly much freer and more informal in its conception than its counterpart, that austere monument of classic and uncompromising steelwork, the New National Gallery that Mies designed for Berlin.

For the true architect or engineer, who thinks in economic and functional terms, there can be no prejudice, no preconceived basic answer to the question 'steel or reinforced concrete?' His motto is 'steel and reinforced concrete'.

By carefully considering the progress of building construction he can derive nothing but benefit from the competition between the two materials. That the possibilities and chances offered by free decision in favour of one or the other form of construction are not always seized and that this factor now still often acts in favour of structural steelwork is due to various reasons. In the present-day context of increasingly exacting demands upon the architect's time and attention, in connection with finance planning, estimating the placing of contracts and the execution of the job, in performing his professional duties and complying with the many regulations and requirements of the public authorities he hardly has the time and energy to go thoroughly into alternative structural solutions and carefully weigh them against one another. Nor can the consulting engineers, especially as they are usually under even greater pressure in preparing their designs and drawings, always produce as wide a variety of alternative designs as might emerge from direct contact with a number of efficient structural steelwork or reinforced concrete firms. The procedure of inviting several such firms to compete for a job often produces surprisingly good ideas in the field of bridge or industrial building construction, but in the field of highrise buildings it is not directly practicable. On the other hand, in architectural competitions – both in the thoroughness of preparation of the designs and in the assessment thereof – the interests of structural design and economy often receive inadequate attention. The practice of building with systems which facilitate the solution of difficulties in normal, frequently-recurring construction jobs, still creates suspicion among many architects and also some of their clients. They are inclined to regard such systems as a possible threat to their professional integrity or as restricting their freedom of design.

For the steelwork designer there is really no fundamental problem in deciding between system building with predesigned and prefabricated standardised components, on the one hand, and the individual 'one-off' design on the other. Any properly designed steel structure will in any case be composed of prefabricated and, as far as possible, uniform components, thus embodying at least some features of system building or system design. Similarly, the architect will not look upon the existing construction systems, if he knows how to select them correctly to serve his purpose, as objectionable restrictions upon his creative ability but will, instead, view them as a stimulus to systematic planning and design and will take advantage of this. It is certainly not the purpose of this book, and especially the sections concerned with steelwork design, to replace any existing whims and conveniently familiar design habits favouring reinforced concrete – admittedly a rather simpler structural medium from the designer's point of view – by a different set of prejudices or biased preferences for steel construction. Instead, it aims to give helpful suggestions to the designer with due regard to material and structural criteria and to stimulate him to think critically both in steel and in concrete.

If there is indeed an overriding consideration which would justify the architect or the engineer in deciding on steel as a general preference, or in a borderline case after due investigation of the alternatives, it would arise from an acute awareness of, and responsibility for, the service life of a building. In Europe, the average expectation of life of buildings erected in major urban centres is already well below the span of a human life. The American multi-storey buildings are designed for a service life of some 25 years. The adaptability and flexibility of layout which are such important aspects in the development and application of structural steelwork can be said to culminate in the fact that a steel-framed structure can be dismantled without excessive cost, noise or dust nuisance, and that the components can be re-used or sold as scrap. Hitherto, the scrap value of steelwork has, as a rule, covered the cost of demolition. From this point of view alone the steel-framed structure represents a favourable method of construction in terms of economy and preservation of the environment.

Architects and building owners must, and gradually will, come to realise that the buildings they erect are not intended as monuments for posterity, but should instead provide, at acceptable cost, an adaptable, light, unobtrusive housing and internal environment to suit the rapidly changing and fluctuating pattern of life, as the best steelwork designers of the last century in fact achieved. Claims to everlasting grandeur can no longer constitute a valid criterion for true architecture; its distinctive characteristic is, rather, the unity of function, design and shape, harmonised with a clear conception of the time scale. Viewed from this angle, steel is an eminently contemporary material, in accord with the spirit of our time: and from the same viewpoint, too, it is perhaps not without deeper significance that such outstanding and, in their day, such contemporary buildings as the Crystal Palace and the great mechanical engineering exhibition hall in Paris no longer exist.

Examples of Multi-Storey Steel-Framed Buildings

WALTER HENN

Co-authors:

Ingo Grün
Gunter Henn
Hans-Georg Köhler

European and German symbols for qualities of steel and rolled sections

	Qualities of steel		Rolled sections		
EURONORM 25	Fe 37	Fe 52	HE-A	HE-B	HE-M
DIN 17 100	St 37	St 52	IPBl	IPB	IPBv

In the following section of this book descriptions are given of 62 projects involving multi-storey steel-framed buildings. They are divided into a number of categories, depending on the function which they perform. This classification results in an approximate grading of the buildings in ascending order of size, ranging from relatively small private houses to some of the world's tallest structures.

Various criteria were applied in choosing the examples. None of the buildings was to be more than about ten years old, while all types of construction were to be represented: rigid frames, simply-supported frames, cantilever systems, suspended structures, etc. Furthermore, the widest possible range of function or use was to be included. Finally, the examples had to constitute a fair international cross-section of modern architecture and structural engineering in steel. Buildings were also included in which the design was governed by the constraints of fabrication or erection or which were of special interest by virtue of the particular method of fire or corrosion protection employed. In addition, certain schemes have been included to demonstrate how the unusual difficulties presented by foundation problems or the need to withstand possible seismic forces can be resolved by appropriate design.

Last but not least, the buildings described had to show beyond doubt that steel is not a material which compels architects to design monotonous grid façades but that, contrary to popular misconception, it offers very considerable scope for novel and imaginative treatment.

With these criteria in mind, a large number of schemes were scrutinised from all over the world. One hundred and twenty of these projects were then selected, all of which could have featured in this collection of examples. The final choice, necessitated by considerations of space in the production of this book, was governed by the view that the buildings should, first and foremost, exemplify architecture in steel suited to European conditions. For this reason, with the exception of a hotel in Tunis, the only non-European buildings included are North American examples.

In order to give the maximum amount of information in the description of each example, it was deemed necessary to arrange the text in a concise form, divided into standardised sections. The data relate to function, dimensions, design features and services. In addition, special advantages offered by structural steelwork, in terms of flexibility of use, efficient prefabrication and rapid erection, have been suitably emphasised in the descriptions. Contrary to general practice in the technical press, both the quantities of materials used and the construction costs, when available, have been quoted. The costs relate to the years indicated. To make a realistic comparison with current construction costs, they must be converted by applying an inflation factor appropriate to the particular country concerned. At the end of each description there are also bibliographical references to enable the reader to obtain further information.

Despite the limitation of space available for illustrations, in every instance the drawings include a plan and cross-section, a diagram showing the structural system, the arrangement of the floor beams and the principal structural details. There is always one general photograph showing the completed building, but particular value is attached to progress photographs as these illustrations give a better idea of the structural steelwork and thus augment the information given in the drawings and in the text.

No.	Example	Place	Country	Architects
	Residential buildings			
1	Private house	Commugny	Switzerland	Annen, Siebold, Siegle
2	Private house	Landskrona	Sweden	Bergström
3	Private house	Stallikon	Switzerland	Wellmann
4	Private house	London	England	Winter and Associates
5	Housing estate	Piombino	Italy	Gorio, Grisotte, Mandolesi, Petrignani
6	Blocks of flats	Rouen	France	Lods, Depondt, Beauclair, Alexandre
7	Housing estate	Taranto	Italy	Nizzoli
8	Block of flats	Brussels	Belgium	Roggen, Liénard
9	Housing estate	Balornock	Scotland	S. Bunton and Associates
10	System building	Stora Tuna	Sweden	Johnsson
	Communal buildings			
11	Mountain refuge	Stelvio Pass	Italy	Conte, Fiori
12	Children's home	Müllheim	Germany	Blaser, Nees and Beutler
13	Day nursery	Berlin-Schöneberg	Germany	Bratz, Hassenstein, Schmidt-Thomsen
14	Students' hall of residence	Paris	France	Parent, Foroughi, Ghiai, Bloc

No.	Example	Place	Country	Architects
	Hotels			
15	Hotel du Lac	Tunis	Tunisia	Contigiani
16	Alpha Hotel	Amsterdam	Holland	Groosman Partners
17	Jolly Hotel	Rome	Italy	Monaco
	Hospitals			
18	Maternity hospital	Sevenoaks	England	Gollins, Melvin, Ward and Partners
19	Group practice clinic	Salt Lake City	USA	Sugden Associates
20	Accident hospital	Vienna	Austria	Hoch
	Schools			
21	Primary school	Berlin-Wittenau	Germany	Hundertmark, Grünberg
22	Physical education college	Magglingen	Switzerland	Schlup
23	Technical college	Brugg-Windisch	Switzerland	Haller
24	High School	Chicago	USA	Mies van der Rohe
	Universities and research centres			
25	Office building for administration of student affairs	Brunswick	Germany	Henn
26	School of architecture	Nanterre	France	Kalisz, Salem
27	Sports academy	Munich	Germany	Heinle, Wischer and Partner
28	Industrial research institute	Düsseldorf	Germany	Hitzbleck, Meyer, Rinne
29	Medical research laboratories	San Francisco	USA	Marquis and Stoller
30	Institute of Technology	Lille	France	Balladur, Tostivint
31	Faculty of Sciences complex	Paris	France	Seassal, Cassan, Coulon, Albert, de Gortchakoff
32	Department of Electrical Engineering	Delft	Holland	van Bruggen, Drexhage, Sterkenburg, Bodon
	Exhibition buildings and passenger handling facilities			
33	Industrial pavilion	Hanover	Germany	Hentrich, Petschnigg and Partner
34	Group of exhibition pavilions	Toronto	Canada	Craig, Zeidler, Strong
35	Motorway restaurant	Montepulciano	Italy	Bianchetti
36	Multi-storey car park	Bremen	Germany	Kaffee HAG, Building Dept
37	Terminal building, Orly	Paris	France	Meyer
	Shopping centres			
38	Supermarket	Interlaken	Switzerland	Wyler
39	Department store	Stockholm	Sweden	Backström, Reinius
40	Shopping centre	Berlin-Steglitz	Germany	Heinrichs, Geiger, Bartels, Schmitt-Ott
	Official buildings			
41	Library	Pantin	France	Perrottet, Kalisz
42	Library	Paris	France	Lods, Depondt, Beauclair, Malizard
43	Federal court	Karlsruhe	Germany	Baumgarten
44	European Court of Justice	Luxembourg	Luxembourg	Conzemius, Jamagne, van der Elst
45	Television centre	Berlin-Charlottenburg	Germany	Tepez
46	Office building for Members of Parliament	Bonn	Germany	Eiermann
	Administrative buildings			
47	Office building for a light engineering works	Stäfa	Switzerland	Dahinden
48	Office building for a tubeworks	Aulnoye	France	Albert, Champetier de Ribes
49	Office building for a structural steelwork firm	Langenhagen	Germany	Wilke
50	Bank and office building	Milan	Italy	Belgiojoso, Peressutti, Rogers
51	Administrative building for an electrical engineering firm	Munich	Germany	Henn
52	Administrative building for an oil company	Rome	Italy	Moretti, Morpurgo
53	Administrative building for an electrical engineering firm	Saint Denis	France	Zehrfuss
54	Office building for an insurance company	London	England	Gollins, Melvin, Ward and Partners
55	Bankrashof office building	Amstelveen	Holland	Vink and van de Kuilen, Klein
56	Office building for pension fund administration	Luxembourg	Luxembourg	Ewen, Kayser, Knaff, Lanners
57	Office building	Puteaux	France	Binoux, Folliasson, Fayeton
58	Office building for an insurance company	Brussels	Belgium	Dufau, Stapels
59	Administrative building for an aluminium concern	San Francisco	USA	Skidmore, Owings and Merrill
60	Tour du Midi	Brussels	Belgium	Aerts, Ramon
61	Head office building for a steel company	Pittsburgh	USA	Harrison and Abramowitz and Abbe
62	World Trade Center	New York	USA	Yamasaki and Associates, Roth and Sons

1 PRIVATE HOUSE IN COMMUGNY, SWITZERLAND

Architectural design: Annen, Siebold, Siegle, Geneva.
Structural design: Balzari, Blaser, Schudel AG, Berne.
Built: 1967.

Function
Two-storey house without a basement. Entrance, staircase, garage and heating plant in the recess at ground level. Upstairs: residential level with covered balcony on south side.

Dimensions
Overall dimensions 15.20 × 11.45 m; height above ground level 5.50 m. Recessed storey with overall dimensions 7.70 × 9.50 m. Storey height 2.47 m, ceiling height 2.27 m.

Structural features
Columns consisting of IPE 200 sections, set in foundation pockets, are located on a 3.75 × 3.75 m square grid with five columns on each longitudinal and four on each transverse grid line. The outer columns are continuous through both storeys and are interconnected longitudinally by IPE 200 beams at roof and floor level, forming a rigid-jointed system. Composite floors, consisting of profiled steel decking with a 10 cm concrete slab, span in the transverse direction; imposed load 200 kg/m². The external boundary of the floors comprises an exposed fascia beam, assembled from angle sections, which extends around the building. The columns and beams are in weathering steel.
Wind forces in the longitudinal direction are absorbed by portal frame action; in the transverse direction they are transmitted by the floors to vertical cross-bracings between the internal columns on the three middle grid lines.
• Façade: Glazed areas on the upper floor consist of storey-height solar heat-resistant glass in steel frames. The cladding consists of timber units, 3.75 × 2.47 m × 14 cm thick, with impregnated vertical strips of fir on the outside, veneered chipboard on the inside, with 10 cm thermal insulation and vapour barrier in between. The recessed bottom storey is of masonry construction.
• Partitions are of stud framework faced with wood, combined with built-in wall cupboards.

Services
Oil-fired central heating and hot water system 30 600 kcal/h total capacity. Extractor fans for bathroom, WC and kitchen.

Areas and volume

gross area	242 m²	area on plan	175 m²
effective floor area	187 m²	volume	1117 m³

Quantities of materials	steel	concrete	reinforcement
total	11.6 t	105 m³	3.5 t
per m³ volume	10.4 kg	0.094 m³	3.1 kg
per m² gross area	47.9 kg	0.434 m³	14.5 kg

Construction cost (1967)

total cost	201 500 Sw.fr.	per m² gross area	832 Sw.fr.
per m³ volume	180 Sw.fr.	per m² eff. floor area	1078 Sw.fr.

Reference
Bauen + Wohnen, 12/1969, p. 438.

South elevation with covered balcony

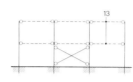

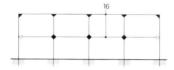

Structural system, longitudinal and transverse sections

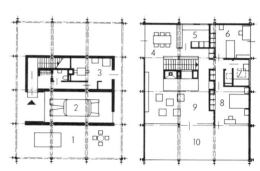

Ground floor, upper floor, scale 1:400

1 Covered area
2 Garage 3 Boiler room
4 Dining area 5 Kitchen
6 Child's room 7 Bathroom
8 Parents' bedroom
9 Living room 10 Balcony

Vertical section through external wall and horizontal section through corner of building

11 Columns IPE 200
12 Fascia beam L 200 × 85
13 Profiled steel decking and concrete slab
14 Angle L 150 × 150
15 Timber wall unit
16 Floor beam IPE 200
17 Haunch plate
18 Pin joint
19 Angle for wind-bracing
20 Diagonals, 45 × 10 mm bars

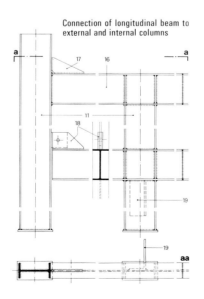

Connection of longitudinal beam to external and internal columns

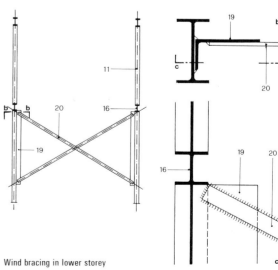

Wind bracing in lower storey

2 PRIVATE HOUSE IN LANDSKRONA, SWEDEN

Architectural design: L. Bergström, Stockholm.
Structural design: A. Larsson, Stockholm.
Built: 1963.

Function

The living room, work room, kitchen, boiler room and garage are in a long low block, to one longitudinal side of which is connected a square block containing the bedrooms and bathroom. The entrance hall forms the link between the two blocks.

Dimensions

Long block 30.70 × 4.20 m, square block 7.80 × 7.80 m on plan. Height above ground level 3.10 m, ceiling height 2.35 m. Both blocks are raised 14 cm above the ground.

Structural features

To facilitate transport and to achieve rapid erection in only four months, the long block was divided into six units. Each unit comprises four corner columns, channel sections U 120, which are interconnected by edge beams, U 220, at roof and floor level on the two longitudinal sides so as to form a rigid framed structure. In the transverse direction timber beams, 5 × 23 cm in section, span between the edge beams at floor and roof level. At the joints between adjacent units their channel-section corner columns are paired with their webs in contact and welded together to form I-sections.
Wind forces are absorbed in the longitudinal direction by rigid frame action, while those acting transversely are resisted by the columns set in 1.20 m deep individual concrete foundations.
• Roof: 15 cm thick thermal insulation inserted between the structural timber beams, lined on the underside with plywood sheeting; above the insulation is an air cavity and tongued-and-grooved boarding with roofing felt.
• Façade: From inside outwards, timber boarding, sealant, 10 cm thermal insulation, waterproof plywood and corrugated asbestos cement sheets, supported by timber studding fixed between the edge beams of the structural steelwork.

Areas and volume

gross area	196 m²	area on plan	196 m²
effective floor area	100 m²	volume	615 m³

Quantities of materials

	steel	concrete	reinforcement
total	8740 kg	7.62 m³	590 kg
per m³ volume	14.2 kg	0.012 m³	1.0 kg
per m² gross area	44.6 kg	0.039 m³	3.0 kg

Construction cost (1963)

total cost	200 000 S.kr.	per m² gross area	1020 S.kr.
per m³ volume	325 S.kr.	per m² eff. floor area	2000 S.kr.

References

Villa Tidskriften Hem i Sverige, 8/1963, p. 279.
Allt i Hemmet, 10/1963, p. 30.
Arkitektur, 10/1963, p. 270.
Detail, 2/1965, p. 171.

View of building

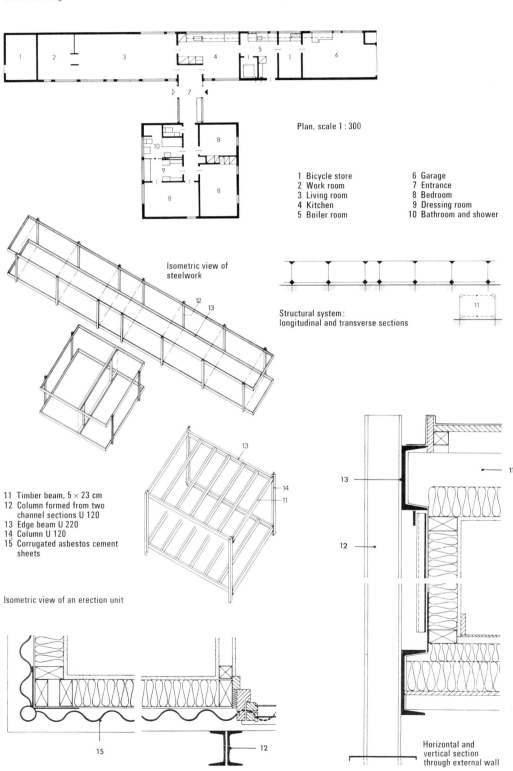

Plan, scale 1:300

1 Bicycle store
2 Work room
3 Living room
4 Kitchen
5 Boiler room
6 Garage
7 Entrance
8 Bedroom
9 Dressing room
10 Bathroom and shower

Isometric view of steelwork

Structural system: longitudinal and transverse sections

11 Timber beam, 5 × 23 cm
12 Column formed from two channel sections U 120
13 Edge beam U 220
14 Column U 120
15 Corrugated asbestos cement sheets

Isometric view of an erection unit

Horizontal and vertical section through external wall

3 PRIVATE HOUSE NEAR STALLIKON, SWITZERLAND

Architectural design: P. Wellmann, E. Wellmann, Zurich.
Structural design: A. Urech, Zurich.
Built: 1971.

Function

Standardised two-storey building which can be used as a dwelling or an office. The upper storey, comprising apartment or office accommodation together with sanitary facilities to suit the requirements of use, has no internal supports but is raised on external steel columns. The residential version has a cantilevered balcony at one end. The recessed bottom storey contains the heating plant and the spiral staircase to the upper floor.

Dimensions

The upper storey is rectangular in plan 16.80 × 8.00 m, storey height 2.80 m, ceiling height 2.40 m. The bottom storey is 2.20 m high with 2.10 m headroom.

Structural features

The elevated structure is supported by four external columns fixed in foundation pockets. It cantilevers at both ends and is composed of four storey-height lattice girders in the external wall planes. Transverse beams IPE 220 and 270 span 7.80 m between the lattice girders of the longitudinal walls at roof and floor level. At both these levels the framework is stiffened with horizontal cross-bracings (40 × 10 mm rectangular section or 40 × 40 × 4 mm angle). Wind forces are transmitted to the foundations through the four fixed-base external columns.
The external columns are rectangular hollow sections, 300 × 150 mm, spaced at 11.20 m centres in the longitudinal direction. The lattice girders are also composed of RHS. The longitudinal flanks are divided into four panels by uprights spaced at 2.80 m centres between the external columns, plus one cantilevering bay at each end of the building. The ends comprise two panels, the uprights being spaced at 3.90 m centres.
• Aluminium cladding consists of 50 mm thick sandwich panels (93 × 280 cm) incorporating plastic windows or fixed glazing set in rubber gasket strips. The cladding panels also serve as fire protection to the external structural steelwork. There are fabric sun-blinds on the outside of the windows.
• Steelwork erected by mobile crane. Main structural work completed in three days; total construction time three weeks.

Areas and volume

gross area	270 m²	area on plan	135 m²
effective floor area	240 m²	volume	480 m³

Quantities of materials	steel	concrete	reinforcement
total	6530 kg	5.5 m³	140 kg
per m³ volume	13.6 kg	0.011 m³	0.3 kg
per m² gross area	24.2 kg	0.020 m³	0.5 kg

Construction cost (1972)

total cost	133 000 Sw.fr.	per m² gross area	493 Sw.fr.
per m³ volume	277 Sw.fr.	per m² eff. floor area	554 Sw.fr.

Reference

Bauen + Wohnen, 11/1972, p. 498.

Layout of upper floor, scale 1:400

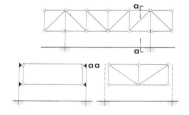

Structural system: longitudinal frame, cross-section and end frame

Part cross-section

Part longitudinal section

Vertical section of a longitudinal outside wall

1. Asbestos cement fascia
2. RHS 100 × 75 mm
3. Transverse beam IPE 240
4. Timber boarding, 27 mm thick
5. Timber joist 200 × 36 mm
6. Timber ceiling
7. Cladding unit, 50 mm thick
8. Diagonal RHS 50 × 40 mm
9. Upright RHS 100 × 50 mm
10. External column RHS 300 × 150 mm
11. Transverse beam IPE 200
12. Foundation pocket

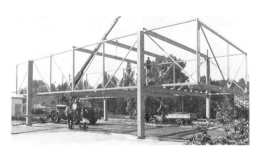

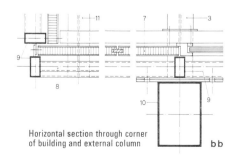

Horizontal section through corner of building and external column

4 PRIVATE HOUSE IN LONDON, ENGLAND

Architectural design: J. Winter and Associates, London.
Structural design: H. Heller.
Built: 1968/1969.

Function
Three-storey dwelling, with entrance, dining room/kitchen, guest room and bathroom on ground floor. Living rooms and bedrooms on the two upper floors.

Dimensions
Rectangular on plan with overall dimensions 11.27 × 6.55 m. Height above ground level 7.51 m, storey heights 2.57 m, ceiling heights 2.44 m in first and second storeys, 2.34 m in third storey.

Structural features
The structural system comprises four transverse three-storey rigid frames (span 6.30 m, spacing 3.66 m), rigidly interconnected in the longitudinal direction by edge beams. Beams and columns are of I-section. The columns are 152 × 152 mm UCs, the transverse beams 254 × 101 mm UBs and the edge beams are 260 × 146 mm UBs. All the connections are welded. The continuous 13 cm thick in-situ concrete slab acts compositely with the transverse beams, the shear connectors being welded cuttings from rolled steel sections. Imposed load on floors 244 kg/m². Wind forces are transmitted in both directions by rigid frame action.
• Façade: The columns and edge beams are externally protected with vermiculite and faced with weathering steel plate, 6 mm thick. The façade bays have fixed glazing and narrow ventilation casements. The sides of the building are in part infilled with panels composed of 5 cm thick foamed polyurethane, faced externally with weathering steel sheet, 2.5 mm thick.

Areas and volume

gross area	207 m²	area on plan	74 m²
effective floor area	187 m²	volume	550 m³

Steel content

total	7600 kg	per m² gross area	36.7 kg
per m³ volume	13.8 kg	per m² eff. floor area	40.6 kg

Construction cost (1968)

total cost	£12 400	per m² gross area	£59.90
per m³ volume	£22.50	per m² eff. floor area	£66.30

Erection of structural steelwork

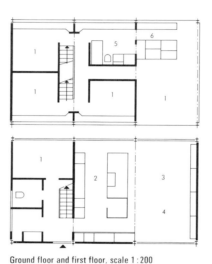

Ground floor and first floor, scale 1:200

References
Architectural Design, 8/1970.
Architects Journal, 8/1970.
Acier-Stahl-Steel, 10/1971, p. 390.
L'Architecture d'Aujourdhui, 163/1972, p. 28.

Beam-to-column connections

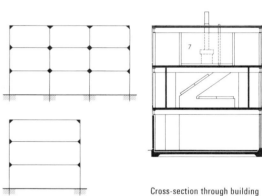

Structural system: longitudinal section and cross-section

Cross-section through building

1 Bedroom 2 Kitchen
3 Dining room 4 Playroom
5 Bathroom 6 Dressing room
7 Living room with fireplace

8 Universal column 152 × 152 mm
9 Weathering steel cladding
10 Ventilation casement
11 Floor slab
12 Universal beam 260 × 146 mm
13 Sliding sash

Isometric view of external wall

5 HOUSING ESTATE AT PIOMBINO, ITALY

Architectural design: P. Gorio, M. Grisotte, E. Mandolesi, A. Petrignani.
Structural design: Columbini.
Built: 1968/1972.

Function

Residential development comprising 508 dwellings in twelve ten-storey blocks of flats, eight two-storey blocks of terraced houses, and twenty-four detached houses.
On each floor of the multi-storey blocks are two flats arranged on each side of a pair of internal cores. All the dwellings have three rooms, kitchen with dining space, bathroom and washroom. Each dwelling also comprises two loggias located by the outer faces of the two cores. One core contains the liftshaft, staircase and services. The other core, which adjoins the bathroom and WC, serves as a light well. The bottom storey is recessed to the cores. Because of the sloping site, the basement is partly above and partly below ground.

Dimensions

High-rise blocks 15.60 × 15.60 m overall. Cores 4.50 × 4.90 m, 2.00 m apart. Storey height 3.20 m, ceiling height 2.70 m, planning grid 1.10 × 1.10 m.

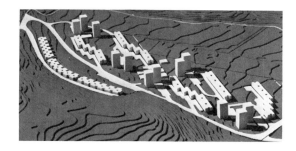

Model of housing estate

1 Vertical circulation core
2 Light well
3 Loggia
4 Living room
5 Kitchen
6 Dining area
7 Children
8 Parents
9 Bathroom

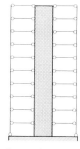

Structural system: section through core

Structural features

Pin-jointed framework comprising external columns, which are interconnected by perimeter beams, and floor beams supported externally by the perimeter beams and internally by the concrete walls of the two stiffening cores. In front of each side of the building there are three columns, at 5.50 m centres, in three lengths HE 180 A, B and M, reducing in section with height. The continuous IPE 300 edge beams, arranged behind the external columns, are connected to the latter by means of short brackets. The IPE 160 floor joists, which span 5.40 m, are at 91 cm centres. They support floors comprising 8 cm hollow tiles and 10 cm reinforced concrete topping. Welded studs were used for shear connection. In order to ensure composite action for both dead and live loads, and to reduce deflections, the floor beams were propped until the concrete had matured.
• Façade: Structural frame has infilling of double leaf construction. The inner leaf, consisting of hollow blocks plastered on both faces, is supported by the concrete floor; the outer leaf is of unrendered red brick which is supported by the top flange of the perimeter beam. Window frames of galvanised steel sections, partial double glazing, external venetian blinds.
• Fire protection: External steelwork unprotected. For fire-fighting: two riser mains with hydrants on the landings.
• Corrosion protection: Two priming coats of synthetic resin with red lead and lead chromate; for the floor beams which come into contact with concrete, synthetic resin with zinc dichromate; two finishing coats.

10 Column HE 180 A, B and M
11 Perimeter beam IPE 300
12 Floor beam IPE 160
13 Bracket
14 Hollow block wall
15 Brick infilling
16 Concrete floor
17 Hollow tiles

Typical floor, scale 1:300

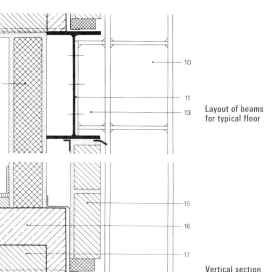

Layout of beams for typical floor

Vertical section through external wall

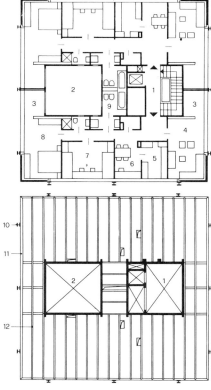

Reference
Acier-Stahl-Steel, 9/1970, p. 373.

63

6 BLOCKS OF FLATS IN ROUEN, FRANCE

Architectural design: M. Lods, P. Depondt, H. Beauclair, M. Alexandre, Paris.
Structural design: L. Robustelli.
Built: 1968/1969

Individual block of flats

Function

This residential development comprises 500 flats in 25 blocks with a total floor area of 38 500 m². The blocks, each of which is a five-storey structure without a basement, are either isolated or arranged in interconnected groups. On the ground floor there are the entrance hall, heating plant, storage space and room for refuse collection. Each upper floor comprises five flats, with three to five rooms, positioned around a central staircase.

Dimensions

The individual blocks are rectangular on plan with overall dimensions 23.70 × 19.20 m; height above ground level 14.60 m. Storey height 2.91 m, ceiling height 2.40 m. Central staircase 9.00 × 4.50 m.

Structural features

Two types of large prefabricated unit were used for the structural steelwork: shear wall assemblies consisting of columns and cross-bracings, extending the full height of the building, and space frame floor units, each corresponding in size to the floor area of one flat. After construction of the foundations, two shear wall units in each direction of the building were erected in such positions that the floor units of the staircase could be connected to their innermost columns. The other floor units are joined to this stiffened staircase and transmit their loads to the shear wall units, to continuous columns provided outside the cladding and to storey-height columns within the flats.

Before erection, the large floor units were assembled from elements of identical size, 3.60 × 2.40 m, each of which comprises a top and bottom steel mesh mat. These mats, which are 30 cm apart, are staggered half a mesh width in relation to one another in both horizontal directions. The joints in the top and bottom mats are interconnected diagonally by round steel bars, 8 mm dia, to form a space frame. The mats themselves consist of 10 mm dia bars at 300 mm centres in both directions.

Each floor element is enclosed by 300 mm deep lattice girders composed of angle sections. The elements are joined to one another by bolting these girders together. After erection in the building these edge girders, interconnected in pairs, form a girder grillage spanning between the columns. The connections to the external columns are made with the aid of brackets. The unit weight of the structural floor system is 47 kg/m². All the loadbearing steel components, except the staircase structure, are of weathering steel. Corrosion protection for the staircase steelwork is provided by zinc-chromate paint.

• Floors: Reinforced lightweight concrete slabs, 90 × 60 × 4 cm thick, laid on Neoprene pads resting on the top steel mat. Imposed load 175 kg/m². Fire-resistant vermiculite ceiling panels are clipped to the bottom mat. This floor construction provides soundproofing superior to the requirements of the local building regula-

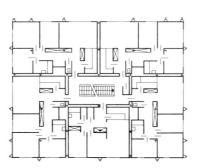

Typical floor, scale 1:500

Layout of apartment-size floor units

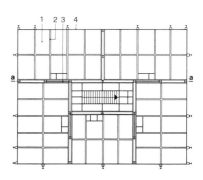

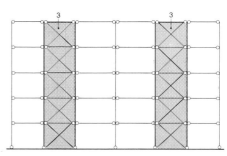

Structural system, section aa

General view of the housing estate

tions. Ducts for services are accommodated in the floor space frame.

• Foundations: Under the columns of the shear wall units there are eight concrete-filled tubular steel piles, 22 cm dia and 7.0 m deep. The other columns have individual foundations.

Services

Temperature control with hot and cold air. The air is conveyed through ducts in the floors to the outlets in the window jambs. Extraction of air from kitchen, bathroom and lavatory by means of extractor fans on the roof.

Area and construction cost

Total residential floor area	38 500 m²
Cost of construction	21 533 673 Fr.fr.

Steel content

Total 1900 t	per m² of residential floor area 49.4 kg

References

Bâtir, 174/1969.
Architecture de Lumière, 18/1969.
Facades Légères et Cloisons Industrialisées, 37–38/1968.

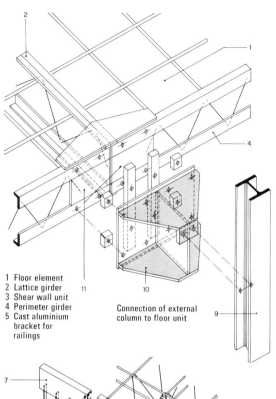

1. Floor element
2. Lattice girder
3. Shear wall unit
4. Perimeter girder
5. Cast aluminium bracket for railings

Connection of external column to floor unit

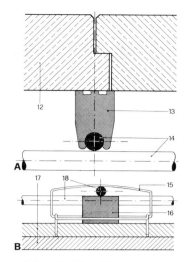

A Assembly of floor slabs
B Attachment of ceiling panels

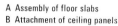

Structural steelwork completely erected

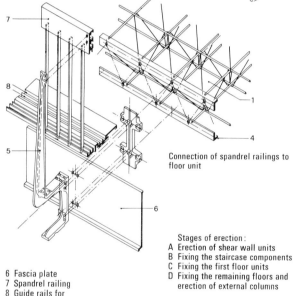

Connection of spandrel railings to floor unit

6. Fascia plate
7. Spandrel railing
8. Guide rails for sliding sashes
9. External columns HE 120 A
10. Bracket
11. Stiffening plate
12. Lightweight concrete slab
13. Neoprene pad
14. Bars of top steel mat
15. Fixing stirrup
16. Spacer block
17. Vermiculite ceiling panel
18. Bars of bottom steel mat

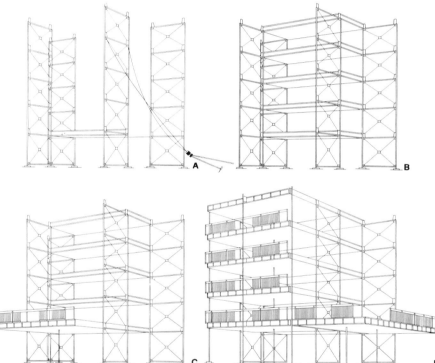

Stages of erection:
A Erection of shear wall units
B Fixing the staircase components
C Fixing the first floor units
D Fixing the remaining floors and erection of external columns

Hoisting a floor unit

Fixing the fire-resistant ceiling panels

65

7 HOUSING ESTATE AT TARANTO, ITALY

Architectural and structural design: Nizzoli, Milan.

Function

This residential development comprises 460 flats, accommodated in point blocks, four or nine storeys high, and long four-storey blocks. The latter, each of which has a structural core in the middle and two flats per floor, are joined to the point blocks with central core and three flats per floor. The buildings are arranged in eight groups separated by open spaces landscaped with lawns and trees and also providing car parking facilities. The buildings are raised off the ground on pilotis, so that the ground floor area is largely open and with unrestricted access.

Dimensions

Each point block comprises three wings, each 10.60 m square on plan, and has overall dimensions of 23.0×23.0 m. Height above ground level 30.5 m. The long blocks have overall dimensions of 24.1×10.8 m on plan and rise to a height of 14.0 m above ground level. Storey height 3.27 m, ceiling height 3.00 m. Headroom on open ground floor space 2.70 m and 3.20 m.

Structural features

Uniform structural system for all the buildings. In the two longitudinal façades and on the centre-line: columns consisting of sections ranging from HE 140 A to HE 180 B, at 3.44 m centres, on tapered reinforced concrete columns in the open bottom storey. In the longitudinal direction the external columns are interconnected by pin-jointed beams, channel sections U 220, while the central columns are rigidly interconnected by HE 180 A beams. In the transverse direction IPE 100 secondary beams are pin-jointed to the columns. All structural connections are bolted. Wind forces are transmitted through the floors to the reinforced concrete cores and to vertical cross-bracings in the end walls of the wings.

- Floors: Reinforced concrete T-beam floor spanning between outside and central beams. Precast floor units, 5.20 m long, 0.75 m wide, 19 cm thick, resting on in-situ reinforced concrete beams. Reinforcement is welded to the top flanges of the steel beams; structural continuity of the concrete beams over the central steel beams is ensured by lapping of the reinforcement. Underfloor heating installed in the floor slabs.
- Façade: Structural steel frame has double-leaf infilling, composed from outside to inside as follows: brick facing, plaster, fibreboard insulation, cavity, perforated blocks set on edge. Window frames consist of galvanised cold-formed sections.
- Foundations on limestone. Pad foundations under the columns, strip foundations under the wind bracings in the ends of the buildings. Bearing pressure on subsoil 3 kg/cm².

Reference
Acier-Stahl-Steel, 7–8/1969, p. 343.

Long block with open bottom storey (top) and erection of a point block

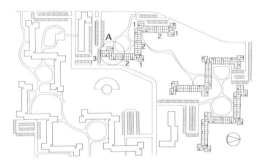

Location plan, scale 1 : 7000

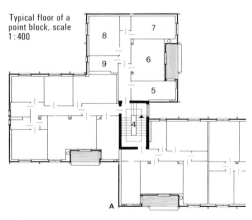

Typical floor of a point block, scale 1 : 400

1 Point block
2 Long block
3 Car parking
4 Staircase core
5 Kitchen
6 Living room
7 Child's room
8 Parents' room
9 Bathroom

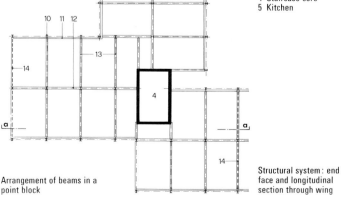

Arrangement of beams in a point block

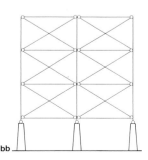

Structural system: end face and longitudinal section through wing

Supports for columns and beam connections to concrete core

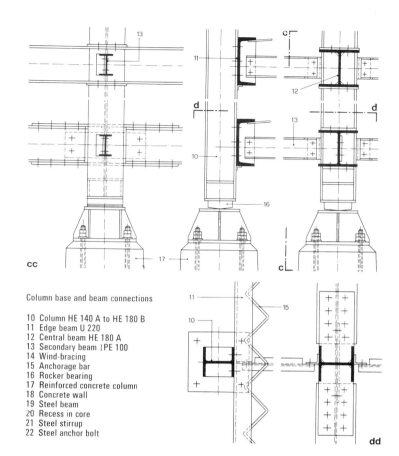

cc

Column base and beam connections

10 Column HE 140 A to HE 180 B
11 Edge beam U 220
12 Central beam HE 180 A
13 Secondary beam IPE 100
14 Wind-bracing
15 Anchorage bar
16 Rocker bearing
17 Reinforced concrete column
18 Concrete wall
19 Steel beam
20 Recess in core
21 Steel stirrup
22 Steel anchor bolt

dd

Point block with adjoining long block

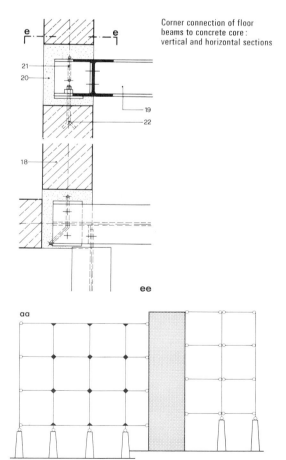

Corner connection of floor beams to concrete core: vertical and horizontal sections

8 BLOCK OF FLATS IN BRUSSELS

Architectural design: M. Roggen, F. Liénard.
Structural design: M. van Wetter, Brussels.
Built: 1966/1967.

Function
High-rise residential building comprising 96 flats accommodated on 16 upper floors, each with six four-room flats. On the ground floor: entrance hall, caretaker's flat and heating equipment. Above the ground floor there is a mezzanine for services. Access to upper floors is provided by means of two lifts, in a concrete core, and two spiral staircases at the ends of a central corridor.

Dimensions
Rectangular on plan with overall dimensions 32.76 × 11.66 m. Height above ground level 53 m. Height of first storey 2.97 m, mezzanine 1.32 m. Storey height 2.78 m, ceiling height 2.48 m, in upper storeys.

Structural features
Columns at 3.60 m centres along the two longitudinal faces and central axis of the building. The inner columns are arranged on both sides of the corridor. In the transverse direction the columns are interconnected by pin-jointed main beams. In the longitudinal direction secondary beams, likewise pin-jointed to the inner columns and the main beams, are spaced at 1.80 m centres.
Floors are stiffened by horizontal cross-bracing installed in the perimeter bays. The columns along the longitudinal faces of the building are rigidly connected to one another by the perimeter floor beams to form a portal frame system extending the full height of the building and serving to resist wind forces in the longitudinal direction. The ends of the building are provided with vertical lattice frames to act as shear walls for resisting the transverse wind forces.
The reinforced concrete core is small as it is not structurally connected to the steel frame and does not have to resist wind forces. The columns comprise HE 220 M and HE 200 B rolled steel sections, which were fabricated in four-storey lengths and bolted together on the site.
The beams along the longitudinal faces of the building, varying from HE 140 A to HE 220 A, are rigidly connected to the columns by means of end plates and high-strength friction-grip bolts. The main beams consist of IPE 240 and the secondary beams of I 140, with bolted connections.
• Floors: Trapezoidal sheet steel troughing units, 8.5 cm in depth, are spot-welded to the secondary beams. The units are covered with 2 cm thick cork, hardboard and linoleum. Imposed load on floor 250 kg/m^2. Suspended ceiling of plasterboard.
• Façade: Curtain walling with transoms at floor and cill level, in combination with vertical posts. The transoms at floor level are connected to the perimeter beams through adjustable attachments and transmit all the façade and wind loads to each floor. Sandwich-type spandrel panels. Window frames of standard steel sections with simple windows horizontally or vertically pivoted.
• Foundations: Concrete piles supported by gravel at 12 m depth. The pile heads are joined together by reinforced concrete beams.

Typical upper floor and ground floor, scale 1 : 400

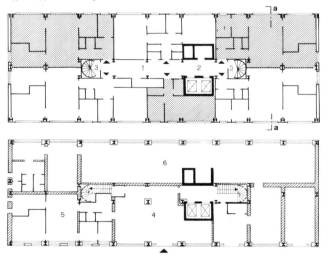

Equipment
Hot water central heating with radiators. Three boilers on ground floor, one of which is for hot water supply; each boiler has capacity of 240 000 kcal/h.

Reference
Acier-Stahl-Steel, 6/1967, p. 263.

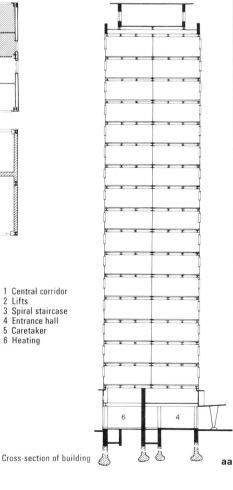

1 Central corridor
2 Lifts
3 Spiral staircase
4 Entrance hall
5 Caretaker
6 Heating

Cross-section of building

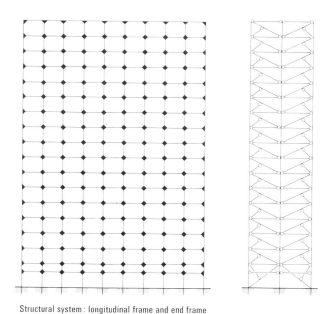

Structural system: longitudinal frame and end frame

Arrangement of beams in typical upper floor

Areas and volume

Gross area	6494 m²	area on plan	382 m²
Effective floor area	5312 m²	volume	19 864 m³

Quantities of material

	steel	concrete
Total	407 t	485 m³
per m³ volume	20.5 kg	0.024 m³
per m² gross area	62.7 kg	0.075 m³

Construction cost (1967)

Total cost	41.7 million B.fr.	per flat	435 000 B.fr.
per m³ volume	2100 B.fr.	per m² eff. floor area	7850 B.fr.

7 Columns HE 220 M and HE 200 B
8 Main beams IPE 240 and HE 240 A
9 Secondary beams I 140
10 Perimeter beams HE 120 A to HE 240 A
11 Free-standing reinforced concrete core
12 Spiral staircases
13 Service shafts

Anchorage of a column base

14 Wind-bracing
15 Stiffening rib
16 35 mm base plate
17 Holding-down bolt, 64 mm dia
18 Steel sleeve
19 Channel section U 140
20 Slotted connecting angle
21 Washers
22 Spandrel panel
23 Horizontal member
24 Steel frame
25 Window cill of steel sheet
26 Sheet steel troughing

Vertical section through façade

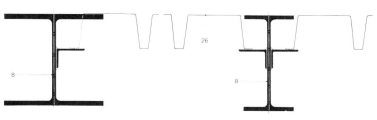

Connection of profiled floor unit to main beams

Detail of rigid connections in longitudinal frame

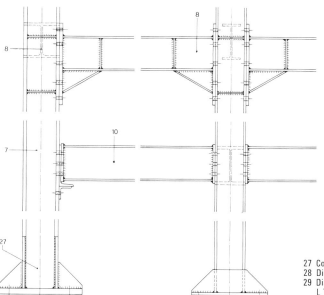

27 Column base
28 Diagonal, HE 180 M
29 Diagonal, two angles L 70 × 70 × 7

Detail of wind-bracing in end frame

9 HIGH-RISE RESIDENTIAL DEVELOPMENT AT BALORNOCK, SCOTLAND

Architectural design: S. Bunton and Associates.
Structural design: W. A. Fairhurst and Partners.

Function

Eight multi-storey blocks comprising a total of 1356 flats for 5600 residents in all. The flats, ranging from one to four rooms, are accommodated in two slab blocks (with 26 and 28 storeys respectively) and in six 31-storey point blocks. The buildings occupy a site of 9 hectares. Each slab block has three vertical circulation cores, while each point block has one core. A core contains two lifts, two staircases and service shafts. Communal facilities in the basements; clothes drying space is provided on the roofs.
The housing development also comprises a shopping centre, school, day nursery and covered car parking space.

Dimensions

Slab blocks 93 × 11 m overall on plan, rising to 77 m and 71 m above ground level. Four point blocks 22.90 × 15.25 m, two point blocks 30.50 × 18.30 m overall on plan; height above ground level 84 m. Storey height 2.70 m, ceiling height 2.45 m, in upper storeys. On ground floor headroom is 2.75 and 3.35 m.

Structural features

The structural steelwork for all the buildings transmits the vertical loads and also the horizontal wind forces. Internal and external columns continue unspliced through three storeys and are interconnected in both directions by floor beams. Beams and columns are of I-section; all connections are bolted. Wind-bracing by vertical lattice systems in the party walls between the flats and by rigid interconnection of floor beams and columns to multi-storey portal frames in the external walls and in the internal walls near the cores. Structural co-operation of the portal frames and lattice bracings is ensured by the rigid floors. To reduce vibration and increase rigidity, the external columns and perimeter beams of the point blocks are encased in concrete up to the 10th floor. For a wind pressure of 150 kg/m^2 the lateral deflection of the buildings is calculated as less than 1/500 of the height of the building.
The concrete casing also provides effective fire protection. Elsewhere, lightweight asbestos casing is used.
• Composite floors of Holorib decking with 12.5 cm thick concrete filling. Continuity reinforcement over the floor beams. Floor slabs connected to the steel beams by means of welded shear studs in pairs at 60 cm centres. Fire protection provided by ceiling of asbestos cement panels attached to the ribbed sheet with 2.5 cm cavity in between. Fire resistance 1 hour. Flooring: in the bathrooms, PVC tiles on 7 cm screed; elsewhere, parquet on glass fibre mats.
• Façade: Storey-height cladding units are fixed to the outside edge of each floor by means of steel angle sections. These units each comprise a timber frame with external cladding of asbestos cement sheets and inner vapour-tight panel with aerated gypsum filling. Windows have wooden frames and horizontally pivoted steel lights.

Structural system: cross-sections through a point block

View of three point blocks

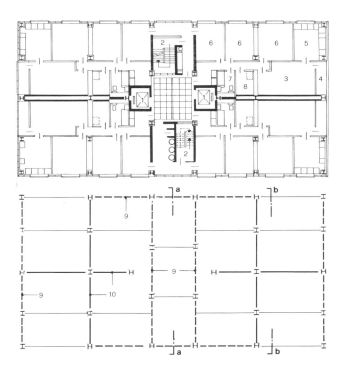

Typical floor in a point block, scale 1:350

1 Lift
2 Staircase
3 Living room
4 Loggia
5 Kitchen
6 Room
7 Bathroom
8 Heating and water tank
9 Portal frame
10 Wind-bracing

Arrangement of beams in a typical floor

Erection of structural steelwork and construction of composite floors

Vertical section showing the connection between the cladding and the floor

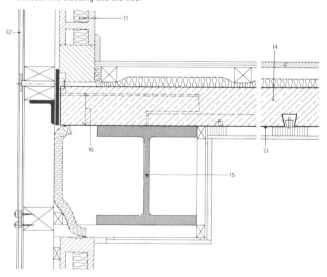

11 Prefabricated wall panel
12 Asbestos cement sheet
13 Holorib decking
14 In-situ concrete floor
15 Perimeter beam
16 Bracket for connection of cladding

Total quantities of materials

steel	concrete	reinforcement
11 341 t	16 350 m³	571 t

Construction cost

Total cost of the buildings £6.25 million

Reference

Acier-Stahl-Steel, 4/1966, p. 173.

- Foundations: In all, 704 reinforced concrete piles, 3.5 m average length, bearing on sandstone and shale. Piles 90 cm dia.

Services

Electric storage heating, infra-red radiant heating and wallmounted radiators. Power supply from the public mains, transformer station in basement. Water storage tanks in basement and on roof, with a smaller tank (225 litres) in each flat.

View of the site with buildings in various stages of construction

View along an upper storey in a slab block, showing transverse wind-bracing

10 SYSTEM BUILDING AT STORA TUNA, SWEDEN

Architectural design: Y. Johnsson, Stockholm.
Structural design: A. Johnson, Stockholm.
Built: 1969/1970.

Function
Two-storey balcony-access type building comprising seven 2/3 room residential flats on each floor. If desired, two adjacent flats can be combined into one 5-room flat. The upper access balcony is reached by means of two external staircases.

Dimensions
Overall dimensions on plan 51.5 × 14.4 m. Height above ground level 6.50 m. Storey height 2.97 m, ceiling height 2.50 m.

Structural features
Building system comprises a structural steel frame and prefabricated lightweight concrete wall and floor units. For acoustic insulation between the flats, in both the vertical and horizontal directions, all the floors and party walls are of two-leaf construction, and the steel frame is in seven sections with double columns at the joints.

Each section of the building comprises four two-storey frames arranged in the longitudinal direction, spanning 7.35 m and spaced at 3.00 m centres. The two outermost frames, consisting of HE 200 A rolled steel sections, are situated in the longitudinal façades and are rigid-jointed. The external frames of adjacent sections are staggered by a column width in relation to one another and are thus located in different planes without direct contact. The U 200 channel-section columns and the HE 220 A beams of the two inner frames are pin-jointed. Here, too, the columns of adjacent segments are not in direct contact with one another.

Wind forces in the longitudinal direction are resisted by the external rigid frames; those acting transversely to the building are resisted by wind-bracings, consisting of angle sections situated in each party wall separating adjacent flats.

- Floors consist of precast reinforced lightweight concrete slabs of 3.00 × 0.60 m, 20 cm thick, imposed load 100 kg/m². Slabs are interconnected by means of glued tongued-and-grooved joints along the longitudinal edges. Floor finish: plastic sheet and two layers of fibre board; suspended ceiling of 13 mm thick plasterboard and 30 mm mineral wool.

Services
Main services installed in a creep-through space under the building. To reduce noise, a separate riser pipe is installed for each flat, consisting of a small-bore pipe accommodated within the party walls between adjacent flats. Service ducts in the ceiling cavities. All pipes mounted on rubber for sound insulation. To reduce noise of baths filling, water enters through a valve in the bottom of each bath.

References
Byggnadsindustrin, 6/1969, p. 69.
Lättbetong, 4/1969, p. 14.
Acier-Stahl-Steel, 9/1969, p. 363.

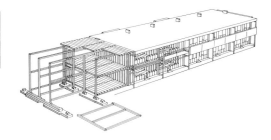
Building components and their sequence of erection

Structural system: longitudinal portal frames and cross-section

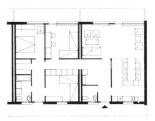
Five-room flat on ground floor

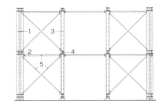
External longitudinal frame and cross-section through the structure

Hoisting the external wall units into position

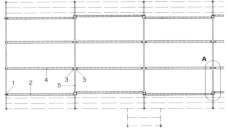
Arrangement of upper floor beams

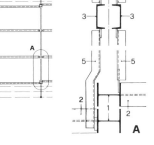

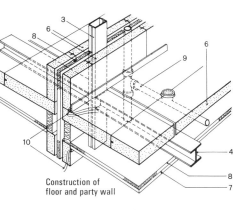

1 Rigid frame column HE 200 A
2 Rigid frame beam HE 200 A
3 Internal column U 200
4 Internal beam HE 220 A
5 Bracing, 70 × 70 × 7 angles and 30 × 5 mm bars (diagonals)
6 Lightweight concrete slab
7 Plasterboard
8 Mineral wool
9 Waste pipe
10 Electric wiring

Construction of floor and party wall

Detail of the balcony façade

Steelwork with staggered arrangement of longitudinal rigid frames

11 MOUNTAIN REFUGE IN THE STELVIO PASS, ITALY

Architectural design: C. Conte, L. Fiori, Milan.
Structural design: C. Pesenti, Bergamo.
Built: 1964/1966.

Function

Refuge and accommodation for mountain climbers in an Alpine pass at an altitude of 3000 m. Grouped around the central staircase are the entrance hall, kitchen and restaurant on the ground floor, lounges and other communal rooms on the first floor and bedrooms and lavatories on the second and third floors.

Dimensions

On plan the building comprises four wings enclosing a square core, the wings projecting farther outwards with increase in height. Overall dimensions of the area covered on plan 25.60 × 25.60 m. Height above ground level 19.10 m. Storey height 3.20 and 4.30 m on the ground floor, 3.35 m on the second floor and 2.00 to 4.30 m, varying with the slope of the roof, on the third floor.

Structural features

The design and construction were governed by very difficult weather and transport conditions. After trial erection in the fabrication shops, in the course of which the exact erection procedure was determined and certain connections were improved, the steelwork was actually erected on the site in forty days. The complete structural work was carried out in three months.
To withstand the high wind and snow loads (600 kg/m^2) the steel structure was designed as a rigid portal frame system in both directions. Each of the four sides of the core comprises two frames with hinged feet, one upon the other, interconnected at their corners by the cantilevering portal cross beams and stiffened by horizontal bracings. The bottom portal frame extends 7.80 m through two storeys but the top portal frame is only one storey in height. The span of the frames is 10.00 m. The floor beams, the cantilevers and the roof structure are attached to this square portal frame system. Two main beams span in each direction across the space between the cross beams of the portal frames and cantilever 5.60 m beyond them. At the points of intersection of these beams in the interior of the structure they are rigidly connected to four columns.
Spanning between the main beams are secondary beams, spaced at 1.50 m centres, to which galvanised steel troughing units are spot-welded which serve to carry the timber floor. Beams also span across the space between the cross beams of the top portal frames and cantilever beyond them. The main beams of the roof structure are connected to the ends of these cantilevers. The four sloping sections of the roof are staggered vertically in relation to one another, so that the steel purlins at the top of the roof are connected at different levels to columns which transmit their load to the beams of the floor below or to the four internal columns. The purlins in turn carry rafters consisting of rolled steel sections.
The ground floor beams are connected to the portal frame columns and to the inner columns and are supported on concrete walls which, in the region of the core, extend up to the ground floor of the building.
• Façade: The cladding consists of timber panels with internal thermal insulation and vertical window openings with double glazing.

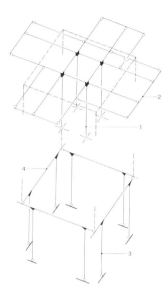

Rigid frame system for the two bottom storeys (shown separated here)

Section through building

1 Internal column
2 Edge beam
3 Portal frame column
4 Cross beam of portal frame
5 Hip rafter
6 Purlins
7 Rafters

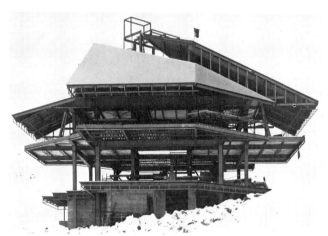

2nd floor with bedrooms, scale 1:500

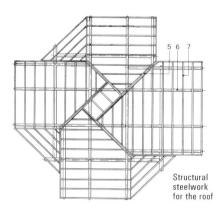

Structural steelwork for the roof

Areas and volume

gross area 1255 m^2 area on plan 510 m^2 volume 4520 m^3

Steel content

total	140 t	per m^2 gross area	111.6 kg
per m^3 volume	31.0 kg		

Construction costs (1966)

total cost	204.8 million lire	per m^2 gross area	163 000 lire
per m^3 volume	45 300 lire		

Reference

Acier-Stahl-Steel, 7–8/1967, p. 325.

12 CHILDREN'S HOME AT MÜLLHEIM, GERMANY

Architectural design: W. Blaser, Nees and Beutler, Basle.
Structural design: Nafz, Müllheim; Gruner and Jauslin, Basle.
Built: 1966/1967.

Function

The scheme comprises seven blocks for 60 children, ten teachers and the warden and his wife. Around a communal building with kitchen facilities are grouped five two-storey residential blocks for children and an administrative building with the warden's flat. In each children's block, dormitory accommodation on the upper floor and living rooms on the ground floor are arranged around a central service core.

Dimensions

Children's blocks square on plan 12.60 × 12.60 m. Height above ground level 5.70 m. Storey height 2.85 m, ceiling height 2.50 m.

Structural features

Each building has a reinforced concrete basement. Above ground level, two-storey columns arranged on a square grid of 4.20 m are joined together in both directions by floor beams. The upper floor is of in-situ concrete construction with embedded IPE 160 floor beams. The roof, which is stiffened by diagonal bracing, consists of lightweight concrete slabs. Wind forces are transmitted through the roof and upper floor to the concrete core of the building. Columns are cruciform in section, comprising two 80 × 80 × 12 angles welded together.
• Façade: External bays have room-height infilling consisting of prefabricated window units, with insulating glass and sliding lights, and wall units of multi-layer construction, faced externally with Perlichron-Eternit.

Areas and volume

gross area	3235 m²	area on plan	1357 m²
effective floor area	1732 m²	volume	10 052 m³

Steel content

total	90 t	per m² gross area	27.8 kg
per m³ volume	9.0 kg	per m² eff. floor area	52.0 kg

Construction cost (1967)

total cost	1.967 million DM	per m² gross area	608 DM
per m³ volume	195 DM	per m² eff. floor area	1132 DM

References

Werk, 2/1968, p. 88.
Bouw, 1969, p. 697.
Detail, 2/1969.

Plan

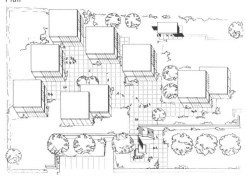

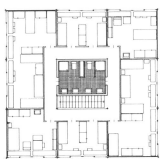
Upper floor with dormitories, scale 1:300

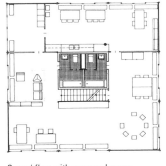
Ground floor with communal rooms

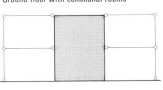
Structural system, section through concrete core

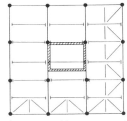
Arrangement of roof beams

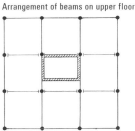
Arrangement of beams on upper floor

1 Wall panel
2 Cruciform column, two angle sections L 80 × 80 × 12
3 Sliding light
4 Steel plate 400 mm × 8 mm
5 Concrete placed after erection of wall

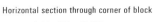

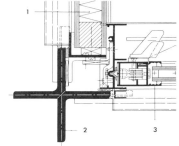
Horizontal section through corner of block

Vertical section through external wall

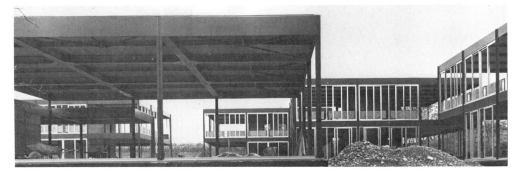
General view showing children's blocks in background and partly completed communal building in foreground

13 DAY NURSERY IN BERLIN SCHÖNEBERG

Architectural design: Bratz, Hassenstein, Schmidt-Thomsen, Berlin.
Structural design: Steelwork contractors.
Built: 1970/1971.

Function

Classrooms and play rooms for 146 children in a two-storey building, without basement. The two storeys are subdivided into three zones for different age groups. On the ground floor: crèche and kindergarten, separated from each other by a central area comprising lavatories and kitchen. In addition there are rooms for staff, office space and heating equipment. Infant school on the upper floor.
Each zone has a separate entrance. A second staircase from the communal room in the kindergarten gives additional access to the upper floor.

Dimensions

The shape of the building on plan corresponds to the internal layout. Planning grid 3.0 × 6.0 m. Ground floor overall dimensions 45 × 30 m; upper floor 30 × 15 m (set back in relation to ground floor). Height above ground level 6.70 m. Storey height 3.11 m, ceiling height 2.51 m.

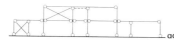
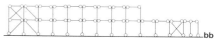

Structural system: transverse and longitudinal sections

Ground floor, scale 1 : 900

1 Entrance to crèche
2 Stairs to infant school
3 Entrance to kindergarten
4 Play area
5 Kitchen
6 Communal room
7 Classroom
8 Staff
9 Boiler room

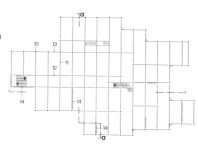

Arrangement of beams in upper floor

Structural features

Pin-jointed steel framework comprising single-storey or (in the set-back upper storey area) two-storey square hollow columns (70 × 70 mm) and I-section floor beams spanning in the transverse direction of the building. Spacing of columns in transverse direction 3.00 m and 6.00 m, in longitudinal direction 3.00 m. Floor beams consist of I 200, IPE 270 or IPE 300, depending on span and imposed load; perimeter beams are U 200 channels. For adjustment and bracing in longitudinal direction: columns interconnected by L 65 × 7 angles with pin joints. The precast concrete slabs at first floor level, which are 3.00 m long and 10 cm thick, are joined together with in-situ concrete over the main beams which are, in turn, provided with stud shear connectors to ensure composite construction. Imposed load 500 kg/m². Roof of 12.5 cm thick lightweight concrete slabs, stiffened during erection by horizontal cross-bracing along the centre. Wind forces transmitted in both directions through floor and roof to vertical cross-bracings between the columns.
• Façade: Spandrels and fascias consist of 17.5 cm thick lightweight concrete slabs mounted on a backing framework of horizontal angles and vertical T-sections connected by means of brackets to the columns. Cladding units are 3.00 m long and 62.5 cm high. Composite windows fitted between the slabs. Sun protection by awnings and external venetian blinds.
• Fire protection: Columns protected with asbestos slabs or concrete. Both storeys have suspended acoustic ceilings with 90 min fire protection.

Areas and volume

gross area	1220 m²	area on plan	914 m²
effective floor area	990 m²	volume	3970 m³

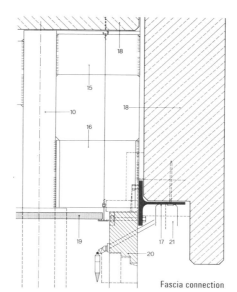

Fascia connection

10 Square hollow section 70 × 70 mm
11 Floor beam
12 Angle section L 65 × 65 × 7
13 Perimeter beam U 200
14 Wind-bracing
15 Bracket ½ IPE 200
16 Bracket IPE 180
17 Cleat ½ IPE 200
18 Lightweight concrete slab
19 Fire-resistant acoustic ceiling
20 Timber window
21 Venetian blind
22 Asbestos slab
23 Plasterboard
24 Galvanised flat bar 40 × 3 mm

Steel content

total	38 t	per m² gross area	31.6 kg
per m³ volume	9.6 kg	per m² eff. floor area	38.4 kg

Construction cost

total cost	1.31 million DM	per m² gross area	1073 DM
per m³ volume	330 DM	per m² eff. floor area	1325 DM

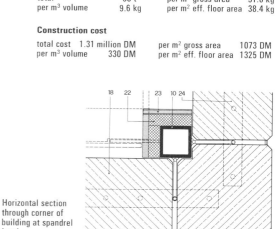

Horizontal section through corner of building at spandrel level

View through lower-storey steelwork

14 STUDENTS' HALL OF RESIDENCE, CITÉ UNIVERSITAIRE, PARIS

Architectural design: C. Parent, Neuilly-sur-Seine; M. Foroughi, E. Ghiai, Teheran.
Structural design: Steelwork contractors.
Built: 1966/1968.

Function

Residential accommodation for students comprising 96 rooms located on eight floors (1st to 4th, 6th to 9th). The fifth floor is open except for the warden's flat and four guest rooms. On the ground floor: entrance hall, refectory, library, offices, infirmary, and caretaker's flat. In the basement: central heating plant, recreation rooms, garages. Vertical circulation: two lifts and a services shaft in the internal reinforced concrete core; on one side of the building there is an external spiral staircase connected by walkways to the various floors.

Dimensions

Ten-storey high-rise building, rectangular on plan, overall dimensions 40.60 × 10.20 m. Height above ground level 38.21 m. Storey height 3.00 m. A low-rise building, which comprises ground floor and basement and is shaped like two cubes displaced in relation to each other, is structurally separate from the main building.

Structural features

Rigid frame system with suspended storeys. Three steel portal frames, spanning 12.90 m at 14.50 m centres, each with two transverse girders, placed at the top and mid-height of the building respectively. At these two levels the portal frames are rigidly interconnected by longitudinal girders on the centre-lines of the portal columns and on the centre-line of the building. The longitudinal girders cantilever 5.80 m beyond the end columns of the portal frames, and the ends of these cantilevers are joined together by edge girders. Cross-beams arranged at 2.90 m centres span between the horizontal members. A block comprising four storeys is suspended from the transverse and longitudinal girders both at the top and mid-height of the building. Suspension is effected by means of hanger rods. The floors of these storeys function as rigid horizontal decks, being braced by lattice girders arranged around the perimeter. The second and seventh floors of the building are connected by horizontal abutments to the portal frame columns.
Wind forces in both directions are distributed uniformly by the floors to the rigid hangers and are transmitted through the floor abutments and upper hanger connections to the main structural system of the building. This system resists the wind forces by rigid portal frame action in both directions. The portal columns are fixed at the base, thus reducing deformation.
• Portal columns, cross girders, longitudinal girders and edge girders are of box-section. Structural connections are welded. Each column was erected in six units. Cross-beams each consist of two HE 340 B rolled steel sections with 150 mm space between them for the bolted attachment of the hangers. The latter consist of HE 140 B, IPE 200 and U 140 sections arranged in four longitudinal rows. They are connected to the portal frame cross girders by means of capping plates and, at each connection, four 45 mm dia threaded bolts which are welded to diaphragms in the box-sections. Hangers have fire-protective casing. The actual portal frame members, being on the outside of the building, require no fire protection.
• Floors: Holorib steel decking on floor beams, with 9 cm in-situ concrete topping provided with reinforcement in upper surface; imposed load 175 kg/m². Ceilings of 16 mm thick rockwool slabs with tongued-and-grooved joints, suspended from blocks in the dovetailed ribs. Fire resistance 30 minutes.
• Each of the six portal frame columns is founded on an open masonry caisson resting on the floor of a disused and reclaimed quarry, 18 m below ground level. These caisson foundations carry a 2.0 m deep concrete perimeter beam which interconnects them and absorbs the fixing moments of the portal frame columns.

Structural system: cross-section

View of building showing continuous balconies in front of the rooms

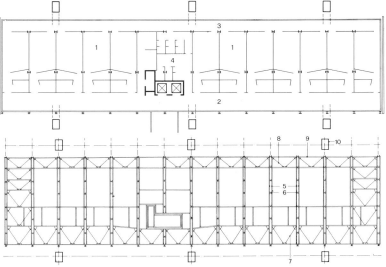

Typical floor with 12 students' rooms, scale 1:400

1 Student's room
2 Corridor
3 Balcony
4 Ablutions

Structural system for a typical upper floor

Areas and volume

gross area	5655 m²	area on plan	810 m²
residential area	3779 m²	volume	18 603 m³

Steel content

total	850 t	per m² gross area	150 kg
per m³ volume	45.7 kg	per m² residential area	225 kg

Construction cost (1969)

total cost	8.93 million Fr.fr.	per m² gross area	1579 Fr.fr.
per m³ volume	480 Fr.fr.	per m² residential area	2363 Fr.fr.

Reference

Acier-Stahl-Steel, 6/1968, p. 275.

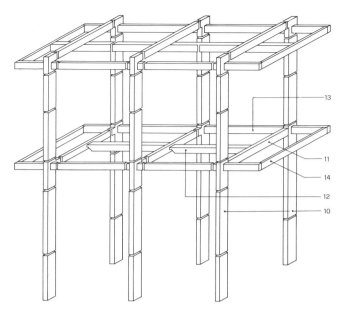
Isometric view of portal frame system

5 Cross beam comprising two channels U 200
6 Hanger
7 Wind-bracing L 70 × 70
8 Wind-bracing L 50 × 50
9 Floor edge beam U 150
10 Portal frame column 830 × 1500 mm
11 Portal frame cross girder 870 × 1200 mm
12 Central girder 830 × 1050 mm
13 Longitudinal girder
14 Edge girder
15 Hanger HE 140 B
16 Hanger U 140
17 Hanger IPE 200
18 Cross beam comprising two HE 340 B
19 Threaded bolts, 45 mm dia
20 Diaphragm
21 Stiffeners HE 180 B

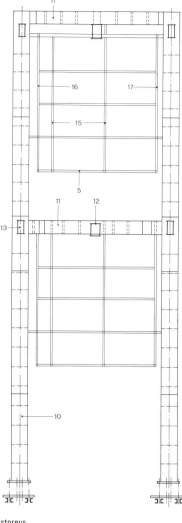
Transverse portal frame with suspension system

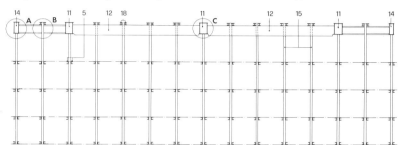
Longitudinal section through structural system for the lower block of four storeys

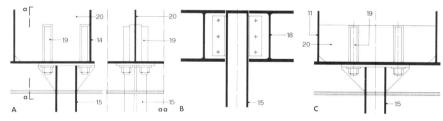
Connection of hangers to: A edge girder, B cross beam, C cross girder of portal frame

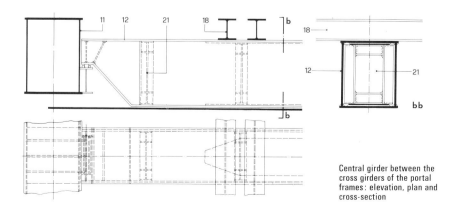
Central girder between the cross girders of the portal frames: elevation, plan and cross-section

Portal frame system with suspended storeys

77

15 HOTEL DE LAC, TUNIS

Architectural design: R. Contigiani, Rome.
Structural design: L. Nerap, Albert and steelwork contractors.
Built: 1969/1972.

Function

Hotel with 416 beds, in ten upper storeys which cantilever progressively outwards like wings. The bedrooms are situated on the south longitudinal side, being reached by way of a corridor at the rear. Emergency staircases projecting obliquely from the main structure are provided at each end. The main stairs, lifts and services are in a concrete core projecting rearwards in the middle of the building.
A low podium block contains the entrance hall, offices and communal rooms. Over the ground floor is a mezzanine with technical services. There is a restaurant on the top floor.

Dimensions

Overall dimensions on plan: 44.40 × 12.50 m on first floor, increasing to 82.80 × 12.50 m on tenth floor. Height above ground level 40.1 m. Each storey cantilevers 2.40 m beyond the one below. Emergency staircases project 6.00 m obliquely from the ends of the building. Storey height 3.20 m, ceiling height 2.95 m.

Structural features

The poor subsoil conditions in the immediate proximity of the sea necessitated a piled foundation. As this involved the construction of 190 reinforced concrete bored piles extending to a depth of 60 m, the designers aimed at keeping the area of the foundation as small as possible. It was this requirement that resulted in the striking shape of the building with its cantilevering wings.
The structural steel frame consists of pin-jointed members including main columns in the two longitudinal façades and one internal row of columns, interconnected by floor beams in the transverse direction. In the longitudinal direction there are continuous reinforced concrete ribbed floors which are prestressed between anchorages installed at the ends of the building. The forces from the cantilevered ends of each storey are transmitted through an inclined strut and the prestressed floors. Under symmetrical loading the horizontal forces balance each other in each floor. Horizontal forces due to symmetrical floor loading and to longitudinal wind are transmitted through diagonal bracings installed in the vertical plane between the internal columns at the centre of the building. This bracing system terminates in two V-shaped struts on the ground floor. Wind forces acting in the transverse direction are transmitted through the floor slabs to diagonal bracings located at four transverse axes and at the two ends of the building.
The sloping cantilevered staircases at the ends are connected to the main structural framework in each storey.
The columns in the longitudinal façades extend down to the foundations, whereas the internal columns and the inclined struts at the cantilever ends terminate in the mezzanine, in which storey-height plate girders and lattice girders span transversely between the external columns. These transverse girders are interconnected by a longitudinal plate girder located under the internal columns. This longitudinal girder is

1 Inclined strut, square hollow section, reducing from 275 × 275 mm to 125 × 125 mm
2 Steel diaphragm plate
3 Internal columns, box-section composed of two channels U 300
4 Longitudinal beam, two L 60 × 60 × 6
5 Cross beam IPB 160
6 Column, welded I-section 350 × 800 mm
7 Diagonal, welded I-section 250 × 350 mm
8 Longitudinal girder, 2285 mm deep plate girder
9 Transverse girder, 2350 mm deep plate girder
10 V-struts, two T 300 × 30
11 V-struts, two T 200 × 20
12 Prestressing cable
13 Diagonal 300 × 18 mm
14 Wind-bracing
15 Prestressed ribbed concrete floor
16 Anchorage
17 Concrete filling
18 Cable duct
19 Moulds for ribbed floor
20 Beam IPE 160

Rear view of the building, showing vertical circulation core in the middle

8th floor, scale 1:700

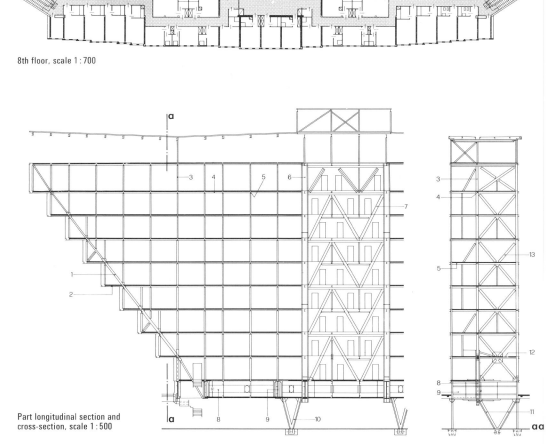

Part longitudinal section and cross-section, scale 1:500

additionally supported on V-struts on each side of the entrance hall which absorb the horizontal thrust exerted by the two inclined struts. Over the entrance two external columns terminate on, and are structurally intercepted by, a plate girder.
The main staircase and the two concrete lift-shafts are structurally separate from the steel frame of the building.
• Fire protection: Two-leaf brick partition walls enclose the columns, floor beams and vertical wind-bracing.

Areas and volume

gross area	8200 m²	area on plan	1400 m²
effective floor area	7000 m²	volume	26 000 m³

Quantities of materials

	steel	concrete	reinforcement
total	620 t	485 m³	100 t
per m³ volume	23.8 kg	0.019 m³	3.8 kg
per m² gross area	75.6 kg	0.059 m³	12.2 kg

Construction cost (1972)

Total cost	15.0 million DM	per m² gross area	1829 DM
per m³ volume	577 DM	per m² eff. floor area	2143 DM

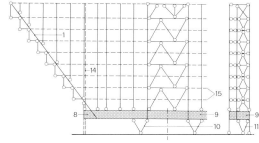
Structural system: part longitudinal section and cross-section

Longitudinal section through anchorage for prestressing cable

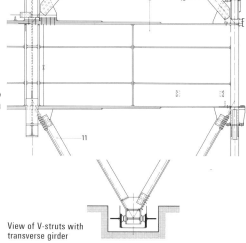
View of V-struts with transverse girder

Plan of the steel diaphragm plate at end of cantilever

21 Steel plate 5 mm thick
22 Welded shear connector 30 × 3 mm
23 Pocket for anchorage
24 Cleat 300 × 30 mm
25 Gusset plate
26 End-to-end connection
27 Web plate 2210 × 16 mm
28 Manhole

Steelwork in course of erection, with inclined column and diaphragm plates at one end of the building
Connection of wind-bracing to a transverse beam; prestressing cables in position between the moulds of the ribbed concrete floor

Inclined column with connection to steel diaphragm plate

Elevation of longitudinal girder

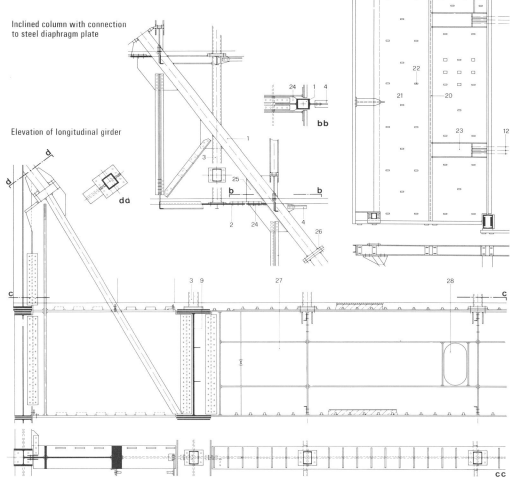

16 ALPHA HOTEL, AMSTERDAM

Architectural design: Groosman Partners, Rotterdam.
Structural design: Raadgevend Ingenieursbureau Ir. E. J. A. Corsmit, The Hague.
Built: 1968/1971.

Function

Hotel with 603 double bedrooms. A low-rise podium is surmounted by two staggered slab blocks connected to an 18-storey central core. The podium accommodates reception, management, restaurants, bars and kitchen. On the first floor of the slab blocks: administrative offices, staff accommodation. On the floors above (11 in one block and 14 in the other) are the bedrooms, situated on both sides of a central corridor.
Vertical circulation comprises four passenger lifts, two goods lifts and a main staircase in the core, together with two emergency staircases, one at the end of each block.

Dimensions

18-storey rectangular core 16.5 × 12.5 m, height above ground level 50 m. 16-storey slab block 68.0 × 15.2 m on plan, 45 m high; 13-storey slab block 48.0 × 15.2 m on plan, 36 m high. Storey height on upper floors 2.65 m, ceiling height 2.52 m; ground floor: 3.92 m and 3.45 m respectively.

Structural features

The slab blocks comprise transverse frames, spaced at 4.0 m intervals, each consisting of two columns interconnected by beams, braced by diagonal members in each storey. Precast prestressed concrete slabs span longitudinally between the beams. As the columns, beams and diagonal bracing are installed in the walls separating the rooms, the construction depth of the floors is merely the thickness of the prestressed concrete slabs. Transverse wind forces are transmitted through the rigid frames; longitudinal wind forces are transmitted by the floor slabs to the reinforced concrete core.
The columns, which reduce in section with height from HE 240 M to HE 220 A, are spaced at 10.60 m centres in the transverse direction of the building. Beams are IPE 270. Joints in columns are welded, with intermediate steel plates. All other structural connections are bolted.
- Floors: The 13 cm thick prestressed concrete slabs, 4.0 × 1.50 m, are connected to the beams in the transverse frames by welding to the bars left projecting from the concrete. Joints filled with in-situ concrete; no floor screed. Imposed load 300 kg/m².
- Erection: The transverse frames for the 16-storey slab block were assembled flat on the ground and then successively raised to the vertical position by means of a hoist on the roof (Porte-des-Lilas system). Floor slabs, spandrel panels and windows were erected in sequence to keep pace with these operations. In this way the structure of one 4 m long section of building was completed every seven days.
The transverse frames for the 13-storey slab block were erected in two-storey sections with the aid of a mobile crane.
- Fire protection: Columns, beams and bracing members incorporated in double-leaf lightweight concrete walls. The beams are protected with sprayed asbestos where they cross the central corridors.

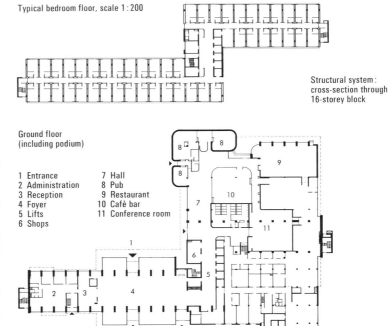

Typical bedroom floor, scale 1:200

Ground floor (including podium)

1 Entrance
2 Administration
3 Reception
4 Foyer
5 Lifts
6 Shops
7 Hall
8 Pub
9 Restaurant
10 Café bar
11 Conference room

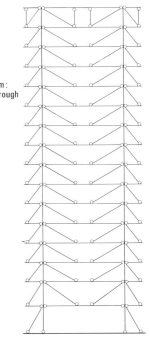

Structural system: cross-section through 16-storey block

- Foundation: Subsoil improved by the dumping of sand; ground-water at 1.50 m depth. Piled foundation comprises 196 in-situ concrete piles, extending to 10 m depth, bearing capacity 125 tons per pile. Pile heads under core are connected by a concrete raft, while under the two slab blocks they are connected to one another by concrete beams.

Services

Natural gas fired central heating plant installed in top storey of core. In all, six boilers with total capacity of 2 550 000 kcal/h. Radiators in bedrooms and administrative offices. Restaurant and bars air-conditioned (two air-conditioning plants and a cooling tower on roof of podium structure).

Areas and volume

gross area	30 043 m²	area on plan	4150 m²
effective floor area	15 707 m²	volume	82 115 m³

Material quantities

	steel	concrete	reinforcement
total	650 t	6155 m³	230 t
per m³ volume	7.9 kg	0.075 m³	2.8 kg
per m² gross area	21.6 kg	0.205 m³	7.7 kg

Construction cost (1970)

total cost	24.1 million D.fl.	per m² gross area	800 D.fl.
per m³ volume	294 D.fl.	per m² eff. floor area	1540 D.fl.
per bed	20 000 D.fl.		

References

Bouw, 1969, p. 2014.
Bouwen met Staal, 1969, p. 38.
Acier-Stahl-Steel, 5/1971, p. 201.

Partly erected slab blocks

Raising a fully assembled transverse frame

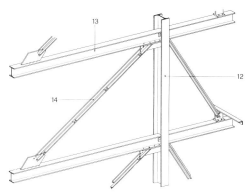
Detail of transverse frame

12 Column HE 240 M to HE 220 A
13 Beam IPE 270
14 Diagonal member composed of two angles
15 Floor slab
16 Core wall
17 Anchor socket
18 Flat bar
19 Reinforcement

Connection of floor slab and cross-beam of a frame to the core

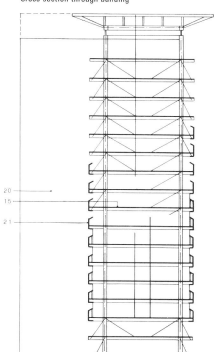
Cross-section through building

20 Core of building
21 Spandrel panel
22 Hoisting equipment
23 Hoisting ropes
24 Temporary bracing for erection
25 Transverse frame
26 Pivoting point

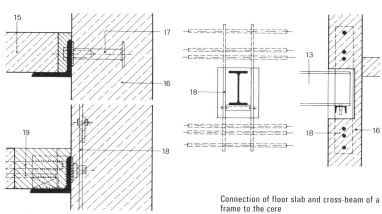
Raising a transverse frame

17 JOLLY HOTEL, ROME

Architectural design: E. Monaco, V. Monaco, Rome.
Structural design: G. Covre, M. Ferretti, Rome.
Built: 1968/1971.

Function

Hotel with 170 bedrooms on seven upper floors. The building is divided into groups, each comprising from two to five bedrooms, arranged on both sides and at the ends of a central corridor. The ground floor, of open construction, comprises the entrance, driveway for cars and parking space. In the first and second basement storeys: reception, lounge, restaurant and kitchen. In the third and fourth basement storeys: technical installations, car park and ancillaries. Vertical circulation facilities comprise three staircases, five passenger and four goods lifts.

Dimensions

Irregular shape on plan with maximum overall dimensions 66.0 × 29.0 m. Height above ground level 25.0 m. Storey height (upper floors) 3.25 m and 3.05 m; ceiling height 3.00 m and 2.80 m respectively.

Structural features

The functional subdivision of the building is clearly shown in its construction. In front of the glazed areas of the individual groups of rooms are columns, each consisting of two channel sections U 240, spaced at 10.35 and 6.90 m centres and interconnected by façade beams comprising two channels U 450. Along the corridors there are internal columns HE 300 B with 600 mm deep welded longitudinal plate girders. I 450 cross beams span between the longitudinal girders and fascia beams in the two outer walls and I 600 beams in the bedroom walls. Stubs of the cross beams, whose top flanges are flush with the top flanges of the fascia beams, are welded to the fascia beams to serve as brackets to support the sun-screens and flower-boxes.

The floors consist of 6.5 cm deep steel sheet troughing with concrete cover of 6.0 cm, supported by secondary beams IPE 160 spaced at 1.76 m centres and spanning between the cross beams. Imposed load 250 kg/m². Wind forces are transmitted through the floors to the reinforced concrete walls of the staircases and liftshafts.

- Façade: Windows have wooden frames and plate glass with high degree of sound insulation. Spandrel panels, 80 cm deep, are faced with safety glass.

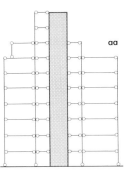
Structural system: cross-section

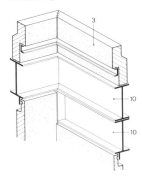
Corner of external wall

Cross-section of building above basement level

1 Bedroom 2 Corridor
3 Precast lightweight concrete unit
4 External column two channels U 240
5 Fascia beam two channels U 450
6 Internal column HE 300 B
7 Longitudinal welded plate girder, 600 mm deep
8 Cross beam I 600
9 Secondary beam IPE 160
10 Fascia beam I 450
11 Sun-screens

Arrangement of beams on part of a typical upper floor

Quantities of materials

	steel	concrete	reinforcement
total	1300 t	1000 m³	1200 t
per m³ volume	19.1 kg	0.147 m³	17.6 kg

Construction cost (1971)

total cost 3.5 milliard lire   per m³ volume   51 470 lire

Reference

Acier-Stahl-Steel, 11/1972, p. 456.

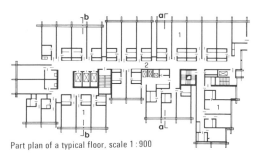
Part plan of a typical floor, scale 1:900

Steelwork in course of erection

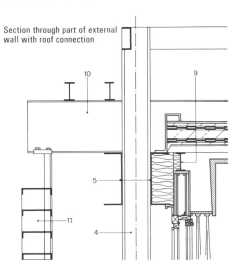
Section through part of external wall with roof connection

18 MATERNITY HOSPITAL, SEVENOAKS, ENGLAND

Architectural design: Gollins, Melvin, Ward and Partners, London; Ch. Scott.
Structural design: Clarke, Nicholls and Marcel.
Built: 1964/1965.

Function

Within a larger hospital complex: two-storey building with facilities for treatment of outpatients on ground floor and maternity wards with a total of 24 beds on the upper floor. Central core contains services, sanitary facilities, lift and two staircases. Ground floor comprises consulting rooms, waiting rooms, staff rooms and surgeries. Upper floor: operating theatres, nurses' rooms, kitchen and wards.

Dimensions

Overall dimensions 55.20 × 24.60 m. Height above ground level 10.25 m. First storey is recessed 1.50 m on three sides, with one shorter side recessed 6.40 m, where entrance is situated. Storey height 3.28 m, ceiling height 2.56 m.

Structural features

Three-bay transverse frames (spans: 9.04 m, 6.50 m, 9.04 m), at 4.80 m centres. Columns are 204 × 204 mm UCs interconnected with 690 mm deep castellated beams, the openings in which are used for service ducts. Structural connections made with high strength friction grip bolts or welded. In the longitudinal direction a 15 cm thick in-situ concrete floor slab spans from frame to frame, the top chords of the castellated beams being embedded in the concrete. Imposed load on floor 293 kg/m². Wind forces are transmitted in the transverse direction by portal frame action, in the longitudinal direction through the floor slab to staircase and through frames in the external walls. The transverse frames are interconnected longitudinally by 178 × 102 mm beams and 178 × 76 mm channels.
• Façade: Closed wall panels and panes of glass fitted directly to the steelwork with Neoprene gaskets. Allowable tolerances for the cladding units ±0.8 mm; for the structural steelwork: ±1.5 mm.

Services

Air-conditioning plant for the maternity wards, ventilation system for the inner rooms. Cold-water tank (5450 litres), for emergency supply, installed on roof.

Areas and volume

gross area	2360 m²	area on plan	1350 m²
		volume	7200 m³

Quantities of steel

total	75 t	per m² gross area	31.8 kg
per m³ volume	10.4 kg		

Construction cost (1965)

total cost	£250 000	per m² gross area	£106
per m³ volume	£34.80		

Reference

Acier-Stahl-Steel, 6/1966, p. 257.

End view, showing main entrance

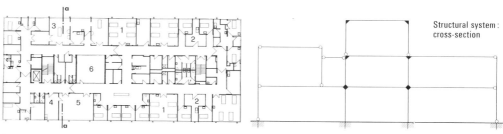
Upper floor, scale 1:900

Structural system: cross-section

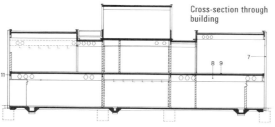
Cross-section through building

Beam connections to internal and external columns

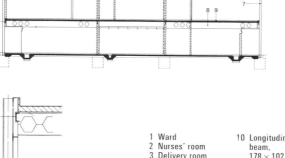
Vertical section through longitudinal wall

1 Ward
2 Nurses' room
3 Delivery room
4 Kitchen
5 Dining room
6 Light well
7 Universal column, 204 × 204 mm
8 Castellated beam
9 Concrete floor
10 Longitudinal beam, 178 × 102 mm
11 Edge beam, 178 × 76 mm channel
12 Wall unit
13 Neoprene gasket
14 Mullion
15 Recessed façade on ground floor

Horizontal sections through external wall at window and at spandrel panel levels

19 GROUP PRACTICE CLINIC IN SALT LAKE CITY, USA

Side facing road, with main entrance

Architectural and structural design: J. Sugden Associates, Salt Lake City.
Built: 1965.

Function

Two-storey building raised on stilts, comprising consulting rooms and communal facilities for fourteen doctors. On the bottom floor, which contains the entrance hall and waiting rooms, five consulting-room/surgery units are grouped around a core with lift, X-ray room and cloakrooms. A dispensary and a cafeteria are also accommodated on this floor. On the floor above are nine consulting-room/surgery units, an optician's shop, offices and commonrooms for doctors and other staff.
Plant is installed in a penthouse on the roof and in the recessed basement under the raised building. An external staircase gives access to the main entrance. Internally, the building has a staircase at each end.

Dimensions

Overall dimensions 43.40 × 18.80 m. Height above ground level 9.40 m. Storey height 3.05 m, ceiling height 2.63 m. Recessed base under the building 12.20 × 12.20 m.

Structural features

The structural steelwork comprises columns at 6.10 m centres in both directions. In the external walls the columns and edge beams are interconnected by diagonal bracings forming trusses two storeys high. At roof and floor level the external columns are interconnected across the width of the building by main beams which are also connected to the internal columns. Secondary beams, extending in the longitudinal direction of the building, are pin-jointed to the main beams. The secondary beams are spaced at 1.53 m centres for the floors and at 3.05 m centres in the roof. They carry steel sheet troughing with 9 cm cover of reinforced concrete, the decking units being welded to the steel beams to provide wind resistance in the floors.
Wind forces in both directions are transmitted through the lattice bracings in the external walls and by the bottom floor slab to the base structure, which has reinforced concrete walls. Floor beams are standard 254 mm deep wide flange and light beams. Edge beams are channel sections with bronze facings. Columns are I-sections, 203 × 203 mm and 153 × 98 mm. Secondary beams are connected to main beams by means of gusset plates welded to the flanges. Columns are spliced in each storey and are welded to the gusset plates. The truss units were prefabricated and transported to the site on low-loaders.
• Façade: Truss bays have infilling of room-height solar heat resistant glass. The panes of glass are fixed by means of bronze sections and Neoprene gaskets to the rear flanges of the columns and with Neoprene gaskets to the bronze facings at floor level. Sun shade is provided by vertical venetian blinds installed on the inside of the glass.
• Foundation: Pad foundations under the columns, strip foundations under the walls of the basement structure and of the staircases.

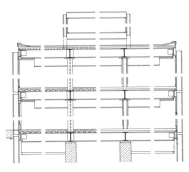

Cross-section through building

Upper floor, scale 1:600

Lower floor

1 Hall
2 Waiting room
3 Consulting-room/surgery unit
4 Administration
5 Optician
6 Nurses' room
7 Entrance hall
8 Cafeteria
9 X-ray room
10 Laboratory
11 Dispensary

Plant

Individually controlled convectors through which hot or cold water is circulated are installed at ceiling height in the rooms and directed at the external glazing.

Areas and volume

gross area	1768 m²	area on plan	815 m²
effective floor area	1027 m²	volume	6050 m³

Quantities of materials

	steel	concrete	reinforcement
total	73 t	219 m³	4.8 t
per m³ volume	12.1 kg	0.036 m³	0.8 kg
per m² gross area	41.3 kg	0.124 m³	2.7 kg

Construction cost (1965)

total cost	$574 000	per m² gross area	$325
per m³ volume	$95	per m² eff. floor area	$559

Reference

Bauen und Wohen, 10/1966, p. 407.

Prefabricated truss units transported by road

Joints in trusses at corner of building

Part of the finished steelwork

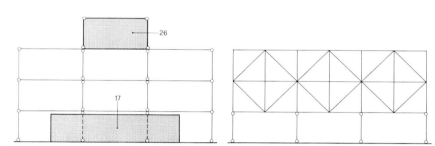

Structural system: cross-section and lattice bracing in end wall

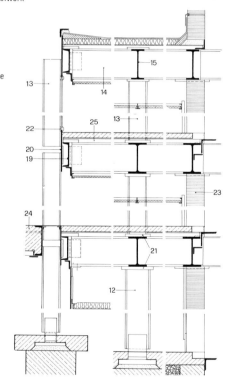

Part cross-section through the structure

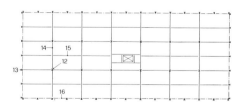

Beam arrangement in roof

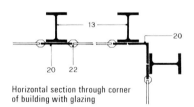

Horizontal section through corner of building with glazing

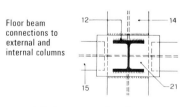

Floor beam connections to external and internal columns

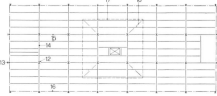

Beam arrangement for lower floor

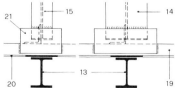

12 Column, I-section
   203 × 203 mm
13 Column, I-section
   153 × 98 mm
14 Main beam, I-section
   254 × 203 mm
15 Secondary beam, I-section
   254 × 102 mm
16 Trusses
17 Base structure
18 Diagonal connecting beam
19 Edge beam, channel section
20 Bronze plate
21 Gusset plate
22 Neoprene gasket
23 Wall of liftshaft
24 Landing of entrance stairs
25 Concrete on steel sheet
   troughing
26 Infilling

85

20 ACCIDENT HOSPITAL IN VIENNA, AUSTRIA

Architectural design: A. Hoch, Vienna.
Structural design: R. Krapfenbauer, Vienna.
Built: 1967/1972.

Function

Hospital for casualty cases, comprising an 8-storey block containing the wards and a 3-storey block containing operating theatres and treatment facilities. The two blocks are positioned at right angles to each other, both being joined to a vertical circulation core which they share in common and which accommodates two staircases and five lifts. In addition, each block has a staircase at the end remote from the core.
Eight-storey block: Central medical attendance and sterilisation facilities on the ground floor, doctors' rooms on first floor, operating theatres on second floor. On the next four floors are the wards with 120 beds in all. Dining rooms and kitchen on seventh floor. Helicopters can land on the flat roof.
Three-storey block: Technical services, bed preparation department and archives on the ground floor. On the first floor: outpatients' department, for 600 patients per day. Operating theatres and therapy rooms for the seriously injured on the second floor. Intensive care unit in a penthouse on the roof.
Both blocks have a basement containing air-conditioning plant and air-raid shelters.

Dimensions

Eight-storey block with overall dimensions of 40.60 × 17.20 m, height above ground level 32.30 m. Three-storey block with overall dimensions of 83.00 × 25.60 m, height above ground level 12.10 m. Storey height/ceiling height: basement 2.70/2.30 m, ground floor 3.70/3.00 m, first and second floor 4.20/3.40 m, third to seventh floor 4.00/3.20 m.

Structural features

The high standard and complexity of the technical equipment in the hospital calls for a form of construction which not only allows the numerous services (ducts, pipes, cables, etc.) to be conveniently installed but which makes it possible subsequently to alter any of these services or install new ones, without disrupting the normal work of the hospital. To meet these requirements, a lightweight structural steel system was developed, comprising Vierendeel girders, I-section columns and double-leaf partition walls, planned on a 1.40 × 1.40 m basic grid. The Vierendeel girders are designed as twin battened girders with 8 cm space in between, corresponding to the cavities of the double-leaf partitions, so that services can run vertically through all walls and floor beams and can also pass horizontally in any direction through the Vierendeel girders. The ceilings and the casings to the columns and walls consist of detachable fireproof panel units.
In both buildings the main girders span transversely between columns consisting of IPE 260. Spans in eight-storey building: 2.00 × 8.40 m; in three-storey building: 7.00 m, 2.80 m, 5.60 m, 2.80 m and 7.00 m. Main girders spaced at 4.20 m centres; secondary girders, in longitudinal direction, spaced at 1.40 m centres. All structural connections are bolted and regarded as pin joints. Main girders: depth 620 mm, chords consisting of two U 120 channel sections, uprights at 700 mm centres consisting of two U 150. Secondary girders: depth 580 mm, uprights at 700 mm centres, chords and uprights each consisting of two 800 mm deep cold-formed channel sections. The battens for the Vierendeel girders comprise short channel sections.
• Floors: 40 mm deep hot-dip galvanised steel decking comprising units 200 mm wide, laid side by side, and a continuous 1.2 mm thick top steel sheet (Lontain system). Over this is laid a soundproofing layer, carrying concrete slabs, screed and PVC floor covering. Imposed load on floors 400 kg/m². Wind forces are transmitted through the floors to vertical bracings in the core walls and in the end walls of the two blocks.

General view of hospital

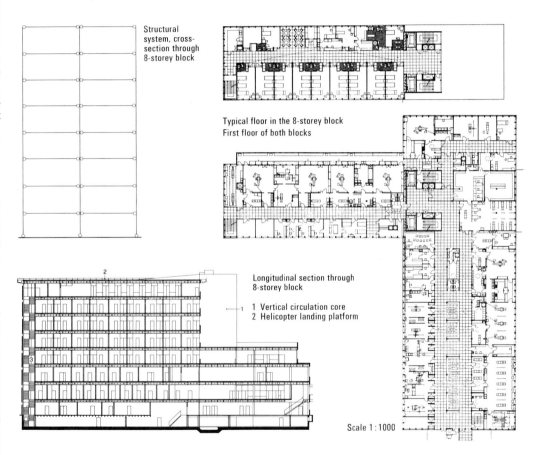

Structural system, cross-section through 8-storey block

Typical floor in the 8-storey block
First floor of both blocks

Longitudinal section through 8-storey block
1 Vertical circulation core
2 Helicopter landing platform

Scale 1:1000

References

Öster. Bau-Zeitung, 13/1968.
Detail, 3/1968, p. 423.
Acier-Stahl-Steel, 4/1968, p. 184.
Der Stahlbau, 1/1969, p. 24.
Fertigbau, 10/1969.
Deutsche Bauzeitung, 10/1970.

Structural steelwork for 3-storey block

Floor construction comprising Vierendeel girders in both directions

All services in floors and walls are accessible

Arrangement of girders in
8-storey block, with adjacent portion of 3-storey block

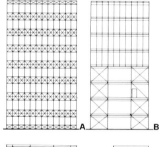
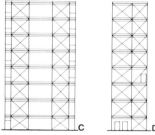
Wind bracings in end wall (A) and in core walls (B, C, D)

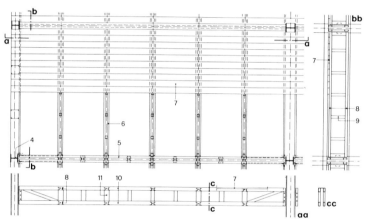
Detail of floor construction

3 Staircase
4 Columns IPE 260
5 Main girder 620 mm deep
6 Secondary girder 580 mm deep
7 Sheet-steel decking units (Longtain)
8 Chord, two U 80 × 40 × 3
9 Upright, two U 80 × 30 × 3
10 Chord, two U 120
11 Upright, two U 150 × 50 × 6
12 Floor construction: screed, concrete slabs, sound-proofing sheet-steel decking units
13 Detachable panel unit
14 Spandrel panel
15 Detachable wall unit
16 Fire encasement
17 Glass-wool filling

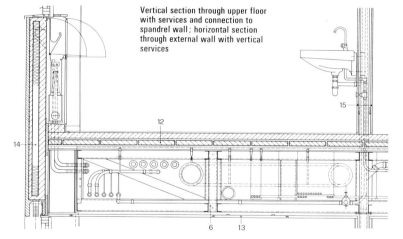
Vertical section through upper floor with services and connection to spandrel wall; horizontal section through external wall with vertical services

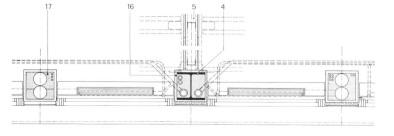
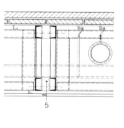
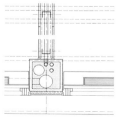

Areas and volume

gross area	16 372 m²	area on plan	3565 m²
effective floor area	8174 m²	volume	61 804 m³

Quantitites of materials	steel	concrete	reinforcement
total	1381 t	3578 m³	199 t
per m³ volume	22.3 kg	0.058 m³	3.2 kg
per m² gross area	84.4 kg	0.219 m³	12.2 kg

Construction cost (1967/1972)

total cost	197 million Au.S.	per m² gross area	12 033 Au.S.
per m³ volume	3187 Au.S.	per m² eff. floor area	24 101 Au.S.

21 PRIMARY SCHOOL AT BERLIN-WITTENAU, GERMANY

Architectural design: D. Hundertmark, H. Grünberg, Berlin.
Structural design: Steelwork contractors.
Built: 1971/1972.

Function
Primary school with 21 classrooms for 800 pupils. A central block with a semi-subterranean basement is connected on two sides by linkways to two classroom blocks comprising three and four storeys respectively. Both classroom blocks are displaced vertically a distance of half a storey in relation to the central block. They contain, on each floor, three or four classrooms disposed around a central vestibule. Two multi-purpose rooms, a staff common-room and a library are accommodated in the central block. Stairs and cloakrooms are located in the two linkways.

Dimensions
Central block 24.40 × 19.60 m; classroom blocks 26.80 × 14.90 m and 26.80 × 17.20 m. Height above ground level 14.50 m. Storey height 3.60 m, ceiling height 3.00 m. Each classroom 66 m².

Structural features
Pin-jointed framework on a 2.40 × 2.40 m grid, with prefabricated wall, cladding and floor units. The HE 120 B columns are continuous through three or four storeys and are interconnected in one direction by HE 340 A floor beams with pin-joints, spanning 7.20 m and 9.60 m and spaced at 2.40 m centres. In the other direction 9 cm thick precast concrete slabs, interconnected by projecting reinforcement and in-situ concrete in the joints, span from beam to beam, with which they act compositely through welded shear connectors. Wind forces are transmitted by vertical cross-bracings and by the reinforced concrete staircase cores.
On the centre-lines of the internal and external walls, secondary frames made of 1.5 mm thick galvanised cold-formed steel sheet are bolted to columns and floor beams. These secondary frames serve as fixings for wall and cladding elements, as door and window frames and as support for services.

Heating
The building is connected to a district heating system. Pumped hot-water heating 90°/70°C. Capacity 534 000 kcal/h.

Areas and volume

gross area	5320 m²	area on plan	2330 m²
effective floor area	2765 m²	volume	22 000 m³

Quantities of materials

	steel	concrete	reinforcement
total	260 t	1570 m³	94 t
per m³ volume	11.8 kg	0.071 m³	4.3 kg
per m² gross area	48.9 kg	0.295 m³	1.8 kg

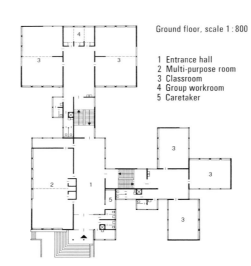

Ground floor, scale 1:800

1 Entrance hall
2 Multi-purpose room
3 Classroom
4 Group workroom
5 Caretaker

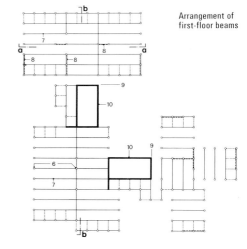

Arrangement of first-floor beams

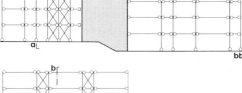

Structural system: longitudinal and transverse sections

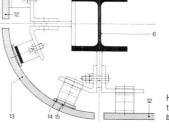

6 Column HE 120 B
7 Floor beam HE 340 A
8 Wind-bracing
9 Expansion joint
10 Reinforced concrete slab
11 Precast concrete slab
12 Asbestos cement cladding panels
13 Shaped corner unit
14 Asbestos cement strip
15 Adhesive compound

Horizontal section through corner of building

Construction cost (1971)

total cost	5.23 million DM	per m² gross area	983 DM
per m³ volume	238 DM	per m² eff. floor area	1892 DM

References
Detail, 1/1970, p. 37.
Bauwelt, 40/1970, p. 1482.
Industrialisierung des Bauens, 8/1972, p. 37.

Structural steelwork with floor slabs laid

22 PHYSICAL EDUCATION COLLEGE, MAGGLINGEN, SWITZERLAND

Architectural design: M. Schlup, Biel.
Structural design: Steelwork contractors.
Built: 1968/1970.

Function

The swimming pool, auditoria, lecture rooms, administrative offices and technical services are accommodated in a four-storey podium block partly recessed into the sloping ground. It is surmounted by a set-back three-storey superstructure containing the main assembly hall and library. The roof of the podium is a public terrace. The entrance to both buildings is at terrace level.

Dimensions

Podium block with overall dimensions of 73.36 × 43.88 m, height 13.20 m. Superstructure with overall dimensions 30.16 × 22.96 m, height above terrace 10.85 m. Storey height 3.45 m, ceiling height 2.45 m.

Structural features

Pin-jointed framework comprising rolled steel columns and 760 mm deep lattice girders in both directions. The 11 cm thick in-situ concrete slab is in composite construction with the girders, the shear connectors being welded studs. Main girders in longitudinal direction of building, spanning 7.20 m; secondary girders in transverse direction, spanning 14.40 m, top chord comprises channel section, bottom chord and diagonal members each consist of two angle sections. Imposed loads on floor 300 and 400 kg/m².
Wind forces transmitted through floor slabs to the concrete core and to concrete retaining walls set in the hillside.
- Façades: Continuous steel mullions of weathering steel, spaced 1.20 m centre-to-centre. Between them are storey-height window units with heat-resistant laminated glass. Spandrel panels are also of weathering steel.
- Foundation: Fissured limestone at 4 m below ground level. Bearing pressure 7.5 kg/cm². At the back of the podium, the walls are anchored into the rock.

Services

Heating installation with associated air-conditioning system for hall of swimming pool and auditoria. Air ducts pass through the openings in the lattice girders.

Lattice girders with stud shear connectors

View from north-west

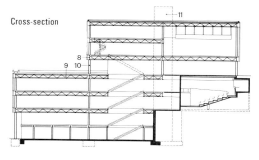

Cross-section

First floor of superstructure, scale 1:1000
Ground floor of podium

1 Auditorium
2 Foyer
3 Swimming pool
4 Changing rooms
5 Equipment issue counter
6 Games equipment store
7 Office

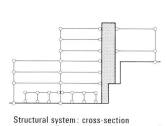

Structural system: cross-section and longitudinal section

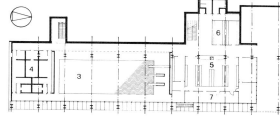

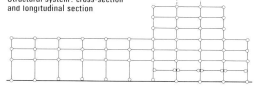

Arrangement of floor beams in podium, scale 1:1500

8 Main girder
9 Secondary girder
10 Column HE 320 B
11 Concrete core
12 Top chord U 180
13 Bottom chord ½ HE 220 B
14 Diagonal pair of angles, L 70 × 70 × 7
15 Gusset plate for secondary girder

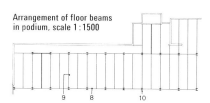

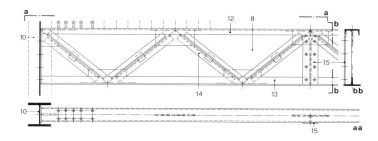

Areas and volume

gross area	10 073 m²	area on plan	3254 m²
effective floor area	6943 m²	volume	44 889 m³

Quantitites of materials

	steel	concrete	reinforcement
total	300 t	3000 m³	170 t
per m³ volume	6.7 kg	0.067 m³	3.8 kg
per m² gross area	29.8 kg	0.298 m³	16.9 kg

Construction cost (1970)

total cost	8.98 million Sw.fr.	per m² gross area	891 Sw.fr.
per m³ volume	200 Sw.fr.	per m² eff. floor area	1293 Sw.fr.

References

Bauen in Stahl, 10/1970, p. 55.
Bauen + Wohnen, 10/1971, p. 437.

23 TECHNICAL COLLEGE AT BRUGG-WINDISCH, SWITZERLAND

Architectural design: B. and F. Haller, Solothurn.
Structural design: Steelwork contractors.
Built: Main and laboratory buildings 1964/1966; assembly hall and refectory 1968/1970.

Function

The college comprises three interconnected buildings serving different purposes. The four-storey main building contains classrooms and administrative offices. The laboratories and auditoria are accommodated in a two-storey building with relatively elaborate technical equipment. The two buildings are linked by a bridge at first-floor level. The assembly hall and refectory are in a single-storey building.
In the main building the rooms are arranged around a central atrium with galleries and two staircases. On the ground floor: entrance hall, administration and staff commonroom. On the three upper floors are classrooms and drawing offices. Galleries are used as promenades during intervals between lessons and for exhibitions. Technical and ancillary services are accommodated in the basement.

Dimensions

Four-storey main building with overall dimensions 54.0 × 54.0 m. Height above ground level 17.0 m. Two-storey laboratory building with overall dimensions 106.8 × 27.6 m. Height above ground level 8.5 m. Assembly hall and refectory is a single-storey structure with overall dimensions 27.6 × 27.6 m. Height above ground level 4.5 m. In all buildings: storey height 4.20 m, ceiling height 3.27 m. Basements under all three buildings, interconnected by two underground passages.

Structural features

Same structural system used in all the buildings. Reinforced concrete basement is surmounted by tubular columns arranged on a square planning grid of 8.80 × 8.80 m. Columns rigidly interconnected in both directions by floor beams, so as to form one-, two- or four-storey rigid frames to resist the wind forces in the respective buildings. During erection, the steelwork was assembled with the aid of temporary bolted connections and subsequently, after final adjustment, all the joints were welded.
• Columns: Storey-height steel tubes, 318 mm diameter. Welded to the upper end of each column are two horizontal gusset plates for connecting the top and the bottom chords of the floor beams. Plastic downpipes for discharging rainwater from the roof are placed inside the columns.
• Floors: Each square floor bay is subdivided by three secondary girders. All the inner floor girders are of lattice construction, 700 mm deep, with top and bottom chords of T-sections and diagonal members of angle sections. Plate girders are used around the perimeter. The floors, which comprise in-situ concrete laid on corrugated steel sheets, are not composite with the supporting girders. Imposed load 500 kg/m^2.

Main building viewed diagonally

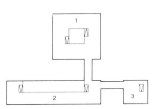

Location plan of the three buildings

1 Main building
2 Laboratory building
3 Assembly hall and refectory
4 Classroom
5 Gallery
6 Atrium

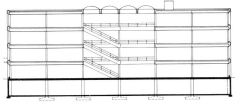

Section through building across atrium

Typical upper floor, scale 1:1000

Arrangement of beams in an upper floor

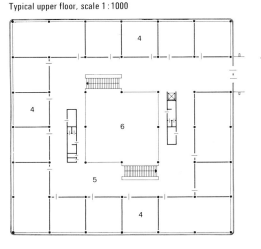

Steelwork being erected with mobile crane

Structural system

Section between atrium and external wall

Section through atrium

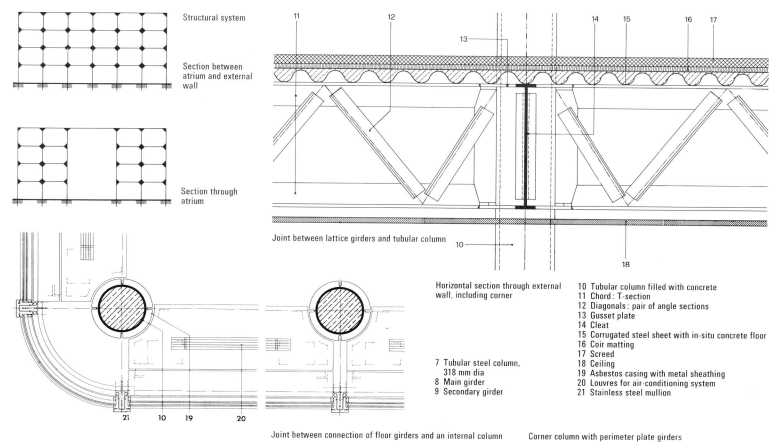

Joint between lattice girders and tubular column

Horizontal section through external wall, including corner

7 Tubular steel column, 318 mm dia
8 Main girder
9 Secondary girder

10 Tubular column filled with concrete
11 Chord: T-section
12 Diagonals: pair of angle sections
13 Gusset plate
14 Cleat
15 Corrugated steel sheet with in-situ concrete floor
16 Coir matting
17 Screed
18 Ceiling
19 Asbestos casing with metal sheathing
20 Louvres for air-conditioning system
21 Stainless steel mullion

Joint between connection of floor girders and an internal column

Corner column with perimeter plate girders

- Façade: Curtain wall with storey-height laminated glazing framed in stainless steel sections. Spandrel sandwich panels externally profiled and internally smooth stainless steel, foamed plastic insulation. Demountable internal partitions can be connected to any mullion. Sun shading by internal venetian blinds.
- Fire protection: Asbestos casing and concrete filling for tubular steel columns. Sprayed asbestos on plate girders. Suspended ceilings comprise fireproof panels.

Services

As the floor girders are lattice members, services can be conveniently installed in both directions in the ceiling space. All rooms fully air-conditioned. The building is connected to a district heating plant.

Ducts and services accommodated in the ceiling space

Detail of corner of curtain wall

Areas and volume

gross area	23 987 m²	effective floor area	16 825 m²
net area	23 141 m²	volume	114 638 m³

Quantitites of steel

total	885 t	per m² gross area	36.9 kg
per m³ volume	7.7 kg	per m² eff. floor area	52.5 kg

Construction cost (1966/1969)

total cost	26.8 million Sw.fr.	per m² gross area	1116 Sw.fr.
per m³ volume	234 Sw.fr.	per m² eff. floor area	1593 Sw.fr.

References

Schweizerische Bauzeitung, 18/1966.
Bauen + Wohnen, 8/1968, p. 297.
L'Architecture Francaise, July/August 1969.
Detail, 1/1969, Konstruktionstafel
Schweizer Journal, 4/1970
Schweiz. techn. Zeitschrift 33/1970.

24 HIGH SCHOOL IN CHICAGO, USA

Architectural design: Office of Mies van der Rohe, D. Lohan, partner in charge, Chicago.
Structural design: Nelson, Ostrom, Baskin, Berman and Associates.
Built: 1971/1973.

Function

School for 3000 pupils, comprising an eight-storey teaching block and a two-storey sports building. The buildings are situated on opposite sides of a street and are interconnected by a footbridge at first-floor level.
Teaching block: The school is subdivided into four groups, each comprising 750 pupils. On the third to sixth floors of the building each group has nine classrooms, a group commonroom, staff commonroom, promenade hall for use during school breaks, and food-service counter. On the first, second and seventh floors there are rooms for specialist subjects which are used by all the groups. The entrance hall and an auditorium with seating capacity for 750 are on the ground floor. In the basement there are rooms for specialist subjects and a kitchen. Technical services are accommodated in a roof penthouse. Vertical circulation facilities comprise two escalators, two lifts and three staircases. The speed and direction of movement of the escalators are varied to suit the requirements at any particular time.
Sports building: Two gymnasia and a swimming pool at ground floor level. Changing rooms and training rooms in the basement. On the upper floor: teaching facilities and technical equipment. Three staircases.

Dimensions

Both buildings are based on a planning module of 1.53 × 1.53 m.
• Teaching block: Overall dimensions 73.44 × 45.90 m, height above ground level 41.80 m. Ground floor: storey height 6.40 m, ceiling height 5.50 m. Basement and upper floors: storey height 3.96 m, ceiling height 3.05 m. Ground floor recessed 4.60 m from overall perimeter of the building.
• Sports building: Overall dimensions 91.80 × 36.72 m, height above ground level 14.00 m. Ground floor: storey height 6.40 m, ceiling height 5.30 m. Upper floor: storey height 4.27 m, ceiling height 3.05 m. Large gymnasium extends through height of both upper storeys. Basement: storey height 3.66 m, ceiling height 2.44 m.

Structural features

In the transverse direction there is a rigid frame system comprising universal columns and universal beams. The columns are arranged on a 9.14 × 9.14 m grid and are spliced every alternate storey. The transverse beams are connected to the columns by means of end plates and high-strength friction-grip bolts. Secondary beams spaced at 3.05 m centres support sheet-steel decking and carry a 12.7 cm lightweight concrete slab, composite action being ensured by stud shear connectors. Pinned joints are used for the connections between the secondary beams and the internal columns and main beams but rigid joints for wind resistance are used for the connections to the columns in the longitudinal façades.
The auditorium on the ground floor is free from internal supports and is spanned in the transverse direction by two box girders, on each of which two of the columns from the upper seven floors are supported. These box girders are 2130 mm deep, span a distance of 27.42 m and are carried by 604 × 604 mm box columns.
The footbridge connecting the two buildings comprises two lattice girders interconnected at floor and ceiling level by cross beams. The lattice girders are spaced at 9.14 m centres and span 38.61 m, the chord members comprising universal columns compounded with plates to form box sections, while the diagonal and vertical web members consist of I-sections.
To allow for temperature movement, the footbridge is supported at each end by suspender bars to which the bottom chords of the lattice girders are attached and which in turn are connected to brackets projecting from the columns of the teaching block and the sports building.
• Façade: Columns and beams are encased in concrete and sheathed externally with steel plate, stud shear connectors being used to ensure connection between the sheet and the concrete. The spandrel is of cavity construction, the intervening space being insulated. The steel framed windows have double glazing.
• Fire protection: The columns are encased in concrete. There is a sprinkler system on all floors.
• Foundations: Both buildings are supported by caissons founded on rock or stable subsoil at 11 m depth.

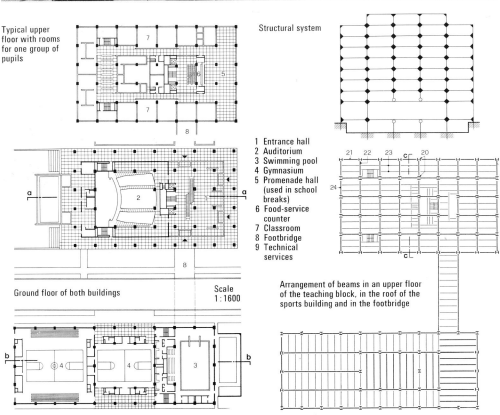

1 Entrance hall
2 Auditorium
3 Swimming pool
4 Gymnasium
5 Promenade hall (used in school breaks)
6 Food-service counter
7 Classroom
8 Footbridge
9 Technical services

Typical upper floor with rooms for one group of pupils

Ground floor of both buildings Scale 1:1600

Structural system

Arrangement of beams in an upper floor of the teaching block, in the roof of the sports building and in the footbridge

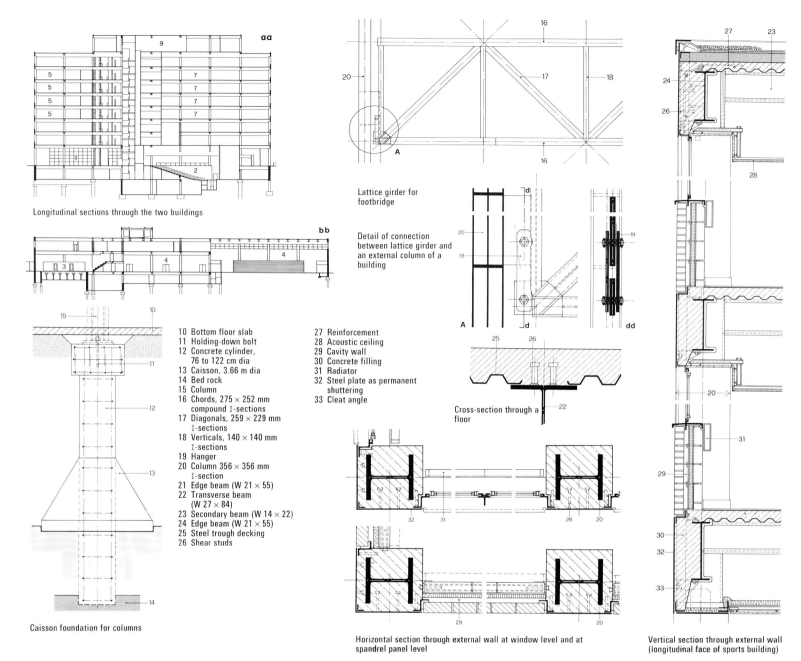

Longitudinal sections through the two buildings

Lattice girder for footbridge

Detail of connection between lattice girder and an external column of a building

10 Bottom floor slab
11 Holding-down bolt
12 Concrete cylinder, 76 to 122 cm dia
13 Caisson, 3.66 m dia
14 Bed rock
15 Column
16 Chords, 275 × 252 mm compound I-sections
17 Diagonals, 259 × 229 mm I-sections
18 Verticals, 140 × 140 mm I-sections
19 Hanger
20 Column 356 × 356 mm I-section
21 Edge beam (W 21 × 55)
22 Transverse beam (W 27 × 84)
23 Secondary beam (W 14 × 22)
24 Edge beam (W 21 × 55)
25 Steel trough decking
26 Shear studs

27 Reinforcement
28 Acoustic ceiling
29 Cavity wall
30 Concrete filling
31 Radiator
32 Steel plate as permanent shuttering
33 Cleat angle

Caisson foundation for columns

Cross-section through a floor

Horizontal section through external wall at window level and at spandrel panel level

Vertical section through external wall (longitudinal face of sports building)

Longitudinal face of the teaching block in course of erection

Services

Teaching block has high-pressure air-conditioning system, augmented by electric radiators fixed behind the spandrels of the external walls. Classrooms in the sports building have low-pressure air-conditioning plus electric radiators. Gymnasium and swimming pool hall have ordinary ventilation and heating.

Areas and volume

gross area	41 000 m^2	area on plan	6849 m^2
effective floor area	27 900 m^2	volume	193 000 m^3

Quantitites of materials

	steel	concrete	reinforcement
total	2700 t	11 614 m^3	308 t
per m^3 volume	14.0 kg	0.060 m^3	1.6 kg
per m^2 gross area	65.9 kg	0.283 m^3	7.5 kg

Construction cost (1972)

total cost	$15.75 million	per m^2 gross area	$384
per m^3 volume	$81.6	per m^2 eff. floor area	$565

25 OFFICE BUILDING FOR ADMINISTRATION OF STUDENT AFFAIRS, TECHNICAL UNIVERSITY OF BRUNSWICK, GERMANY

Architectural design: W. Henn, Brunswick.
Structural design: K. Pieper, Brunswick.
Built: 1968/1969.

Function

Office building for a staff of 50, in individual offices. Basement contains store rooms, commonrooms and a laundry. On ground floor: entrance hall, offices, caretaker's quarters. Upper floor: offices.

Dimensions

Two-storey building with basement, overall dimensions 30.25 × 15.25 m, height above ground level 6.48 m. Storey height 3.06 m, ceiling height 2.60 m. Basement partly above ground level.

Structural features

In transverse direction: three-storey rigid frames, comprising three 5 m bays spaced at 5 m centres, so that the columns are arranged on a square grid. In the longitudinal direction: temporary underfloor beams as an aid for erection, subsequently removed after the floors were concreted. Wind forces transmitted transversely by rigid frame action, longitudinally by rigid interconnection of two transverse frames.
• Columns: External columns are rectangular box-section members (200 × 240 mm, 8 mm plate thickness) placed 65 mm outside the plane of the façade. Internal columns are rolled sections (HE–A 160 to 260); transverse beams for the rigid frames are I 60 to 320. The bottom beams are pin-jointed to the external columns in order to reduce the bending moments in these columns in the bottom storey. Rigid frames erected with the aid of fitted and high-strength friction-grip bolts.
• Floors: Reinforced concrete ribbed floors span between transverse beams; imposed load on floors 500 kg/m^2.
• Façades: Timber cladding units strengthened by vertical steel sections, attached to the floors by means of anchor rails. Composition of cladding units (from outside to inside): horizontal spruce boarding, foil, laminated thermal-insulating filling 7 cm, vapour barrier, incombustible plywood, 16 mm. The units are 13.5 cm thick and have a thermal resistance value $k = 1.85$ m^{2}h°C/kcal. Horizontally pivoted windows comprise aluminium frames with double glazing. Horizontally projecting aluminium sun-baffles are fitted between external columns on south face.
• Foundation: Subsoil consists of fine and medium sand to 2.00 m below the surface. Permissible bearing pressure 4 kg/cm^2. Columns have pad foundations; basement walls supported on strip footings.

Services

Connection to district heating system through a heat exchanger. Heating capacity installed in the building 130 000 kcal/h; panel radiators. Inner rooms ventilated by ducted air delivered by fan, capacity 3000 m^3/h; extraction by fan on roof, capacity 3200 m^3/h.

End view with external columns exposed along longitudinal face

Structural system: cross-section and longitudinal section

Cross-section through building

Upper floor with central corridor and individual offices, scale 1 : 500

1 External column 340 × 200 mm
2 Temporary erection beam
3 Beam for stiffening in longitudinal direction
4 Internal column HE-A 160 to 260
5 Transverse beam for rigid frame
6 Ribbed reinforced concrete floor

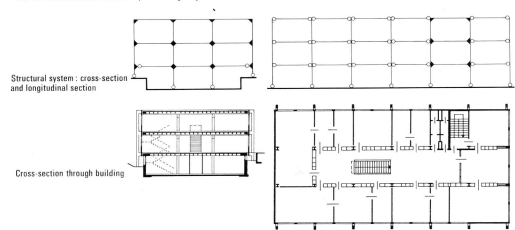

Connections between cross beam and external column

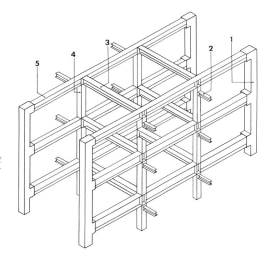

Isometric view of the two rigidly interconnected transverse frames

Areas and volume

gross area	1685 m^2	area on plan	500 m^2
net area	1435 m^2	circulation areas	378 m^2
effective floor area	939 m^2	volume	4845 m^3

Quantities of materials

	steel	concrete	reinforcement
total	45 t	623 m^3	33 t
per m^3 volume	9.3 kg	0.128 m^3	6.8 kg
per m^2 gross area	26.7 kg	0.370 m^3	19.6 kg

Construction cost (1969)

total cost	1.40 million DM	per m^2 gross area	831 DM
per m^3 volume	289 DM	per m^2 eff. floor area	1491 DM

References

Bauen + Wohnen, 1/1971, p. 23.
Bauen in Stahl, 27/1972.

26 SCHOOL OF ARCHITECTURE IN NANTERRE, FRANCE

Architectural design: J. Kalisz, R. Salem, Bagnolet.
Structural design: Steelwork contractors.
Built: 1971/1972.

Function

School of architecture for 1000 students in a five-storey building, characterised by its polygonal shape on plan and by the elevational treatment comprising recessed and protruding features in the various storeys. The ground floor comprises six lecture rooms, studio, restaurant, administrative offices and technical services, laid out as a continuous interconnected space. A covered patio is used for exhibitions and information purposes.
In the upper storeys, over each lecture hall, are associated drawing offices, classrooms, model-making workshops and other facilities. There are thus six groups of integrated facilities, each with its own spiral staircase disposed around a cylindrical concrete core. In addition, there are one internal and two external double-flight staircases. The cores accommodate cloakrooms and vertical service shafts.

Structural features

Pin-jointed framework with columns HE 300 A spaced at distances which, depending on the size of the spatial subdivisions of the building, are equal to once or twice the modular dimension of the planning grid. Floor beams IPE 500 have spans of 5.85 or 11.70 m in the interior, while those in the external walls span 5.85 or 8.28 m, depending on whether they coincide with grid lines or are diagonals. The columns have shop-welded brackets for supporting the floor beams, which are bolted to them on site. Floors consist of precast concrete slabs with in-situ concrete in the joints. Wind forces are transmitted through the floors to the six cylindrical slip-formed concrete cores, each 3.50 m in diameter.
- Façade: Window areas have double glazing (storey-height sheets, 90 and 180 cm wide) in aluminium frames. Non-glazed external wall areas have infilling consisting of 90 cm wide storey-height panels with external aluminium sheet and internal painted steel sheet, with a layer of insulating material sandwiched between them.
- Foundations: The site is a reclaimed quarry. Piled foundation: timber piles extending to depth of 20 m.

Services

Low-pressure air-conditioning system. Heating by radiators connected through a heat exchanger to district heating plant.

References

Technique et Architecture, Dec. 1971, p. 134.
Architecture d'Aujourd'hui, 160/1972, p. 80.
Architecture Francaise, 357–358/1972, p. 22.
Revue Bâtir, 13/1972.
Forum, 1/1972.
Planen + Bauen, 11/1972.

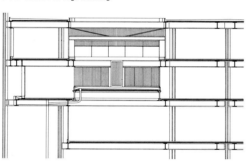

Ground floor, scale 1:1500

1 Commonroom
2 Lecture hall
3 Large hall
4 Reinforced concrete core
5 Restaurant
6 Administration
7 Entrance hall
8 Exhibition and information
9 Technical services
10 Cladding panel
11 Mullion
12 Fixed glazing

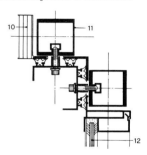

Horizontal section through corner of external wall

Cross-section through building

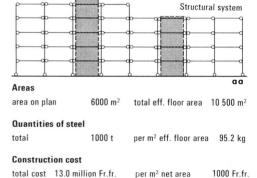

Details of exposed structural system

Structural system

Areas
area on plan 6000 m² total eff. floor area 10 500 m²

Quantities of steel
total 1000 t per m² eff. floor area 95.2 kg

Construction cost
total cost 13.0 million Fr.fr. per m² net area 1000 Fr.fr.

27 SPORTS ACADEMY IN MUNICH, GERMANY

Architectural design: E. Heinle, R. Wischer and Partner, Stuttgart.
Structural design: Leonhardt, Andrä and Boll, Stuttgart.
Built: 1971/1972.

Function

Built as an offshoot of the Olympic Games, this complex of buildings is a training academy for sports teachers which is also used for students' sports. Connected to a central building is a hall of residence with 45 individual rooms for participants in training courses. To the rear is a quadrangle and open-air theatre. Around this group are arranged other buildings containing halls for athletics, games, gymnastics and physical training; also, a set of six smaller halls for physical training, fencing, boxing, judo, wrestling and table tennis.

In the four-storey central building the rooms are arranged around a central main assembly hall in the first two storeys, and around a patio in the two upper storeys. The assembly hall accommodates 500 persons, with tiered seating and a platform. Changing rooms, store rooms and technical services are accommodated in the bottom storey. The first floor, the external walls of which are set back from the general façades, comprises the entrance hall, reception and exhibition areas. The two top storeys contain three auditoria (seating accommodation for 75 or 150 persons), library with reading room, work rooms, conference rooms, refectory and administration offices.

Access to the upper floors is provided by three staircases, one of which is in an elongated rectangular core which in addition contains a lift and cloakrooms. The lift machinery and other technical services are accommodated in a penthouse superstructure over this core.

Dimensions

Central building 50.40 × 43.20 m overall, patio 21.60 × 21.60 m. Planning grid 7.20 × 7.20 m. Storey height – headroom: 4.25 m – 2.50 and 3.09 m in the bottom storey; 4.69 m – 3.40 m in the second storey; 3.90 m – 2.96 m in the top two storeys.

Structural features

Pin-jointed framework on square grid of rolled steel beams in composite construction with precast concrete floor slabs.

The columns, consisting of HE 240 B sections, are interconnected in one direction by beams HE 400 B spaced at 7.20 m centres. Supported on these beams and spanning in the other direction are IPE 300 sections forming secondary beams, spaced at 2.40 m centres, in composite construction with the precast concrete floor slabs.

The joints between the slabs are filled with in-situ concrete to form rigid floors which are provided with horizontal bracing only where they are pierced by large openings. Wind forces are transmitted through the floor slabs and four vertical cross-bracing systems arranged in partition walls between adjacent columns.

- Floors: 7.20 × 2.40 m precast concrete slabs, 10 cm thick. Protruding steel stirrups are looped around stud shear connectors welded to the secondary beams. Imposed load on floors 300 and 500 kg/m². Ceilings consist of coffered metal panels (60 × 60 cm) with air inlet and outlet grilles for the ventilation system. Services are installed in the ceiling space.
- Façade: Two-storey-height mullions, of rectangular hollow section, are suspended from the edges of the roof and are connected with sliding bearings to the edges of the floor slabs. The cladding bolted externally to these mullions (which are spaced at 1.80 m centres) comprises window units, sheet-steel sandwich panels along perimeter of floor, and stove-enamelled aluminium spandrel panels. Cladding is secured with the aid of vertical sub-frame components. The latter, as well as the floor panels and the window frames, are of weathering steel. Seals are formed with neoprene gaskets. To prevent rust stains, rainwater is discharged through horizontal channels to the vertical sub-frame members. Windows have bronze-tinted fixed glazing and bottom-hinged opening lights in the top and bottom zones.
- Foundation: Pad foundations under the columns extend 1.35 m below ground level; reinforced concrete raft under the core. Gravel subsoil, with water table below base of foundation. Calculated bearing pressure on soil 4.0 kg/cm².

Central building as seen from quadrangle

A model of the complex showing sports halls, central building, hall of residence and smaller halls

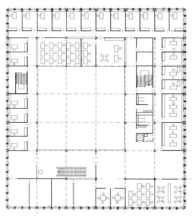
Second floor, scale 1:900

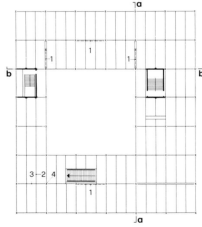
Arrangement of beams in an upper floor

1 Wind-bracing
2 Secondary beam IPE 300
3 Main beam HE 400 B
4 Column HE 240 B

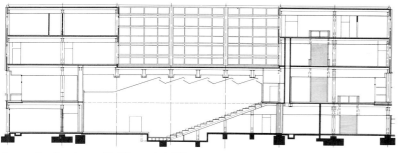

Cross-section through central building, scale 1:500

Longitudinal section through central building

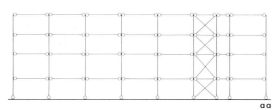
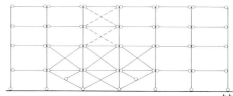

Structural system, longitudinal section and cross-section

Areas and volume (central building)
gross area	7700 m²	area on plan	2180 m²
net area	6300 m²	volume	35 000 m³

Steel quantities
Total	680 t	per m² gross area	88.3 kg
per m³ volume	19.4 kg	per m² eff. floor area	108.0 kg

Construction cost (1970)
total cost	12.53 million DM	per m² gross area	1633 DM
per m³ volume	358 DM	per m² net area	1990 DM

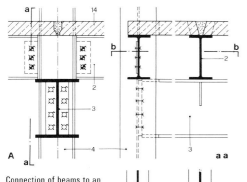

Connection of beams to an outer column

Connection of mullions to edges of roof and floors

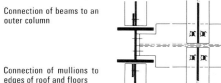

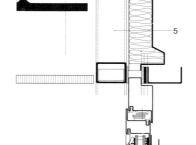

Vertical section through façade

Horizontal section through façade

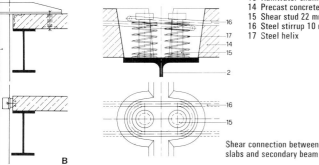

5 Mullion 70 × 50 mm
6 Spandrel panel
7 Window unit
8 Sub-frame component
9 Lipped gasket
10 Fireproof cladding to column
11 Neoprene
12 Flat steel bar
13 Rainwater channel
14 Precast concrete slab 10 cm thick
15 Shear stud 22 mm dia
16 Steel stirrup 10 mm dia
17 Steel helix

Shear connection between precast concrete floor slabs and secondary beam

Detail of the façade

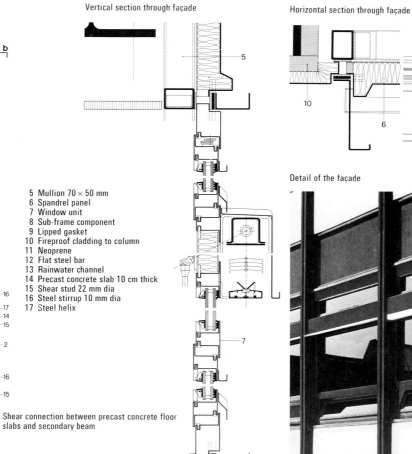

Services

Central heating with radiators, connected to a district heating system. Total capacity (including hall of residence) 560 000 kcal/h. Low-pressure ventilation for entrance hall and rooms occupied by staff. Low-pressure air-conditioning for assembly hall and lecture halls, capacity 370 000 kcal/h.

References
Bauen + Wohnen, 7/1972, p. 344.
Detail, 4/1972, Konstruktionstafel.
Deutsche Bauzeitschrift, 8/1972, p. 1387.

28 INDUSTRIAL RESEARCH INSTITUTE FOR THE VEREIN DEUTSCHER EISENHÜTTENLEUTE, DÜSSELDORF, GERMANY

Architectural design: F. Hitzbleck, J. Meyer, H. Rinne, Düsseldorf.
Structural design: G. Lewenton, E. Werner, L. Schwartz, Duisburg
Built: 1970/1972.

Function

Research laboratories and administrative offices in a three-storey building with basement under its entire area on plan. On the ground floor and the two upper floors are laboratories and offices; boiler room and stores are in the basement. More laboratories are accommodated in an adjacent single-storey building.
Vertical circulation in the three-storey building comprises a lift and a single-flight staircase; central corridors in the longitudinal direction.

Dimensions

Rectangular on plan, overall dimensions 13.20 × 51.00 m. Height above ground level 10.60 m. Internal core 4.80 × 10.80 m. Storey height 3.20 m, ceiling height 2.75 m.

Structural features

Reinforced concrete basement surmounted by two inner and two outer rows of columns, the latter located in front of the longitudinal façades, which extend continuously to the full height of the three storeys and are interconnected in both directions by floor beams forming a pin-jointed framework.
In-situ concrete floor slabs (imposed load 500 kg/m^2) transmit the wind forces in both directions to the concrete core. Internal columns consist of HE 180 B sections; external columns comprise 180 × 180 × 9 mm square hollow sections in weathering steel.
• Fire protection: The external columns are filled with water and are connected to one another at top and bottom to form closed circuits. In the event of fire the columns are cooled by circulating water. A storage cistern ensures a constant water level, extra water being obtained from the mains when necessary. An anti-freeze compound has been added to the water.
Internal columns are protected by asbestos cladding, floor beams by intumescent paint (fire resistance 30 min).
• Façade: Band of weathering steel sheet cladding at floor level with room-height units comprising windows and spandrel panels along longitudinal sides. Sun protection by external blinds. The end walls comprise closed panels with masonry backing.
• Foundations: Subsoil is fine to medium sand; bearing pressure 3.5 kg/cm^2. Strip footings under the basement walls, raft under the core.

References

Bauwelt, 40/1970, p. 1489.
VDEh Stahl und Eisen, 22/1970, p. 1234.
Acier-Stahl-Steel, 10/1971, p. 385.
Der Bauingenieur, 5/1971, p. 193.

Ground floor with central corridor and individual rooms, scale 1:800

1 Square hollow section 180 × 180 mm
2 Steel plate 143 × 20 mm cross-section
3 Weathering steel cladding
4 Inlet pipe 50 mm dia
5 Floor beam I 360

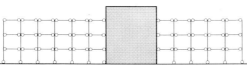

Structural system: cross-section and longitudinal section

Areas and volume

gross area	2800 m^2	area on plan	700 m^2
effective floor area	1910 m^2	volume	8660 m^3

Quantities of materials

	steel	concrete	reinforcement
total	100 t	950 m^3	55 t
per m^3 volume	11.5 kg	0.110 m^3	6.4 kg
per m^2 gross area	35.7 kg	0.339 m^3	19.6 kg

Construction cost (1971)

total cost	2.8 million DM	per m^2 gross area	1000 DM
per m^3 volume	324 DN	per m^2 eff. floor area	1470 DM

Steel frame with temporary bracing during erection

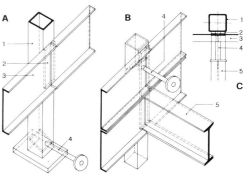

Pipe connections to external columns
A Column base, B Column head, C Horizontal section

Cooling system for columns

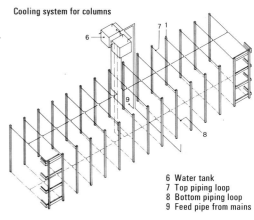

6 Water tank
7 Top piping loop
8 Bottom piping loop
9 Feed pipe from mains

Column cooling water pipes on the roof

29 MEDICAL RESEARCH LABORATORIES, UNIVERSITY OF CALIFORNIA, SAN FRANCISCO, USA

Architectural design: Marquis and Stoller.
Structural design: E. Elsesser.
Built: 1965/1966.

Function

A two-storey central corridor system extending parallel to the slope of the site is flanked by a pair of two-storey blocks on the lower side and by one such block on the upper side. No basement. Both storeys accommodate medical research laboratories with associated offices. There is provision for subsequent extension by lengthening the central corridors and the addition of more buildings.

Dimensions

All the buildings are rectangular on plan, overall dimensions 15.25×9.15 m. Height above ground level 5.70 to 11.60 m. Storey height 3.67 m, ceiling height 2.46 m. Planning grid 3.05×3.05 m. Width of corridor 2.30 m.

Structural features

On account of the slope of the site each building is raised on four columns of unequal length, spaced at 9.15 m centres in both directions. At both roof and first floor level there are longitudinal Vierendeel girders which cantilever 3.05 m beyond the columns at each end and which have been left exposed in the façades and on the inside of the corridor. Longitudinal beams consisting of 457×203 mm I-sections, also exposed, are provided at ground floor level. All the connections between the columns and the longitudinal girders and beams are rigid. Transversely, the longitudinal members are interconnected by six Vierendeel girders or rolled steel beams, which cantilever over the full width of the corridor.
Columns are 305×305 mm I-sections (12 WF). Vierendeel girders are 920 mm deep, with uprights at 1525 mm centres; chords and uprights are 152×152 mm I-sections (6 WF).
Floors stiffened by horizontal diagonal bracing. Wind forces transmitted by portal frame action.
• Floor: Timber boards and joists. No suspended ceiling.
• Fire protection: Sprinkler system, in central corridors and staircases only.

Services

Main services extend vertically through apertures in the corridor floors, horizontally through the open panels in the Vierendeel girders; branch connections to the individual laboratories. Fresh air intake from underneath the buildings (elevated on stilt-like columns), air outlets on the roof.

References

Architectural Record, 4/1965, p. 32.
Progressive Architecture, 7/1967, p. 169.
Daily Pacific Builder, 3/1967, p. 1.
Architecture/West, May 1965, p. 19.
Design in Steel, 1967.

Structural system: longitudinal section and cross-section

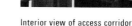

Entrance floor, scale 1:700

1 Access corridor
2 Laboratory
3 Office

Interior view of access corridor

4 Services
5 Vierendeel girder
6 Floor beam I-section 456 mm deep
7 Column I-section 305×305 mm
8 Top and bottom chords I-section 152×152 mm
9 Uprights I-section 152×152 mm
10 Site joint
11 Timber joist floor
12 Aperture in floor
13 Fresh air inlet
14 Air outlet

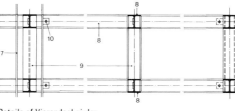

Details of Vierendeel girder

Areas and volume

gross area	930 m²	area on plan	558 m²
effective floor area	745 m²	volume	2970 m³

Construction cost (1964)

total cost	$311 000	per m² gross area	$334
per m³ volume	$105	per m² eff. floor area	$417

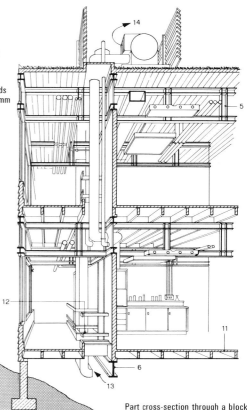

Part cross-section through a block

30 INSTITUTE OF TECHNOLOGY, UNIVERSITY OF LILLE, FRANCE

Architectural design: J. Balladur, B. Tostivint, Paris.
Structural design: Steelwork contractors.
Built: 1968/1969.

Function

A complex of thirteen university buildings, ranging from one to three storeys in height, seven of which provide teaching facilities for 1200 students. Two three-storey central buildings accommodate a library, medical aid centre, administrative offices and rooms for the teaching staff. Each of these two structures is connected to a building containing an auditorium and commonrooms. The other one-storey and three-storey buildings contain computer equipment, laboratories and special research facilities. Some of the buildings have basements for technical installations. Vertical circulation arrangements comprise lifts in some of the buildings and staircases at or near their ends. Buildings are interconnected by covered ways.

Dimensions

All the buildings are 17.38 m wide, with lengths ranging from 32.80 to 79.50 m. Storey height 3.30 m, headroom 3.10 m to floor slab or 2.85 m to floor beams. Overall dimensions of auditorium buildings 23.40 × 11.40 m, headroom 4.15 to 8.00 m.

Structural features

Four longitudinal rows of columns RHS 120 × 120 mm and 180 × 130 mm, just inside the two longitudinal façades and on each side of the central corridor, respectively. Column spacing: 3.60 m in the longitudinal direction; 7.08, 2.65 and 7.08 m in the transverse direction. The columns flanking the corridor are fixed at the base and are rigidly connected by transverse IPE 240 or 180 beams in each floor. Welded plate girders, 278 mm deep, link the façade columns with these multi-storey rigid frames. In the longitudinal direction the columns are connected to one another by channel sections UPF 50 × 50 × 4. Wind forces are transmitted in the transverse direction through the floor slabs to the transverse portal frames, and in the longitudinal direction to vertical cross-bracings between the internal columns. The transverse frames were prefabricated in the works, using welded connections. The wind-bracings consist of channel sections UPF 80 × 50 × 4.
- Floors: The floor beams are in composite construction with a 12 cm thick in-situ concrete slab spanning in the longitudinal direction, the shear connectors being welded studs. Imposed load on floors: 250 kg/m² in lecture rooms, 300 kg/m² in laboratories.
- Façade: Storey-height units, 3.30 × 1.80 m, erected between vertical sections attached to the structural columns. Windows have double glazing. Spandrel sandwich panels comprise two metal sheets with polyurethane thermal insulation in between, the external sheet being plastic coated. Sun shading by means of slatted roller blinds, fixed externally.
- Corrosion protection: Steel components have been given two protective coats of glycerophthalin with zinc chromate.
- Fire protection: Columns and floor beams encased in Asbestolux with 10 cm cavity. Roof framework is protected by suspended ceiling of mineral wool boards.
- Foundations: Strip footings under the basement walls. The buildings without basements have open caisson foundations with longitudinal capping beams. Foundation depth up to 3.50 m on loadbearing strata of silt and firm clay.

Services

Central heating plant in basement of the central building. Heat supplied by long-distance pipeline from university's own heating system. Capacity 4 million kcal/h. Laboratories have additional air heating.

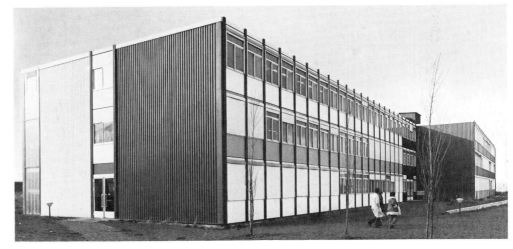

Faculty of food technology

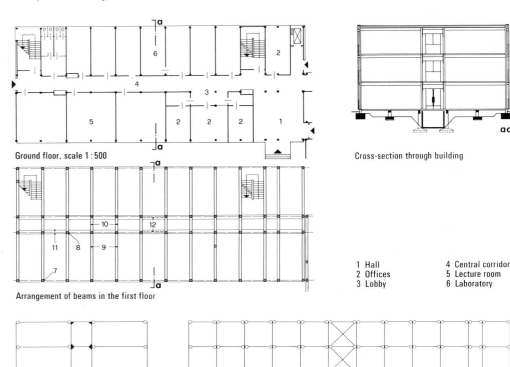

Ground floor, scale 1:500

Arrangement of beams in the first floor

Cross-section through building

1 Hall
2 Offices
3 Lobby
4 Central corridor
5 Lecture room
6 Laboratory

Structural system: cross-section and longitudinal section

Areas and volume

| gross area | 20 600 m² | volume | 60 000 m³ |

Quantities of steel

| total | 650 t | per m² gross area | 31.6 kg |
| per m³ volume | 10.8 kg | | |

Construction cost (1969)

| total cost | 17.0 million Fr.fr. | per m² gross area | 825 Fr.fr. |
| per m³ volume | 283 Fr.fr. | | |

References

Chantiers coopératifs, June 1970.

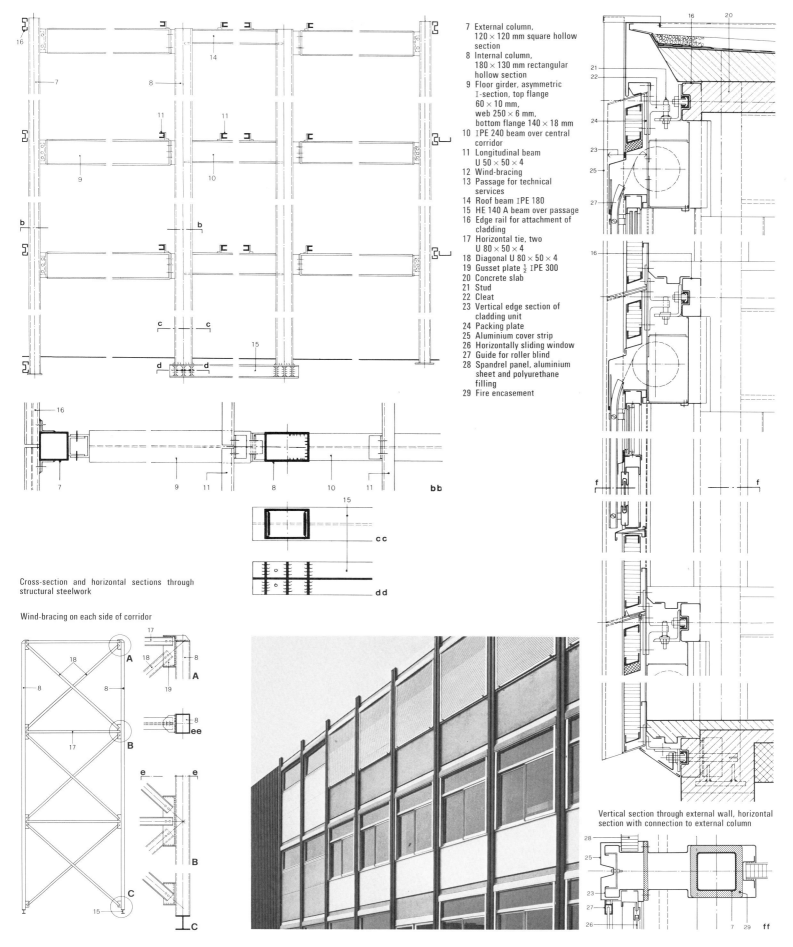

7 External column, 120 × 120 mm square hollow section
8 Internal column, 180 × 130 mm rectangular hollow section
9 Floor girder, asymmetric I-section, top flange 60 × 10 mm, web 250 × 6 mm, bottom flange 140 × 18 mm
10 IPE 240 beam over central corridor
11 Longitudinal beam U 50 × 50 × 4
12 Wind-bracing
13 Passage for technical services
14 Roof beam IPE 180
15 HE 140 A beam over passage
16 Edge rail for attachment of cladding
17 Horizontal tie, two U 80 × 50 × 4
18 Diagonal U 80 × 50 × 4
19 Gusset plate $\frac{1}{2}$ IPE 300
20 Concrete slab
21 Stud
22 Cleat
23 Vertical edge section of cladding unit
24 Packing plate
25 Aluminium cover strip
26 Horizontally sliding window
27 Guide for roller blind
28 Spandrel panel, aluminium sheet and polyurethane filling
29 Fire encasement

Cross-section and horizontal sections through structural steelwork

Wind-bracing on each side of corridor

Vertical section through external wall, horizontal section with connection to external column

101

31 FACULTY OF SCIENCES COMPLEX, UNIVERSITY OF PARIS, FRANCE

Architectural design: R. Seassal, U. Cassan, R. Coulon, E. Albert, C. de Gortchakoff, Paris.
Structural design: Steelwork contractors.
Built: 1958/1973.

Function

Complex of buildings with 400 000 m² of effective floor area for the Faculty of Sciences, University of Paris. The site, measuring 333 × 275 m, is subdivided into rectangular bays, five in each direction. At the intersections of the grid lines are located vertical cylindrical circulation cores, of reinforced concrete construction, connected to the five-storey buildings which are raised on pilotis. The complex thus comprises 22 courtyards, including the large courtyard at the main entrance. The last-mentioned courtyard, which takes up four bays, is dominated by the administrative tower block, sited diagonally to the axes of the remainder of the complex.

Some of the courtyards have basements beneath them, thus providing accommodation for a computer centre, a particle accelerator and an underground car park. At ground level there are open spaces under the buildings and also auditoria which extend down into the basement. The five upper floors of the shorter buildings comprise lecture halls and rooms for tutorials. Associated institutes and laboratories occupy the five upper floors of the respective adjacent longer buildings. Technical services, storage facilities and archives are accommodated in two basements. Each of the 33 vertical cores contains two spiral staircases, a passenger lift, a goods lift and shafts for services. An underground railway station is located near the main entrance.

Dimensions

Overall dimensions: longer buildings 53.6 × 18.0 m, shorter buildings 41.6 × 18.0 m. Height above ground level 22.50 m. Storey height of upper storeys 3.40 m, ceiling height 2.94 m. On ground floor: storey height 5.50 m, ceiling height 3.50 m. The vertical cores are 9.50 m diameter.

Structural features

The reinforced concrete basements are surmounted by the cylindrical cores. The latter are linked by structural steel frames consisting of two rows of tubular columns, exposed outside the longitudinal façades of each building and interconnected in the transverse direction by floor beams. Each upper floor comprises a central longitudinal beam resting on pin-jointed columns which are supported on transverse fish-belly girders spanning over the entire width of the ground floor. Every alternate external column is intercepted and carried by an edge girder at first floor level. Thus, while the tubular columns are spaced at 3.00 m centres at ground level, they are, like the transverse floor beams, at 1.50 m centres in the storeys above.

Wind forces are transmitted through the floors to the concrete cores, to which the floor beams are connected. Resilient connections between the ends of the beams and the anchor plates embedded in the concrete absorb wind forces elastically and also allow up to 8 mm change in

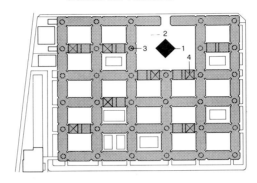

Location plan: 1 Tower block 2 Main courtyard 3 Vertical circulation core 4 Auditorium

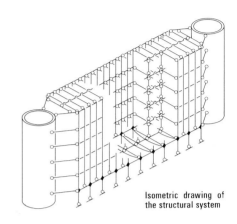

Isometric drawing of the structural system

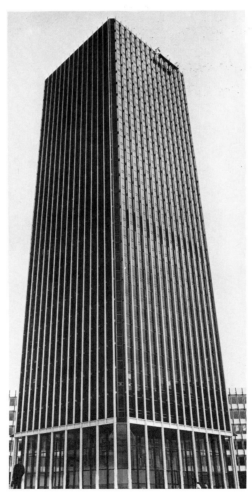

Administrative tower block

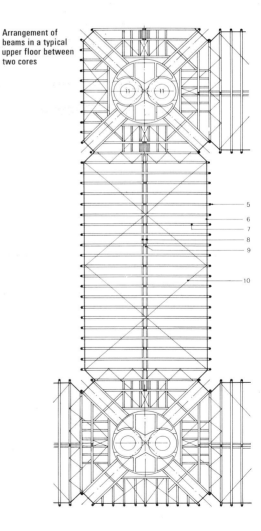

Arrangement of beams in a typical upper floor between two cores

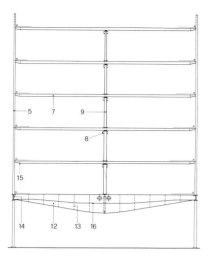

Cross-section through one of the buildings

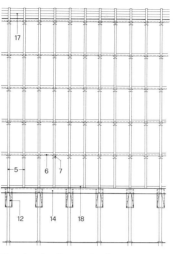

Part elevation of external wall

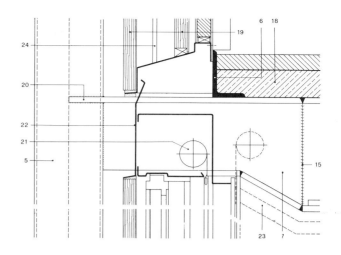

Vertical section through façade showing beam-to-column connection

Areas and volume (not including tower block)

gross area	205 578 m²	area on plan	34 263 m²
effective floor area	158 395 m²	volume	788 049 m³

Quantities of steel (not including tower block)

total	22 400 t	per m² gross area	109 kg
per m³ volume	28.4 kg		

5 External column, 219 or 168 mm dia
6 Edge beam
7 Transverse beam HE 300 A
8 Central longitudinal beam, two U 260
9 Central column
10 Temporary bracing (during erection)
11 Spiral staircase
12 Fish-belly girder
13 Diaphragm
14 Edge girder HE 600 A
15 Site joint
16 Suspended ceiling
17 Parapet
18 Concrete slab
19 Marble slab
20 Top flange of floor beam
21 Sun-blind
22 Steel sheet
23 Soffit
24 Guideway

length in the building in consequence of thermal expansion.

The tubular columns are 219 mm dia × 25 mm thick on the ground floor and 168 mm dia × 6.3 and 4.5 mm thick in the upper storeys. As fire protection and in order to prevent sound transmission, the tubular columns were filled with fine concrete at the time of erection. The edge girders at first floor level are HE 600 A sections. The fish-belly girders are welded box sections, 700 mm wide and 1760 mm deep at mid-span, 250 mm wide and 500 mm deep at the ends. The transverse beams are HE 300 A, while the central longitudinal beam in each upper floor is composed of two U 260 channels and the internal columns are I-sections diminishing in size with height.

- Floors: The 7 cm thick in-situ concrete floor slabs, which transmit the wind forces to the cores, are in composite construction with the transverse beams, the shear connectors being welded studs. Imposed load 500 kg/m². The suspended ceilings are provided with radiant heating.
- Façade: The storey-height framed components, 1.50 m wide, are attached to the transverse beams. The spandrels comprise 3 cm thick marble slabs, an intervening air space and heat-insulating panels on the inner face. The windows run in vertical guides attached to the mullions and, when opened, they slide down in front of the spandrel panels. The vertical joints between the cladding units are faced with 30 cm wide marble slabs and the horizontal joints have sheet-metal flashings.
- Foundations: The upper subsoil strata of sand, gravel and marl have low loadbearing capacities. Piled foundations: Piles extending down to loadbearing limestone at 36 m depth. Under each vertical core there are 12 reinforced concrete bored piles, 42 cm dia.

Reference

Acier-Stahl-Steel, 5/1967, p. 199.

Erecting the steelwork between two cores

Façade detail

Fish-belly girders spanning the ground floor

32 DEPARTMENT OF ELECTRICAL ENGINEERING, TECHNISCHE HOGESCHOOL, DELFT, HOLLAND

Architectural and structural design: J. P. van Bruggen, G. Drexhage, J. J. Sterkenburg, A. Bodon, Rotterdam.
Built: 1964/1969.

Function

This is a 23-storey slab block containing lecture rooms, laboratories and demonstration rooms. Rooms are situated on both sides of a central corridor, a flexible internal layout being possible. On all floors continuous 1.00 m wide balconies extend along both longitudinal faces of the building and are enclosed by external glazing. The various technical services for the laboratories are accommodated on these balconies, in front of the spandrel panels. These services are directly accessible at all times and can therefore be easily replaced or extended. Antenna equipment for experimental purposes is fixed on the roof, which is designed as a flat terrace surrounded by a 2.00 m high parapet. On the ground floor are the entrance hall and access to the lecture rooms in an annexe. The technical services are accommodated in the basement and top storey.
Vertical circulation: four passenger lifts and a service lift, together with staircases at the two extreme ends of the building.

Dimensions

Shape on plan: two longitudinally displaced rectangles. Overall dimensions 82.50 × 17.50 m. Height above ground level 89.7 m. Basement: storey height 4.50 m, headroom 4.20 m. Ground floor: storey height 5.20 m, headroom 4.45 m. Upper storeys: storey height 3.75 m, headroom 3.25 m.

Structural features

The structural steel frame consists of columns and floor beams in both directions, in combination with five pairs of reinforced concrete shear walls, the latter being located at the four ends of the building and also beside the liftshaft. The shear walls combine with the steelwork to stabilise the building in the transverse direction. The building is stiffened in the longitudinal direction by the rear wall of the shaft in which the four passenger lifts are accommodated side by side. The columns are I-sections with individual loads up to 500 t. Column spacing: 4.05 m in longitudinal direction; 5.87 m, 2.91 m and 5.87 m transversely. The HE 300 A cross-beams carry a 10 cm thick in-situ concrete floor slab designed for 500 kg/m² imposed load. Fire protection of columns: 3 cm thick perlite-gypsum plaster on expanded metal.
• Foundation: Piled foundation grillage comprising 686 piles extending to 12 m depth, 100 t capacity per pile. Basement designed as torsionally rigid reinforced concrete box.

Services

Induced-draught ventilation in the rooms. Outer corridors ventilated by fans. Total capacity: 30 000 m³/h on west side, 9000 m³/h on east side.

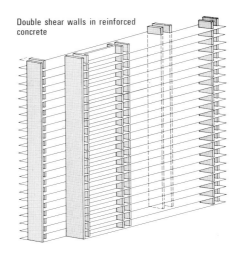
Double shear walls in reinforced concrete

1 External glazing
2 Outer corridor
3 Services
4 Internal glazing
5 Induced-air ventilator
6 Concrete floor
7 Column 340 × 340 mm
8 Edge beam
9 Precast concrete slab
10 Floor beam HE 300 A
11 Longitudinal beam
12 Reinforced concrete shear wall

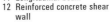

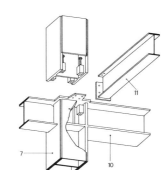

Structural system: cross-section

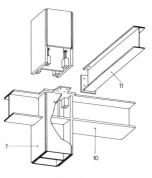

Detail of joint in transverse frame

Areas and volume

gross area	30 620 m²	area on plan	1250 m²
effective floor area	14 635 m²	volume	116 120 m³

Quantities of materials

	steel	concrete	reinforcement
total	1650 t	15 209 m³	1496 t
per m³ volume	14.2 kg	0.131 m³	12.9 kg
per m² gross area	54.0 kg	0.497 m³	48.8 kg

Construction cost (1969)

total cost	31.4 million D.fl.	per m² gross area	1025 D.fl.
per m³ volume	270 D.fl.	per m² eff. floor area	2145 D.fl.

References

Bouw, 2/1968, p. 42, and 23/1971, p. 904.
Bouwen met Staal, 2/1968, p. 2.

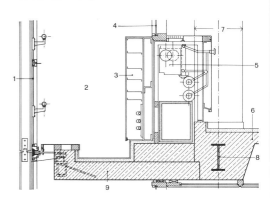

Vertical section of spandrel with outer corridor

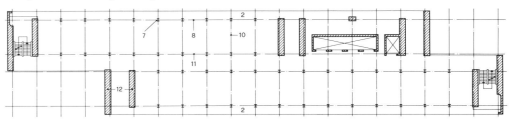

Arrangement of beams in a typical storey, scale 1:650

33 INDUSTRIAL PAVILION AT THE HANOVER FAIR, GERMANY

Architectural design: Hentrich, Petschnigg and Partner, Düsseldorf.
Structural design: Dölling, Mühlheim.
Built: 1970/1971.

Longitudinal side of the shorter block

Function

Two rectangular pavilion blocks displaced parallel to each other, comprising 690 m² of exhibition space in all. The basement contains an auditorium for 200 persons, canteen, kitchen, store room and technical services. The ground floor is free from internal supports and provides exhibition space, including 123 m² with two-storey headroom for displaying large objects. The upper floor contains a conference room and interview rooms.
Main stairs located between the two blocks. In addition, spiral staircase between ground floor and upper floor. Vertical services installed in hollow columns, one of which contains a service lift.

Upper floor, scale 1:800 1 Main staircase 2 Interview rooms 3 Conference room 4 Visitors' room 5 Lounge 6 Space above ground floor

6 Steel trough decking
7 Transverse beam IPE 450
8 Top and bottom chords of Vierendeel girder
9 Stainless steel cladding
10 Steel column 1800 mm dia
11 Upright of Vierendeel girder
12 Pin-joint

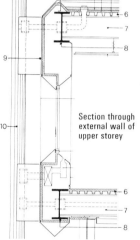
Section through external wall of upper storey

Dimensions

Overall dimensions of the two blocks 49.0 × 10.8 m and 35.0 × 10.8 m. Height above ground level 8.50 m. Lower storey height 4.57 m, ceiling height 3.82 m. Upper storey height 3.83 m, ceiling height 2.75 m.

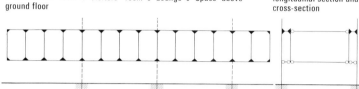
Structural system: longitudinal section and cross-section

Structural features

The structure is characterised by the arrangement of the upper storeys, which are suspended from three rows of tubular steel columns, so that they span over the ground floor and cantilever a distance of 7.0 m and 14.0 m, respectively, at the ends. The upper storey of each block comprises two Vierendeel girders integral with the longitudinal façades and interconnected by cross beams at roof and floor level. The girders are pin-jointed to brackets on the columns, which are spaced at 14 m centres in both directions. Horizontal lattice bracing in longitudinal direction at roof level; transverse bracing in the end faces of the upper storeys. Steel trough decking serves to stiffen the upper floors. Wind forces are transmitted by the eight external columns fixed in foundation pockets in the reinforced concrete basement structure.
The 1800 mm dia tubular columns are of welded construction, from 15 mm thick plate, stiffened with internal diaphragms. The Vierendeel girders, which are 4.24 m deep, comprise welded I-section chords, 620 mm deep, and HE 600 M uprights spaced at 3.50 m centres. Owing to transport restrictions the Vierendeel girders were brought to the site in pieces, field joints being made at the mid-point of the depth of the uprights. Transverse beams IPE 450, spaced at 1.75 centres, bolted connections. Imposed load on upper floor 275 kg/m².

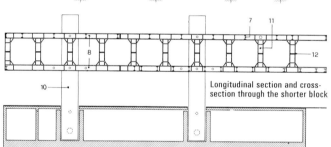
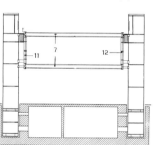
Longitudinal section and cross-section through the shorter block

• Façade: Ground floor glazed; Vierendeel girders in upper storeys have stainless steel cladding; opening windows, provided with double glazing.

Services

Air heating system operated with fresh air during exhibitions, at other times with recirculation of air, minimum temperature +5°C, air delivery rate 40 000 m³/h. Gas-fired boiler has capacity of 750 000 kcal/h. Ducts and flues are accommodated in the tubular steel columns, which are internally accessible.

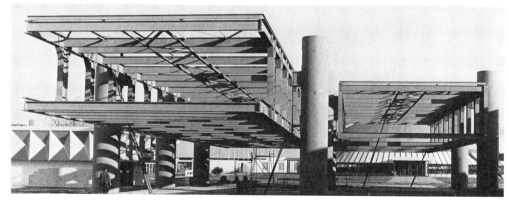
The steelwork for the upper storeys of the two blocks is suspended between cylindrical steel columns

References

Acier-Stahl-Steel, 1/1972, p. 1.
Bauen + Wohnen, 12/1971, p. 560.
Deutsche Bauzeitschrift, 4/1972, p. 613.
Detail, 4/1971, Konstruktionstafel.
Zentralblatt für Industriebau, 7/1971, p. 246.
Element + Fertigbau, 8/1971.

Areas and volume

gross area	3150 m²	area on plan	1050 m²
effective floor area	2250 m²	volume	12 500 m³

Quantities of materials	steel	concrete	reinforcement
total	400 t	1600 m³	150 t
per m³ volume	32.0 kg	0.128 m³	12.0 kg
per m² gross area	127.0 kg	0.508 m³	47.6 kg

34 GROUP OF EXHIBITION PAVILIONS IN TORONTO, CANADA

Architectural design: Craig, Zeidler, Strong, Toronto.
Structural design: Gordon Dowdell Associates, Scarborough.
Built: 1968/1971.

Pavilion with bridge linking it to the shore

Function

Exhibition and recreation grounds laid out on three artificial islands in Lake Ontario. In a bay enclosed by the islands is a group of five pavilions or pods, built on stilts in the water and so arranged that the bridges linking the pavilions to the islands and to one another are joined to the corners of these structures, which are square on plan.
One of the pavilions contains a restaurant and a hall for parties, receptions, etc. The four other pavilions contain 3000 m^2 of exhibition floor space. Visitors walk through the buildings at various levels on footbridges which intersect them in the two diagonal directions. External staircases are provided at the corners and ramps along the sides of the pavilions. The principal access to the pavilions from the mainland is gained by way of a two-storey bridge, the bottom storey of which is glazed. Service ducts and also film projectors for audio-visual display purposes are provided over the two diagonal footbridges inside the pavilions.

Dimensions

Overall dimensions of the pavilions 27.5 × 27.5 m; height 9.72 m. Headroom inside pavilions generally 7.30 m, but 2.30 m over the footbridges. The pavilions are erected at a height of 10.30 m above water level on 32 m high columns.

Structural features

At the centre of each pavilion are four 762 mm dia tubular columns arranged in a square with a side length of 3.55 m. At 20.0 m above water level, four pairs of lattice girders are connected to these columns. The lattice girders extend to the four corners of the square structure and their ends are tied back by inclined cable stays to the tops of the columns protruding above roof level. The top chords of the lattice girders are interconnected by a peripheral edge girder and by floor beams running longitudinally and transversely.
The lower-level floor beams, which are interconnected to form a grillage, are attached on the inside to the four tubular columns, while the outer ends are suspended by vertical hangers from the ends of the lattice girders and additionally supported by inclined cables in the façades. The intersections of the floor beams are supported by externally exposed trussed bracings.
The rectangular hollow section beams at top and bottom level carry steel trough decking with 13 cm in-situ concrete slab. Imposed load on floors 488 kg/m^2. Wind forces are transmitted through the floor slabs and through horizontal cross-bracings between the pairs of lattice girders to the four tubular columns, which are set in open caisson foundations and are interconnected by vertical bracings.

Aerial view of the exhibition grounds and pavilions

Diagrams showing the various functional levels

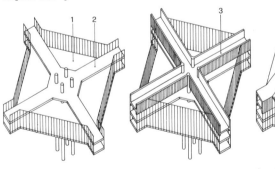
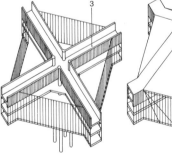

1 Exhibition area
2 Footbridge
3 Service bridge
4 Roof promenade

- Lattice girders: 15.50 m long, 3.64 m deep. Top chord comprises broad flange beam 254 × 254 mm, bottom chord RHS 254 × 153 mm, diagonals RHS 153 × 102 mm. Ducts accommodating various services are installed between the lattice girders, which are spaced 3.55 m apart. The footbridges inside the pavilions are suspended from the lattice girders.
- Façade: Single glazing fitted in aluminium frames; closed surfaces comprise 9 cm thick panels with metal sheet on the outside and thermal insulating board on the inside.

5 Exhibition pavilion
6 Restaurant
7 Access bridges
8 Cinema in spherical dome

Location plan, scale 1:8000

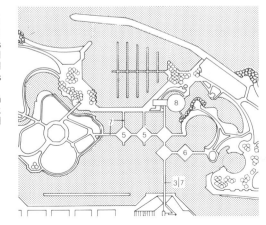

References

Architectural Forum, 7–8/1971, p. 31.
L'Architecture d'Aujourd'hui, 162/1972, p. 54.

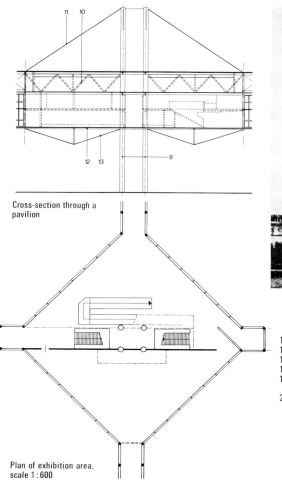

Cross-section through a pavilion

Plan of exhibition area, scale 1:600

Structural system: diagonal section

9 Tubular columns 762 mm dia
10 Pairs of lattice girders
11 Cable stays
12 Floor beams
13 Truss ties
14 Edge beam, RHS 305 × 203 mm
16 Floor beam I-section 254 × 254 mm
17 Inclined tube 102 mm dia
18 Aluminium frame
19 Square hollow section 254 × 254 mm
20 Mezzanine floor
21 I-section wind-bracing

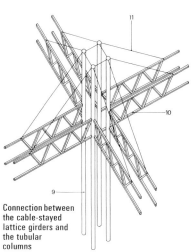

Connection between the cable-stayed lattice girders and the tubular columns

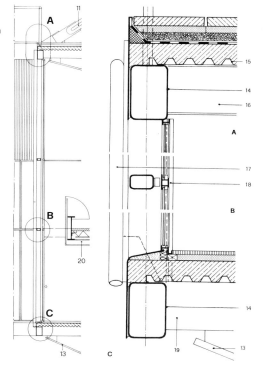

Vertical section through, and details of external wall

Areas and volume

gross area	11 400 m²	area on plan	6233 m²
exhibition area	3000 m²	volume	77 533 m³

Quantities of materials

	steel	concrete	reinforcement
total	1315 t	1220 m³	360 t
per m³ volume	16.9 kg	0.016 m³	4.6 kg
per m² gross area	115 kg	0.107 m³	31.6 kg

Construction cost

total cost	$6.1 million	per m² gross area	$533
per m³ volume	$78		

35 MOTORWAY RESTAURANT NEAR MONTEPULCIANO, ITALY

Architectural design: A. Bianchetti, Milan.
Structural design: Steelwork contractors.
Built: 1966.

Function

Two-storey elevated structure spanning a motorway. On the upper floor there is a restaurant for 120 people together with the kitchen, store and staff rooms. On the floor below are a supermarket, store rooms, cold storage facilities and cloakrooms.
On both sides of the motorway there is access through a core comprising two lifts and two staircases. Adjacent to one of the cores is a low building in which the heating and air-conditioning plant is housed.

Dimensions

Overall dimensions of the superstructure: 79.50 × 15.00 m. Height above ground level 16.15 m. Clearance over motorway 5.50 m. Storey height – ceiling height: 4.38 – 3.80 m on first floor; 4.45 – 2.75 m on second floor of the superstructure.

Structural features

The striking appearance of this elevated restaurant is largely due to the two single-leg rigid frames which support the building and which are in weathering steel. The main girder of each of these frames cantilevers 17.50 m rearwards from the tapered leg, while in the other direction it extends across the width of the motorway, rests on the concrete core on the opposite side and cantilevers 14.00 m beyond it. At roof level the girders are interconnected at intervals of 4.00 m by cross girders from which the two-storey superstructure is suspended. The depth of main girder is 3800 mm over the leg, decreasing to 1700 mm at each end; width of flange 900 mm. Each frame is of welded plate construction, the webs being stiffened by welded plates at each cross girder connection, while the leg is pin-jointed at its base to a concrete footing. The cross girders are normally plate girders with a maximum depth of 1410 mm at mid-span, but the cross girder over the concrete core on which the longer arms of the main girders of the frames are supported is of box-section.
The transverse beams in the upper floor of the superstructure are 400 mm deep rolled steel sections which are suspended from each cross girder by two inner hangers HE 120 A and two outer hangers HE 160 B, the latter being in the two longitudinal façades. The transverse members for the floor below are lattice girders, 13.36 m long and 1300 mm deep; these are supported by two outer hangers HE 160 B only, located in the longitudinal façades. Both floors have longitudinal secondary beams IPE 220, spaced at 2.60 m centres, carrying steel trough decking laid transversely and filled with in-situ concrete. Imposed load 550 kg/m². The roof and floors are stiffened with horizontal cross-bracings. Wind forces are transmitted to the two concrete cores.

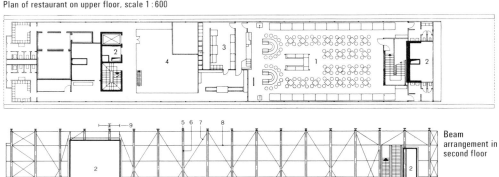

Plan of restaurant on upper floor, scale 1:600

Beam arrangement in second floor

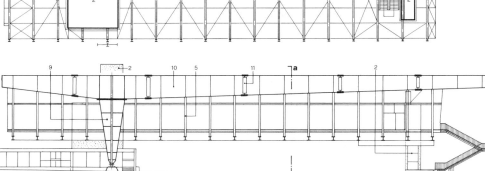

Longitudinal elevation

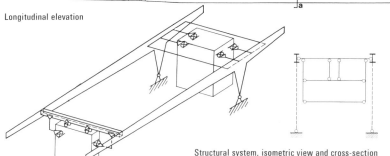

Structural system, isometric view and cross-section   Cross-section

Areas and volume

gross area	2240 m²	area on plan	950 m²
effective floor area	1500 m²	volume	11 260 m³

Quantities of steel

total	206 t	per m² gross area	92.9 kg
per m³ volume	18.5 kg		

Construction cost (1966)

total cost	650 million lire	per m² gross area	290 000 lire
per m³ volume	57 800 lire	per m² eff. floor area	433 000 lire

References

Costruzioni Metalliche, 4/1967.
Acier-Stahl-Steel, 4/1971.

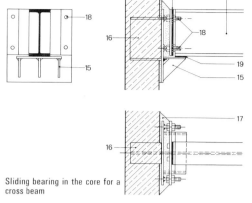

Sliding bearing in the core for a cross beam

1 Restaurant
2 Vertical circulation core
3 Kitchen
4 Store room
5 Hanger HE 160 B
6 Cross-beam 420 mm deep
7 Wind-bracing
8 Longitudinal beam IPE 220
9 Leg of frame
10 Main girder
11 Erection splice
12 Plate girder
13 Hanger HE 120 A
14 Lattice girder
15 Bearing bracket
16 Embedded I-section
17 Mortar packing
18 Adjusting bolts
19 Teflon sliding surface
20 Stiffener
21 Edge beam HE 220 A
22 Top chord $\frac{1}{2}$ IPE 500
23 Bottom chord $\frac{1}{2}$ IPE 400
24 Diagonal and vertical members composed of two angles L $70 \times 70 \times 6$

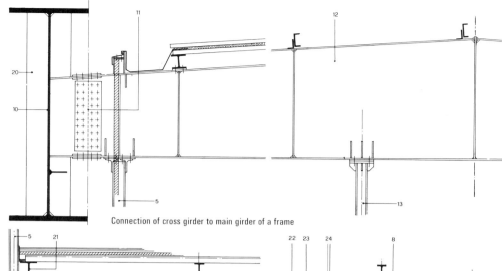

Connection of cross girder to main girder of a frame

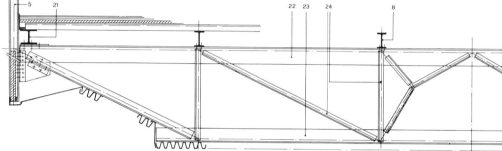

Lattice girder supporting the lower floor, with hanger connection

Services under the floor over the supermarket

Rigid frames in course of erection

Hinged bearing at foot of rigid frame

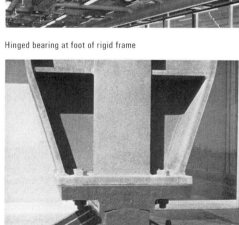

Erection of the suspended steelwork

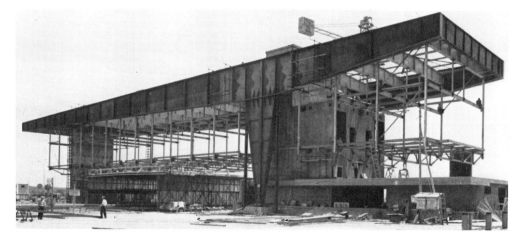

109

36 MULTI-STOREY CAR PARK IN BREMEN, GERMANY

Architectural design: Building Department, Kaffee HAG, Bremen.
Structural design: Steelwork contractors.
Built: 1970/1971.

Function

Multi-storey car park with 748 parking bays on seven floors for use by the staff of an industrial firm. On the ground floor: parking bays with their own entrance and exit, for use by managerial staff. At semi-basement level, with ramped access, is a car-washing plant, repair shop and service station.
On the six upper floors the parking bays are positioned at right angles to two access roads. Vehicles entering the park use a semi-helical ramp built at one end of the building; a second semi-helical ramp, protruding from one of the longitudinal faces, is used by vehicles leaving. Staircases are accommodated in two cores, one of which additionally contains two lifts.

Dimensions

Overall dimensions 78.06 × 32.77 m. Height above ground level 20.80 m. External radius of the semi-helical ramps 11.60 m, radius of lane centre-line 9.05 m, lane width 3.00 m, gradient 9%.
Storey height – headroom: 3.80 – 3.07 m and 2.65 – 2.10 m on ground floor; 2.70 – 2.10 m and 2.65 – 2.10 m on first floor; 2.65 – 2.10 m on second to fifth floors; 2.95 – 2.23 m and 2.82 – 2.10 m on sixth floor.

Structural features

Layout comprises, on each upper floor, two 6.00 m wide access roads flanked on each side by 5.00 m deep rows of parking bays. Correspondingly to this layout the structural system has been designed as a pin-jointed framework comprising a row of columns in each of the two longitudinal façades, a row of columns extending along the longitudinal centre-line, and floor beams spanning 16.00 m from the central to the external columns. Columns are spaced at 2.50 m centres longitudinally, corresponding to the width of a parking bay. External columns consist of HE 180 A and B, internal columns of HE 300 A and B, floor beams of IPE 450 (IPE 500 over the car-washing plant, because of additional loads due to the suspended ceiling and ventilation plant). Floors are of composite construction comprising 10 cm thick precast concrete slabs (2.50 × 8.07 m) spanning between the floor beams and interconnecting the structural steelwork in the longitudinal direction. Composite action is ensured by stud shear connectors and in-situ concrete in the joints. Steel beams were given a camber – by propping them up at mid-span – prior to concreting, so that composite action was achieved for dead as well as live loads. Imposed load on floors 350 kg/m². Wind forces are transmitted through floors to vertical diagonal bracing in the end and longitudinal walls.
Semi-helical ramps are structurally separate from the main building. The ramp structure comprises inner and outer columns, radially placed floor beams and an in-situ concrete deck slab. Horizontal forces due to wind and traffic are transmitted by three groups of bracing erected between the outer columns.
• Façade: Spandrel panels of 8 mm thick asbestos cement, 1.20 m high. Above these are horizontal louvre-type slats. The end walls and the two semi-helical ramps have cladding consisting of plastic-coated sheet-steel troughing units.
• Fire protection: Car-washing plant is separated from rest of building by fire walls. Floor above it is provided with a suspended ceiling and thus conforms to fire resistance class F 90 (90 min). The car park areas have no structural fire protection. Adjacent buildings are protected by a fire wall at one end of the car park. Riser mains for fire-fighting in the two staircase cores.
• Corrosion protection of the structural steelwork: sandblasting, two priming coats of zinc chromate paint, two plastic-based finishing coats.

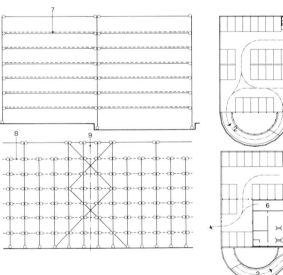

Structural system: cross-section and part longitudinal section

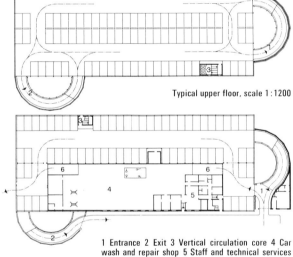

Typical upper floor, scale 1:1200

Ground floor

1 Entrance 2 Exit 3 Vertical circulation core 4 Car wash and repair shop 5 Staff and technical services 6 Ramp

Areas and volume

gross area	20 880 m²	area on plan	2985 m²
effective floor area	17 834 m²	volume	57 160 m³

Quantities of materials

	steel	concrete	reinforcement
total	780 t	870 m³	197 t
per m³ volume	13.6 kg	0.015 m³	3.5 kg
per m² gross area	37.4 kg	0.042 m³	9.5 kg

Construction cost (1971)

total cost	5.2 million DM	per m² gross area	249 DM
per m³ volume	91 DM	per m² eff. floor area	292 DM

Reference

Merkblatt Stahl, 211/1972, p. 44.

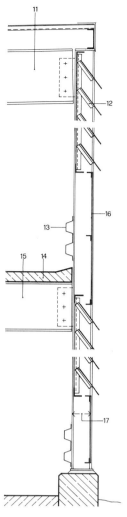

Vertical section through external wall

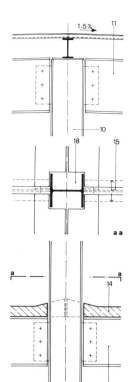

Internal column with beam connections

7 Composite floor
8 Concrete slab
9 Centre-line of building
10 Internal column HE 300 A and B
11 Roof beam IPE 500
12 Louvre wall with asbestos cement slats

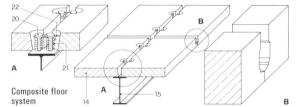

Composite floor system

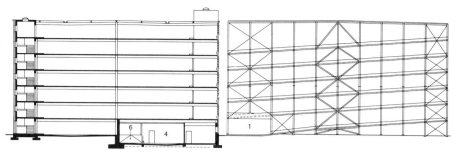

Cross-section through building

External wall of semi-helical entrance ramp (developed elevation)

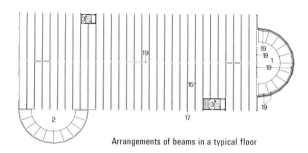

Arrangements of beams in a typical floor

13 Guard rails
14 Precast concrete slab 10 cm thick
15 Floor beam IPE 450
16 Asbestos cement spandrel panel
17 External column HE 180 A and B
18 Recess in floor slabs
19 Wind-bracing
20 Shear stud
21 Helical reinforcement
22 Reinforcement loop

Car park before erection of external cladding

Parking spaces as seen from the semi-helical entrance ramp

Interior view of the semi-helical exit ramp

37 TERMINAL BUILDING AT ORLY AIRPORT, FRANCE

Architectural design: M. Meyer, Paris.
Structural design: Kuhn, Saarbrücken.
Built: 1968/1969.

Function

Airport passenger terminal for dealing with eight long-range aircraft simultaneously at two pierheads, these being situated on each side of a main building to which they are connected by two-storey link buildings. The latter each terminate at one corner of the pierhead structure which is square on plan and of which the three other corners have short projecting structures from which access to and from the aircraft is provided by telescopic gangways. Other gangways are connected to the four sides of the pierhead structure.

The pierheads are of three-storey construction plus a basement under the whole area of the building. There are baggage handling facilities and customs on the ground floor; passenger circulation facilities, booking offices and waiting rooms are on the first floor. These two storeys are connected to the main terminal building by two walkways for passengers and a belt conveyor for baggage. The basement, which accommodates technical equipment and store rooms, is linked to the main building by an underground passage. The recessed top storey contains technical equipment, the public having access to the roof terrace around it.

Dimensions

The square pierhead buildings measure 45.20 × 45.20 m overall. Height above ground level 12.30 m. The projecting structures at the corners are trapezoidal on plan, 11.00 m long and 6.00 m wide at the end. The link buildings are 17.40 m wide. Ground floor and first upper storey: storey height 4.40 m, headroom 2.95 m. Projecting edge of roof 3.70 m wide, extending all round the building.

Structural features

The reinforced concrete basement is surmounted by a steel frame comprising two-storey high columns which are rigidly connected in both directions by floor beams. All wind forces are transmitted by portal frame action.

Columns are spaced on 8.28 m square grid and consist of HE 300 A sections, some parts of which are strengthened with welded plates. All columns are fixed at the base. The HE 650 A main floor girders are rigidly connected to the columns by means of end plates, structural continuity of the four girders meeting at a joint being further ensured by 25 mm thick gusset plates bolted to their top and bottom flanges, the columns extending continuously through holes formed in the gusset plates. Girders on the first floor cantilever 1.90 m, those of the second floor 6.40 m, beyond the external columns; the ends of the girders are interconnected by edge beams.

In one direction, secondary IPE 600 girders are connected midway between the main girders. In the other direction, IPE 160 floor beams at 1.04 m centres carry a floor consisting of 3 cm thick precast concrete slabs and 4 cm thick in-situ concrete topping. Imposed load 500 kg/m².

• Fire protection: A sprinkler system is installed in the cavity between the structural floor and the suspended ceiling which comprises perforated sheet-aluminium profiled units. If a fire breaks out, water issues from the perforations in the ceiling.

• Foundations: The subsoil is fine to medium sand, the water table being below foundation level. Pad footings under columns, strip footings under basement walls; bearing pressure on soil 2.5 kg/cm².

Services

Hot water heating supplied by central heating plant for the airport. Fan convectors installed in the ceilings. Air-conditioning plant installed in top storey.

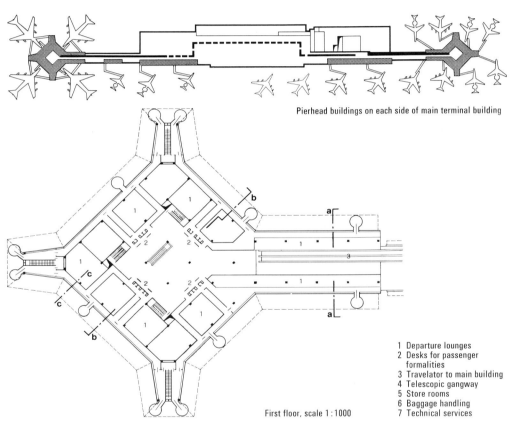

Pierhead buildings on each side of main terminal building

1 Departure lounges
2 Desks for passenger formalities
3 Travelator to main building
4 Telescopic gangway
5 Store rooms
6 Baggage handling
7 Technical services

First floor, scale 1:1000

Areas and volume

gross area	10 300 m²	area on plan	3600 m²
		volume	41 500 m³

Quantities of materials	steel	concrete	reinforcement
total	850 t	11 100 m³	210 t
per m³ volume	20.5 kg	0.270 m³	5.1 kg
per m² gross area	82.5 kg	1.080 m³	20.4 kg

8 Floor beam IPE 160
9 Precast concrete slab 3 cm thick
10 In-situ concrete 4 cm thick
11 Main girder HE 650 A
12 Column HE 300 A
13 Gusset plate
14 Secondary girder IPE 600
15 Roof beam I 300
16 Hole in girder
17 Cantilever
18 Diaphragm

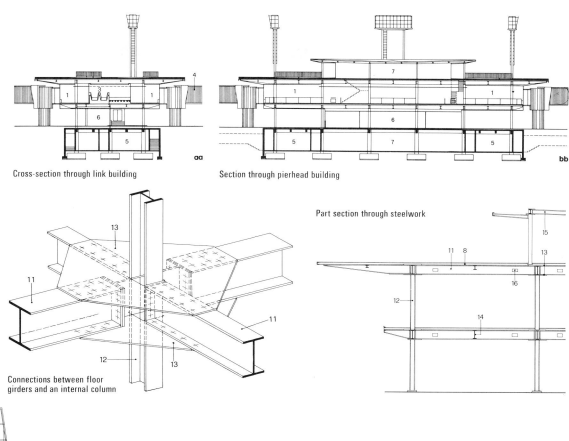

Cross-section through link building

Section through pierhead building

Part section through steelwork

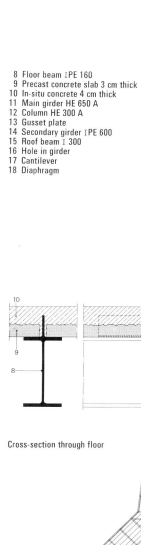

Cross-section through floor

Connections between floor girders and an internal column

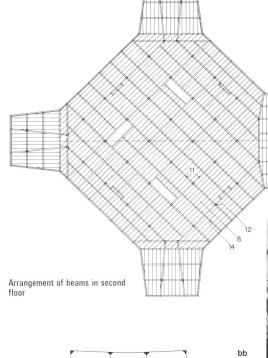

Arrangement of beams in second floor

Details of overhang of roof

Structural system

Structural steelwork with floors partly concreted

113

38 SUPERMARKET AT INTERLAKEN, SWITZERLAND

Architectural design: F. Wyler, Bern.
Structural design: C. Walder and H. von Gunten, Bern.
Built: 1964/1965.

Function

Supermarket with 1350 m² sales area and a restaurant for 120 people accommodated in a two-storey building. Store rooms, cold storage, technical services and cloakrooms are in the basement, which is designed to serve also as an air-raid shelter. The self-service shop on the ground floor has an arcaded entrance from the street; at the rear of the building is the goods reception bay with a platform lift whereby vehicles weighing up to 25 t can be lowered to basement level. The recessed upper storey comprises a restaurant and kitchen. On the roof terrace in front of the restaurant there is room for a further 250 people in the open air.
Vertical circulation facilities comprise two staircases and three goods lifts in the interior of the building. There is also an escalator and one outside staircase leading up from the arcade.

Dimensions

Rectangular on plan with recessed corners, overall dimensions 55.80 × 35.70 m. Height above ground level 11.50 m. Planning grid 5.85 × 5.85 m. The upper storey, set back from the perimeter, is square on plan, 23.40 × 23.40 m. The edge of the roof projects 5.55 m on all sides. Storey height – ceiling height: 5.50 – 3.80 m on ground floor; 4.50 – 2.30 to 4.10 m on upper floor.

Structural features

The reinforced concrete basement (mushroom floor slab and columns with simple bases) is surmounted by a composite framework. A reinforced concrete floor is supported on 28 external columns HE 220 M and nine internal columns HE 300 M (column spacings 5.85 and 11.70 m). This floor is rigidly connected to the internal columns by means of anchor bars and capping plates and is supported along its edges by brackets attached to the external columns. The floor slab is enclosed by an edge beam channel section UNP 400.
Wind forces are resisted by portal frame action and are transmitted through the framework and the floor slab to the concrete core containing one of the staircases. The striking roof structure for the restaurant on the upper floor comprises square mushroom-shaped units, each of which consists of a downward-tapering column carrying a cap in the shape of an inverted pyramid. Each of these mushroom columns consists of four HE 180 B sections joined together cruciform-fashion, the inner flanges being welded to one another to form a box section in which the rainwater downpipe from the roof is fitted. Each mushroom cap comprises four hip rafters fabricated from IPE 400 sections and rigidly bolted to the outer flange of the four members forming the column. The mushroom hip rafters, which taper towards their ends, are interconnected by purlins and edge beams. The purlins and edge beams are stressed, not only in bending by the roof loading, but also in tension, in consequence of the geometric deformation of the mushroom head. The purlins carry lightweight concrete slabs with cork insulation, the soffit being clad with aluminium panels.

View of the front arcade from the street

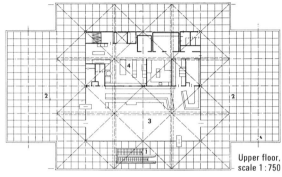

Upper floor, scale 1:750

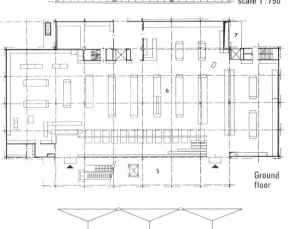

Ground floor

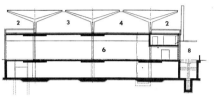

Cross-section through building

1 Escalator
2 Cafe terrace
3 Restaurant
4 Kitchen
5 Arcade
6 Self-service shop
7 Goods reception bay
8 Vehicle lift

Plan of mushroom floor structure

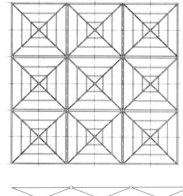

Structural system: longitudinal section and cross-section

At their free ends, hip rafters of adjacent caps are pin-jointed to each other, thus forming three-pinned portal frames extending diagonally in relation to the plan of the building and serving to transmit wind forces to the floor slab.
• Foundation: 40 cm thick reinforced concrete foundation slab. As the water table is 2.50 m above the soffit of the basement floor, the basement is constructed as a watertight concrete tank. Subsoil is gravel. Calculated bearing pressure 2.0 kg/cm².

Areas and volume

gross area	5204 m²	area on plan	1756 m²
effective floor area	3683 m²	volume	22 702 m³

Quantities of materials

	steel	concrete	reinforcement
total	122 t	3014 m³	370 t
per m³ volume	5.4 kg	0.133 m³	16.3 kg
per m² gross area	23.5 kg	0.579 m³	71.1 kg

Construction cost (1965)

total cost	4 795 000 Sw.fr.	per m² gross area	921 Sw.fr.
per m³ volume	211 Sw.fr.	per m² eff. floor area	1302 Sw.fr.

References

Deutsche Bauzeitung, 12/1967, p. 966.
Werk, 4/1968, p. 240.
Detail, 2/1970, p. 268.

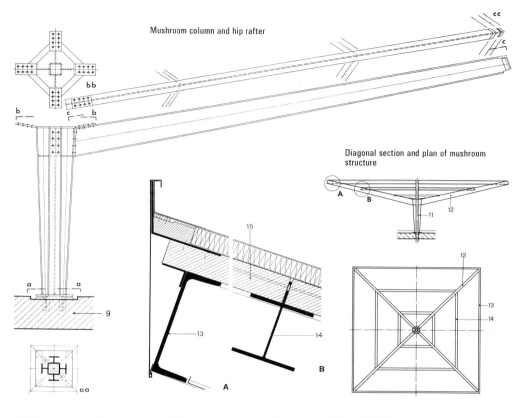

Mushroom column and hip rafter

Diagonal section and plan of mushroom structure

9 Reinforced concrete floor slab
10 Steel column
11 Mushroom column composed of four HE 180 B
12 Tapering mushroom hip rafter, obliquely cut and welded IPE 400
13 Edge beam U 300
14 Purlin HE 180 A
15 Lightweight concrete slab
16 External column HE 240 A
17 Bracket
18 Internal column HE 300 M

Photographs: Mushroom column in upper storey
Corner column and parapet for terrace
Steelwork for a mushroom roof
Erection of a mushroom hip rafter

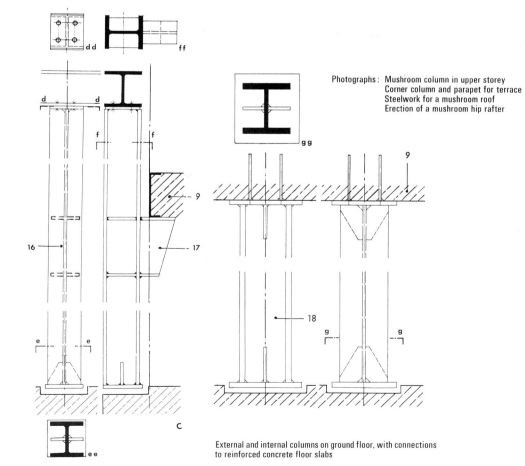

External and internal columns on ground floor, with connections to reinforced concrete floor slabs

115

39 DEPARTMENT STORE IN STOCKHOLM, SWEDEN

Architectural design: S. Backström, L. Reinius, Stockholm.
Structural design: Jacobson and Widmark AB, Stockholm.
Built: 1962/1964.

Function

Department store with five storeys above ground and four in the basement, 17 000 m² sales area, 1000 employees. Parking accommodation for 350 cars is provided on the two lowest basement levels. Food department and storage facilities occupy the next floor of the basement, and on the top basement floor are goods reception, customers' accounts department and sales counters. The ground floor and the first three upper floors are also occupied by sales counters. On the fourth floor are the administrative offices and a staff restaurant. Spiral ramps give access to the underground car park and goods reception. Sales floors are interconnected by escalators, stairs and lifts. There are entrances to the building on all four sides, the main entrance being combined with an entrance to the underground railway station.

Dimensions

Approximately rectangular on plan, overall dimensions 133 × 40 m. Height above ground level 25 m. Storey heights: 2.90 m in underground car park, 3.95 m in two upper basement storeys, 5.50 m on ground floor, 4.70 m in upper storeys; ceiling height 2.55 m in car park, 3.45 m in upper storeys.

Structural features

Steel columns consist of 500 × 500 mm I-sections, fixed at the base, spaced at 10.65 m centres in the longitudinal direction of the building, and 8.80 to 14.60 m (depending on the internal layout of individual floors) in the transverse direction. The two lowest basement storeys have in-situ concrete mushroom floors. On every transverse grid line in the floors above is a pair of continuous main girders, comprising 640 × 300 mm I-sections spaced 820 mm apart, which are supported on brackets on each side of the columns and are simply supported by them. Precast prestressed concrete floor slabs span longitudinally between these girders. Wind forces are transmitted in both directions by vertical bracings composed of I-section members. Erection of the building proceeded simultaneously in both the vertical and horizontal directions, erection and finishing operations on the upper storeys followed those on the storeys below as construction proceeded in the longitudinal direction of the building.

• Floors: Precast prestressed concrete T-beam slabs, 9.50 × 1.50 m, web depth 30 cm, slab thickness between webs 4 cm. The concrete floor units rest on reinforced concrete seating units which in turn are carried by the bottom flanges of the main girders and are fixed to the webs of the latter by means of welded anchor bars. Continuous mesh reinforcement mats were laid over the precast slabs and embedded in 8 cm in-situ concrete topping. Upper floors designed for imposed load of 400 kg/m².

• Façade: The perimeter floor beams carry storey-height reinforced lightweight concrete panels provided with rock-wool thermal insulation and external brick facing. Provision has

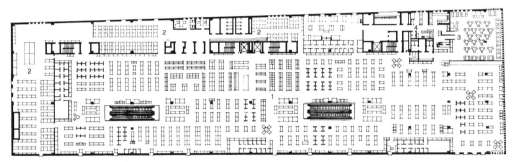

First floor, scale 1:1000

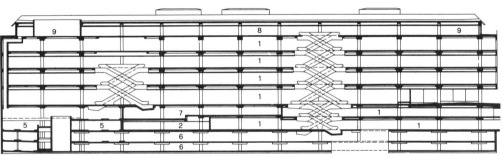

Longitudinal section through building

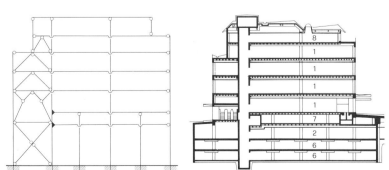

Structural system, cross-section   Cross-section through building

1 Sales area
2 Store rooms
3 Cafeteria
4 Kitchen
5 Two-lane helical ramp
6 Underground car park
7 Delivery of goods
8 Staff restaurant
9 Services

been made for subsequent installation of windows.
- Foundations: Fissured rock with groundwater at 6.40 m depth. The excavations for the building were taken down to a depth of 13.40 m, but the rock below was stabilised by cementation to a further depth of 5 m. External walls sealed with tanking of plastic sheet.

Services

Full air-conditioning for shopping floors, restaurant and administrative offices: six air-changes per hour, air delivery rate 508 000 m³/h, cooling capacity 3 325 000 kcal/h. The other parts of the building are served by three separate ventilation systems.

Areas and volume

gross area	49 000 m²	area on plan	5040 m²
selling area	17 000 m²	volume	200 800 m³

Quantities of materials

	steel	concrete	reinforcement
total	3750 t	16 600 m³	1250 t
per m³ volume	18.7 kg	0.083 m³	6.2 kg
per m² gross area	76.6 kg	0.339 m³	25.5 kg

Construction cost (1964)

total cost	57 million S.kr.	per m² gross area	1162 S.kr.
per m³ volume	284 S.kr.	per m² selling area	3350 S.kr.

References

Acier-Stahl-Steel, 9/1969, p. 358.
Arkitektur, 11/1964.

Main beams across the building with precast concrete seating units to support the floor slabs

Laying the precast prestressed concrete floor slabs

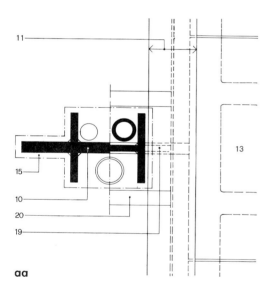
aa
Horizontal section through a column

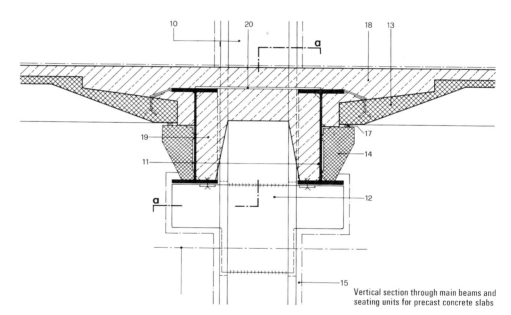
Vertical section through main beams and seating units for precast concrete slabs

Construction proceeding both in horizontal and vertical directions

10 Column, 500 × 500 mm I-section
11 Main beams, pair of 640 × 300 mm I-sections
12 Welded bracket plate
13 Precast prestressed concrete slab
14 Seating unit
15 Fire encasement
16 Suspended ceiling
17 Neoprene strip
18 In-situ concrete filling
19 Bearing stiffener
20 Batten plate

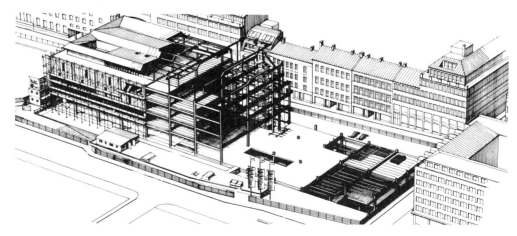

40 SHOPPING CENTRE IN BERLIN-STEGLITZ, GERMANY

Architectural design: G. Heinrichs, S. Geiger, F. Bartels, C. Schmitt-Ott, Berlin.
Structural design: Steelwork contractors and S. Polony, Berlin.
Built: 1968/1970.

Function

Shopping centre for 200 individual traders, together with restaurants, entertainment and amusement facilities and various customer services. The main entrance is located on a busy urban square. At the rear of the shopping centre, which is accommodated in a building with six storeys above ground and two basement floors, is a seven-storey car park with accommodation for 400 cars. The upper basement floor of the main building contains a bazaar-like complex with shops, a teenagers' café, a supermarket (1600 m² floor area), services and store rooms. In the lower basement are further store rooms and rented parking accommodation. On the ground and three upper floors the shops, comprising a total effective area of 13 000 m², are grouped around two covered light-wells surrounded by galleries which are reached by escalators installed between the light-wells. In addition, on the ground floor, at the rear, there is a weekly market which formerly used to be held in the open air on this site. A continuous open concourse on the fourth floor is partly covered by the fifth floor which comprises entertainment facilities, offices and staff rooms. Access to the various floors in the shopping centre is gained by means of staircases on the ground floor and the centrally located escalator system. Furthermore, a staircase and one or two lifts are provided in each of the eight cores spaced along the two longitudinal faces of the building.

Dimensions

Overall dimensions 162.0 × 54.0 m. Height above ground level 24.0 m. Length of commercial part 119.4 m. Storey height – ceiling height: 4.29–3.33 m on ground floor; 3.96–3.00 m on the other floors.

Structural features

The multi-storey car park has in-situ concrete flat-slab floors. The main building, containing the shopping centre, has pin-jointed steel framework with main beams running transversely (6.60 m spans), composite beams in longitudinal direction (11.0 m spans) and columns ranging from HE 500 M to HE 140 B. The main beams, which cantilever up to 5.00 m beyond the columns at the internal light-wells, comprise HE 360 M and HE 400 B.
To facilitate the installation of services, lattice girders were chosen as the longitudinal members, spaced at 2.20 m centres and spanning between the transverse main beams. The bottom chords of the lattice girders consist of tees split from HE 180 B sections; the web members consist of pairs of angle sections with horizontal gusset plates welded to the upper ends. These gusset plates serve as bearings for 5.50 × 2.20 m precast concrete slabs, 10 to 16 cm thick, and also provide for shear connection. Beneath the slabs there are plates with stirrups welded to them. These plates under the slabs are secured to the gusset plates on the lattice girders by means of high-strength friction-grip bolts, thus ensuring composite action, with the concrete slabs acting as the compression flange for the composite girder. Floors designed for imposed loads of 500 kg/m² and 1000 kg/m². In order to ensure composite action between the steel and concrete in carrying the dead weight, the girders were given a camber (with the aid of props, subsequently removed) before the joints between the slabs were concreted and before the bolts were tightened. Wind forces are transmitted through the floors to the eight concrete cores spaced along the two longitudinal faces.

- The curtain walling, which comprises aluminium spandrel panels at floor level and storey-height glazing, is fitted between mullions at 2.20 m centres. Sun protection is provided by external louvre blinds.
- Foundations: Subsoil is compact sand, water table below base of foundation, which is 8.5 m below ground level. Pad footings under the columns; vertical circulation cores are supported on reinforced concrete foundation slabs.

References

Deutsche Bauzeitung, 10/1970, p. 836.
Deutsche Bauzeitschrift, 3/1971, p. 355.
Detail, 5/1971, Konstruktionstafel.
Bauen + Wohnen, 8/1971, p. 351.
Acier-Stahl-Steel, 1/1972, p. 30.

End façade with main entrance

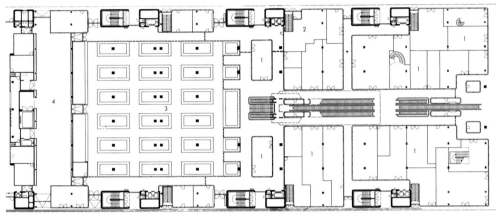
Part plan of ground floor, scale 1 : 1000 (adjacent car park not shown)
1 Shops 2 Restaurant 3 Weekly market 4 Incoming goods

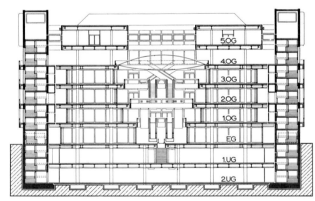
Cross-section through the complex in the light-well region

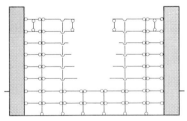
Structural system: cross-section

Erection of lattice girders and floor slabs

Erection of structural framework, proceeding in longitudinal direction

Areas and volume

gross area	68 000 m²	car parking area	10 000 m²
effective floor area	39 000 m²	area on plan	9400 m²
sales area	21 500 m²	volume	276 000 m³

Quantities of materials

	steel	concrete	reinforcement
total	2900 t	9150 m³	600 t
per m³ volume	14.0 kg	0.044 m³	2.9 kg
per m² gross area	51.5 kg	0.163 m³	10.7 kg

Construction cost (1970)

total cost	60 million DM	per m² gross area	1065 DM
per m³ volume	288 DM	per m² eff. floor area	2069 DM

5 Cruciform column, 3 I-sections
6 Main beam HE 340 B
7 Main Beam HE 600 B
8 Lattice girder
9 Precast concrete slab
10 Vertical circulation core
11 Temporary bracing
12 Temporary props
13 In-situ concrete slab
14 Floor beam I 300
15 Bottom chord ½ HE 180 B
16 Diagonal, 2 L 50 × 50
17 Welded T-section cleat
18 Gusset plate
19 Steel stirrup
20 HSFG bolt
21 Edge beam IPE 450
22 Sliding bearing for attachment of cladding
23 Aluminium panel
24 Asbestos panel as thermal insulation

Beam arrangement at ground level

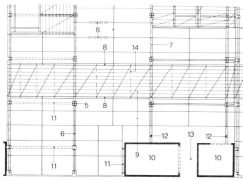

Part cross-section of structural steelwork

Elevation of lattice girder

Shear connection between lattice girders and floor slabs: cross-section and plan

Vertical section through external wall at fifth floor

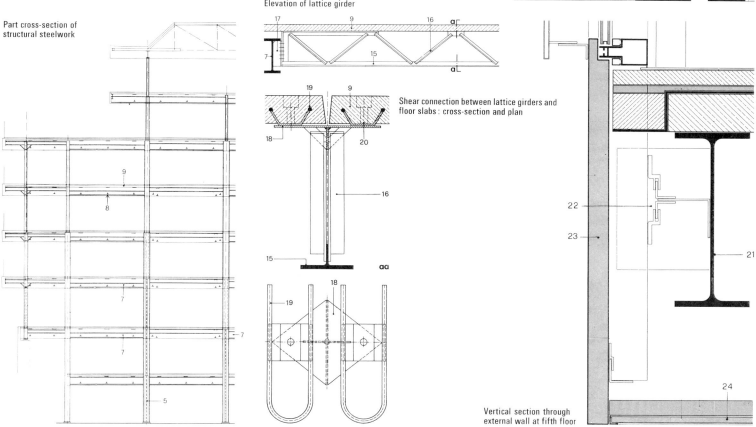

41 LIBRARY IN PANTIN, FRANCE

Architectural design: J. Perrottet, J. Kalisz.
Structural design: M. Kostanjevac.
Built: 1970/1972.

Function

This public library is accommodated in a two-storey building which is structurally and functionally subdivided into five equal units. The central unit comprises the entrance hall and caretaker's flat on the ground floor, with administrative offices on the upper floor. The other four units are connected in pairs to two opposite sides of the central unit. On the ground floor they contain the lending library for adults, discothèque, store, workshop and cloakrooms, while on the upper floor there are the lending library for children, reading room and multi-purpose room.
Under the central unit is a basement with central heating plant and ancillary rooms. Underground passages with services link the basement to the other units. The main staircase is in the entrance hall; there is also a staircase for staff use, while two spiral staircases are located on the outside of the building.

Dimensions

Overall dimensions 38.90 × 26.30 m. Height above ground level 8.20 m. Each unit measures 12.60 × 12.60 m on plan. The upper storey of each unit has bay windows on the three detached sides. Storey height and ceiling height 3.15 and 2.60 m, respectively, on ground floor. The upper storey has 2.70 m headroom on the outside increasing to 4.25 m on the inside.

Structural features

Each of the five units composing the building is supported by four intersecting rigid portal frames. The frames of adjacent units comprise columns shared in common, while on the free sides the columns are located at the quarter points of these sides and stand 60 cm in front of the façade. The portal frames span 12.60 m and are 6.30 m apart: their beams consist of IPE 300, their columns are HE 200 B which are pin-pointed at the base.
In one direction the upper floor comprises IPE 300 beams pin-jointed to the portal frame columns and supported at the centre by a simple HE 140 A column. In the other direction IPE 300 beams span only between these floor beams and the portal frame columns. The edge beams are IPE 300.
These beams support a reinforced concrete slab, 18 cm thick, enclosed at its outer edge within a U 230 channel section. Imposed load 400 kg/m². Composite action with the steel beams is ensured by shear connectors in the region of the supports.
Each unit has a roof in the form of a 1.75 m high pyramid set diagonally so that the inclined surfaces are parabolic hyperboloids. A cold-formed edge beam is connected through brackets to the columns of the portal frames and is, at the centre of each face of the building, tied back to the intersection points of the upper portal frame beams. The four hip rafters, each composed of two U 200 channel sections, converge at the apex of the roof, their lower ends being supported on the edge beams. The horizontal thrust exerted by the hip rafters is resisted by 70 mm dia tubular ties which interconnect the lower ends of the members. The edge beams and hip rafters support timber beams which in turn carry rafters and battens.
• Foundation: Open foundation on firm marl at depth of 8.0 m.

Reference

L'Architecture d'Aujourd'hui, 162/1972, p. V.

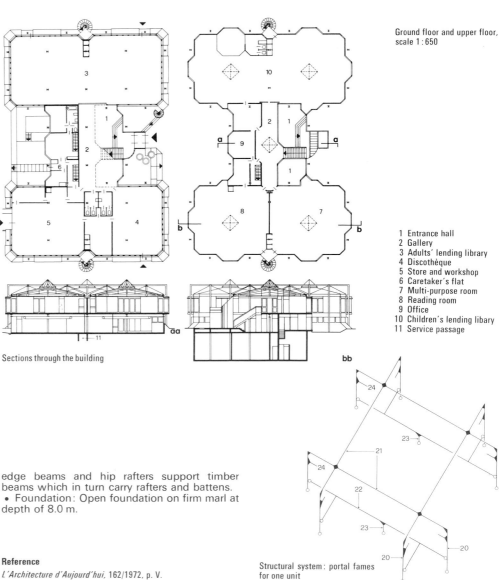

Ground floor and upper floor, scale 1:650

Sections through the building

1 Entrance hall
2 Gallery
3 Adults' lending library
4 Discothèque
5 Store and workshop
6 Caretaker's flat
7 Multi-purpose room
8 Reading room
9 Office
10 Children's lending libary
11 Service passage

Structural system: portal fames for one unit

Details of roof steelwork with rooflight surround

External portal frames with roof framework

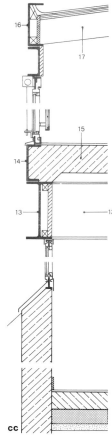

Section and plan of roof steelwork, detail of suspension system

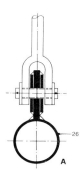
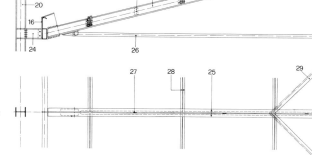

Areas and volume

gross area	1700 m²	area on plan	1000 m²
effective floor area	1180 m²	volume	5500 m³

Quantities of materials

	steel	concrete	reinforcement
total	75 t	800 m³	35 t
per m³ volume	13.6 kg	0.145 m³	6.4 kg
per m² gross area	44.1 kg	0.471 m³	20.6 kg

Construction cost (1971)

total cost	2.21 million Fr.fr.	per m² gross area	1300 Fr.fr.
per m³ volume	402 Fr.fr.	per m² eff. floor area	1873 Fr.fr.

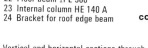

25 Hip rafter, two U 200
26 Tubular tie, 70 mm dia
27 Filler plate
28 Rafter
29 Rooflight surround, cold formed section
30 Rooflight
31 Hanger, 34 mm dia

12 Floor beam IPE 300
13 Edge beam IPE 300
14 Surround to floor slab U 230
15 In-situ concrete floor
16 Edge beam of roof structure, cold-framed section
17 Timber purlin
18 Sandwich panels
19 Mullion U 100 × 50
20 Portal frame column HE 200 B
21 Portal frame beam IPE 300
22 Floor beam IPE 300
23 Internal column HE 140 A
24 Bracket for roof edge beam

Vertical and horizontal sections through the external wall at the corner of the building

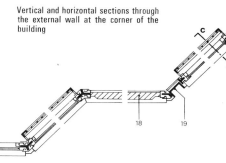

Stub connection between portal frame beam and column

Corner in the top storey of the building

42 LIBRARY IN PARIS, FRANCE

Architectural design: M. Lods, P. Depondt, H. Beauclair, Paris; H. Malizard, Boulogne-sur-Seine.
Structural design: L. K. Wilenko, Paris.
Built: 1966/1968.

Function
Library building with ten storeys above ground and two basement floors. The upper basement contains a car park, while the central library stock of about half a million books is kept in the lower basement. The entrance hall and reception desks are on the ground floor; special libraries, reading rooms and offices are accommodated on the upper floors. Connecting passageways linking the library to a five-storey ancillary building are provided on the first and fourth floors.

Dimensions
Overall dimensions 47.81 × 24.05 m. Height above ground level 34.83 m. Storey height — headroom: 4.03 – 3.77 m on ground floor; 2.97 – 2.56 m on upper floors.

Structural features
The portal frame system, which extends through every storey and in both directions, has been obtained by rigid connections between all columns and beams. The columns are arranged in four longitudinal and seven transverse rows, on a 7.90 × 7.50 m grid. The outer columns on the longitudinal sides are exposed outside the façades and consist of HE 300 B sections, with site joints at every third floor. Internal columns vary from HE 600 M to HE 360 B, progressively reducing in section with height. The floor beams in the transverse direction are HE 280 A, and in the longitudinal direction HE 260 A. The edge beam is a channel section U 270. All the structural connections are welded.
The almost square floor bays comprise a 14 cm thick continuous in-situ concrete slab. Imposed load 200 to 600 kg/m^2. Composite action between the concrete slab and the steel beams was ensured with stud shear connectors. All the steel beams were prestressed by giving them a preliminary camber by means of inclined props with screw jacks, these props being removed after the slab had matured. Because of the rigid beam-to-column connections, the columns also participate in the prestressing, which enables the steelwork to be more fully utilised structurally, so that better economy is achieved. As a result, a 44.8% saving in the overall weight of the steelwork was achieved.
- Façade: The cladding consists of 1.25 m wide storey-height units set between the edges of the floors above and below. The units are aluminium double frames with continuous insulating glass on the inside and with perforated sheet aluminium folding fins for sun protection on the outside. These fins can be opened in pairs, being swung sideways by actuation of a manual crank mechanism.
- Fire protection: Internal columns and floor beams have sprayed asbestos encasement.
- Foundations: Reinforced concrete slab on limestone 6 m below ground level. Permissible bearing pressure 10 kg/cm^2, design bearing pressure 8 kg/cm^2.

1 Library 2 Office 3 Room for meetings 4 Microfilm 5 Passage to ancillary building 6 Exhibition hall 7 Pool

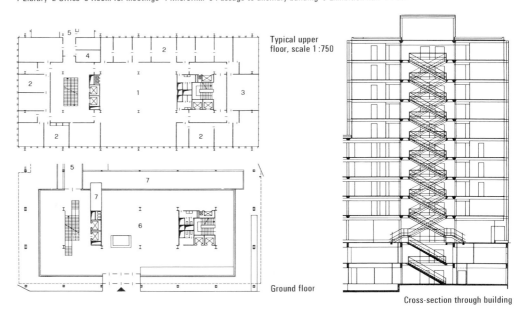

Typical upper floor, scale 1:750

Ground floor

Cross-section through building

Areas and volume

gross area	15 805 m^2	area on plan	1150 m^2
effective floor area	14 360 m^2	volume	47 735 m^3

Quantities of materials

	steel	concrete	reinforcement
total	600 t	2162 m^3	128 t
per m^3 cubic content	12.6 kg	0.046 m^3	2.7 kg
per m^2 gross area	38.0 kg	0.137 m^3	8.1 kg

Construction cost (1968)
Steelwork and floor slabs 2.67 million Fr.fr.

Reference
Bâtir, 185/1970.

Arrangement of beams in fourth floor

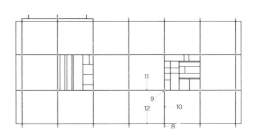

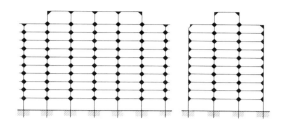

Structural system: cross-section and longitudinal section of rigid steel frame

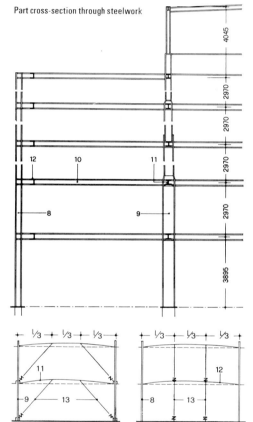

Part cross-section through steelwork

8 External column HE 300 B
9 Internal column HE 600M to HE 360 B
10 Transverse beam HE 280 A
11 Longitudinal beam HE 260 A
12 Edge beam U 270
13 Prestressing props

Method of prestressing the floor beams

Erection of steelwork

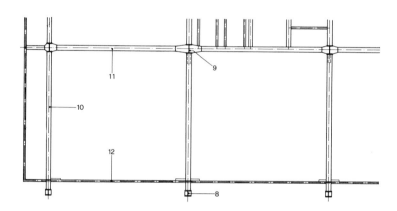

Layout of beams at a corner of the building

Detail of façade

Anchorage of prestressing props at base of column

Prestressing props for floor beams at inner column

123

43 FEDERAL COURT, KARLSRUHE, GERMANY

Architectural design: P. Baumgarten, Berlin.
Structural design: K. J. Peschlow, Berlin.
Built: 1965/1969.

Function

This complex consists of five buildings connected to one another by glazed passages:
- Court building: A three-storey structure with basement under its entire area. Entrance and plenary sessions on ground floor; rooms for reception, the press, and litigants on first floor; courtroom and council chamber on second floor. Car parking facilities are provided in the basement.
- Judges' building: This comprises three storeys surrounding an inner courtyard; ground floor has open layout with two staircases leading to the two upper floors, which provide office accommodation for the judges of the two Senates.
- Library building: This is a single-storey structure with two basement floors, the latter accommodating press archives, bookbinding workshop and storage capacity for 300 000 books. The catalogue room, two reading rooms and ancillary rooms are on the ground floor.
- Administrative building and dining hall: These are both single-storey buildings with basements.

Dimensions

- Court building: Overall dimensions 29.50 × 29.50 m. Height above ground level 16.60 m. Storey height 3.90 m on ground and first floors, 7.50 m on second floor. Ceiling height 3.10 and 6.00 m respectively.
- Judges' building: Overall dimensions 43.25 × 42.00 m. Height above ground level 14.40 m. Storey height 3.90 m, ceiling height 3.10 m.
- Library building: Overall dimensions 28.00 × 28.00 m. Height above ground level 7.20 m. Ceiling height 3.50 m.

Structural features

- Court building: The courtroom, free from internal supports, is on the second floor. The roof structure is a beam grillage supported by eight external columns. The structural steelwork for the two bottom storeys comprises columns located on a 7.00 m square grid. Wind forces are transmitted through these two storeys by the floor slabs to concrete shear walls. Similarly, to stiffen the roof framework, the eight external columns are connected by brackets to the floors below. The roof grillage comprises four 2000 mm deep intersecting plate girders together with edge girders and cross-bracing.
The continuous external columns consist of HE 500 M sections. In the two bottom storeys the internal columns and floor beams are of I-section. The floors comprise precast concrete slabs. Imposed load 500 kg/m^2.
- Judges' building: The perimeter row of columns is located inside the external façade of the building. At the corners of the courtyard there are four twin columns exposed outside the façade and interconnected by lattice girders of which the depth corresponds to the two upper storeys. Vertical loads are transmitted by the floor beams to the outer columns and through the lattice girders to the four courtyard columns. The horizontal forces are transmitted through in-situ concrete floor slabs to two concrete cores. Horizontal cross-bracings stiffen the roof framework to resist wind loads.
Columns and beams consist of rolled steel sections; the diagonal members of the lattice girders are bundles of 26 mm dia high tensile bars, the number of which varies from panel to panel, depending on the forces involved.
- The complex is located in a seismic zone. It has therefore been designed to resist additional horizontal loading equivalent to 7.5% of the vertical loads and the foundations have been designed for 50% more than the calculated bearing pressure. As a precaution against air raids a debris load of 1000 kg/m^2 has been taken into account and entrances have been provided in the external walls of the basements.
- Roofs: Ventilated, internally insulated roofs with covering of aluminium sheet and multilayer roofing felt with gravel covering.
- Façades: Storey-height timber-framed units with fixed glazing, sliding and casement windows. Spandrel areas have ventilated timber cladding or aluminium sheeting with standing seams, together with cast aluminium slabs. External louvre blinds provide sun protection.
- Fire protection with intumescent paint or sprinkler system.
- Foundations: Strip and pad foundations on sand and gravel. For earthquake resistance: continuous flexible foundation slab and shear-resisting connections between columns and walls.

Services

Buildings are connected to a steam district heating system; heat exchangers for water heating. Air extractors in the window units.

View of the court building, with library on the left and pavilion on the right

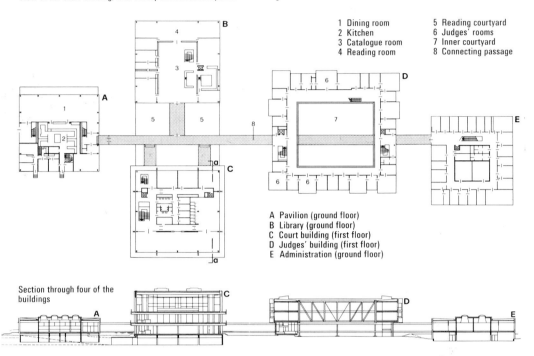

1 Dining room
2 Kitchen
3 Catalogue room
4 Reading room
5 Reading courtyard
6 Judges' rooms
7 Inner courtyard
8 Connecting passage

A Pavilion (ground floor)
B Library (ground floor)
C Court building (first floor)
D Judges' building (first floor)
E Administration (ground floor)

Section through four of the buildings

Areas and volume

gross area	14 323 m^2	area on plan	6771 m^2
effective floor area	9635 m^2	volume	55 000 m^3

Quantities of materials

	steel	concrete	reinforcement
total	1261 t	11 000 m^3	1050 t
per m^3 volume	22.9 kg	0.200 m^3	18.5 kg
per m^2 gross area	88.0 kg	0.768 m^3	73.5 kg

Construction cost (1969)

total cost	17.06 million DM	per m^2 gross area	1191 DM
per m^3 volume	310 DM	per m^2 eff. floor area	1771 DM

References

Die Bauverwaltung, 11/1969, p. 604.
Bauwelt, 48/1969, p. 1714.
Glasforum, 1/1970, pp. 3 and 36.
Deutsche Bauzeitschrift, 4/1970, p. 631.

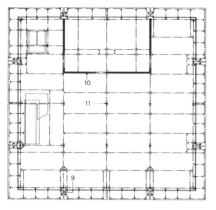

Beam arrangement in second floor of court building

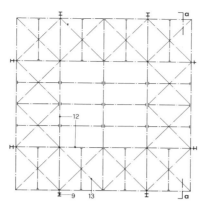

Roof framing

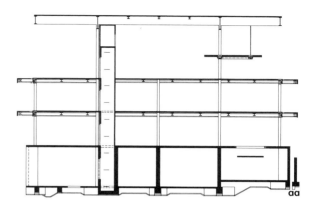

Section through court building

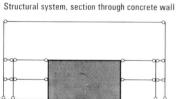

Structural system, section through concrete wall

Courtroom

Judges' building

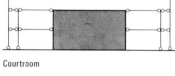

Structural system: section showing two-storey lattice girder

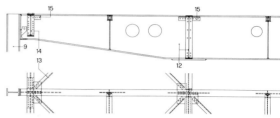

Plan and elevation of plate girder for roof

9 External column HE 500 M
10 Reinforced concrete wall
11 Internal column HE 300 B
12 Plate girder 2000 mm deep
13 Horizontal bracing
14 Edge girder 1000 mm deep
15 Angle cleat for connection of horizontal bracings

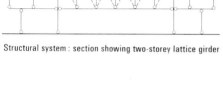

Part of beam arrangement above 2nd floor

16 Twin column
17 Vertical member, HE 360 B, of two-storey lattice girder
18 Concrete core
19 Edge beam IPE 270
20 High tensile bars, 26 mm dia
21 Top chord HE 360 B
22 Centre-line of high tensile bars
23 Anchor block
24 Timber framed roof
25 Ceiling
26 Timber-framed unit
27 Timber sheathing
28 Louvre blind
29 Aluminium sheet with standing seams
30 Cast aluminium slabs
31 Edge beams HE 800 M

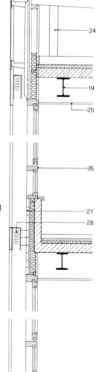

Section through external wall

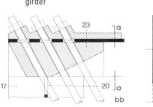 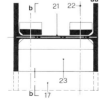

Anchorage of high tensile bars at top of end panel of lattice girder

Inner courtyard of judges' building with connecting passage

View of judges' building

44 EUROPEAN COURT OF JUSTICE, LUXEMBOURG

Architectural design: J. Conzemius, Luxembourg; F. Jamagne and M. van der Elst, Antwerp.
Structural design: Schroeder and Associates, Luxembourg.
Built: 1967/1972.

Function

Six-storey building containing courtrooms and administrative offices. On ground floor: conference hall and general offices. First floor: foyer, library, office of the Clerk, and three courtrooms (extending through two storeys). Second floor: administrative offices. Third floor: offices for president and other judges grouped around two light-wells. The fourth and fifth floors are set back from the perimeter of the building; they contain a restaurant, the caretaker's flat and services.

Archives, ancillary rooms and air-conditioning plant are accommodated in the two basement storeys. An underground car park for 260 vehicles is provided beneath the forecourt. The public entrance is located on one of the long faces of the building and stairs provide access to the foyer on the first floor. The service entrance is on the opposite side. Vertical circulation facilities comprise five passenger lifts and one service lift, together with two staircases, installed in two outer and two inner reinforced concrete cores.

Dimensions

Overall dimensions 108.0×48.5 m. Height above ground level 28.0 m. Recessed top storeys 43.1×27.0 m. Ground floor is raised 1.25 m above the surrounding site. Storey height and ceiling height on ground floor and three upper floors 3.70 and 2.65 m respectively; on fifth floor: 4.31 and 4.04 m. The two basement storeys have storey heights of 3.15 and 2.95 m, with corresponding headroom of 2.80 and 2.65 m.

Structural features

The exterior of this building is characterised by the three-storey columns placed 1.30 m in front of the façades and by the projecting ends of the floor beams, likewise designed as twin sections. The third floor cantilevers a further 1.80 m out beyond the external columns. Internally, there are four reinforced concrete cores interconnected and surrounded by pin-jointed steelwork laid out on a 5.40 m square grid and extending to a height of four or six storeys. Main beams, spaced at 5.40 and 10.80 m centres, extend transversely and vary in span, 5.40, 10.80 or 16.20 m, depending on the type of room or hall involved. Secondary beams, at 2.70 m centres, run in the longitudinal direction of the building.

On the two long sides, transverse beams connect the external columns to this internal framework. These beams are spaced at 5.40 m centres and have a span of 8.20 m. They are in fact two HE 400 A sections, while each external column comprises two HE 300 B sections at 700 mm centres. The two outer rows of internal columns are similar. All the double columns are fixed at the base. At the ends of the building the floor beams span between the external columns and the outer reinforced concrete cores. Perimeter beams, each comprising two HE 400 A, spaced 900 mm apart, interconnect the floor beams and serve also to support the emergency balconies. Floors: steel trough decking with 10 cm in-situ concrete slab. Imposed load 250 to 1000 kg/m², depending on utilisation. Horizontal cross-bracings transmit wind forces transversely and longitudinally to the concrete cores and to the four rows of fixed-base double columns along the longitudinal sides of the building.

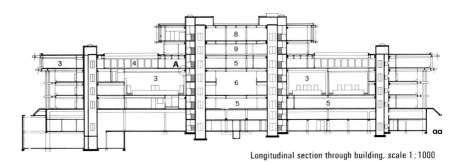

Longitudinal section through building, scale 1:1000

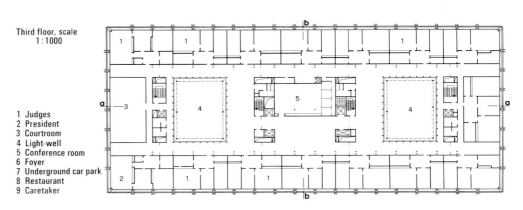

Third floor, scale 1:1000

1 Judges
2 President
3 Courtroom
4 Light-well
5 Conference room
6 Foyer
7 Underground car park
8 Restaurant
9 Caretaker

- Fire protection: All loadbearing steel components inside the building are encased in 4 to 6 cm thick asbestos cement plaster. In addition, there are hydrants and a sprinkler system.
- Façade: Storey-height cladding units composed of steel sections with casement windows and Thermopane glazing. At floor and spandrel levels there are sandwich panels. All the exposed cladding and structural steelwork is in weathering steel.
- Foundations: Pad footings under the columns, slabs under the cores. The subsoil is sandstone.

Services

Circulated hot water heating; three boilers with 3 million kcal/h capacity. Ventilation for archives room and underground car park. Other rooms are air-conditioned: high-pressure system in outer, low-pressure in inner rooms, air delivery rate 54 000 and 107 000 m³/h respectively.

Areas and volume

gross area	31 000 m²	area on plan	5000 m²
effective floor area	12 500 m²	volume	150 000 m³

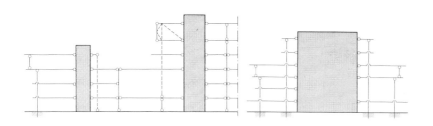

Structural system: part longitudinal section and cross-section

Quantities of materials	steel	concrete	reinforcement
total	2600 t	20 000 m³	1500 t
per m³ volume	17.3 kg	0.133 m³	10.0 kg
per m² gross area	83.9 kg	0.645 m³	48.4 kg

Construction cost (1972)

total cost	485 million B.fr.	per m² gross area	15 645 B.fr.
per m³ volume	3233 B.fr.	per m² eff. floor area	38 800 B.fr.

Façade with exposed double columns and emergency balconies

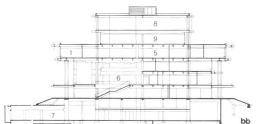

Cross-section through building

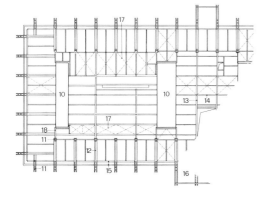

Arrangement of beams in first floor

Beam supported on bracket of concrete core

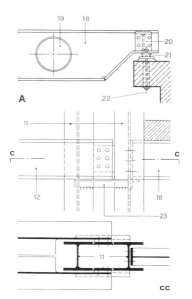

10 Concrete core
11 Double column, two HE 300 B
12 Transverse beam, two HE 400 A
13 Main beam HE 700 B
14 Secondary beam IPE 400
15 Perimeter beams two HE 400 A
16 Canopy over main entrance
17 Wind-bracing
18 Floor beam HE 700 A
19 Hole in beam web
20 Welded cleat for secondary beam
21 Locating pin
22 Anchor bolt
23 Seating angle
24 Sandwich panel
25 Ceiling
26 Steel trough decking
27 Mullion

Internal double column with connections to a transverse beam and a beam linked to the core

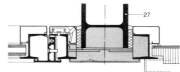

Horizontal and vertical section through façade

Structural steelwork around an outer concrete core

Connection of floor beam to outer concrete core

Part of the completed structural steelwork

45 TELEVISION CENTRE, BERLIN-CHARLOTTENBURG, GERMANY

Architectural design: R. Tepez, Berlin.
Structural design: G. Treptow, Berlin.
Built: 1966/1971.

General view of the television centre; the seven-storey wing of the main building is on the left

Function

This high-rise building, raised on pilotis and comprising 13-storey and 14-storey wings set obliquely to each other, is surrounded by lower buildings containing television studios, workshops, stores and technical services. At their junction, the two wings of the main building are attached to a 20-storey core structure which towers above them.

The main building provides office accommodation for managerial and administrative staff, programme directors, editors, etc., a total of 320. Archives are on the 6th and 7th floors, with air-conditioning plant on the 8th floor. On the 14th floor is a conference room with access to the roof terrace on the adjacent wing. In the cores are three lifts, a staircase and shafts for services; its six top storeys contain technical equipment and include the terrace for parabolic aerials and directional radio.

Joined to the 13-storey north wing is a 7-storey wing in which various further technical facilities, together with editing studios, synchronising studios, etc., are accommodated. This wing is served by two vertical cores, located at the junction with the 13-storey wing and at the other end, respectively, and containing lifts and stairs.

Dimensions

Length of 14-storey wing 28.3 m, 13-storey wing 33.3 m, 7-storey wing 58.7 m. Width of all three wings is 14.18 m. Heights above ground level respectively 56.30, 50.50 and 29.80 m. The tower is 78.20 m in height. Storey height 3.65 m, headroom 3.00 m. Planning grid 1.25×1.25 m.

Structural features

To meet the complex requirements of the services and to achieve maximum flexibility of layout — bearing in mind the possibility of subsequent change in the use of available space — it was necessary to have rooms and internal areas free from supports. By keeping the wings relatively narrow and using composite construction, it was possible to design the floors to span the full width of the building and carry an imposed load of 500 kg/m².

The columns on the longitudinal faces, which are interconnected by pin-jointed IPE 600 and IPE 750 edge beams, are spaced at 7.50 m centres and consist of box-sections composed of HE–M rolled sections with cover plates. In the transverse direction, the floor beams are spaced at 2.50 m centres and consist of 375 mm deep asymmetrical welded plate girders which are in composite construction with 12 cm thick precast concrete slabs.

Stud shear connectors protrude into recesses formed in the edges of these slabs and are enclosed by looped reinforcing bars projecting from the concrete. Before the in-situ concrete was poured into the joints, the floor beams were provided with temporary support at their third-span points to achieve composite action for both dead and live loads. The pin-jointed framework is stiffened by vertical lattice bracings in the end walls and by the longitudinal walls in the three vertical cores in the other direction.

- Façade: The spandrel units, which are attached to brackets on the columns, are 16 cm thick sandwich panels consisting of concrete and cement-bonded wood chips, 1.65 m high. In front of these, with an intervening air-space between, there are aluminium facing panels coated with an anti-drumming agent. Windows have fixed double or, in places, triple glazing, with edge damping for additional acoustic insulation in special rooms. Solar heat is controlled by means of fixed louvres in the upper part of the windows and externally fitted louvre blinds, centrally adjusted for each face of the building and monitored by a storm warning system.
- Fire protection: Sprayed asbestos cement encases the columns and the floor beams adjacent to the corridors; other parts of the floors conform to German fire resistance class F 90 (90 min), this fire rating being achieved by means of suspended ceilings which also perform an acoustic function. Wind-bracings are encased with 6 cm thick gypsum slabs.

Areas and volume

gross area	16 285 m²	area on plan	1771 m²
effective floor area	8499 m²	volume	65 364 m³

Material quantities

	steel	concrete
total	2520 t	3048 m³
per m³ volume	38.6 kg	0.047 m³
per m² gross area	154.7 kh	0.187 m³

Construction cost (1970)

total cost	21.3 million DM	per m² gross area	1308 DM
per m³ volume	326 DM	per m² eff. floor area	2506 DM

References

Detail, 6/1971, Konstruktionstafel.
Bauwelt, 37/1971, p. 1524.
Deutsche Bauzeitschrift, 12/1971, p. 2493.
Acier-Stahl-Steel, 6/1972, p. 262.

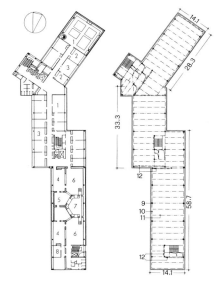

6th floor, scale 1:1500 — Arrangement of beams in a typical floor

1 Archives
2 Film cutting room
3 Offices
4 Services
5 Producers' room
6 Synchronising studio
7 Anechoic room
8 Announcers' room

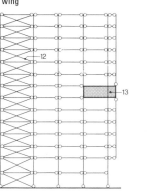
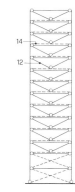

Structural system: longitudinal section and cross-section of south wing

Connection of floor beams and wind-bracing to an external column

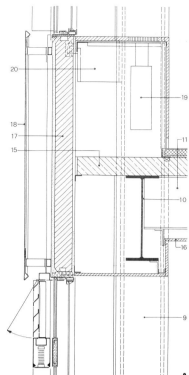

Section through external wall at spandrel panel level

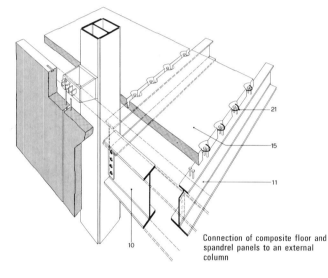

Connection of composite floor and spandrel panels to an external column

9 External column consisting of HE-M sections strengthened with plates
10 Edge beam IPE 600
11 Floor beam, welded I-section, 375 mm deep
12 Wind-bracing
13 Plate girder
14 Composite floor
15 12 cm thick precast concrete slab
16 Fireproof acoustic ceiling
17 Lightweight concrete spandrel panel
18 Aluminium panel
19 Air-conditioning unit
20 Bracket for attachment of spandrel panels

Cross-section through 7-storey wing

21 Reinforcing bars
22 Anechoic room (double wall construction)

Erection of south wing

Erection of north wing with wind-bracing at junction with south wing

Plate girders in the south wing, supporting the columns in the end wall

46 OFFICE BUILDING FOR MEMBERS OF PARLIAMENT, BONN

Architectural design: E. Eiermann, Karlsruhe.
Structural design: Steelwork contractors.
Built: 1966/1969.

Function

Offices and committee rooms for members of the Bundestag of the Federal Republic of Germany in a 30-storey building raised on pilotis. From the 3rd to the 17th floor there are 446 individual members' rooms; from the 19th to the 28th floor there are 114 committee rooms, five conference rooms and 19 two-storey halls. There are plant rooms on the 18th and 29th floors, while the latter floor also comprises a restaurant and café.

The storeys with individual offices each contain a waiting area for visitors adjacent to the vertical circulation core which, itself, contains twelve passenger lifts, one goods lift, two staircases and cloakrooms.

From the basement to the third storey there is a podium which is visually and functionally separate from the high-rise block. One longitudinal side of this low-rise building is set forward from the base of the high-rise block, while its other three sides are recessed beneath it. It accommodates, in its four storeys, a reception hall, post office, telephone and teleprinter exchange, general administrative offices and, in the basement, technical services.

Dimensions

Overall dimensions: 48.0 × 33.0 m. Height above ground level 109 m. Upper storeys generally have a storey height of 3.20 m and a ceiling height of 2.50 m, while the halls have corresponding heights of 6.40 and 5.40 m. Core size overall 30.8 × 8.3 m. Planning grid 3.75 × 3.75 m.

Structural features

The rectangular concrete core braces the building and supports the inner ends of the HE 450 B floor beams, which are spaced at 3.75 m centres and span from the core walls to the external columns, which are of 650 × 650 mm square box section. The core is not situated in the centre of the building and the span of the beams on the two narrower faces and on one of the longitudinal faces is 7.50 m. On the other longitudinal face the distance from the core to the external columns is 15.00 m, which is halved by an additional row of internal columns carrying HE 500 B longitudinal beams. Above the 18th floor of the building these internal columns have been moved 3.75 m inwards towards the core in order to obtain more depth for the halls. These columns are supported on a storey-height lattice girder in the 18th storey.

At third floor level every alternate external column is intercepted by, and supported on, a box girder 2000 mm in depth. The columns under this girder, spaced at 7.50 m centres, have a length of 13.20 m on three sides of the building, that is, where the podium building is recessed under the main high-rise block.

- Floors: The floor beams carry 12 cm thick precast concrete slabs spanning 3.75 m. Imposed load 500 kg/m². For structural and fire protection reasons it was necessary to ensure continuity of the floor slabs. To this end, the slabs were recessed at the top to leave projecting reinforcement exposed at the supports. Additional reinforcement was inserted here and the joint completed with in-situ concrete. Lateral stability of the top flanges of the steel beams is ensured by welded cleats which project between the ends of the slabs. Fire protection of the beams is provided by a sprayed coating (Pyrok) and suspended ceilings. The webs of the beams comprise reinforced apertures for services accommodated in the ceiling space.

- Façade: Longitudinal edge beams for supporting the floor slabs and spandrel units span between brackets bolted to the external columns. Storey-height teak window units with insulating glass are attached to the sheet-metal cladding of the floor edges and to mullions erected in front of the external columns. Closed areas of the façades have a cladding of precast concrete slabs. Concrete units laid on the projecting brackets provide sun protection and serve also as external walkways for window-cleaning. Further sun protection is provided by polyester louvres attached to the mullions.

- Foundations: The loadbearing subsoil consists of gravel: permissible bearing pressure 10 kg/cm². The water table is below foundation level. Under the core a 3.0 m deep foundation raft, 31.0 × 13.5 m, is provided, while the columns stand on strip foundations 2.0 m wide and 3.0 m deep.

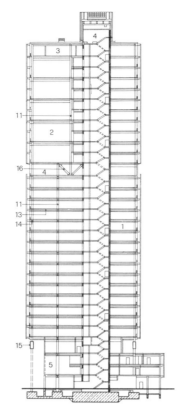

Cross-section of the building taken through the service core

1 Member's office 3 Restaurant
2 Committee room 4 Services

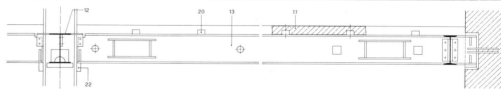

Floor beam details

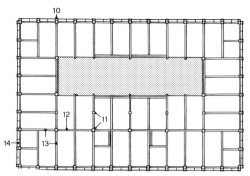
Arrangement of beams in 19th floor

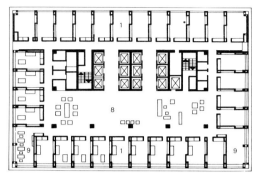
Typical floor with individual offices, scale 1:800

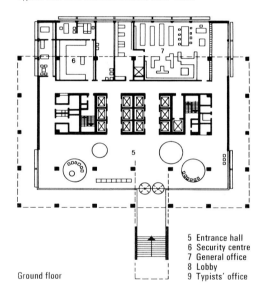
Ground floor

5 Entrance hall
6 Security centre
7 General office
8 Lobby
9 Typists' office

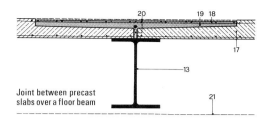
Joint between precast slabs over a floor beam

10 External column
11 Internal beam
12 Longitudinal beam HE 500 B
13 Floor beam HE 450 B
14 Edge beam IPE 270
15 Fascia girder, 770 × 2000 mm box-section
16 Lattice girder
17 Precast concrete slab 12 cm thick
18 Mesh reinforcement
19 In-situ concrete
20 Cleat to ensure lateral stability of beam
21 False ceiling
22 Bracket
23 Longitudinal façade beam
24 Spandrel unit
25 Precast concrete component
26 Polyester louvres
27 Mullion
28 Teak window frame
29 Sheet-metal cladding

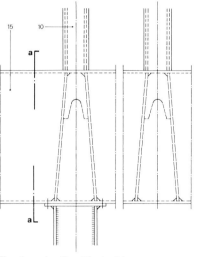
Elevation and section of fascia girder

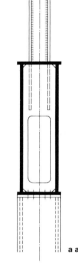
Erection of external columns with brackets for attachment of cladding

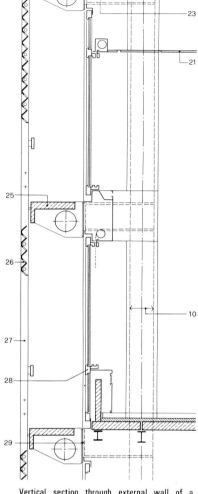
Vertical section through external wall of a committee room

Structural system: cross-section

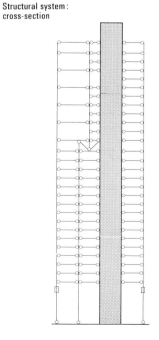

Areas and volume

gross area	44 000 m²	technical services	2810 m²
effective floor area	29 000 m²	area on plan	1855 m²
per occupant	24 m²	volume	150 300 m³

Quantities of materials

	steel	concrete	reinforcement
total	4500 t	18 200 m³	1200 t
per m² gross area	102 kg	0.414 m³	27.3 kg
per m³ volume	29.9 kg	0.121 m³	8.9 kg

Construction cost (1969)

total cost	47.3 million DM	per m² gross area	1075 DM
per m³	315 DM	per m² eff. floor area	1631 DM

References

Der Stahlbau, 8/1970, p. 225.
Schweissen + Schneiden, 3/1971.
Architektur und Wohnwelt, 6/1972.
Deutsche Bauzeitschrift, 9/1972, p. 1583.

47 OFFICE BUILDING FOR A LIGHT ENGINEERING WORKS, STÄFA, SWITZERLAND

Architectural design: J. Dahinden, Zurich.
Structural design: M. Corrodi, Stäfa; B. Möller, Zurich.
Built: 1962/1964.

Function

Administrative offices for a staff of 120 accommodated in two three-storey blocks of equal size. Because of the sloping site they are staggered vertically in relation to one another by half a storey height and are linked by a structure containing stairs and a lift. Reception, a board room, a conference room, a lecture hall and the accounts department are all on the ground floor. Above, there are drawing offices and individual offices for senior staff. The plant room, the archives and air raid shelters are all situated at the two basement levels.

Dimensions

The overall dimensions of both blocks are 19.20 × 12.80 m. Height above ground level 14.50 m. Storey height 3.16 m, ceiling height 2.70 m.

Structural features

Three box columns extending the full height of the blocks stand in front of each longitudinal side of a block and two similar columns stand in front of each transverse side. At each floor level, short brackets extend inwards to support the floor beams. In the transverse direction, the columns on either side of the blocks are interconnected by floor beams to form three-storey rigid frames. Longitudinally, a central beam joins the three transverse beams to one another at each floor level and is rigidly connected to two box columns inside the block and supported at its ends by an edge beam. The columns and foundations are so designed that the blocks can, if required, be subsequently extended upwards by the addition of three storeys. Wind forces are transmitted in both directions by portal frame action to a reinforced concrete core in each block.

- Columns stand on reinforced concrete basement walls. The external columns are welded box sections, 300 × 200 mm, with plate thicknesses varying from 8 to 20 mm. These columns stand at a distance of 8 cm from the façades. They are filled with concrete provided with reinforcement, about 30% of the load being carried by the filling. Internal columns are 300 × 280 mm welded box sections.
- Floors: The transverse and edge beams are I 160 sections, while the central beams in the longitudinal direction are twin I 160 sections. The bottom flanges of the floor beams are compounded with 12 mm thick welded plates. Short cuttings of joist section were welded to the top flanges as shear connectors to ensure composite action with the 20 cm thick in-situ concrete floor slabs. Imposed load 300 kg/m².
- Façades: Storey-height glazing with 1.24 m wide double glazing separated by 7 cm wide aluminium mullions. The edges of the floors are faced with 47 cm deep aluminium panels with internal rock wool insulation, 6 cm thick. Sun protection by vertical louvres on the inside of the windows and the use of solar heat resistant glass, with thin metallic coating, on the outside sheet of the double glazing.
- Foundations: The subsoil consists of loam and alluvial sand deposits over rock located at a depth varying from 6.5 to 8.5 m, dipping from west to east. The west block is founded directly on the rock, whereas an open caisson foundation was used for the east block, concrete-filled tubes, 1.75 m diam, being founded on the rock. The calculated load on a column is 100 t for the existing three-storey building but, should the blocks be extended upwards by a further three storeys, the load will increase to 170 t.

Services

Full air-conditioning with high- and low-pressure installations. The high-pressure appliances are fitted inside the windows. The air outlets from the low-pressure system are provided in the suspended ceilings.

References

Architektur + Wohnform, 74th year, Vol. 1, p. 40.
L'Architecture d'Aujourdhui, 6–7/1965, p. 42.
Bauen + Wohnen, 1/1965, p. 32.
Deutsche Bauzeitung, 3/1965.

View from south-west

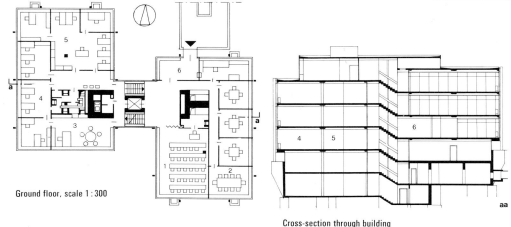

Ground floor, scale 1 : 300

Cross-section through building

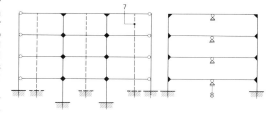

Structural system: longitudinal and cross-sections through a block

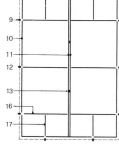

1 Lecture room
2 Conference room
3 Board room
4 Secretaries' office
5 Accounts office
6 Entrance hall

Arrangement of beams on a typical floor.

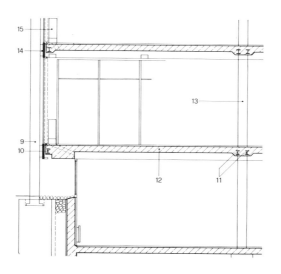

Part cross-section

Part of beam arrangement at a corner of a building

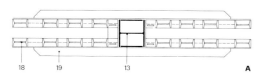

Plan of central beam at an internal column

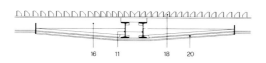

Joint in transverse beam at central beam

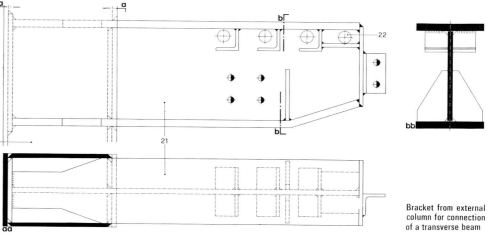

Bracket from external column for connection of a transverse beam

7 Transverse frame
8 Bearing of transverse beam on centre beam
9 Concrete filled external column, welded box-section 300 × 200 mm
10 Edge beam I 260
11 Central beam, two I 160
12 In-situ concrete floor 20 cm thick
13 Internal column, welded box-section 300 × 280 mm
14 Aluminium panel with thermal insulation
15 Air-conditioning appliance
16 Transverse beam I 260 (bottom flange strengthened with welded plate 300 × 12 mm)
17 Secondary beam I 160
18 Shear connectors
19 Splice plate 75 × 16 mm
20 Splice plate 250 × 10 mm
21 Bracket, welded I-section 310 mm deep
22 Hole for floor reinforcement
23 Reinforcement cage
24 Spacer

Cross-section through external column

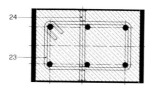

Areas and volume

gross area	2369 m²	area on plan	565 m²
effective floor area	1714 m²	volume	8403 m³

Quantities of materials

	steel	concrete	reinforcement
total	82.0 t	1315 m³	92.8 t
per m³ volume	9.8 kg	0.156 m³	11.0 kg
per m² gross area	34.6 kg	0.555 m³	39.2 kg

Construction cost (1964)

total cost	3.03 million Sw.fr.	per m² gross area	1279 Sw.fr.
per m³ volume	361 Sw.fr.	per m² eff. floor area	1768 Sw.fr.

48 OFFICE BUILDING FOR A TUBEWORKS IN AULNOYE, FRANCE

Architectural design: E. Albert, A. Champetier de Ribes, Paris.
Structural design: J. L. Sarf, Paris.
Built: 1959/1960.

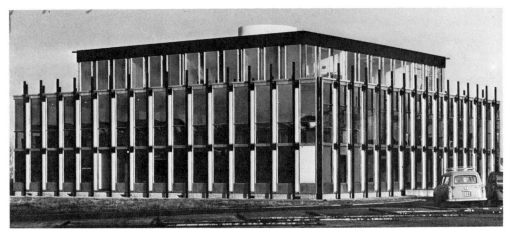

Function

This three-storey building, square on plan, contains the design and administrative offices for a tubeworks. The ground floor comprises the entrance hall, a lecture hall seating 80 people, a library, the archives and ancillary rooms. On the first floor there are a drawing office and individual offices for a total staff of 60, while on the second floor there are a reception hall, senior executives' rooms, a conference room and lounges. A shallow basement used as a plant room extends under part of the building. The upper floors are reached by a spiral staircase arranged around a central cylindrical core containing the cloakrooms.

Dimensions

Overall dimensions of the building 24.12 × 24.12 m. Diameter of the cylindrical core 4.00 m. Height above ground level 12.20 m. The second floor is set back 3.25 m from the perimeter on all four sides of the building. Storey height — headroom to underside of floor beams: 2.78–2.32 m on ground floor, 3.53–3.07 m on first floor, 3.13–2.62 m on second floor. Basement has 1.19 m headroom.

Diagonal view of the building and (below) model of the structural framework

Structural features

Four tubular columns, 216 mm diam, interconnected by IAP 400 beams, are arranged at the corners of a square, 7.50 × 7.50 m, around the central reinforced concrete core. IAP 200 floor beams extend from these beams to the four sides of the building on the first and second floors. These floor beams are connected to 100 × 100 mm square hollow sections located 20 cm in front of the façades of the building. In the region of the corners of the building the floor beams are supported at their inner ends by IAP 400 beams which extend diagonally from the respective internal columns to the 171 mm dia external columns located at the four corners of the building. Rainwater downpipes are accommodated inside the external columns. The square hollow sections along the flanks of the building are filled with concrete.

The 8 cm thick in-situ concrete floor slabs are in composite construction with the floor beams, studs being used as shear connectors. Imposed load 300 kg/m². There are no suspended ceilings, the steel floor beams being exposed beneath the floor slabs.

Wind forces are transmitted through the landings of the spiral staircase from the floors to the concrete core.

The four internal columns extend the full height of the building. At the top they carry four intersecting IAP 450 roof beams which cantilever 4.50 m on all sides and are interconnected at their ends by a perimeter beam. For the transmission of wind loads the roof structure is connected to the core by intermediate beams which span between the edge beams and the main beams and are embedded in the concrete wall of the core.

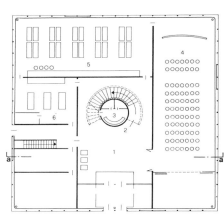

Ground floor, scale 1:450

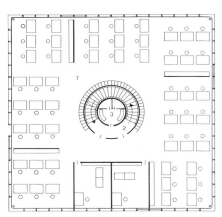

First floor

• Façade: The first two storeys are clad with continuous 1.10 m wide units, each fitted between two external columns and fixed to the outside edges of the floor slabs. The recessed top storey has similar units but, of course, only one storey high. The units are all provided with double glazing, the outer sheet being metal-coated for solar heat resistance. The 40 cm wide vertical casement between the window frames is of sandwich construction.

Heating

Floor and ceiling radiant heating through heating coils in the floor slabs. Heat is supplied as steam from the adjacent factory and passed through heat exchangers.

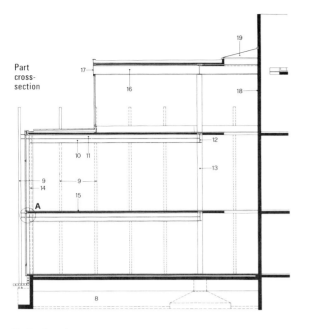
Part cross-section

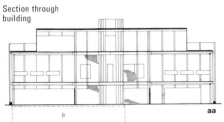
Section through building

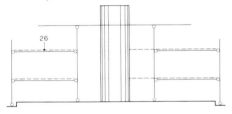
Structural system: section in front of core

1 Entrance hall
2 Spiral staircase
3 Core with cloakrooms
4 Lecture hall
5 Library
6 Archives
7 Drawing office
8 Shallow basement for services

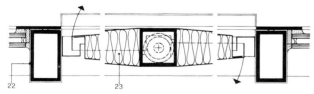
Vertical and horizontal sections through ventilation casement

Areas and volume

gross area	1492 m²	area on plan	606 m²
effective floor area	1216 m²	volume	5215 m³

Construction cost (1960)

per m³ volume   233 Fr.fr.   per m² eff. floor area   1000 Fr.fr.

Reference
Acier-Stahl-Steel, 7–8/1963, p. 333.

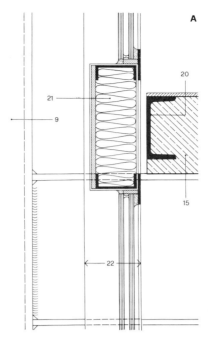
Vertical section through external wall at floor area

9 Square hollow section 100 × 100 mm
10 Diagonal floor beam IAP 400
11 Floor beam IAP 200
12 Main beam IAP 400
13 Tubular column 171 mm dia
14 Corner column
15 Floor slab
16 Roof beam IAP 450
17 Edge beam, welded channel section, 750 mm deep
18 Concrete core
19 Roof light
20 Channel section as floor surround
21 Transom for window components
22 Mullion, RHS 70 × 40 mm
23 Ventilation casement
24 RHS 60 × 40 mm
25 Plate connecting window and casement
26 Composite floor

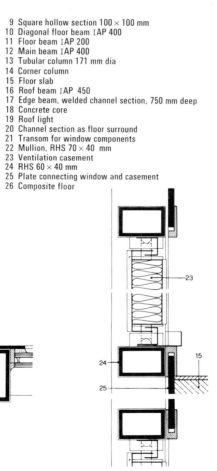

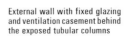
External wall with fixed glazing and ventilation casement behind the exposed tubular columns

Connection between floor beam and external column

Connection between floor beam and core

49 OFFICE BUILDING FOR A STRUCTURAL STEELWORK FIRM IN LANGENHAGEN, GERMANY

Architectural design: H. Wilke, Hanover.
Structural design: Steelwork contractors.
Built: 1971/1972.

Function

This five-storey building, which is circular in plan, comprises offices for the staff of 300 employed by a firm of structural steelwork contractors. On the first to third floors there are individual offices and a drawing office which are arranged around a central service core. This core contains a spiral staircase surrounded by circular galleries with four entrances per floor to the offices and with access to and from stores, cloakrooms and a lift. The top floor, which is smaller in diameter than the lower floors, incorporates the documentation centre, the printing and duplication department, the staff lounge and technical services. The ground floor, which is even more recessed, contains the entrance hall, telephone exchange, interview rooms and exhibition space. There is parking space for 40 cars at ground level under the overhanging part of the building.

Dimensions

The diameter of the building varies, being 38.0 m for the first three upper storeys, 28.0 m for the recessed ground floor and 33.0 m for the top floor. The diameter of the core is 14.9 m and the height above ground level is 18.7 m. The storey height is 3.61 m on the ground, first and second floors, 3.94 m on the third floor and 3.54 m on the top floor. The ceiling height on every floor is 2.82 m.

Structural features

The shape of the building logically leads to a structural frame comprising radial floor beams with columns in concentric rings. On the first, second and third floors, 24 radial beams IPE 400 extend from an inner ring of 24 columns, each consisting of two U 120 channel sections, and cantilever over the intermediate ring of simply-supported storey-height HE 150 B columns to an outer ring of hangers consisting of 50 × 30 mm flat bars. These hangers, which enable the upper floors to overhang the ground floor, are attached to the ends of compound HE 650 beams on the fourth floor. The roof beams of the top storey span between the inner ring of columns and the columns of an outer ring set back a distance of 2.50 m from the circumference of the building.
The inner columns were tied back by anchor bars and concreted into the cylindrical wall of the core after erection of the steelwork. Each upper floor of the building is stiffened by an annular tie-bar interconnecting the intermediate columns and by horizontal lattice bracing fitted between the radial floor beams.
• Floors: The radial beams carry 14 cm thick precast concrete slabs. Joints are formed with in-situ concrete filling. Imposed load 350 kg/m²; on fourth floor: 500 kg/m². Floor slabs have a screed topping with built-in plug sockets. Ceilings consist of mineral fibre acoustic tiles between radial bars and comprise light fittings and perforations for air extraction.
• Façade: In each sector, between two hangers, there are four bays, 1.24 m wide. The secondary framework comprises mullions and transoms with sandwich spandrel panels and double glazing. The 1.51 m high panels comprise plastic coated aluminium sheets with intervening insulating material. Sun protection is provided by the fin-shaped mullions and by tinted glass.
• Fire protection: The columns are clad with asbestos silicate slabs while the floor beams are protected with sprayed asbestos.
• Foundation: Subsoil is sand with 2 kg/cm² permissible bearing pressure. Water table is 70 cm below base of foundation. There are 80 cm deep strip footings under the two rings of columns.

Services

Warm air heating, capacity 400 000 kcal/h. Ventilation system with possibility of operating with recirculated air, capacity 60 000 m³/h. Second storey is partly air-conditioned.

Upper floor with open-plan office, scale 1:700

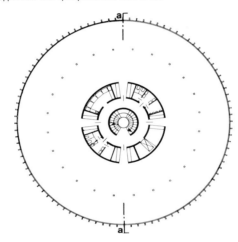

Structural system for a floor

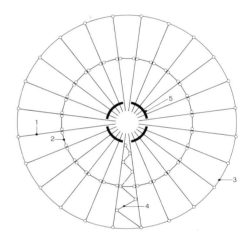

Structural system, radial section

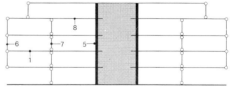

Areas and volume

gross area	4896 m²	area on plan	1134 m²
effective floor area	4000 m²	volume	18 530 m³

Quantities of materials

	steel	concrete	reinforcement
total	265 t	1180 m³	12.0 t
per m³ volume	14.3 kg	0.064 m³	0.6 kg
per m² gross area	54.1 kg	0.241 m³	2.5 kg

Construction cost (1971/1972)

total cost	3.0 million DM	per m² gross area	613 DM
per m³ volume	162 DM	per m² eff. floor area	750 DM

References

Element + Fertigbau, 5/1971.
Zentralblatt für Industriebau, 8/1971, p. 288.
Acier-Stahl-Steel, 12/1972, p. 514.
Bauwelt, 33/1972, p. 1262.
Deutsche Bauzeitschrift, 2/1972, p. 203.

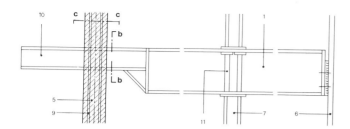

Radial beam with connections to concrete core, inner column and hanger

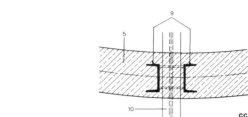

Section through building

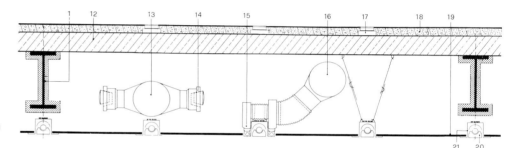

Cross-section through floor

9 Core column, two U 120
10 Cantilever beam extension I 200
11 Web stiffener
12 Precast concrete slab
13 Exhaust air duct
14 Exhaust air valve
15 Air inlet
16 Air duct
17 Plug socket set in floor
18 Screed
19 False ceiling
20 Strip lighting
21 Perforation for air extraction

1 Radial beam IPE 400
2 Annular tie-bar
3 Edge section
4 Lattice bracing
5 Concrete core
6 Hanger, 50 × 30 mm section
7 Simple column HE 150 B
8 Compound floor beam IPE 700

Arrangement of radial beams on columns which were subsequently embedded in the wall of the concrete core

Structural framework with radial beams, precast concrete slabs and external hangers

50 BANK AND OFFICE BUILDING IN MILAN, ITALY

Architectural design: L. Belgiojoso, E. Peressutti, E. Rogers, Milan.
Structural design: I. Tricario, Milan.
Built: 1967/1969.

Function

Seven-storey office building, with semicircular façade, situated on a corner site between two streets at their junction with a public square. The elevational treatment harmonises with the architectural features of the church in the immediate vicinity of the building. In addition, the rounded shape makes for efficient traffic flow.

The ground floor and mezzanine are set back to form an arcade, 8.70 m high, extending round the front. These floors and the first floor are occupied by a bank. The other floors contain individual offices arranged on both sides of a central corridor. Two basement levels provide parking facilities for 40 cars.

Two vertical circulation cores, each containing a spiral staircase and three lifts, are accessible through separate entrances on each side of the main banking hall and are interconnected on each upper storey by a central corridor. In addition, there is a lift connecting the main hall and the first floor.

Dimensions

Shape on plan almost semicircular, radius 20.4 m, actually formed by a polygonal perimeter. Height above ground level 31.3 m. Storey and ceiling heights on ground floor 5.52 and 4.57 m respectively; on upper floors: 3.50 and 3.05 m.

Structural features

The arcades are designed as portal frames, surmounted by a pin-jointed framework stiffened by two concrete cores, the frames conforming to the polygonal outline on plan. Each horizontal portion of the cross beams in the portal frames supports two radial beams, 5.90 m apart, which cantilever 1.0 m beyond the cross beams and carry the columns of the storeys above.

These external columns, which are HE 300 B sections, are connected at each upper floor by HE 300 M transverse beams to a longitudinal beam, cranked on plan, which is supported on the intermediate columns. These radial transverse beams are interconnected by secondary beams (HE 200 A and IPE 240) at 2 m centres. A generally similar arrangement of main beams, secondary beams and edge beams occupies the space between the cranked longitudinal beam and the rear façade of the building. The beams are stiffened by horizontal cross bracings. All the structural connections are bolted and are treated as pin-joints for the purpose of structural analysis.

- The arcade portal frames are composed of welded I-sections, span 11 m, height 8.35 m. Web depth of columns 600 mm, of horizontal members 710 mm. Each portal frame column has a 60 mm thick base plate held down by four 42 mm dia bolts to the reinforced concrete walls of the basement.
- Floors: Floor beams carry steel trough decking, plastered on the underside and carrying a layer of tar-bonded crushed stone, several layers of tar felt and 4 cm reinforced concrete topping. Imposed load 350 kg/m^2.
- Roof structure: Cranked purlins and straight rafters (parallel to the external façade) of I-section. Roof covering consists of 0.8 mm copper sheet on softboard.
- Fire protection: Sprayed asbestos on internal columns and floor beams. External columns unprotected.

Services

Low-pressure air-conditioning system together with hot water central heating fed by three oil-fired boilers, total capacity 1.26 million kcal/h; air-conditioning plant in basement.

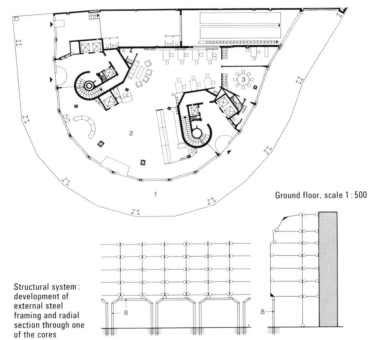

Ground floor, scale 1:500

Structural system: development of external steel framing and radial section through one of the cores

1 Arcade
2 Main banking hall
3 Conference room
4 Offices
5 Central corridor
6 Underground garage
7 Ramp to garage

8 Arcade portal frame
9 Anchor bolt 42 mm dia
10 Portal frame column, welded I-section, flange 400 × 55 mm, web 600 × 25 mm
11 Site joint
12 Portal frame cross beam, welded I-section, flange 600 × 25 mm, web 710 × 25 mm
13 Purlin IPE 220
14 Rafter IPE 120
15 Transverse beam HE 300 M
16 Secondary beam HE 200 A
17 Longitudinal beam HE 300 M
18 External column HE 300 B
19 Internal column, welded I-section 400 × 340 m
20 Wind-bracing, flat 80 × 8 mm

Areas and volume

gross area	7780 m^2	volume	36 157 m^3

Quantities of materials

	steel	concrete	reinforcement
total	826 t	2105 m^3	212 t
per m^3 volume	22.9 kg	0.058 m^3	5.9 kg
per m^2 gross area	106.1 kg	0.270 m^3	27.2 kg

Construction cost (1969)

total cost	1.2 milliard lire	per m^2 gross area	154 252 lire
per m^3 volume	33 188 lire		

Reference

L'architettura, 176/June 1970, p. 76.

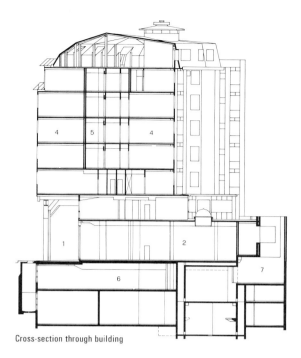
Cross-section through building

Interior view showing radial transverse beams extending from external columns to a concrete core

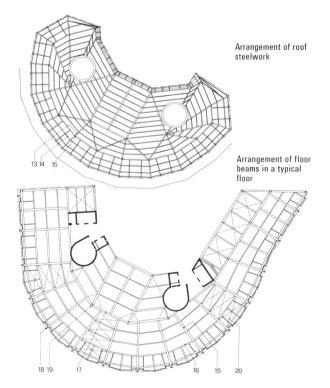
Arrangement of roof steelwork

Arrangement of floor beams in a typical floor

Erection of the steelwork: steel trough decking being laid on floor beams

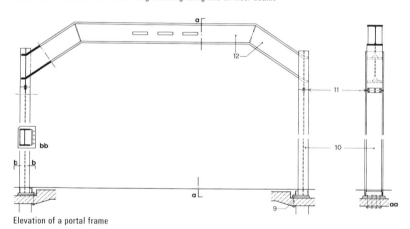
Elevation of a portal frame

External columns terminating on the two-storey arcade portal frames which support them

51 ADMINISTRATIVE BUILDING FOR AN ELECTRICAL ENGINEERING FIRM IN MUNICH, GERMANY

Architectural design: W. Henn, Brunswick, in collaboration with the Building Department of Osram GmbH, Munich.
Structural design: G. Scholz, Munich.
Built: 1964/1966.

South elevation with main entrance

Function

Administrative building for a staff of 850 of whom 610 are in offices. Basement: exhibition space, archives, store, technical services. Ground floor: entrance hall, kitchen, dining room, post room and general services. First to fifth floors: large open-plan office and eight individual offices on each floor. Data processing equipment on first floor; conference facilities on fifth floor.

The main vertical circulation core, located along one side of the building, comprises four passenger lifts, a goods lift, main staircase, service shafts and cloakrooms. A secondary core with emergency stairs is located in the interior. In addition, there are three hydraulic service lifts operating between the basement and ground floor levels in the canteen and general services area.

Structural system

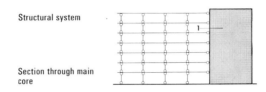

Section through main core

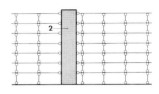

Section through secondary core

Dimensions

Square on plan with overall dimensions 50.30 × 50.30 m, suited to requirements of large open-plan offices. Height above ground level 26.65 m. Basement: storey height 4.00 m, headroom 2.50 to 3.00 m. Ground floor: storey height 4.50 m, ceiling height 3.50 m. Upper floors: storey height 4.00 m, ceiling height 3.00 m.

Structural features

Tubular steel columns, 298 mm dia, on a planning grid of 7.85 × 7.85 m, are interconnected in one direction by main beams; in the other direction these beams and columns are joined to one another by secondary beams. At the perimeter of the building the floor beams cantilever 1.50 m. The joints at the intersections of the floor beams are of rigid construction with cleated connections. Wind forces are transmitted in both directions through the floor slabs to the two reinforced concrete cores.

- Floors: Main and secondary beams are welded I-sections, 680 and 440 mm deep respectively. Secondary beams carry precast reinforced concrete slabs, 3.92 × 1.96 m × 12 cm thick. Imposed load 500 kg/m². Ceilings composed of soundproof aluminium panels.
- Façade: Aluminium curtain walling attached to the cantilever ends of the floor beams. Windows have Thermopane glazing. Aluminium spandrel panels backed by precast concrete with thermal insulation. Sun protection by means of externally fitted aluminium louvre blinds operated electrically by central control for each façade. In the individual offices independent control of the blinds is also possible. In addition, the whole sun protection system is connected to wind gauges and gale warning monitors on the roof.
- Roof: The flat roof is protected against solar heat by slabs of concrete laid on a bed of sand.
- Fire protection: Steel columns encased in hard gypsum plaster; floor beams sprayed with asbestos 3 cm thick. Spandrel walls and window lintels behind curtain wall are of concrete. In the open-plan offices there is one ionisation fire detector per 150 m² of floor area.
- Foundation: Subsoil is alluvial gravel of varying density, so that in places it was necessary to provide compaction. Water table is 4 m below ground level. Columns have individual footings; concrete foundation slabs under the concrete cores.

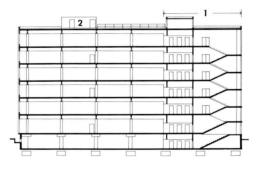

Section through building at main core

Services

The building is connected to the city's district heating system. Low-pressure hot water system for supplying heat to the air-conditioning plant and for heating the ancillary rooms in the main core. Open-plan offices air-conditioned by high-pressure installation with convectors at the perimeter, combined with low-pressure system for the inner zones. Total air delivery rate 263 500 m³/h. Minimum air intake rate into system in summer: 75 m³ per person per hour.

Typical upper floor, scale 1:850

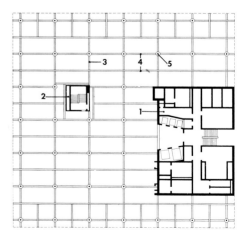

Arrangement of beams in an upper floor

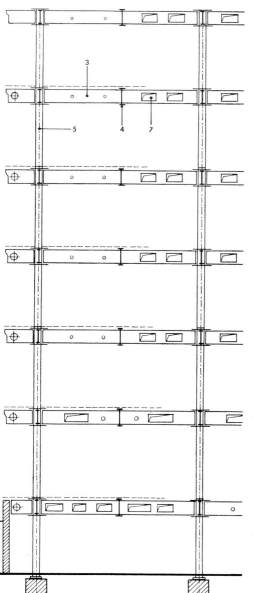

Details of structural steelwork

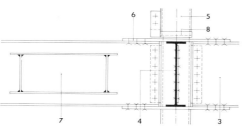

Connection of floor beams to tubular column; opening for services formed in web of main beam

1 Main core
2 Secondary core
3 Main beam 680 mm deep
4 Secondary beams 680 and 440 mm deep
5 Tubular column, 298 mm dia
6 Gusset plate
7 Opening in beam
8 Sleeve around column
9 Casing to air-conditioning convector
10 Floor slab
11 Spandrel slab
12 Cladding attachment
13 Insulating panel
14 8 mm aluminium plate
15 Mullion
16 Sun-blind casing
17 Curtain lighting
18 Ceiling

Laying the precast concrete slabs

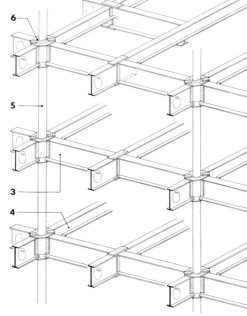

Steelwork at corner of building, with cantilevering floor beams for attachment of cladding

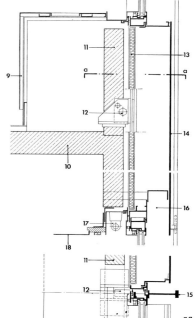

Vertical and horizontal section through façade

Steelwork being erected, with the first and second floor spandrel panels already fixed

Areas and volume

gross area	18 290 m²	area on plan	2586 m²
net area	16 360 m²	core area	484 m²
effective floor area	12 311 m²	volume	74 137 m³

Quantities of materials

	steel	concrete	reinforcement
total	1112 t	4071 m³	222 t
per m³ volume	15.0 kg	0.055 m³	3.0 kg
per m² gross area	60.8 kg	0.223 m³	12.1 kg

Construction cost (1966)

total cost	20.83 million DM	per m² gross area	1139 DM
per m³ volume	281 DM	per m² eff. floor area	1692 DM
per workplace	24 506 DM	per m² net area	1273 DM

References

R. Hohl: *Bürogebäude international*, Stuttgart, 1968.
G. Feuerstein: *New Directions in German Architecture*, New York, 1968.
Deutsche Bauzeitschrift: *Baufachbuch No. 4*, Gütersloh, 1969.
Bauen + Wohnen, 2/1964, p. 72.
Baumeister, 10/1966, p. 1157.
Bauwelt, 4/1966, p. 95.
Deutsche Bauzeitschrift, 1/1967, p. 51.
Bouw, 37/1968, p. 1289.
Bauen in Stahl, 13/1969, p. 73.

52 ADMINISTRATIVE BUILDING FOR AN OIL COMPANY IN ROME

Architectural design: L. Moretti, V. Morpurgo, Rome.
Structural design: A. Bolocan, G. Dori, G. Caccialanza and steelwork contractors.
Built: 1965/1966.

Function

Head office building for an oil company, comprising two similar buildings situated symmetrically in relation to a busy street. Both buildings are T-shaped on plan, each comprising a long seven-storey main block and a four-storey block set transversely at one end.
Each building has a three-level basement, at the bottom level of which is parking space for 500 cars. On the level above is the air-conditioning plant, while the top basement contains a canteen, an auditorium with seating capacity for 300 persons, a post office and ancillary rooms. The building itself accommodates large open-plan offices, individual offices, conference rooms, archives, a telephone exchange, a computer centre and rooms for medical services. On the centre line of the longer block in each building are two vertical circulation cores comprising, in all, nine passenger lifts, one goods lift, three staircases and cloakrooms.

Dimensions

Overall dimensions: main blocks 23×141 m, transverse blocks 23×50 m. Height above ground level 26.9 m. On ground floor: storey height 4.70 m, ceiling height 4.00 m; on upper floors: 3.70 and 3.00 m respectively.

Structural features

Pin-jointed structural framework comprising simple storey-height columns arranged on a 9.00×9.00 m planning grid. HE 500 B main beams are continuous over the columns in the longitudinal direction; HE 220 A secondary beams span between the main beams and are spaced at 2.25 m centres. The main beams cantilever out 1.50 m on the 1st to 4th floors and 3.00 m on the 5th to 7th floors at the ends of the buildings. The secondary beams cantilever out a distance of 1.50 m along the longitudinal sides. In both directions the building derives rigidity to resist wind forces from two concrete cores on the centre line. The basements are in reinforced concrete.
• Columns are I-sections which, in the bottom storeys, have been strengthened with welded plates to form 340×240 mm box sections and filled with concrete as protection against corrosion.
• Floors: To ensure structural continuity, the secondary beams are rigidly connected to the central main beams with high-strength friction-grip bolts. To resist the support moments and reduce deflections, the secondary beams are haunched, the haunches being formed from cuttings of beam sections welded to the bottom flange. Secondary beams carry 9.00 m long sheet-steel decking units, spanning longitudinally, with 3.5 cm concrete filling. These units are 60 mm deep and 630 mm wide. To obviate shrinkage cracks and to ensure load distribution (imposed load 350 kg/m^2), the concrete is reinforced with steel fabric. Additional stiffening of the floors is provided by horizontal bracing in the edge bays and in the bays adjacent to the concrete cores.
In the area between the cores is an expansion joint which extends down into the foundations. In the main beams the expansion joint is formed with web splice plates having slotted holes which permit longitudinal movement and yet allow shear to be transmitted across the joint.
• Façade: Fixed sun protection fins of plastic-coated aluminium placed externally to the façades give vertical emphasis. The windows have solar heat resistant glass in aluminium frames. Spandrel panels comprise two plates of safety glass with thermal insulation sandwiched between.
• Foundations: The subsoil comprises loam, sand, gravel and limestone strata. Piled foundations comprise 62 cm dia piles, 20 to 25 m long, with a bearing capacity of 60 t per pile.

View of the two buildings with a street running between them

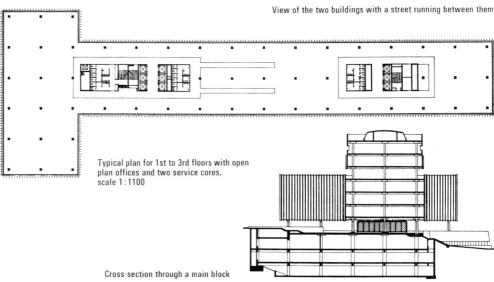
Typical plan for 1st to 3rd floors with open plan offices and two service cores, scale 1:1100

Cross-section through a main block

Structural system: cross-section and longitudinal section

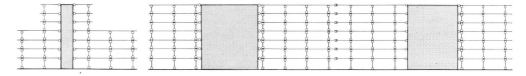

Areas and volume (for one building)

gross area	48 168 m^2	area on plan	3643 m^2
effective floor area	37 300 m^2	volume	189 810 m^3

Quantities of materials (for one building)

	steel	concrete	reinforcement
total	1869 t	19 220 m^3	2256 t
per m^3 volume	9.8 kg	0.101 m^3	11.9 kg
per m^2 gross area	38.8 kg	0.399 m^3	46.8 kg

References

Acier-Stahl-Steel, 3/1966, p. 125.
Detail, 5/1968, p. 929.

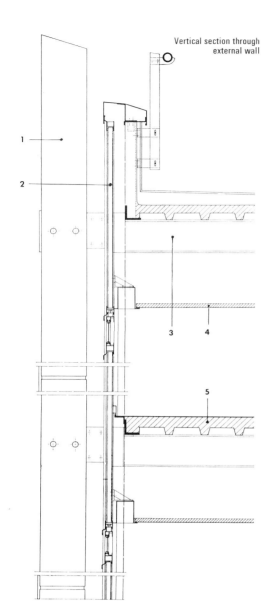

Vertical section through external wall

Part of steelwork on completion of erection

Connection of main and secondary beams to an internal column

Interior view of an upper storey before pouring of concrete floor

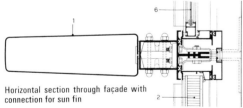

Horizontal section through façade with connection for sun fin

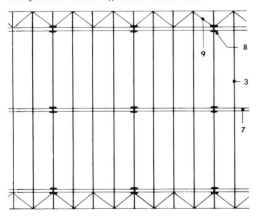

Arrangement of beams in a typical floor in a main block

1 Fixed fin for sun protection
2 Spandrel panel
3 Secondary beam HE 220 A
4 False ceiling
5 Steel trough decking with concrete floor
6 Fixed glazing
7 Main beam HE 500 B
8 Column 340 × 240 mm
9 Horizontal bracing
10 Welded haunch
11 Edge section enclosing floor slab

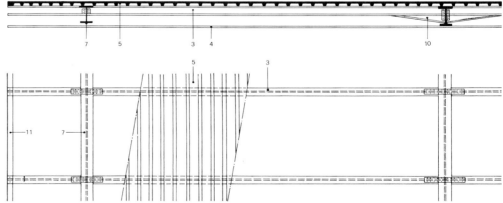

Floor framework, cross-section and plan

53 ADMINISTRATIVE BUILDING FOR AN ELECTRICAL ENGINEERING FIRM IN SAINT DENIS, FRANCE

Architectural design: B. Zehrfuss, Paris.
Structural design: J. Prouvé, L. Fruitet.
Built: 1969/1970.

Function

This suspended structure, which is square on plan, provides 4100 m² of effective floor area on seven floors. The entrance hall with reception desk and exhibition space are at ground floor level, while the services are in the basement. The seven upper floors contain office space, without internal supports, arranged around a central core in which there are three lifts, a spiral staircase, cloakrooms and a shaft for services.

Dimensions

The overall dimensions of the upper floors are 28.80 × 28.80 m, but on the recessed ground floor they are 21.60 × 21.60 m. The core is 9.80 × 9.80 m square on plan, the height above ground level being 38.20 m. The ceiling height in the entrance hall is 4.50 m. The upper storeys are 3.56 m high with a ceiling height of 2.70 m.

Structural features

The main beams of the upper floors are supported at their inner ends by the reinforced concrete walls of the core, while their outer ends are attached to hangers suspended in front of the façades. The tensile forces induced in the hangers are in turn absorbed by diagonal ties on the roof, which transmit these forces to the concrete core. Wind forces are transmitted through the solid floors to the core.

- Floors: From each face of the core to the hangers at the perimeter there are three welded I-section main beams, 700 mm deep, which are interconnected by I-section transverse beams, 300 mm deep. The beams are also interconnected by an edge beam in the plane of the façade, 1.05 m behind the hangers. Between the main and transverse beams is a system of I 140 secondary beams, spaced at 1.20 m centres and supporting 8 cm thick concrete slabs. The imposed load for the upper floors is 250 kg/m². Floors are stiffened by diagonal bracings in the bays adjacent to the core walls.
- Vertical hangers: Pairs of round steel bars, 45 mm dia, placed 1.05 m in front of the façade and spaced at 4.80 m centres. The bars are connected to the projecting ends of the floor beams, in a staggered pattern.
- Diagonal ties: Each tie comprises two 80 mm dia steel bars anchored to a square framework of I-section beams in the concrete core.
- Façades: The aluminium alloy cladding panels, which measure 2.86 × 1.20 m, are fitted between the projecting floor beams. The panels comprise 30 mm thick foamed polyurethane thermal insulation, with a vapour barrier on the inside. The windows have double glazing, tinted grey.

Services

This building is connected to an existing building by a service passage. The upper storeys are fully air-conditioned with high-pressure units at each window. The ground floor has a low-pressure air-conditioning system.

Diagonal view of completed building

The floors span between the core and the external hangers

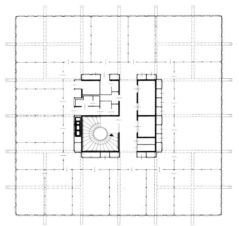

Typical floor   Arrangement of floor beams

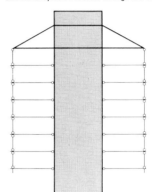

Structural system: section through the core

1 Main beam of welded I-section, 700 mm deep
2 Transverse beam, I 300
3 Pressed steel edge beam
4 Secondary beam, I 140
5 Hanger, two 45 mm dia bars
6 Horizontal wind-bracing
7 Concrete core
8 Diagonal tie, two 80 mm dia bars
9 Anchor frame

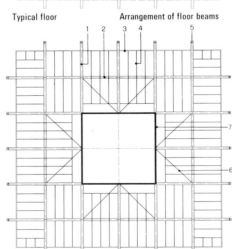

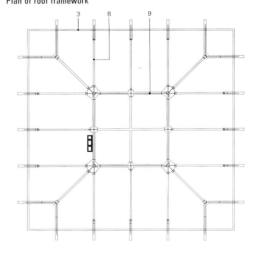

Plan of roof framework

Areas and volume

gross area	6084 m²	area on plan	676 m²
effective floor area	4800 m²	volume	17 800 m³

Quantities of materials

	steel	concrete
total	360 t	1700 m³
per m³ volume	20.2 kg	0.096 m³
per m² gross area	59.2 kg	0.279 m³

References

Acier-Stahl-Steel, 10/1971, p. 398.
Detail, 1972, Konstruktionstafel.

Attachment of lower end of raking stays to edge of roof

Attachment of raking stays to anchorage in the core

Attachment of upper end of raking stay

10 Connection to core
11 Connections for wind-bracing and transverse beam
12 Bearing plates

Attachment of lower end of raking stay

Vertical section through suspension system

Suspension of upper floors

Steelwork for the floors

Attachment of floor beams to hangers

145

54 OFFICE BUILDING FOR AN INSURANCE COMPANY IN LONDON, ENGLAND

Architectural design: Gollins, Melvin, Ward and Partners, London.
Structural design: Scott, Wilson, Kirkpatrick and Partners, London.
Built: 1964/1969.

Function

The building is a 30-storey suspended structure containing 26 100 m² of office space for a staff of 3000. The entrance hall extends through a height of two storeys. From the 2nd to the 13th and from the 16th to the 27th floors are open-plan offices and smaller individual offices. There are two-storey high plant rooms at mid-height and at the top of the building. On the 12th floor there is a data processing centre. A restaurant and kitchen are accommodated on the upper basement floor while there are goods delivery facilities and parking space for 178 cars on the floor below, and the two bottom floors of the four-storey basement contain store rooms and vaults.
There are twelve lifts and two staircases in the central core.

Dimensions

Overall dimensions 37.50 × 37.50 m. Height above ground level 118 m. Storey height generally 3.66 m, ceiling height 2.69 m. Entrance hall 7.80 m high. The central service core is rectangular on plan, 22.85 × 15.25 m.

Structural features

In each storey, the floor beams are supported at their inner ends in pockets formed in the concrete core, while their outer ends are attached to steel hangers in the façades. At about mid-height and at the top of the building the hangers transmit their loads, from the twelve storeys below these levels to a cantilever frame built out from the core. Wind forces are also transmitted to the core, through the floor slabs.
• Floors: Castellated floor beams, supported internally by the four walls of the core, are interconnected by a lattice girder all round the façades and attached by HSFG bolts to the hangers. The floor beams are 685 mm deep I-sections spaced at 3.90 m centres and spanning 11.40 or 8.00 m.
The 13 cm thick lightweight concrete floor slabs are in composite construction with the floor beams, welded studs being used as shear connectors. Imposed load on floors 488 kg/m². The hangers are flat bars, 228 mm wide, increasing in thickness with height from 19 to 51 mm.
• Cantilever frames: The hangers in the façades are suspended from edge girders, those parallel to the two longer sides of the core being plate girders, while those in the other direction are two-storey high lattice girders. From these edge girders the vertical forces exerted by the hangers are transmitted through diagonal ties and horizontal struts (four on each longer side and two on each shorter side of the core) to the core of the building. The horizontal forces transmitted by the cantilever frames to each side of the core balance one another, the diagonal ties being interconnected by prestressed high-tensile steel ties linking their upper points of attachment, while the struts transmit their thrust through the floor slabs within the core. The vertical forces

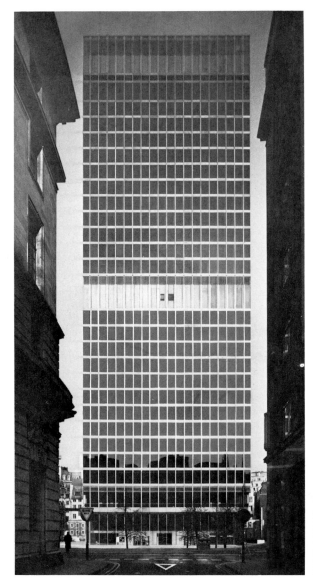

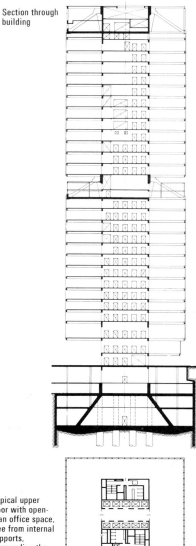

Section through building

Typical upper floor with open-plan office space, free from internal supports, surrounding the vertical core, scale 1:1200

Stages of construction:

A Concrete the twelve cylinder piles and the diaphragm wall for the future excavation.
B Cast columns on piles. Excavate within diaphragm wall, construct slab and commence casting core walls.
C Core in progress. Cast perimeter section of intermediate basement slab to strut diaphragm walls.
D Core continued. Complete excavation and cast remainder of raft and walls.
E After completion of core, erect steelwork from the top downwards, then cast the concrete floors.

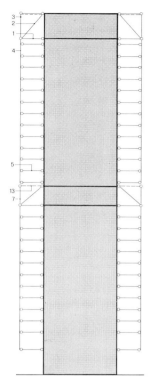

Structural system: section through core in longitudinal direction

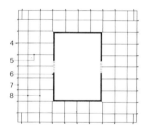

Beam arrangement in a typical floor

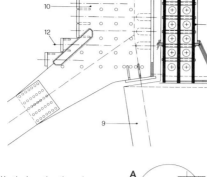

Vertical section through upper cantilever frame

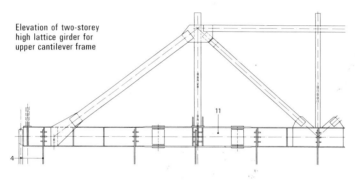

Elevation of two-storey high lattice girder for upper cantilever frame

1. Lattice strut
2. Diagonal tie, six 762 × 46 mm plates
3. Lattice girder, two storeys deep
4. Hanger, flats 228 × 51 to 19 mm
5. Castellated beam, 685 mm deep
6. Concrete core
7. Longitudinal lattice girder, chords L 101 × 89, diagonals L 63 × 51
8. T-section longitudinal beam
9. Bearing bracket
10. Anchor block
11. Welded I-section, 1220 mm deep
12. Centre lines of prestressing bars
13. Stabilising girder

Erection of lower cantilever frame after completion of steelwork for the storeys in the top half of the building

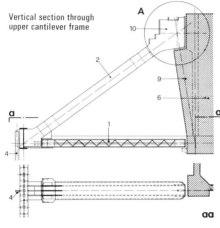

Part of steelwork for lower cantilever frame

Attachment of floor beams to hangers

from the ties are transmitted to the core through the special brackets provided.

The bottom chords of the two-storey high lattice girders are of welded I-section, 1220 mm deep, with plate thicknesses up to 62 mm. The top chords and vertical posts are universal sections and the diagonals are made from plate. The diagonal ties in the cantilever frames are each composed of six plates, 762 × 46 mm, which are 8.70 m long on the longitudinal sides of the core and 6.10 m on the transverse sides. At its lower end, each tie is bolted through three gusset plates to the horizontal strut and to the perimeter girder. Each strut is a lattice member with angle section lacing bars.

- Façade: The mullions consist of extruded anodised aluminium sections spaced at 1.95 m centres, every second mullion enclosing a hanger. The glazing extends over the full depth of a room with aluminium bands at floor level.
- Fire protection: The cantilevering steelwork in the plant rooms, including the perimeter girders, is encased in lightweight concrete to give four hours' fire resistance. The hangers were shot-blasted, zinc-sprayed and painted with one coat of zinc-chromate primer at the works.

After erection of the window wall panels, the hangers were given a further coat of zinc-chromate paint and two coats of micaceous iron-ore paint. The floor beams were shot-blasted and painted with zinc-chromate primer before delivery to the site.

Areas

gross area	53 500 m²	area on plan	6500 m²
effective floor area	25 600 m²		

Quantity of steel

total	2540 t	per m² gross area	47.5 kg

Construction cost

total cost	£8.0 million	per m² gross area	£150
per person	£2667	per m² eff. floor area	£312

References

The Structural Engineer, 4/1967, p. 143.
Acier-Stahl-Steel, 6/1969, p. 269.
Der Stahlbau, 2/1968, p. 41.
Architectural Review, 6/1970, p. 420.
Design, 11/1970.

55 BANKRASHOF OFFICE BUILDING IN AMSTELVEEN, HOLLAND

Architecture design: Vink and Van de Kuilen, Klein.
Structural design: A. J. Uilenreef N.V.
Built: 1968/1970.

Function

This 13-storey office building, square on plan, is built around a central core containing a staircase, two lifts, cloakrooms and a service shaft. On the ground floor there are the entrance hall, shops, offices and ancillary rooms. From the 1st to the 11th floor there are 4150 m² of office space, free from internal supports. The 12th floor comprises store rooms, heating plant and lift machinery. The basement contains further store rooms and the boiler for central heating.

Dimensions

22.0 × 22.0 m overall. Height above ground level 47.6 m. Core 7.20 × 6.80 m. Storey height generally 3.60 m, but 4.02 m on ground floor, headroom 3.00 m.

Structural features

The floor beams in the upper storeys are supported on the inside by the concrete walls of the central core and on the outside through simple bolted connections to the external columns in the façades. Wind forces are transmitted through the composite floors to the core and to vertical bracings in the external walls, the cladding for which appears as a vertical band on each of the sides of the building. The columns, spaced at 3.60 m centres, consist of HE 260 A sections at the base of the building, reducing to HE 180 A at the top. These columns are supported by short lengths of I 400 beams embedded in the top of the walls of the reinforced concrete basement.
On each side of the core there are three floor beams, spaced at 3.60 m centres and spanning 7.20 m. Beams located on the line of two of the core walls are HE 280 A sections which serve to carry the floor beams of the square corner bays; all the other beams are HE 260 A. Each pair of beams having their joint at a corner of the core is connected to vertical steel plates provided with welded anchor bars which are embedded in the concrete. The intermediate beams are supported in recesses in the core walls. The columns are continuously linked together by a perimeter beam.
- Floors: 15 cm thick precast lightweight concrete slabs with 5 cm in-situ concrete topping are supported by the floor beams, with which they act compositely by means of welded shear studs and in-situ concrete in the joints. Imposed load 400 kg/m².
- Fire protection: On their outer side the columns are cased with fireproof slabs, fitted behind the façade cladding, while the inner surfaces of the columns are protected by sprayed asbestos applied to expanded metal lathing.
- Façade: Hardwood window frames. Spandrel panels consist of 12 mm thick softboard with bonded aluminium sheet. Secondary framework for cladding consists of horizontal steel sections at spandrel and lintel level. Sun protection by internally fitted blinds.
- Foundation: Sand stratum under a 70 cm thick layer of peaty topsoil; water table 30 cm below ground level. Piled foundation; basement constructed as a watertight tank. Under the core are 60 piles with 33 more under the external columns. Piles 7.50 m long, 35 × 35 cm in section, enlarged to 50 × 50 cm at the tip; toe pressure 35 kg/cm².

Services

Gas-fired central heating with capacity of 635 000 kcal/h. Upper floors have radiators and are ventilated at a rate of three air-changes per hour.

Structural steel frame temporarily braced before the concreting of the floors

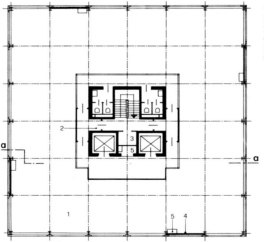

1 Open-plan office
2 Fire refuge area
3 Shaft
4 Unglazed strip with wind-bracing
5 Vertical service duct
6 Composite floor of precast slabs

Structural system: section through core

Typical floor, scale 1:350

Areas and volume

gross area	6976 m²	area on plan	1169 m²
effective floor area	4988 m²	volume	26 690 m³

Quantities of material

	steel	concrete	reinforcement
total	275 t	1055 m³	106 t
per m³ volume	11.3 kg	0.040 m³	4.0 kg
per m² gross area	39.4 kg	0.151 m³	15.2 kg

Construction cost (1967)

total cost	4.15 million D.fl.	per m² gross area	595 D.fl.
per m³ volume	116 D.fl.	per m² eff. floor area	832 D.fl.

References

Bauw, 10/1970, p. 380.
Bouwen met Staal, 11/1970, p. 4.
Acier-Stahl-Steel, 1/1971, p. 2.

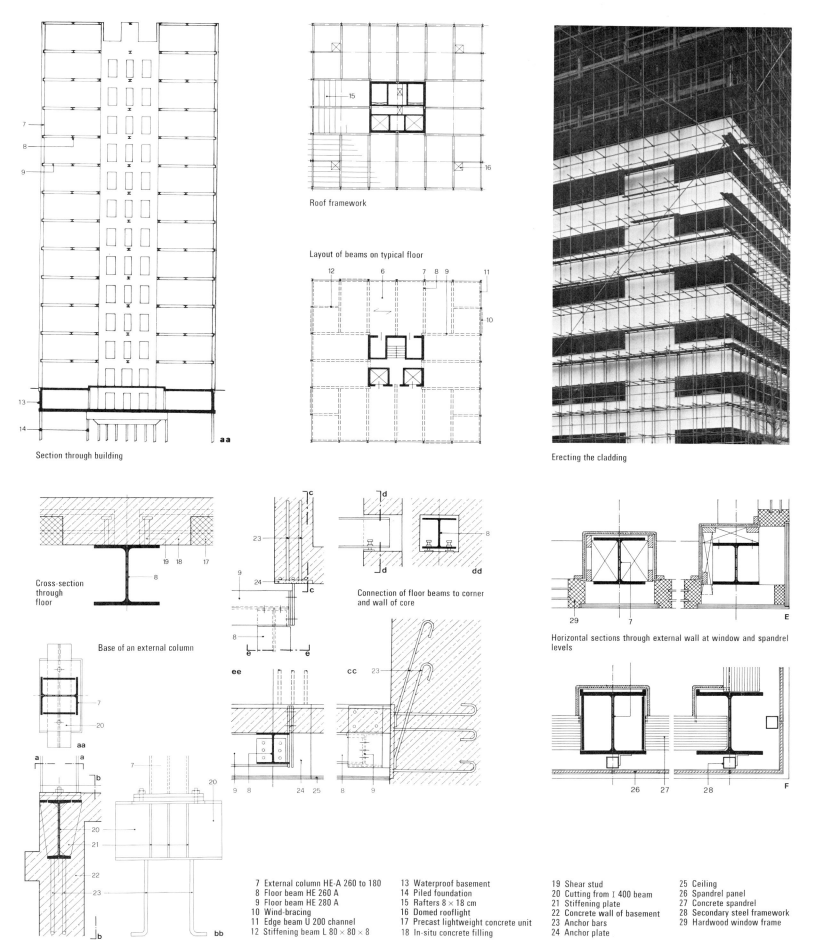

Section through building
Roof framework
Layout of beams on typical floor
Erecting the cladding
Cross-section through floor
Base of an external column
Connection of floor beams to corner and wall of core
Horizontal sections through external wall at window and spandrel levels

7 External column HE-A 260 to 180
8 Floor beam HE 260 A
9 Floor beam HE 280 A
10 Wind-bracing
11 Edge beam U 200 channel
12 Stiffening beam L 80 × 80 × 8
13 Waterproof basement
14 Piled foundation
15 Rafters 8 × 18 cm
16 Domed rooflight
17 Precast lightweight concrete unit
18 In-situ concrete filling
19 Shear stud
20 Cutting from I 400 beam
21 Stiffening plate
22 Concrete wall of basement
23 Anchor bars
24 Anchor plate
25 Ceiling
26 Spandrel panel
27 Concrete spandrel
28 Secondary steel framework
29 Hardwood window frame

56 OFFICE BUILDING FOR PENSION FUND ADMINISTRATION IN LUXEMBOURG

Architectural design: Ewen, Kayser, Knaff, Lanners, Luxembourg.
Structural design: L. Gehl, C. Klein, A. Loewen, J.-P. Strauss, Luxembourg.
Built: 1972.

Function

Head office of a private staff pension fund, occupying an eight-storey rectangular building and a six-storey L-shaped building. Both blocks have a triple-bay internal layout with offices arranged on each side of a narrow inner core comprising staircases, lifts, cloakrooms and services. In the basements there are heating and ventilation plant, refuse incineration and vacuum cleaning equipment.
Between the blocks is a two-level underground car park for 150 vehicles. At ground level there is a courtyard.

Dimensions

Overall dimensions of the rectangular building 29.40 × 16.80 m, height above ground level 28.0 m. Side lengths of the L-shaped building 56.30 and 40.50 m, width of ends 16.80 m, height above ground level 18.90 m. Storey height on ground floor 3.60 m, ceiling height 3.20 m; on upper floors 3.03 and 2.60 m respectively.

Structural features

On each side of the long internal concrete core is a pin-jointed structural framework comprising external columns, floor beams and edge beams. The external columns, consisting of HE 300 B sections, spaced at 5.40 m centres, are located 17 cm in front of the façade, being provided with short brackets by means of which the channel section edge beams, U 150 × 37, interconnect the columns. The floor beams HE 260 B, spaced at 2.70 m centres, are pin-jointed to the columns and edge beams and are supported in the interior of the building by the core walls, which are provided with bearing recesses for the purpose. The floor beams support 10 cm thick in-situ concrete slabs laid on Holorib steel decking spot-welded to the beams. Imposed load 350 kg/m². Wind forces are transmitted through the floors, which are stiffened by horizontal cross-bracings, to the concrete core and to the end walls which are of masonry construction.

- Façade: In each bay of the structural framework there are four deep aluminium window frames, with double glazing, set between the prominently exposed mullions. The two centre windows in each bay have fixed glazing; the two outer windows can be slid open. Along the outer edges of the floor there are box-section aluminium units in which louvre sun-blinds are installed.
- Fire protection: The floor beams and Holorib decking are protected by sprayed asbestos.
- Foundations comprise pad footings and strip footings on rock. Bearing pressure 5 to 10 kg/cm².

Services

The building is fully air-conditioned. Convector units deliver air which has been appropriately heated or cooled, depending on the external temperature. The underground car park is ventilated and heated with exhaust air from the offices. Heating capacity of the plant is 3.0 million kcal/h, cooling capacity 1.2 million kcal/h, air circulation rate 100 000 m³/h.

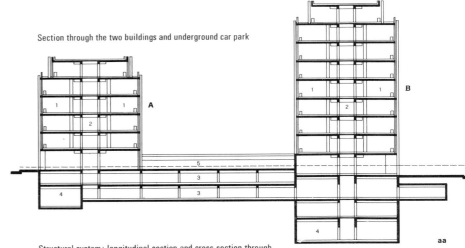

Section through the two buildings and underground car park

Structural system: longitudinal section and cross-section through building

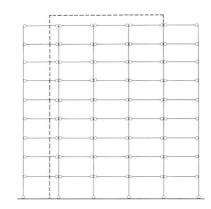

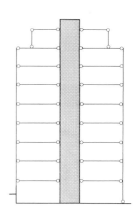

Areas and volume

gross area	17.540 m²	area on plan	2332 m²
effective floor area	15 960 m²	volume	58 563 m³

Quantities of materials

	steel	concrete	reinforcement
total	600 t	4300 m³	310 t
per m³ volume	10.2 kg	0.073 m³	5.3 kg
per m² gross area	34.2 kg	0.245 m³	17.7 kg

Typical floors for both blocks, scale 1:900. 1 Individual office
2 Service core 3 Car park 4 Technical services 5 Courtyard

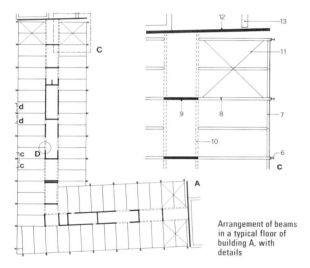

Arrangement of beams in a typical floor of building A, with details

Part of the structural steelwork, showing connection to the concrete core

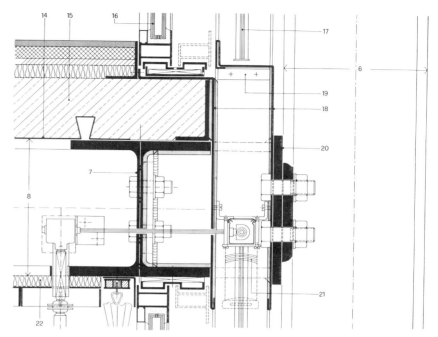

Vertical section through façade with connection between beam and column

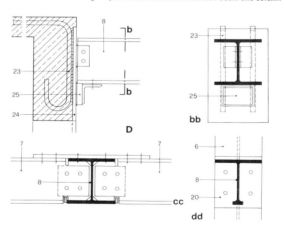

Connection of floor beam to core, cc connection of edge beam to floor beam, dd connection of floor beam to external column

Façade detail

6 External columns HE 220 B
7 Edge beam HE 220 B
8 Floor beam HE 220 B
9 Reinforced concrete core wall
10 Reinforced concrete beam as bearing for floor beams
11 Wind-bracing L 80 × 80 × 7
12 Party wall
13 Old building
14 Holorib decking
15 In-situ concrete slab 10 cm thick
16 Sliding window
17 Guide for louvre blind
18 Aluminium sheet
19 Angle section L 100 × 50 × 6
20 End plate of floor beam
21 Louvre blind
22 Acoustic ceiling
23 Reinforcement
24 Anchor plate 15 mm thick
25 Bearing cleat

151

57 OFFICE BUILDING IN PUTEAUX, FRANCE

Architectural design: J. Binoux, M. Folliasson, Boulogne-sur-Seine; J. Fayeton, Paris.
Structural design: Steelwork contractors.
Built: 1966/1968.

Function

This is a 13-storey block of offices, with a floor area of 7400 m², rising above a podium structure containing a car park for 220 cars, a restaurant and technical services. The first twelve storeys in the block contain offices, but on the 13th floor there are conference rooms and a board room, while on the 14th floor there are additional technical services.
Vertical circulation facilities comprise four lifts in the concrete service core and a staircase enclosed in an internal lightweight concrete fire-resisting shaft. The lifts are accessible from all the floors in the podium.

Dimensions

The high-rise block is rectangular on plan, 43.5 × 18.0 m overall. Height above ground level 47.8 m. On the upper floors the storey height is 3.00 m and the ceiling height 2.60 m, but on the ground floor the storey height is 5.60 m and the headroom under the exposed girders is 3.60 m. Between this block and the podium building there is a 5.90 m high circulation area.

Structural features

The high-rise building is supported on two rows of concrete columns, 9.00 m apart, which reach up into the circulation area from the podium beneath. The columns support six transverse girders, normally at 6.00 m centres but 9.00 m at the concrete core, which cantilever 4.80 m out on both sides and serve to take the loads from the storeys above. These transverse girders, which are of welded steel box-section, are 0.90 m wide and 2.00 m deep between the concrete columns, at the position of which they are interconnected by longitudinal girders, 1.00 m deep. Neoprene cushions between the concrete columns and the transverse girders ensure efficient load distribution. At one end of the building two additional longitudinal girders interconnect the last two transverse girders and project a distance of 5.10 m. The cantilevering ends of these girders and of the transverse girders along the two longitudinal sides of the building are welded to a continuous edge girder comprising an HE 1000 M rolled section. The cantilevers were given an initial upward deflection of as much as 13 mm to compensate for the subsequent deformation resulting from the dead weight of the completed building.
The substructure supports the pin-jointed framework of the slab block, consisting of external columns, two rows of internal columns and floor beams. Wind forces are transmitted longitudinally through the floor slabs to the concrete core, 8.00 × 7.00 m, and transversely through the floor slabs to the core and also to the end wall, which is constructed in concrete. The columns are spaced in the longitudinal direction at 3.00 m centres and, in the transverse direction, at 7.90, 3.40 and 7.90 m. The floor beams, which comprise pairs of channel sections, straddle the I-section columns and, where they terminate at the concrete core, they are embedded in it.

- Floors: In-situ concrete, 9 cm thick, with light top reinforcement, laid on Holorib decking, which serves as permanent shuttering and bottom reinforcement, in composite construction with the floor beams. Imposed load 300 kg/m².
- Façades: Lightweight concrete spandrel panels attached to the steel mullions and faced with opaque glass. Window frames of anodised aluminium, also fixed to the mullions. Fixed glazing with louvre blinds which can be lowered either externally or internally, as desired.
- Foundations: The subsoil is sand. The end wall and core are founded on concrete slabs 4.00 m below ground level. The concrete columns are supported on transverse concrete raft foundations, interconnected by longitudinal beams and exerting a bearing pressure of 5 kg/cm². The bearing pressure under slab footings is 4 kg/cm².

Services

Induced-draught air-conditioning system with a convector unit in each bay of the façade. Temperature of air delivered to the system is 12°C in summer, 55°C in winter. Heating installation fired with natural gas. Cooling towers erected on the roof of slab building. Cooling units are accommodated in the basement of the podium block. Boiler capacity 2 × 1.4 million kcal/h. Horizontal services are installed in the space between the ceilings and the floors above.

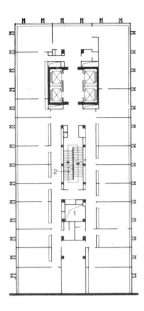

11th floor with offices on each side of circulation zones, scale 1:600
1 Air-conditioning shaft
2 Double staircase with fireproof walls
3 Car parks
4 Restaurant
5 Technical services

Longitudinal façade of high-rise building with podium block in foreground

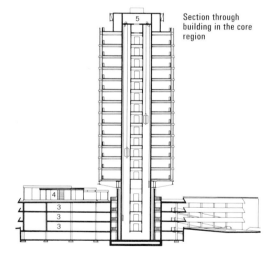

Section through building in the core region

Areas and volume (slab block only)

gross area	10 930 m²	area on plan	783 m²
effective floor area	7400 m²	volume	33 500 m³

Steel content (slab block only)

total	1000 t	per m² gross area	91.5 kg
per m³ volume	29.9 kg	per m² eff. floor area	135.1 kg

Construction cost (1968)

total cost	17.97 million Fr.fr.	per m² gross area	937 Fr.fr.
per m³ volume	332 Fr.fr.	per m² eff. floor area	974 Fr.fr.

Reference

Acier-Stahl-Steel, 7–8/1969, p. 307.

Part of the façade

Substructure on reinforced concrete columns

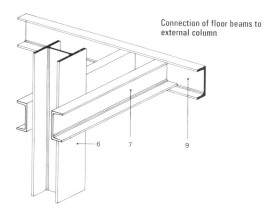

Connection of floor beams to external column

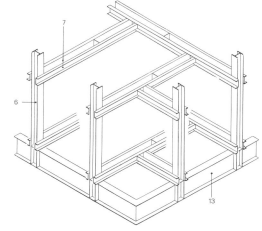

Steel framework supported on substructure at a corner of the building

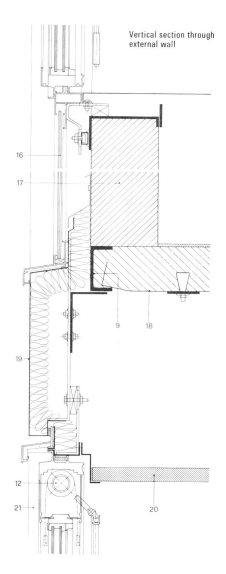

Vertical section through external wall

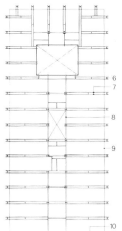

6 External column HE 260 A
7 Floor beam, two U 300
8 Longitudinal beam IPE 100
9 Floor edge beam U 100
10 Reinforced concrete end wall
11 Movable partition

12 Louvre blind
13 Edge girder HE 100 M
14 Transverse girder of substructure, 2000 × 900 mm box section
15 Reinforced concrete column 90 × 90 cm

16 Enamelled glass
17 Lightweight concrete spandrel panel
18 Holorib steel decking
19 Sheet-metal cladding with thermal insulation
20 Ceiling
21 Mullion

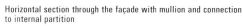

Arrangement of beams in an upper floor

Steelwork in course of erection

Horizontal section through the façade with mullion and connection to internal partition

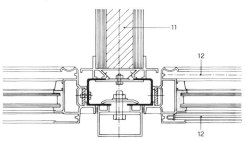

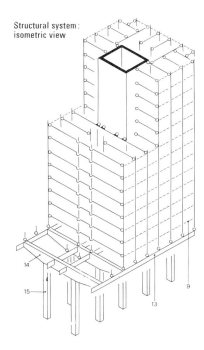

Structural system: isometric view

153

58 OFFICE BUILDING FOR AN INSURANCE COMPANY IN BRUSSELS, BELGIUM

Architectural design: P. Dufau, Paris; R. Stapels, Brussels.
Structural design: Société d'Etudes Verdeyen et Moenaert, Brussels.
Built: 1968/1971.

Function

Head office premises for an insurance company, accommodating a staff of 2500. A square podium block is surmounted by an eight-storey superstructure in the shape of a symmetrical cross on plan. The four wings of the building extend from a central core and are supported on a substructure raised above the podium. The top storeys are set back from the general plane of the façades. The first seven floors of the cruciform superstructure accommodate offices situated on both sides of a central corridor in each wing. On the eighth floor are the board and conference rooms.

The podium block contains a two-storey high entrance hall, an auditorium, a computer centre and various ancillary rooms. The archives and technical services are in two basement storeys.

Access to the upper floors is provided by eight lifts in the central core and by four staircases at the tips of the wings.

In a square annexe there is parking accommodation for 750 cars on four basement levels, together with a restaurant at ground level which communicates with the roof terrace of the podium block and is also connected to the cafeteria under the wings of the superstructure.

Dimensions

Central core 18.5 × 18.5 m. Length of each wing 35.5 m, width 19.5 m. Height above ground level 48.0 m. Storey height 3.50 m, ceiling height 3.00 m. Podium block has overall dimensions of 88.0 × 88.0 m. Annexe 54.0 × 54.0 m.

Structural features

The four wings of the multi-storey superstructure are supported over the reinforced concrete podium block by a substructure, which, under each of the four wings, comprises two longitudinal rows each consisting of three reinforced concrete columns extending down through the podium to the foundation. These columns measure 1.44 × 1.44 m in cross-section and are spaced at 10.50 m centres in both directions. Each row of columns carries a longitudinal capping member comprising a pair of reinforced concrete girders, each 2.20 m deep and 1.25 m wide, spaced 1.44 m apart and joined together over each column. Built into these twin longitudinal concrete girders are ten pairs of transverse steel girders spaced at 3.50 m centres and comprising two welded channel sections, depth 1350 mm, flange width 485 mm, spaced 300 mm apart. Each of these transverse girders supports an eight-storey rigid frame consisting of two external columns HE 300 B, arranged in front of the longitudinal faces of the wings, and two internal columns HE 300 A, all of which are rigidly interconnected at each upper floor by cross beams with spans of 8.05, 2.90 and 8.05 m. All these cross beams comprise twin channels U 300 which straddle the columns.

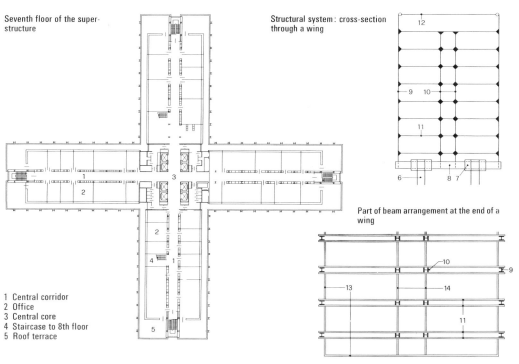

Seventh floor of the superstructure

Structural system: cross-section through a wing

Part of beam arrangement at the end of a wing

1 Central corridor
2 Office
3 Central core
4 Staircase to 8th floor
5 Roof terrace

In the other direction, the rigid frames are linked by I 20 edge beams just behind the longitudinal façades and by I 20 floor beams between the internal columns. The cross beams of the frames in each storey carry an in-situ concrete floor slab, imposed load 400 kg/m². The eighth storey is free from internal columns; the roof being supported by welded plate girders (depth 1020 mm, span 19.00 m). All the externally exposed steelwork is in weathering steel.

Wind forces are transmitted transversely through the multi-storey rigid frames, and in the longitudinal direction through the reinforced concrete floor slabs to the central concrete core.

• Façade: Double glazing of gold-tinted glass is fitted between mullions and transoms consisting of steel sections. The low spandrel walls

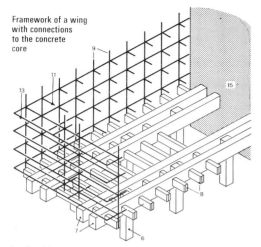

Framework of a wing with connections to the concrete core

Steelwork in top storeys

Connection of floor beams to external column

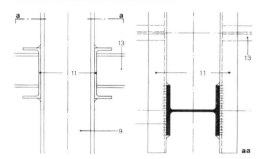

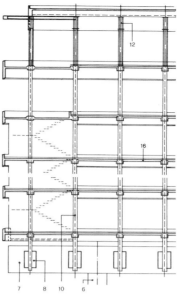

Part longitudinal section through a wing

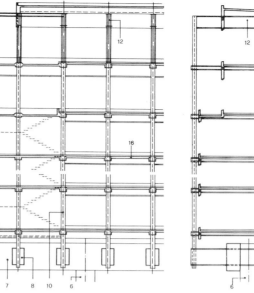

Cross-section through a wing

6 Reinforced concrete column 1.44 × 1.44 m
7 Reinforced concrete girder 2.20 × 1.25 m
8 Transverse girder consisting of two channel sections 1350 × 485 mm
9 External column HE 300 B
10 Internal column HE 300 A
11 Cross beam, two U 300
12 Welded plate girder, 1020 mm deep
13 Edge beam I 20
14 Longitudinal beam I 20
15 Concrete core
16 In-situ concrete floor slab
17 Tinted glass
18 Weathering steel cladding
19 Double glazing

Vertical section and details of external wall

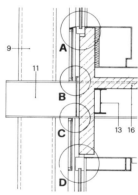

and window heads are clad with bronze-tinted glass, so that the overall effect from the exterior is of glazing extending the full height of each storey. The edges of floors are faced with weathering steel sheet.

• Foundation: The subsoil is clay interspersed with peat, the water table being at ground level. For this reason the building has a piled foundation comprising 1170 piles, interconnected at their heads by a 2.20 m deep reinforced concrete beam grillage as a precaution against differential settlement. To safeguard against buoyancy uplift, the grillage is closed at top and bottom with reinforced concrete slabs and the cavities within this structure are filled with sand.

Areas and volume (including annexe)

area on plan	9350 m²	parking space	25 500 m²
gross area	82 500 m²	office space	49 900 m²
volume	330 000 m³	space per person	20 m²

Quantities of materials

	steel	concrete	reinforcement
total	2582 t	49 000 m³	8430 t
per m³	7.8 kg	0.148 m³	25.5 kg
per m² gross area	31.3 kg	0.594 m³	102.0 kg

References

Acier-Stahl-Steel, 3/1971, p. 97.
La Technique des Travaux, 1971, p. 123.

Erecting the steelwork and concreting the floor slabs

59 ADMINISTRATIVE BUILDING FOR AN ALUMINIUM CONCERN IN SAN FRANCISCO, USA

Architectural and structural design: Skidmore, Owings and Merrill.
Built: 1968.

Function
High-rise building, headquarters of the Aluminum Company of America, with 37 000 m² of office space on 24 upper floors. Technical services on the 25th floor. The entrance, two storeys above street level, is reached from a landscaped plaza and low-rise restaurant buildings. Under the plaza are three storeys accommodating service rooms, goods reception facilities and parking space for 1300 cars. Two escalators provide a link between street level and plaza level. Vertical circulation facilities in the high-rise building are in the central core containing five express lifts, five ordinary lifts and two staircases.

Dimensions
Rectangular on plan, overall dimensions 62.30 × 31.15 m. Height above street level 116 m. Storey height 3.96 m, ceiling height 2.74 m. Height of entrance hall 4.90 m.

Structural features
The building is located in an earthquake zone, the horizontal seismic forces being substantially greater than the wind forces. To cope with them, the structure is provided with diagonal bracings, left externally exposed as characteristic features of the elevational treatment.
In front of each façade are lattice panels comprising continuous columns and two sets of diagonal members. This lattice system stiffens the building against lateral deflection and can absorb relatively large seismic shocks. Additional stiffening of the building is provided by rigid frames in the central core which is so designed, in compliance with local regulations, that it can resist 25% of the seismic forces.
The external lattice system consists of welded box-sections with plate thicknesses up to 100 mm. The diagonals are 410 × 410 mm in cross-section, being arranged at a distance of 360 mm from the façade. The columns, which are 960 × 510 mm in cross-section, are incorporated into the façade. Each corner column is virtually a pair of columns, connected at right angles to one another.
- Floors: Floor loads are supported in the interior by the core walls and at the periphery of the building by the external columns and by hangers. The latter are connected, at every sixth floor, to the intersection points of the diagonal elevation of the building. Floor edge beams positioned just behind the façade are connected to the columns and hangers by means of brackets. The floors span without intermediate columns between these edge beams and the walls of the core. Each floor comprises main and secondary beams and a 6.5 cm thick in-situ concrete slab on steel trough decking. Imposed load 390 kg/m².
- Curtain wall with mullions and transoms of aluminium sections attached to the edges of the floor. Mullions spaced at 1.57 m centres, transoms at 1.98 m. Infilling consists of double glazing; spandrel panels are of dark-tinted safety glass.

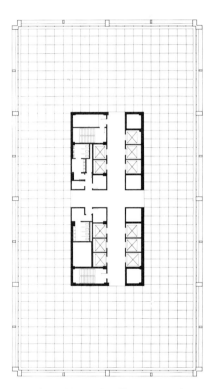

Typical floor with open-plan office space around vertical service core

Cross-section of building

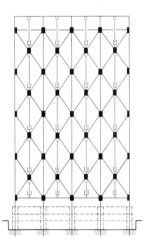

System for the external loadbearing structure

- Fire protection: External columns, diagonals and suspenders are provided with fire encasement clad with aluminium sheet. All the upper storeys have a sprinkler system.
- Foundation: Below plaza level the columns of the high-rise building are embedded in reinforced concrete walls, 51 cm thick. Piled foundation comprises 566 bored piles extending down to rock at 43 m depth. Each pile has a capacity of 200 tons. To resist ground-water pressure a reinforced concrete raft, 30 cm thick, is provided.

Services
Air-conditioning with high-pressure system in the window zones and low-pressure system in the interior.

Areas and volume

gross area	54 846 m²	area on plan	2161 m²
effective floor area	37 281 m²	volume	216 199 m³

Construction cost (1968)

total cost	$ 15.2 million	per m² gross area	$ 277
per m³ volume	$ 70.3	per m² eff. floor area	$ 408

References

Progressive Architecture, 12/1968, p. 84.
Detail, 1/1970, Konstruktionstafel.

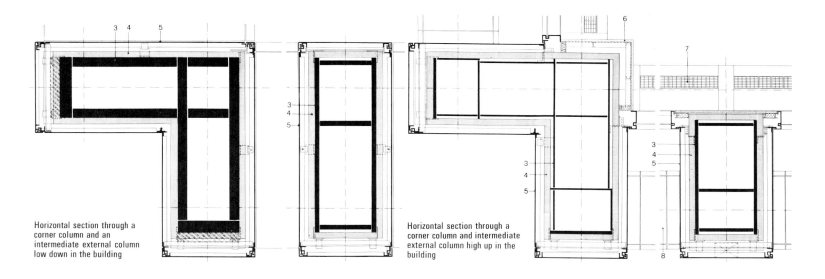

Horizontal section through a corner column and an intermediate external column low down in the building

Horizontal section through a corner column and intermediate external column high up in the building

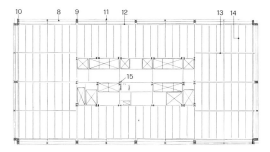

Arrangement of beams in a typical upper floor

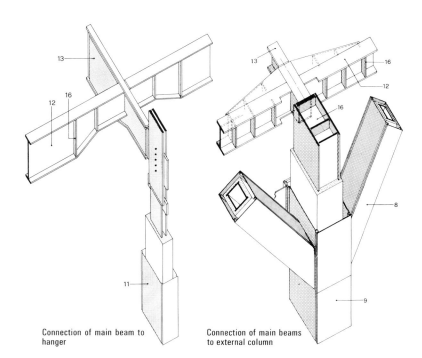

Connection of main beam to hanger

Connection of main beams to external column

1 Plaza level	6 Internal plastering	12 Edge beam, 915 mm deep
2 Reinforced concrete low-rise structure (columns of high-rise block are embedded in the walls of this structure)	7 Convector	13 Main beam, 915 mm deep
	8 Diagonal member (outside dimensions 65 × 65 cm)	14 Secondary beam
	9 External column (outside dimensions 121 × 74 cm)	15 Core column
3 Steel plate up to 100 mm thick	10 Corner column (outside dimensions 175 × 175 cm)	16 Stiffener
4 Fire encasement		17 Mullion
5 Cladding	11 Hanger (outside dimensions 81 × 31 cm)	18 Spandrel panel

Vertical section through spandrel panel on an upper floor

Portion of the longitudinal façade above plaza level

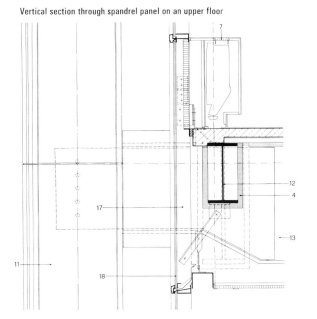

60 TOUR DU MIDI, BRUSSELS, BELGIUM

Architectural design: R. Aerts and P. Ramon, Brussels.
Structural design: A. Lipsky, Brussels.
Built: 1962/1966.

Function

High-rise administrative building and two low-rise subsidiary buildings accommodating the offices of a pension insurance institution. The 37-storey tower block provides 41 000 m² of office space for 3500 people. Technical services are installed in the basement and on the top floor. Vertical circulation facilities comprise eleven lifts and a staircase in the central core: a second staircase is located adjacent to one of the external walls. The subsidiary buildings contain two halls with counters for transacting day-to-day business with the public, together with a restaurant, conference room, data processing equipment and parking space for 300 cars on three basement floors.

Dimensions

The tower block is square on plan with 38.50 × 38.50 m overall dimensions. Height above ground level 149.5 m. Storey height 3.80 m, ceiling height 3.72 m. Central core 19.70 × 19.70 m overall.

Structural features

The upper floors of the building cantilever out on all four sides of the square core, which carries all the vertical and horizontal loads. The core comprises four 700 × 700 mm corner columns (high yield stress steel, grade A 52) joined together by reinforced concrete walls. Connected to the core corner columns at each floor are four continuous steel girders extending across the full width of the building and cantilevering out 9.40 m on each side of the core. For reasons of fabrication and erection, these deep girders could not be so designed as to interpenetrate; for this reason they have been placed parallel to one another on each floor, and in alternate directions at every successive floor. Each floor is enclosed by concrete encased I-section edge beams, 900 mm deep, these beams being attached to the ends of the cantilever girders on each floor. These girders terminate at two of the façades. Along the other two façades the edge beams are either suspended from the cantilever girders of floor above (which run in the other direction) or are supported by those of the floor below. In this way, each two successive storeys form a structural unit. Wind forces are transmitted through the floor slabs to the central core.

• Structural system for the floors: The cantilevered main girders are Preflex plate girders, 1235 mm deep at the core, diminishing in two steps to 430 mm at their ends. Each girder is 38.26 m long and weighs 39 t; the web thickness varies from 8 to 15 mm; the steel is grade A 52. Due to the cantilevering of both ends of the girder, the whole of the bottom flange is in compression. To assist the steel to resist the compressive forces, the bottom flange is encased in concrete which acts compositely with it and also provides fire protection for the steel. To compensate for deflection due to the dead weight of the floors, the girders were cambered prior to erection. By means of hydraulic jacks, such steel girders are given curvature in the opposite direction to their subsequent deflection under working conditions and, in this preflexed condition, the bottom flange is embedded in concrete, which is allowed to harden before the jacks are released. When the steel tends to spring back, it is prevented from doing so by the concrete, which thus acquires a precompression, while the girder as a whole retains a camber of sufficient magnitude to compensate for dead weight deflection.

In the floor areas between the cantilever girders and the edge beams are precast concrete beams spaced at 1.80 m centres. These beams carry an in-situ concrete slab, 8 cm thick, which acts compositely with them. Imposed load 600 kg/m².

• Foundations: To safeguard the building against tilting, the base of the core is widened by means of buttresses composed of steel girders. There are two butttesses at each corner, aligned in the direction of the core walls. The buttresses extend from the foundation (at a depth of 9 m) to the first cantilevered floor above the ground; they are interconnected by horizontal ties at the base. The raft foundation, 54.10 × 54.10 m, is designed as a grillage comprising four intersecting reinforced concrete girders, 5 m deep, with intermediate girders, 3 m deep.

The subsoil is sandy clay; bearing pressure 2.2 kg/cm². Jacking points are provided under the core to compensate for differential settlement.

Erection of the Preflex cantilever beams, in alternate directions at each successive floor, followed by the concreting of the floor slabs.

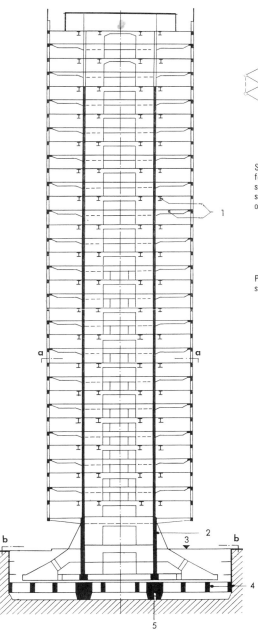

Vertical section through building

Typical upper floor with office space around the vertical circulation core, scale 1:600

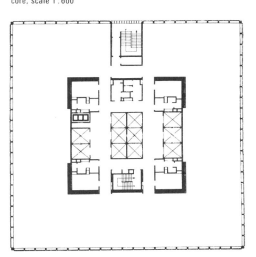

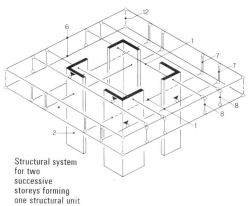

Structural system for two successive storeys forming one structural unit

Plan of roof structure

Steel buttresses at base of core

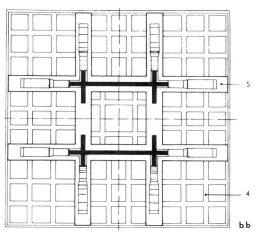

Plan of foundation

1 Main girder
2 Core comprising steel columns embedded in concrete
3 Entrance level
4 Grillage foundation raft
5 Reinforced concrete girder under core wall and buttresses
6 Edge beam
7 Strut, 83 mm dia steel tube
8 Tie, 50 mm dia steel tube
9 Precast concrete secondary beam
10 In-situ concrete floor slab
11 Ventilation shaft
12 Corner post
13 Mullion
14 Air-conditioning unit
15 Glass cladding
16 Ceiling
17 Air-extraction grille

Areas and volume

gross area	85 630 m²	area on plan	1480 m²
effective floor area	63 400 m²	volume	355 800 m³

Quantities of materials

	steel	concrete	reinforcement
total	5778 t	25 000 m³	4100 t
per m³ volume	16.2 kg	0.070 m³	11.5 kg
per m² gross area	67.4 kg	0.292 m³	47.9 kg

Construction cost (1966)

total cost	1400 million B.fr.	per m² gross area	16 349 B.fr.
per m³ volume	3935 B.fr.	per m² eff. area	22 082 B.fr.

References

Architecture, 83/1968, p. 144.
Detail, 5/1968, p. 937.
Der Stahlbau, 2/1968, p. 39.
La Maison, 1/1967, p. 15, and 1/1968, p. 40.

Part cross-section through an upper storey

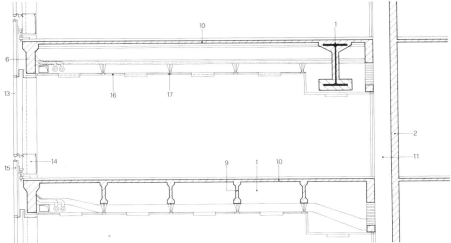

61 HEAD OFFICE BUILDING FOR A STEEL COMPANY IN PITTSBURGH, USA

Architectural design: Harrison and Abramowitz and Abbe, New York.
Structural design: Skilling, Helle, Christiansen, Robertson, New York; Edwards and Hjorth, New York.
Built: 1967/1970.

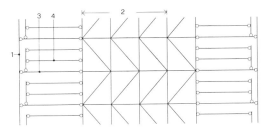

Structural system: part cross-section (above); isometric view showing primary floor beams between core and external columns (below)

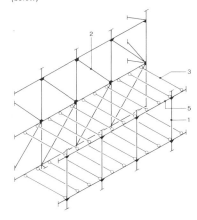

Function

The upper floors of this 64-storey head office building for the United States Steel Corporation contain 145 000 m² of office space. On the ground floor are the entrance hall and two banks of lifts. On the first floor is a hall and a further four banks of lifts. There are four basement levels, the bottom two of which contain an underground car park for 650 vehicles. On the top basement floor are two restaurants, an auditorium, shops, and loading bays for goods vehicles. Another restaurant is located on the top floor of the building. There is a helicopter landing platform on the roof.

The central core contains 54 lifts, three staircases, cloakrooms and service shafts. Successive blocks of eight storeys are each served by eight lifts connecting them to the entrance hall. The basement levels are reached by means of escalators from the entrance hall.

Dimensions

On plan the building has the shape of an equilateral triangle with re-entrant corners, the length of side being 67.32 m. Height above ground level 256 m. The central core is also triangular with sides of 48.30 m, the distance from the façades to the walls of the core being 13.80 m. Storey height on upper floors generally 3.61 m, ceiling height 2.59 m. Height of entrance hall 10.40 m. Storey height of top basement 6.70 m, second basement 7.30 m, parking areas 3.36 m. The dimensional design of the building has been based on a 1.32 m module.

Structural features

The structural system for this high-rise building is determined essentially by the method of transmitting the wind forces to the foundations. The walls of the triangular core are formed with lattice panels which are shear-connected to each other at the corners and whose columns are rigidly fixed to the foundation structure, thus forming a triangular tube functioning as a vertical cantilever which absorbs and transmits all horizontal forces.

At the top of the building a rigid space frame 'hat' connects the core to the 21 external columns which are located 91 cm in front of the façades. The connections between these columns and the space frame are designed to transmit both tension and compression. Thus the top of the building can undergo lateral displacement, but no rotation, so that its deflections are reduced and extreme movements due to temperature variations are also prevented. At every third floor the external columns are joined to one another by horizontal stubs projecting from the façade, which are rigidly connected to them. The beams supporting these floors, known as primary floors, are connected directly to these stubs. The beams for the intermediate floors, called secondary floors, are connected at their outer ends to two-storey high columns which stand, just inside the façades,

1 External column 733 × 712 mm
2 Lattice panel of core wall
3 Primary floor beam, I-section 686 mm deep
4 Secondary floor beam
5 Spandrell girder 1190 × 368 mm

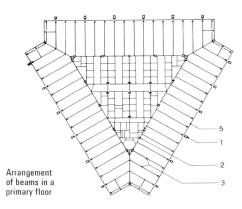

Arrangement of beams in a primary floor

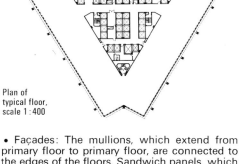

Plan of typical floor, scale 1:400

on the beams for the primary floors. The inner ends of all the floor beams are supported in the core walls.

The external columns and the horizontal stubs in front of the façades are of box-section. These columns, which are spaced at 11.90 m centres, are 733 × 712 mm in section at the base; the breadth remains constant all the way to the top of the building, but the depth progressively diminishes; the plate thicknesses are up to 100 mm. The horizontal stubs are 1190 mm deep. The floor beams, columns supporting the secondary floors and diagonals in the core walls are all of I-section. All externally exposed steel components are in Cor-Ten steel.

• Floors: The floor beams support steel trough decking with a 6.4 cm thick lightweight concrete slab. Suspended ceilings comprise lighting and ventilation units.

• Façades: The mullions, which extend from primary floor to primary floor, are connected to the edges of the floors. Sandwich panels, which are clad with Cor-Ten sheet on the outside face and galvanised steel sheet on the inside, are fitted between the mullions and alternate with bronze-tinted solar heat rejecting glass. Bottom-hinged lights can be opened for cleaning.

• Fire protection: The 18 external box columns, which are filled with water, are subdivided into four units varying from 14 to 18 storeys high, each unit forming a closed circulation system in combination with a vented storage tank. Antifreeze and corrosion-inhibiting agents are added to the water. The beams and internal columns are protected with asbestos.

• Foundations: Subsoil consists of shaly sandstone at 22 m, rock at 37 m depth. Water table

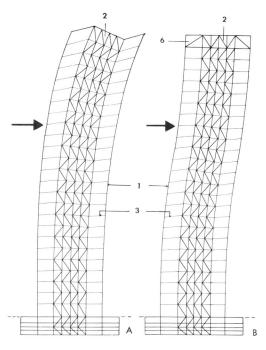

Deflection of the structural system due to wind load:
A Without resistance of external columns
B With resistance of external columns, through space frame in top storey

6 Space frame 'hat'
7 T-1 steel
8 Cor-Ten steel
9 Ex-Ten steel
10 A 36 steel
11 Vented storage tank
12 Upper ring main
13 Lower ring main
14 Column, 152 × 152 mm I-section
15 Stub
16 Air-conditioning duct with air diffuser
17 Lighting unit with aperture for air extraction
18 Cor-Ten steel cladding for spandrel girder
19 Mullion
20 Sandwich panel, Cor-Ten steel sheet on outside
21 Steel trough decking with in-situ concrete floor slab
22 Angle cleat

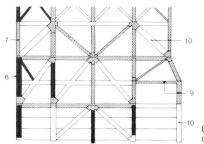

Grades of steel in the bottom storeys of the core structure

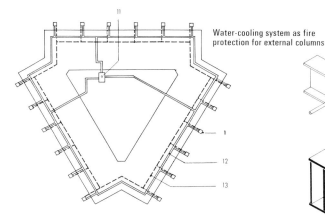

Water-cooling system as fire protection for external columns

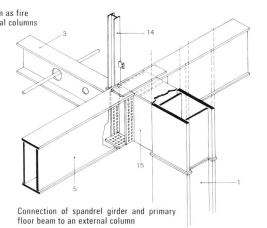

Connection of spandrel girder and primary floor beam to an external column

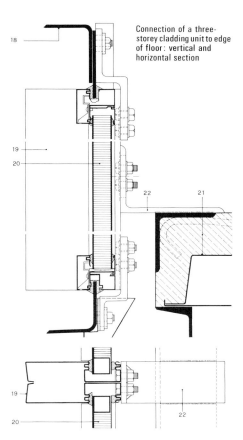

Connection of a three-storey cladding unit to edge of floor: vertical and horizontal section

8.50 m above lowest basement level. External walls of basement anchored to resist water pressure. Bottom slab, resting on the sandstone, is 3.65 m thick under the core and 1.50 m thick outside this zone. Total load transmitted by the building is 28 800 tons; predicted total settlement is 1.12 m.

Interior view showing arrangement of ceilings and partitions

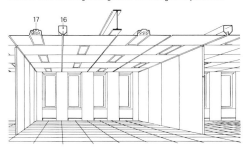

Areas and volume

gross area	270 000 m²	gross area of a floor	3829 m²
area on plan	12 140 m²	net area of a floor	2790 m²

Quantities of steel

total	40 000 t	per m² gross area	148 kg

References

Der Stahlbau, 10/1970, p. 298.
Acier-Stahl-Steel, 9/1967.

62 WORLD TRADE CENTER, NEW YORK, USA

Architectural design: M. Yamasaki and Associates, Troy, Michigan; E. Roth and Sons, New York.
Structural design: Skilling, Helle, Jackson, Seattle.
Built: 1966/1973.

Function

The complex comprises two tower office blocks, each 110 storeys high, together with four 8-storey and 10-storey ancillary buildings which are grouped around the towers and enclose a plaza. The complex provides for a total staff of 50 000 people but it also receives up to 80 000 visitors daily. It accommodates the premises of business firms, insurance companies, banks and public authorities and includes a hotel.

The 110 upper floors contain open-plan offices, free from internal supports, each with 2900 m² of effective floor area. On five basement floors are stations served by underground railway and rapid-transit lines, underground parking space for 2000 cars, and service facilities of various kinds. Below these, on a sixth basement floor and in four double-height storeys distributed over the height of the building, are air-conditioning plant and other technical services.

Each of the tower blocks has 100 passenger lifts and four goods lifts. A lobby on the ground floor and two skylobbies, on the 44th and 78th floor respectively, subdivide the building into three circulation zones. Each skylobby is connected by eleven or twelve lifts to the ground floor. From each of the three lobbies 24 local lifts give access to the floors above. Furthermore, five express lifts go non-stop from the ground floor to the 107th and 110th floors. As three local lifts, one above the other, can operate in the same shaft, only 56 shafts are needed to accommodate the 104 lifts in the building (the liftshafts occupy 13% of the area of each floor). Maximum transit time, including change of lift, is 2 minutes. In an emergency, a fully occupied tower block (assuming it to contain 55 000 people, including visitors) can be emptied in 5 minutes.

Dimensions

Two towers, each 411 m high, 63.5 × 63.5 m square on plan, core 24 × 42 m. Storey height 3.66 m, ceiling height 2.62 m. Height of entrance hall 22.3 m.

Structural features

The structural design of the two towers is determined by the method of absorbing and transmitting the wind forces. On each of the façades a Vierendeel girder type wall is formed by 59 box-section columns (spaced at 1.02 m centres) which are rigidly connected to spandrel panels at each floor level. At the corners of the building these walls are interconnected to transmit shear, so that, together with the floors of the building, they form a torsionally rigid framed tube which is fixed to the foundations and transmits all wind loads. The floors span without intermediate columns between the external columns and the core, the 44 box-section columns of which have to carry vertical loading only.

External framework and façade: The external columns are of constant overall cross-section,

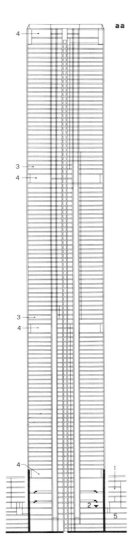

Vertical section through a tower block
1 Ancillary building 2 Plaza level
3 Skylobby 4 Technical services
5 Underground car park

The tower blocks viewed from the Hudson River

450 × 450 m. The spandrel panels interconnecting them comprise steel plates, 1.32 m deep. 12 m above the entrance level the columns are combined in groups of three to form single base columns, spaced at 3.05 m centres and with an overall cross-section of 800 × 800 mm.

The wall thickness and grade of steel in the external columns are varied in successive steps in the upward direction: wall thickness decreasing from 12.5 to 7.5 mm, yield point of the steel from 70.0 to 29.5 kg/mm². To ensure that the floors remain plane, i.e., free from warping distortion, the external columns are so designed that the stresses, and therefore the strains, produced in them by vertical loads are equal to those produced in core columns (mild steel with yield point of 24 kg/mm²). The reserve stress capacity in the external columns which is provided by the progressively graded qualities of steel serves to absorb wind load. The design value adopted for wind pressure over the entire height of the building is 220 kg/m². The calculated maximum deflection at the top of the building is 28 cm.

The external framework was erected using prefabricated three-storey units, each comprising columns interconnected by spandrel panels. These units, ranging in weight from 22.3 to 6.0 t, were fitted together, alternately staggered in one storey heights, and spliced with high-strength friction-grip bolts.

The external cladding to columns and spandrels consists of aluminium sheet. The window openings, 1.98 × 0.48 m, are infilled with bronze-tinted solar-heat rejecting glass fitted into aluminium frames and sealed with Neoprene gaskets. Automatic window cleaning is by means of rotating brushes guided along rails fitted on the column cladding.

- Floors: Composite floors comprise 900 mm deep bar joists (spaced at 2.04 m centres and braced transversely by secondary joists) and a 10 cm thick lightweight concrete slab laid on steel trough decking as permanent formwork. Composite action between the concrete and the steelwork is ensured by extending the diagonal web members of the joists through the steel decking and embedding them in the slab. Dead weight of floor 50 kg/m², imposed load 488 kg/m².

Each upper floor comprises 32 prefabricated units spanning between core and external

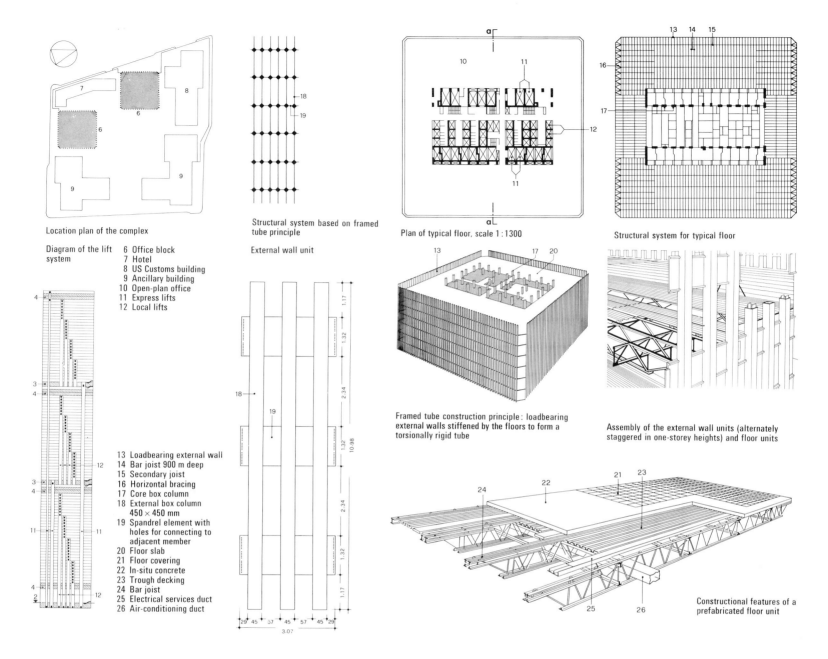

Location plan of the complex

Diagram of the lift system
6 Office block
7 Hotel
8 US Customs building
9 Ancillary building
10 Open-plan office
11 Express lifts
12 Local lifts

Structural system based on framed tube principle

External wall unit

Plan of typical floor, scale 1:1300

Structural system for typical floor

13 Loadbearing external wall
14 Bar joist 900 m deep
15 Secondary joist
16 Horizontal bracing
17 Core box column
18 External box column 450 × 450 mm
19 Spandrel element with holes for connecting to adjacent member
20 Floor slab
21 Floor covering
22 In-situ concrete
23 Trough decking
24 Bar joist
25 Electrical services duct
26 Air-conditioning duct

Framed tube construction principle: loadbearing external walls stiffened by the floors to form a torsionally rigid tube

Assembly of the external wall units (alternately staggered in one-storey heights) and floor units

Constructional features of a prefabricated floor unit

columns. These units are of two sizes: 18.3 × 6.0 m along the longitudinal faces of the core and 10.7 × 4.0 m along the transverse faces. Additional beams are provided to strengthen the four corner bays.
Oscillations due to wind are absorbed by viscoelastic shock absorbers installed between the floor beams and the external columns. Electricity and telephone cables and air-conditioning ducts are incorporated in the floor units, being installed prior to erection.
• Fire protection of the steelwork is provided by 3 mm thick sprayed vermiculite plaster. The core has been designed as a safety zone with emergency stairs for escape and with hydrants for active fire-fighting operations. Water for extinguishing fires is available in tanks, each of 18 500 litres capacity, which are installed on the technical services floors. Additional riser pipes are installed in the core.
• Foundations: Rock with a permissible bearing pressure of 39 kg/cm^2 occurs at a depth of 22.5 m. The excavation for the basement and foundations, area 440 000 m^2, is enclosed by a 90 cm thick reinforced concrete diaphragm wall anchored back into the surrounding ground.

The column foundations comprise two-layer grillages which transmit the loads through a concrete base slab, 2.1 m thick, to the underlying rock.

Services

The whole complex of buildings is fully air-conditioned. In the tower blocks, air-conditioning units are installed in front of the windows around the perimeter of the floor. The cooling plant at bottom basement level obtains cooling water from the adjacent Hudson River at a rate of 330 000 litres/min. There are seven cooling units, each with a capacity of 21 000 Mcal/h. The cooled water (at a temperature of 3.3°C) is distributed through a total of 20 distributing stations.
The top part of each tower (59th to 110th floor) has a high-pressure air-conditioning system comprising two distributing stations on the 75th floor and two on the 108th. On the floors below and in the ancillary buildings a low-pressure system is provided, with a total of 16 distributing stations located in the basement and on the 7th and 41st floors. Heating is by means of steam generated in the central air-conditioning plant.

Areas and volume (per tower)

gross area	418 000 m^2	area on plan	4032 m^2
effective floor area	319 000 m^2	volume	1 754 000 m^2

Quantities of steel (structural steelwork in one tower)

total	78 000 t	per m^2 gross area	186.6 kg
per m^3	44.5 kg	per m^2 eff. floor area	244.5 kg

References

Architectural Forum, 4/1964, p. 119.
Engineering News-Record, 9/1964, p. 36; 11/1971.
Der Stahlbau, 11/1964, p. 350; 4/1970, p. 123.
Der Bauingenieur, 9/1965, p. 373; 11/1967, p. 421.
Bauwelt, 32/1966, p. 909.
Acier-Stahl-Steel, 12/1966, p. 556; 6/1970, p. 273.

Erection of structural steelwork completed

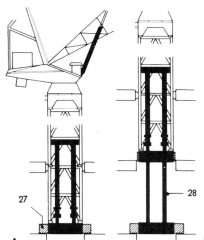
Climbing crane with hydraulic lifting device:
A Crane-supporting shaft secured in working position
B Crane being raised by extension of the stilts

Horizontal section through an external column with window frame connection

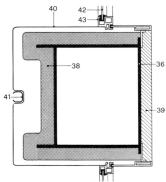

Erection operations with four climbing cranes over the liftshafts at the corners of the core

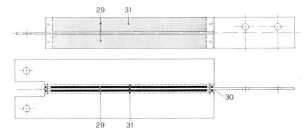
Shock absorbers between floor girders and external wall: plan and elevation

Plan of foundation for a tower block, scale 1:1700

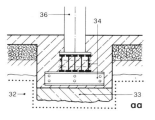

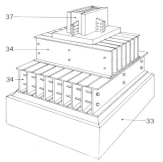
Grillage foundation under a core column

Foundations under external columns and core columns

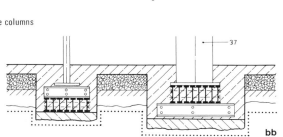

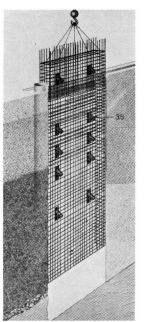
Assembling the reinforcement cage for a section of the diaphragm wall

27 Laterally extendable brackets
28 Stilts
29 Visco-elastic pads
30 High tensile steel wire
31 Steel plates
32 Rock
33 Concrete base slab
34 Grillage beams
35 Sleeves for steel anchor ties
36 External column
37 Core column
38 Vermiculite plaster
39 Special fire-resistant plaster
40 Aluminium sheet
41 Guide rail for window-cleaning equipment
42 Special tinted glass
43 Aluminium window frame

164

Principles of Design and Construction

HANSJÜRGEN SONTAG

Co-authors:

Peter Cziffer
Anneliese Zander

European and German symbols for qualities of steel and rolled sections

	Qualities of steel		Rolled sections		
EURONORM 25	Fe 37	Fe 52	HE-A	HE-B	HE-M
DIN 17 100	St 37	St 52	IPBl	IPB	IPBv

Characteristics of a steel structure	
Properties of a structural steel frame	165
Design appropriate to the material	165
Column arrangement	166
Height of buildings	167
Bracing	167
Space-enclosing construction	168
Vertical circulation and services	168
Flexibility of layout	168
Construction procedure	170
Fire and corrosion protection and sound insulation	170
Economy of a steel structure	
Cost comparison	171
Criteria for the choice of materials	171
Dimensional co-ordination	
Modular co-ordination	172
Tolerances	175
Deformations	176
Steel content	172
Industrialisation	172
Building systems	172

Fundamentals of Planning

Many different materials and methods, with widely varying characteristics, are available for the construction of the loadbearing framework of a building. The advantages of building materials or systems can be fully exploited only if their distinctive properties and features are taken into account at the design stage, that is, the design must really do justice to the material. This is especially true of buildings in which structural steelwork is used. In the following treatment of the subject they will be referred to as multi-storey steel-framed buildings.

Although virtually any constructional problem can be solved by steelwork, an optimum solution can hardly be expected if the building was originally designed for the particular properties and peculiarities of some other material. This is true for both the overall design and the detailing of a multi-storey steel-framed building.

PROPERTIES OF A STRUCTURAL STEEL FRAME

In design:

- large spans
- with columns of small section,
- great building heights and high loadbearing capacity
- combined with low dead weight of the structure,
- structural systems in which openings can easily be provided
- to simplify the installation of services.

In construction:

- prefabrication and erection of the components,
- hence, shorter construction time,
- close dimensional tolerances,
- therefore, ease in fixing the cladding,
- erection independent of weather conditions,
- modest demand on space on the site,
- dry construction.

In use:

- greater flexibility
- because of limited number of internal supports,
- adaptability of framework to change of use

- thus increasing the effective life of the building,
- ease in dismantling or demolition.

These beneficial properties of structural steelwork increase the economy of the building.

DESIGN APPROPRIATE TO THE MATERIAL

To enumerate the various features which enable the design of a multi-storey steel-framed building to do full justice to the constructional material, it is necessary to start with the properties of the supporting structural system itself. The subsequent main constructional work, the cladding and the services should be suited to this.

The structural system

Basic components

The basic components of a steel structure are hot-rolled steel sections. These comprise thin-walled slender elements which possess high loadbearing capacity in conjunction with small overall dimensions and low weight. While it may be said that they represent a limited range of types, they can nevertheless be combined into a wide variety of structural components which may be readily connected together.

The structural steel frame

The system that performs the loadbearing function — the structural steel frame — is assembled from these relatively slender steel members. The frame is thus a skeleton which does not have a space-enclosing function to perform, although it does provide support for such space-enclosing elements as floors, walls and partitions.

Fabrication

The steelwork is usually prefabricated in workshops. Fabrication uses industrialised methods, often involving electronically controlled production lines which can operate economically even without mass production. However, with structural steelwork mass production techniques can be advantageous, particularly from the point of view of design.

Influence of structural design

The following structural criteria are of major importance in the overall design of the building: the location of the columns, and particularly

- the type,
- the arrangement and
- spacing of the supports and thus
- the spans of the floor structure, the bracing of the building, and particularly
- the type of stiffening components and
- their arrangement.

The space-enclosing system

Requirements applicable to the system

The optimum use of the favourable properties of structural steelwork can be more readily achieved if the space-enclosing system is contrived so that it is:

- composed of prefabricated units to ensure that the inherent speed of erection of structural steelwork results in a short overall construction time;
- light, so that the overall weight of the building is minimised;
- adaptable to flexible internal layout of the floor space in the steel-framed building and to modification of its components;
- designed to suit the specific properties and features of the structural steel frame;
- amenable to economical fire protection.

Components involved

The following components should be compatible with the structural steel frame:

- the floors,
- the roof,
- the external walls,
- the internal walls and
- the vertical circulation system.

With careful co-ordination it is possible to achieve maximum economy in the construction of the building as a whole.

Characteristics of a steel structure — Design appropriate to the material • Column arrangement

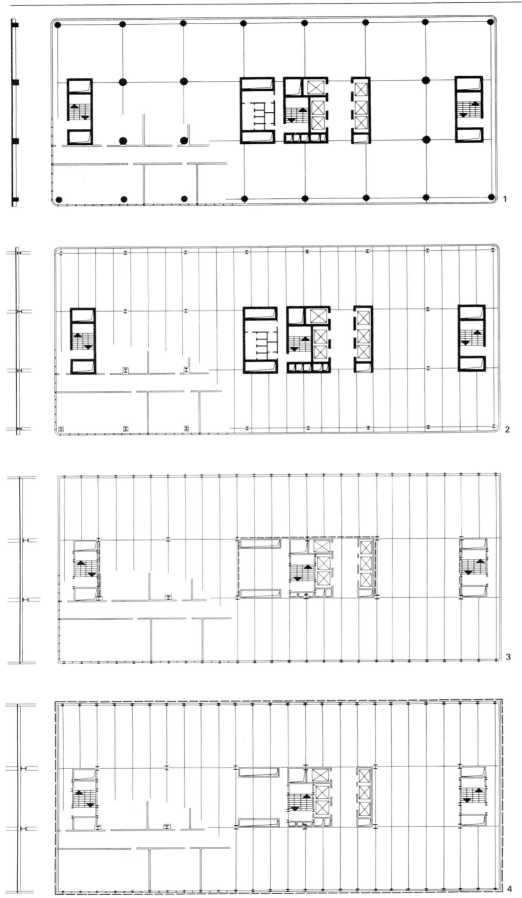

Installation of plant

The open arrangement of the structural steel frame facilitates the installation and subsequent modification of services and equipment in both the vertical and horizontal directions. The structural system and the layout of these services are to some extent interdependent. Steel frames offer many and varied possibilities for the installation and attachment of all kinds of services and are therefore especially convenient in buildings in which central heating or air-conditioning plant has to be provided.

Example

To illustrate the interdependence of design and structure, four variants in planning a typical floor in a 15-storey building will be considered.
1 Plan of a block for a polytechnic in Hamburg (architects: Kallmorgen, Riecke and Karres). Concrete frame with circular columns, 0.40 to 0.80 m dia, spaced at 8.40 m centres in both directions: external columns just within the façade; vertical circulation and cloakrooms in concrete cores; transverse ducts, pipes, etc., pass through the main beams; external spandrel walls integral with the floors.
2 Structural system replaced by a steel frame without changing the design, except that the cross-sections of the columns are reduced.
3 Design based on a structural steel frame: closely-spaced external columns with dimensions varying from 16 cm to 27 cm square; building is stiffened by vertical lattice bracings (four in transverse, one in longitudinal direction); staircases, lift and service shafts enclosed within fire-resisting walls; reduction of the number and cross-sectional dimensions of the internal columns; by arranging the main beams under the floor beams, space for transverse horizontal services is provided; the external wall, designed as a system of mullions and cladding, can be attached to the external columns.
4 Optimum solution as a multi-storey steel-framed building; in combination with spandrel beams, the external columns form rigid frames or, with external diagonals, lattice bracing to stiffen the building; the internal space can now be freely subdivided, the vertical circulation and services being independent of the structural system.

COLUMN ARRANGEMENT

Nature and arrangement of columns

The vertical loads on a multi-storey steel-framed building are transmitted to the foundations through steel columns and sometimes through concrete walls. As a rule, the columns are located at the intersection points of a grid. The properties of a structural steel frame are best suited to an orthogonal grid, especially one which is rectangular rather than square (p. 188), as the former usually provides a more economical solution. Nonetheless, a skew grid may sometimes be advantageously adopted (p. 197).

Column cross-sections

Even when cased for fire protection, steel columns have smaller cross-sectional dimensions than concrete columns of the same load-bearing capacity (p. 226). As they occupy less space, a more favourable ratio between total area on plan and effective area is obtained.

Height of buildings • Bracing

Characteristics of a steel structure

External columns

Closely spaced external columns in contact with the external walls are especially characteristic of steel buildings (p. 201). The advantages of such a column arrangement are:

• the columns are of very small section, thus occupying the minimum of usable space;
• the external cladding can be attached to them, so that separate mullions can be eliminated (p. 309);
• the columns provide connections for internal partitions on every planning grid line of the building.

Low weight

Because of the comparatively low weight of structural steelwork there is, in a steel-framed building, a favourable ratio between imposed and dead loads. There are also fewer loads to be transmitted than in a concrete structure. As a result, savings are made in the cost of the structural system and of the foundations. With difficult subsoil conditions such savings can be considerable.

Long spans

With steel beams it is possible to use long spans, with widely spaced columns, thus providing flexibility of internal layout. Steel beams are economical for column spacings ranging from 6.00 to 18.00 m, and in special cases up to 30.00 m.

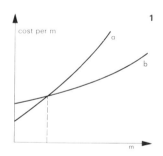

A cost comparison with other methods of construction, as in Figure 1, shows the cost of the steel beam b to increase more gradually with increasing span than that of the concrete beam a. The location of the intersection of the two curves depends on the magnitude of the loading and on the depth of the beams. For short spans the concrete beam is usually cheaper, for long spans the steel beam has the advantage.

With storey-height lattice girders it is economically possible to span distances from 30.00 to 60.00 m. Even longer spans are practicable with modern buildings constructed on the bridge principle, with girders the depth of which corresponds to the full height of a building (p. 207). Structural steelwork can also be used to intercept and transmit heavy loads from cantilevered portions of buildings (p. 203) or from buildings as a whole (p. 202) or to provide the supporting members of suspended structures (p. 205).

HEIGHT OF BUILDINGS

Structural steel systems can be readily applied to buildings of any height.

Single-storey buildings

The frames of single-storey buildings have only to support the relatively small roof loads. In a light steel structure with a steel-sheet roof, the ratio of dead weight to imposed load is therefore small.

Multi-storey buildings

In the case of buildings which comprise several storeys but are not high enough to rank as high-rise buildings, a steel frame in combination with a suitable roof structure is an economical system, especially if long spans are required and the building contains extensive plant and equipment. The great majority of buildings, residential and non-residential, come within this category.

High-rise buildings

In such buildings, the relatively small section of steel columns is a major advantage, especially if they are closely spaced external columns which merge into the façades. The low weight of the structural steel frame also has important influence on cost.

BRACING

All buildings have to be stiffened to resist wind loading, but in certain parts of the world they have to be given additional bracing to withstand earthquake and, in certain special cases, other horizontal forces. The stiffening system must:

• transmit these forces to the foundation and
• limit the horizontal deflection of the building.
With tall buildings
• special consideration must be given to the possible effects of oscillation due to wind.
The magnitude of the horizontal force exerted by wind depends on:
• the wind speed,
• the aerodynamic shape of the building and
• the surface configurations of the façade.
The value of wind speed adopted in design depends on:
• the height of the building,
• its geographical location (p. 209).

Methods of stiffening

A structural steel frame can be braced in the following ways:

• rigid frame systems (p. 210),
• vertical triangulated bracing systems (p. 213) or
• concrete walls, either as individual shear walls or combined as cores (p. 217).

The choice of the correct method of bracing is of major importance to the structural design and may even govern the whole design concept of a high-rise building. It affects:

• the utilisation of the building,
• its economy,
• its external appearance,
• the construction procedure.

Effect on use of building

Stiffening by means of lattice bracing or solid diaphragm walls creates fixed points in the building, thus restricting the freedom of internal layout and arrangement of the circulation system. The methods of stiffening and their disposition within the structure are therefore of major influence on the design. A lattice panel can be much more readily penetrated than a concrete wall. A combination of the stiffening or bracing structure with the vertical circulation facilities is an obvious possibility, but with structural steelwork it frequently does not provide the ideal solution. Often, it is more advantageous to install the bracing in the external walls, as this eliminates structural restrictions upon the freedom of internal layout.

Effect on economy

In each individual case a decision must be made whether a building can be stiffened more economically by means of steel bracing or by concrete walls or cores. Cores are heavy, require more elaborate foundations, and may, for tall buildings, have to be constructed with thicker walls and more reinforcement than necessary for the fire protection of the services within the cores. Vertical steel bracing is therefore usually the less expensive solution, especially when arranged in large bays or panels.

Effect on appearance

If the bracing is located in the external walls of the building and takes the form of rigid frames (p. 212) or systems with externally exposed diagonal members (p. 216) the pattern of force transmission is clearly revealed: but the building may present a rather unusual appearance. Nevertheless, because of its economy, this form of construction is likely to be of major importance in the future. This method allows those high-rise buildings which are square, rectangular or circular on plan to be constructed as very rigid tubes.

Effect on construction procedure

From the erection point of view concrete cores are foreign bodies in the structural steel frame. The speed with which the core can be constructed by slip-forming may be misleading as the overall construction time comprises the time for setting up and subsequently dismantling the formwork. Steel erection usually does not start until the core has been completed. In addition, the differences in the tolerances applicable to concrete and steel and the problems associated with connecting the steel components to the concrete core may give rise to difficulties and delays (pp. 217, 267).
Construction usually proceeds more smoothly and easily if the bracing system is of steel and is erected with the rest of the steelwork.

Stiffening of high-rise buildings

In the construction of high-rise buildings, the stiffening or bracing members may constitute a substantial proportion of the total cost: even in a 20-storey building they may comprise 30% of the overall weight of structural steel. Some conclusions can be drawn from a comparative assessment of American high-rise buildings

167

Characteristics of a steel structure

erected in the last ten years, **1**. The broken curve **a** indicates the steel consumption, in kg/m² of floor area in each storey, plotted against the height of the building, this being the steel serving solely to transmit the vertical loads, that is, for the columns and beams only. The continuous curve **b** indicates the actual total steel consumption. The difference between the

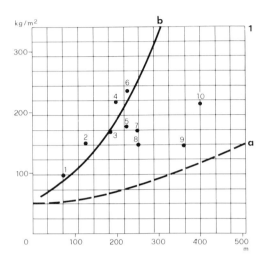

1 Gateway Center
2 Equitable Insurance, Chicago
3 First National Bank, Seattle
4 Civic Center, Chicago
5 Toronto Dominion Center
6 Chase Manhattan Bank
7 First National Bank, Chicago
8 US Steel, Pittsburgh
9 John Hancock Center, Chicago
10 World Trade Center, New York

two curves therefore represents the quantity of steel needed (per m² of floor area) for stiffening the structure. It should be noted that the buildings with the higher weights of steel are portal framed structures of the older type, whereas the more recent buildings 8, 9 and 10 have their stiffening elements located in the external walls and thus require much less steel to perform this function. (See also: 'Hochhäuser in Stahlkonstruktion in den USA', Detail, 3/1969.)

SPACE-ENCLOSING CONSTRUCTION

In considering the foregoing requirements for the space-enclosing elements of a building, the choice of the most appropriate type will be influenced by various criteria, the most important of which are given below.

Floors

The following methods of construction are commonly employed for floors supported by steel beams:

- in-situ concrete slab on formwork;
- precast concrete slabs (p. 278);
- steel decking with in-situ concrete (p. 276).

Composite action between the concrete slabs and the steel beams, which is possible in all three methods, makes for greater economy of construction (p. 256). Appropriate procedures have been developed for all these types of floor.

Economy in the design of the floor steelwork from the point of view of fire protection is an important consideration (p. 279). Many well tried forms of construction are also available to meet these requirements. There are, for example, efficient false ceiling systems which satisfy the criteria of fire protection, as well as performing their other functions (pp. 281–284).

The space between the false ceiling and the structural floor can be utilised to accommodate various services (p. 285). Cables can also be installed in steel decking.

The construction depth of the floor structure as a whole has a direct effect on the height of the building.

The interrelationship of problems of design, the construction of the structural system, the space-enclosing features and the plant and services emerges very clearly in connection with the choice of the floor construction method.

Roofs

Multi-storey steel-framed buildings generally have flat roofs. Lightweight sheet-steel roofing and normal or lightweight concrete are available for their construction (pp. 299–303).

External walls

In accordance with the essential character of steel-framed buildings, the external walls are non-loadbearing and are merely attached to the supporting frame, usually at each storey. All the usual materials and forms of construction are employed. Metal cladding systems are generally best suited to fulfil the requirement of low weight, and their use is facilitated by the dimensional accuracy of the steel frame (pp. 304–317).

Internal walls

Internal walls and partitions of light construction can also be most appropriately constructed in steel. The choice of type of construction depends on the method of fixing to the structural frame and the magnitude of the deflections which could possibly occur in the floor structure (pp. 318–326).

Fire-resisting walls

These walls, serving to subdivide the interior of the building into fire compartments, must fulfil special requirements with regard to strength and fire resistance rating.

VERTICAL CIRCULATION AND SERVICES

Vertical circulation

In its most comprehensive sense, this concept denotes all the facilities for the vertical transportation and movement of people and goods – staircases, escalators and lifts – while the term 'services' denotes hot and cold water pipes, gas supply pipes, power cables, telephone cables, air-conditioning ducts, etc. The shafts in which these facilities and services are accommodated are referred to as vertical

Space-enclosing construction

circulation cores. If the core walls, which are in any case necessary to comply with fire regulations, are also to be used as structural bracing (p. 217), they must be appropriately designed for that function. Such walls, generally of concrete, constitute foreign bodies in the steelwork, as already stated.

On the other hand, if the structural frame is stiffened by means of triangulated bracing or rigid frames, the vertical circulation facilities and services are passed through holes formed in the successive floors and are enclosed within light fire-resisting walls supported on those floors. The flights of stairs, consisting of prefabricated steel or concrete units, are installed between the beams of the floor structure.

Horizontal services

Simple and convenient routing of the horizontal pipes, cables, ducts, etc., is very important, bearing in mind the high proportion of the overall cost of a modern building that may be concentrated in its technical services. This is more particularly true of buildings with air-conditioning plants (high-rise buildings and buildings covering a large area on plan) or buildings requiring a wide variety of services (laboratories and hospitals).

As the horizontal distribution zones are nearly always located within the floor structure, the steel beam floor is very advantageous because it provides ample space to accommodate these services. This problem will be examined in greater detail in the following chapters (see more particularly pp. 192, 241).

The services can be passed through suitable apertures in the beams or girders, lattice girders and castellated beams being especially advantageous in providing great freedom of layout in this respect. Any holes required in the web of plate girders and rolled sections are normally cut at the fabrication stage but, in exceptional cases, the holes may be made after erection.

If a floor comprises steel beams in two directions, one above the other, then two superimposed zones are available for the services, enabling them to cross one another without interpenetration. A change in direction of the services naturally involves a change in level.

The greater freedom offered for the installation and layout of services in a steel-framed floor is especially advantageous where services have to be subsequently augmented or modified.

FLEXIBILITY OF LAYOUT

An even stronger incentive in favour of the more extensive use of steel in multi-storey buildings arises from the need for convertibility of buildings. As a result of the rapid rate of technological evolution in every field of human activity and the increasingly exacting requirements to be fulfilled in terms of available space and quality of equipment, it not infrequently occurs that the internal layout and facilities originally provided in a building soon prove to be inadequate or no longer fully appropriate. A building can attain a longer useful life if it is flexible in the sense that it can, by structural conversion, be readily adapted to changing requirements.

Depending on the extent of the conversions involved, the following degrees of flexibility may be distinguished:

Flexibility of layout

Characteristics of a steel structure

First degree of flexibility

The utilisation of the floor space is altered by rearrangement of space-separating elements, for example, folding screens or banks of cupboards which can be moved by the users of the building themselves, **1**.

Second degree of flexibility

In this case, altering the internal layout involves dismantling and re-erecting the partitions. This is practicable only if the latter are non-loadbearing. This requirement can only be fulfilled by framed buildings in which the loadbearing and space-enclosing functions are separate and are performed by different sets of components. The columns of a framed building may, however, present an objectionable obstruction to internal rearrangement. Wide column spacings and long floor spans therefore improve the flexibility of layout. Flexibility in altering and rearranging the services is also essential. In a building provided with movable partitions, these must of necessity be installed in the floors. Hence it is desirable to provide a floor in which the services can be suitably installed and subsequently modified, as required. These conditions can be very efficiently satisfied in steel-framed structures, **2**.
It is necessary to distinguish between the following different types of partition:

2.1 movable prefabricated units;
2.2 demountable walls with re-usable components;
2.3 walls which can be demolished, but whose materials cannot be re-used.

Third degree of flexibility

Modification of the loadbearing structure becomes necessary, for example, in the following circumstances:

• strengthening the structure to carry heavier loads;
• increasing the spans by the removal of internal supports;
• increasing the height with additional storeys;
• extending a building;
• demolishing certain parts of a building.

These modifications can be carried out easily, inexpensively and without excessive disturbance to the existing building if it is steel-framed.

Increasing the spans

Suppose that the intermediate support under a beam has to be removed, **3**. To resist the increased bending moment, the beam is provided with flange plates a and the column, consisting of a rolled section, is strengthened by means of welded plates b which turn it into a box-section. The beam-to-column connection is strengthened by the addition of a bracket c. The web of the rolled steel beam is generally adequate to carry the increased shear forces; if not, it too must be strengthened. In many cases the foundations can carry heavier loading than that for which they were originally designed, as the subsoil will by now have had time to consolidate under them.

Increasing the height of a building

It is possible to continue normal use during the construction of one or more additional storeys in a steel-framed building, as the roof has to be disturbed only at the points where column splices have to be made. The rest of the roof can be left in position until the extra storeys have been completed and roofed in. The individual members of a structural steel frame are relatively light and can be erected on top of the existing structure with the aid of tall building cranes. The use of prefabricated floors on such jobs is particularly to be recommended in order to reduce construction time.

Conversion of staircases

If the walls of the stairwell do not also perform a structural function in stiffening the building, the staircases in a multi-storey steel-framed building can, without excessive difficulty or inconvenience, be moved to a different position, provided that this is permissible under the relevant building regulations.

Fourth degree of flexibility

This comprises structures which — except for the foundations — can be completely dismantled and whose components can be re-used for other purposes and in combination with other structures. For this, the space-enclosing elements and the plant, as well as the actual structural frame, must be composed of demountable standardised components, **4**. Demolition of such a building does not involve the destruction of more than a small number of parts, for example, the roof covering, floor finishes, services, etc. The foundations or basements are either backfilled or used for other structures. Constructional systems which comply with these requirements are available for external and internal walls and for ceilings. The following are examples of structural systems based on the re-use of components:

• the Mero space-frame system comprising tubular members assembled by bolting (p. 255, **4**),
• the Krupp-Montex system, as applied to university building construction (p. 181 and p. 211, **12**), as well as other systems usually restricted to buildings intended for specific purposes (see also p. 172).

Fifth degree of flexibility

The final link in this chain of reasoning on the adaptation of an existing building to meet changing circumstances is represented by demolition to make way for a new building when structural conversion is not economically possible. Thus, in the USA, it has been found that high-rise buildings have an average service life of less than 30 years.
When considering the type of structural system to adopt for a projected new building, it makes economic sense to include ease of demolition in the comparison of costs, **5**. Steel-framed buildings can be demolished quickly, at moderate cost and without undue nuisance from noise and dust. The dismantled parts of the steel frame can be used again or can at least be sold for scrap, the proceeds from which will help to reduce the cost of demolition.

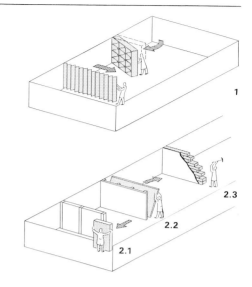
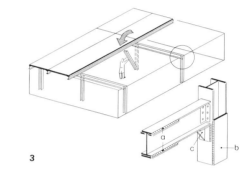
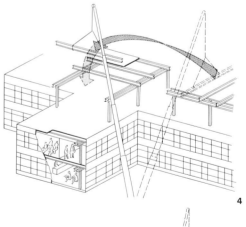
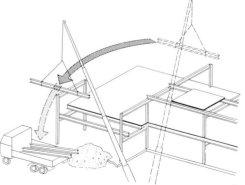

Characteristics of a steel structure

Speed of construction

With many projects speed of construction is of major economic importance. It can be achieved by means of the following measures:

- early detailed planning of the structure;
- subdivision of the structure into units which can be suitably prefabricated and erected;
- precise planning and scheduling of all construction operations.

A short overall construction time results if these principles are applied not only to the structural system of the building but also to its cladding and fittings.

Short erection time

A steel structure provides ideal conditions from the point of view of speedy erection, for it consists of prefabricated units which are assembled with the aid of mobile cranes. The site connections are usually made by bolting and are therefore immediately effective in transmitting loads. Temporary bolting is also sometimes used to assemble the components for welded site joints.
The high dimensional accuracy of structural steelwork is another factor that makes rapid erection possible. With one crane and one gang of erectors, it is possible to erect from 200 to 300 tonnes of steel per month. With a steel consumption of, say, 15 kg/m^3, the monthly output of a crane can thus amount to about $20\,000 \text{ m}^3$ of cubic content. By working more than one shift per day and by using several cranes, this rate of erection can of course be increased many times over (pp. 347–352).

Construction irrespective of weather conditions

Steelwork erection is largely independent of the prevailing weather and of the season. It does not necessitate interruption of building operations in winter or elaborate protective measures for continuing in cold weather, as are necessary for concrete construction in some climates.

Speed of completion

If suitable prefabricated forms of construction are chosen for the space-enclosing components of a structural steel frame – floors, stairs, external walls (including basement walls), roof and internal walls – the entire structure above foundation level can be erected independently of weather and season. In this way, speed of erection leads naturally to speed of completion. Prefabricated secondary components offer the additional advantage that 'dry' construction methods are employed, thus excluding moisture from the building.

Short overall construction time

The speed with which multi-storey steel-framed buildings can be put up is often a decisive factor in choosing steel in preference to other types of construction, even if the cost involved happens to be higher. With major projects it is thus possible, with steel, to save months on the duration of the job (pp. 341, 347). Steel gives an even more important time-saving advantage when it makes possible the use of the winter months for erection.

Less space needed on site

Steelwork erection requires very little space on the building site itself, as the delivery of materials to the job can be so controlled that intermediate storage is not required. Usually, all that is needed is space for the cranes and standing room for the transport delivering steelwork. On urban sites with cramped conditions this is often a decisive consideration in the choice of a steel structure.

Protection of the environment

Steelwork erection is not a noisy operation. Also, it is clean, there is very little waste material and no dust is produced. For these reasons, structural steelwork goes a long way to meeting the requirements of environmental protection. The same is true of prefabricated secondary components. When a building has to be pulled down, the advantages that steelwork offers in this respect, too, are obvious, the structural members being dismantled with little noise or dust.

Remote sites

Steel is virtually the only material suitable for major constructional work on remote or inaccessible sites, for example, in arctic regions or in mountainous country. Helicopters can be used to carry individual loads to otherwise inaccessible locations.

FIRE AND CORROSION PROTECTION AND SOUND INSULATION

It is often asserted that the cost of compliance with the requirements for fire and corrosion protection and sound insulation of a steel structure make it an uneconomical form of construction. This is a fallacy. The cost involved in these protective measures is generally overestimated. There are modern inexpensive ways of fulfilling the requirements imposed.

Structural fire protection

Measures to achieve adequate structural fire protection are necessary in all buildings, whatever the constructional material used. The cost involved by the direct protection of steelwork is relatively low. It certainly does not make a steel building uneconomical. What is important is to achieve correct design of the steel structure, secondary components and finishes. Quite often, space-enclosing components such as floors or walls can, without additional cost, serve also to provide fire protection for the steelwork. The problem of fire protection in general is dealt with in pp. 327–336. Suitable solutions for designing the various components in question are indicated in other appropriate chapters.

Corrosion protection

As steelwork inside buildings with normal atmospheric humidity is unlikely to rust, corrosion protection inside multi-storey steel-framed buildings is not a real problem. Complete protection against corrosion is necessary only for those components which are exposed to the external air or which are in rooms with high humidity levels. There are well-tried and reliable methods of providing such protection. For further particulars, see pp. 357–359.

Sound insulation

Occasionally the opinion is voiced that a building with a structural steel frame behaves less favourably from the point of view of sound insulation than other buildings. To this it can be said that the acoustic properties of a building depend almost entirely on the acoustic properties of the space-enclosing components, such as floors, roofs, external walls and partitions. The properties of the structural frame are only of minor significance in this respect. A steel frame can behave no differently from a concrete frame with the same type of cladding and other secondary features, since the components of the steel frame itself cannot vibrate freely, being rigidly interconnected by the floors. The floor beam systems can vibrate only with the floor as a whole. Because of their fixity or lateral restraint at each floor of the building, the columns have only very low natural frequencies. As far as noise transmission is concerned, there is little to choose between steel and concrete as the velocity of propagation of sound waves is not very different in the two materials: 4000 m/s in concrete, 4900 m/s in steel.
If the components of the steel frame also perform a space-enclosing function, for example, if the webs of columns or floor beams are located in the planes of partitions (p. 322), intermediate sound-insulating layers are needed.

Reference: Moll, *Bauakustik*, Verlag Ernst & Sohn.

COST COMPARISON

In judging the economy of the material used for the structural frame of a building the cost of the frame must not be compared in isolation with the cost of a structural frame built from another material. Instead, all the factors influencing economy must be taken into consideration, for example:

- cost of land,
- cost of financing,
- cost of the main structural frame,
- effect of the space-enclosing features,
- effect of the plant and services installed,
- how these factors affect the utilisation of the building,
- cost of subsequent alterations,
- cost of demolition (in the case of short-lived buildings).

Cost of land

Because of the smaller cross-section of its members a structural steel frame occupies less space on plan and, for equal area on plan, provides more effective floor space than a concrete-framed building. Thus, with steel, the cost of land per m^2 of effective floor area is lower. As the dead weight of a structural steel frame is low, cheap land of low loadbearing capacity can be used while the cost of the foundations can be kept to a minimum.

Sound insulation • Cost comparison • Criteria for the choice of materials Economy of a steel structure

Cost of financing

Short construction time makes for lower financing charges. As steel-framed construction often results in a substantial saving of time in comparison with other methods, the corresponding saving in the cost of financing the project can be considerable and often justifies incurring higher expenditure in constructing a steel frame.

The saving on financing can show itself in various ways:

Earlier earnings from rents

Shorter construction time means that the building can be put into service earlier and thus begin to pay its way from an earlier date.

Saving on rents paid to third parties

When a commercial or industrial organisation is to occupy its own premises, earlier availability of the latter saves the cost of rent or other expenditure in securing accommodation for its staff and activities in rented premises pending the construction of its own building.

Shorter interruption of activities

In the case of structural conversions or buildings erected as replacements for demolished buildings, shorter construction times reduce the period of interruption of normal activities and therefore cut the costs or the loss of earnings associated with such interruption.

Elimination of winter risk

In climates with cold winters, concrete construction involves the risk that any period of building inactivity allowed for in the cost calculations will turn out to be longer if the winter happens to be exceptionally severe. With structural steelwork this risk is largely eliminated, as erection can proceed virtually all the year round.

Construction costs

Problems associated with cost comparisons

To be valid, a cost comparison between variants of a particular structural member in different materials must include the effects of its characteristics on other members and on the utilisation of the structure. What should or should not be included in a comparison of the economy of different materials cannot be laid down in general terms. Each case must be considered individually, as some examples will serve to show:

Steel columns – concrete columns

The steel column requires fire protection, therefore an additional trade is involved in making the comparison. Despite encasement, the steel column has a smaller cross-section than its concrete equivalent; it can be accommodated within the thickness of external or internal walls and thus occupies less effective floor space, or none at all. The external wall claddings can be fixed directly to the steel columns, thus eliminating mullions or a secondary framework. Consequently, the façade construction is cheaper and more effective use can be made of the interior space.

Steel beams – concrete beams

When secondary steel floor beams are supported on top of the girders or main beams, the construction depth of the floor structure is greater than that of an equivalent concrete floor. As a result, the storey height, the overall height of the building and its cubic content are increased. On the other hand, the steel floor structure provides more space for the installation of services, so that savings in cost may be made.

The steel beam requires fire protection. The suspended ceiling can, however, be utilised to perform that function. The ceiling may be needed under a concrete floor as well, in which case it need not be fire-resisting and may therefore be somewhat cheaper. With a suitable fire-protecting ceiling, the steel beams in the floor cavity need not be cased and can conveniently be used to support services, which can subsequently be attached directly to them. Thus, in making the comparison, the effects on the construction depth of the floors and on convenience in installing the services must be considered.

Spans

Longer floor spans make for greater flexibility of internal layout and utilisation of a building. In a steel-framed building they involve less extra cost in comparison with a concrete structural frame. The reduced number of columns associated with longer spans saves internal space and allows a smaller area on plan.

Scope of cost comparisons

To take account of the effect of the choice of structural system, a cost comparison may comprise:

• individual structural members (such as beams), or
• larger assemblies (such as floors);
• the whole structural frame, or
• the complete carcase of the building;
• the complete carcase with all the space-enclosing features, or
• the finished building.

Optimum structural systems

If the various parameters can be freely chosen in designing a building, there is, in every individual case, a different optimum structural system for every material and for every method of construction. There are buildings which can be erected more cheaply in steel, and others for which concrete is the less expensive material. But the solution in steel will have different characteristics from those of the concrete structure. Before a valid decision can be made, a number of alternative solutions must be investigated. A valid comparison is, however, possible only if each solution has optimum characteristics for the material it embodies.

Utility value analysis

Inclusion of considerations of utilisation in the cost comparison is possible only through utility value analysis. This allows values which are not directly comparable to be compared by expressing them in monetary terms. In other words, answers are sought to the following questions: 'What can I afford?' or 'What can I allow: greater comfort or greater freedom of internal layout or shorter construction time to cost?'

Cost of subsequent alterations and demolition

In far-sighted planning, the cost of subsequent structural alterations to meet changed requirements, in whatever degree of flexibility, should be included in the calculation of costs. For a steel structure this cost is more easily assessable, and generally lower, than for other structures.

The cost of subsequent demolition is not normally included in the initial costing unless the building is intended as a more or less temporary structure. Demolishing a steel structure is less expensive, especially when demountable construction systems are used.

CRITERIA FOR THE CHOICE OF MATERIALS

The following check list can provide some guidance in deciding between steel and concrete. Structural steelwork is, as a rule, the more economical alternative when one or more of the following requirements have to be fulfilled:

• long floor spans,
• great height of building,
• low weight of building in view of poor subsoil,
• elaborate services to be installed in each floor,
• large imposed loads,
• small cross-sectional dimensions of columns,
• flexibility of internal layout,
• open-plan system for offices, etc.,
• possibility of altering the structural frame,
• short construction time,
• erection during the winter,
• close tolerances for finishings and fittings,
• erection on a restricted site with little or no storage space,
• limited requirements for fire protection.

171

Economy of a steel structure

A measure of the economy of a steel structure is provided by the quantity of steel that goes into its construction, expressed as a weight per m² of floor area or per m³ of cubic content. It depends on many factors, for example:

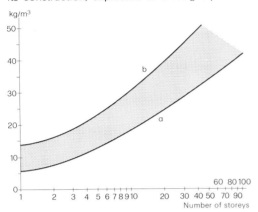

- number of storeys,
- imposed load,
- column spacing in both directions,
- construction depth of the floors,
- method of stiffening or bracing the building,
- grade of steel chosen,
- the type of floor structure.

The above diagram provides approximate guidance. In it, the weight of steel in kg/m³ of cubic content has been plotted against the number of storeys. The lower curve a is valid for low loads and short spans, while curve b corresponds to high values of the imposed loads and long spans. An accurate estimate of the steel content can be made with the aid of the design charts for columns and beams provided on pp. 229–231, 246–247, 251, 259.

INDUSTRIALISATION

The structural steel components of a multi-storey building can for the most part be fabricated on fully or semi-automated production lines in up-to-date fabricating shops. In this field of construction industrialisation has already made considerable progress. It is noteworthy that this type of automated production is entirely independent of the standard types or series of rolled steel sections used in structural steelwork. The machines used for the work are automatically adjustable in such a way that a whole range of different sections can be dealt with in succession without any loss of time. The program of machining and fabricating operations is fed to the machines through the medium of magnetic tape or punched cards. It is interesting to note that the mass production of substantial numbers of identical components in steelwork effects savings as much at the design stage as at the actual fabricating stage. The automated procedures are facilitated by the fact that the number of different rolled sections in terms of their shapes and dimensions is relatively small. Yet this relatively limited range of basic elements does not place any appreciable restriction upon the designer's freedom and scope.

The automation of fabrication and the mechanisation of erection result in a substantial saving in man-hours per tonne of steel erected. With ever-rising wages, structural steelwork is bound to become relatively more and more advantageous in comparison with other methods of building construction, as it has made the transition from wage-intensive to capital-intensive production. Thus it exemplifies the principle that the cost of building can be stabilised or indeed lowered by suitably limiting the number of basic types and by mechanising and automating the production processes.

BUILDING SYSTEMS

There is no clear-cut definition of a 'building system'. Usually, it denotes a range of widely adaptable standard components which can be assembled in many combinations to produce buildings of different kinds and for different purposes. The number of types of components and their variants differs from one system to another.

The units are fabricated in jigs or on production lines, sometimes automated. The object of building systems is to reduce the time required for the design and construction of buildings and also to reduce their cost.

A distinction must be drawn between 'closed' systems, which comprise complete buildings, and 'open' systems, which can be incorporated as sub-systems into any kind of building. There are systems which are essentially structural and systems which provide finishes and fittings. Most structural systems have their preferred finishing systems to be used in combination with them.

Structural steel systems are highly adaptable and versatile. Quite often, the construction method, and not the component, is standardised, for example, the type of beam arrangement, the column types, the structural connections. In such systems the length and the section or size of the members can be freely chosen, so that they can be economically suited to the given spans, loads, construction depths, services to be installed, etc. This freedom of choice is possible without excessive cost and effort if electronic computer programs and system drawings have been prepared. Despite this freedom available to the designer, it is desirable to base the dimensions on the international modular system (10, 30, 60 cm) if only for the very good reason that this facilitates the installation of the finishing elements.

Many structural steel systems are commercially available. The buildings are often offered on a turn-key basis, that is, fully erected and ready to use, by the supplier firms. Mostly, such systems have been developed to fulfil special requirements, such as schools, universities, residential buildings, office buildings, multi-storey car parks, hospitals, etc. As a rule, however, they can also be used for other purposes and are 'multi-storey building systems' in the true sense. The sheer number of systems available and their increasingly rapid development rule out any detailed discussion here.

DIMENSIONAL CO-ORDINATION

Modular co-ordination

Results of preparatory work by the ISO (International Standards Organisation) and the IMG (International Modular Group) are contained, inter alia, in the German Standard DIN 18 000. In the explanatory comments accompanying the standard, the purpose and function of modular co-ordination are explained as follows:

'The development of building technology, the rationalisation of building, the increasing use of prefabricated components, and the advancing industrialisation of building construction generally, call for a binding agreement on the standardisation of components and on universally valid rules for the co-ordination of structural dimensions with one another.

'In addition, such an agreement serves to limit the variety and diversity of components and also furthers the aim that manufacturers produce products which are, without loss and without requiring subsequent modification, capable of being combined, interchanged and used for different purposes.

'An agreement of this kind, possessing international validity, exists in what is known as modular co-ordination. Its application makes possible the standardisation of components and equipment and thus makes their mass production for an extended market a practicable proposition. It also facilitates the designer's work in producing designs that are suitably geared to available manufacturing facilities for the components used. It is a prerequisite for the rationalisation of design and manufacture and it helps the construction industry in the calculation and costing of buildings. A modular co-ordination system forms the basis for further standardisation, including the introduction of a standard system of tolerances, a system of preferred principal dimensions, and — with a view to a further reduction in the range and diversity of component types — agreements on preferred dimensions for particular groups of building components and products.'

In the introduction to the same DIN and in the accompanying commentary it is also stated:
'In order to interrelate the constructional dimensions, modular co-ordination makes use of modules, namely: the basic module, the multimodules and the submodules.

'Co-ordinating dimensions, that is, structural dimensions which are to be suitably inter-related and which determine the location of components relative to one another, should be modular dimensions, that is, whole multiples of the modules.

'The basic module is designated by M. Its magnitude is

$$M = 100 \text{ mm}$$

'The multimodules are whole multiples of the basic module, namely:

$$3 M = 300 \text{ mm}$$
$$6 M = 600 \text{ mm}$$

'The submodules are obtained by division of the basic module by whole numbers and serve solely for forming structural dimensions which are smaller than 3 M.

'The modular dimensions are so-called controlling dimensions, comprising the dimensions of the actual components together with their joints.'

The 6 M multimodule, that is the dimensional unit of 600 mm, is a very useful dimension for a planning grid. It is eminently suitable as the basis of the dimensional system for structural

Modular co-ordination

Dimensional co-ordination

components to be produced by industrialised methods, as it comprises the important prime numbers 2, 3 and 5 ($600 = 2^3 \times 3 \times 5^2$) and is thus divisible in many different ways without involving decimal fractions of millimetres. This facilitates calculation and makes it possible to utilise production machines whose working dimensions are generally adjustable in increments of whole millimetres. The divisibility of the number 600 is illustrated in **1**.

For steel structures themselves, compliance with a modular co-ordination system is not very important, as the fabrication of steel components is not dependent on fixed increments. For the space-enclosing components of steel structures, however, it is usual to employ materials which are prefabricated and which should, for this reason, conform to modular dimensions. In that case, of course, the steelwork to which such components are fixed should similarly comply.

The above-mentioned DIN also lays down the following requirements regarding the application of modular co-ordination:

'A three-dimensional grid should be employed for the planning of modular co-ordinated structures. The lines and planes of this grid should, as a rule, be mutually perpendicular and be spaced at modular distances in relation to one another (**2**).

'The spacings of the lines and planes in a modular grid can be unequal for the different directions of the co-ordinate system (**3**).

'All the components of the building should be incorporated in this grid. For each component a modular space should be determined within which that component, including its joints, is accommodated (**4**).'

requisite for industrialisation), issued in February 1972 by the Länderarbeitsgemeinschaft Hochbau (Interprovincial Working Group on Building Construction) of the Federal Republic of Germany, has the following to say:

'The interior planning grid should be 0.60×0.60 m. The layout grid has the unit dimension

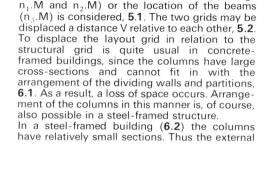

2

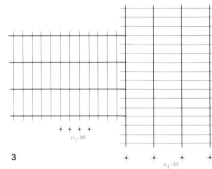

3

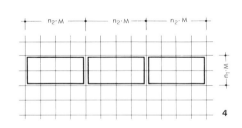

4

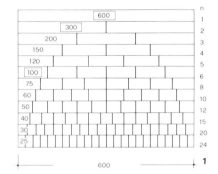

1

The structural grid may, for a steel-framed structure, mean different things, depending whether the location of the columns (spacings $n_1.M$ and $n_2.M$) or the location of the beams ($n_3.M$) is considered, **5.1**. The two grids may be displaced a distance V relative to each other, **5.2**. To displace the layout grid in relation to the structural grid is quite usual in concrete-framed buildings, since the columns have large cross-sections and cannot fit in with the arrangement of the dividing walls and partitions, **6.1**. As a result, a loss of space occurs. Arrangement of the columns in this manner is, of course, also possible in a steel-framed structure.

In a steel-framed building (**6.2**) the columns have relatively small sections. Thus the external

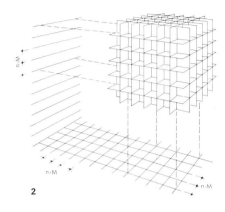

5.1   5.2

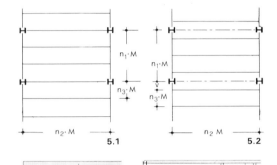

6.1   6.2

Structural and interior planning grids

In planning a structure it may be advantageous to use grids with different grid dimensions, namely, a structural grid and an interior planning grid. These grids may coincide, or they may be displaced relatively to each other; in the latter case the amount of displacement, V, should be a whole modular dimension.

With regard to this subject, the *Forderungskatalog zur Standardisierung im Hochschulbau als Voraussetzung für die Industrialisierung* (*Catalogue of requirements for standardisation in university building construction as a pre-

of 1.20 m, is superimposed on the basic planning grid and is supplemented by 0.60 m. The structural grid is also superimposed on the interior planning grid and its dimensions are multiples of those of the layout grid. Preferred dimensions are 7.20 and 8.40 m.'

A similar *Catalogue of requirements* is being compiled for school buildings in the Federal Republic. If prefabricated internal components are to be employed, it is suggested that the requirements applicable to university buildings should also be applied to the design of other multi-storey structures.

columns can be located on the grid points of the beam grid (which corresponds to the spacing of the windows), so that the partitions can be connected directly to them. In that case, for the external columns, the structural grid coincides with the layout grid. The internal columns may be similarly accommodated on the layout grid or alternatively — if their combination with the partitions is not desired — they may have their own grid, as in a concrete-framed building. The external columns can generally be provided with the appropriate features to allow modular partition units to be connected to them.

173

Dimensional co-ordination

Modular co-ordination

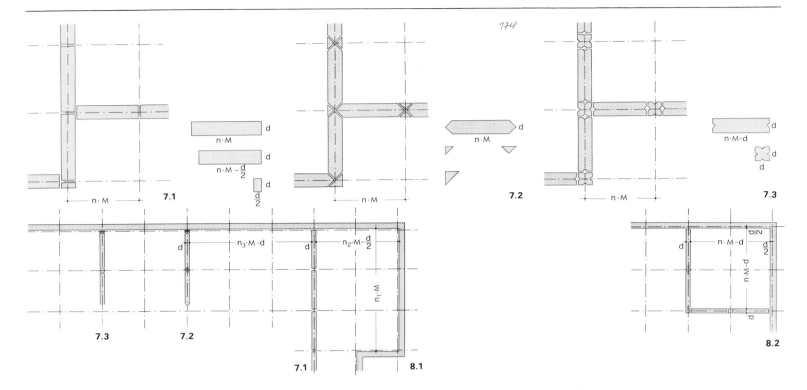

For each component, it is necessary to define the dimensions which are critical in fabrication and to ensure co-ordination of assembly with the other components in the structure.
Nevertheless, it is not essential and sometimes not possible to ensure that all the dimensions are modular. Non-modular dimensions and floor thicknesses result in non-modular layout and size of room (**8.1** and **8.2**).
It is often advantageous to provide the ceilings with a modular grid of fixtures for demountable partitions, so that the latter can be installed in any position between window units.

Modular partitions

In principle, the following possibilities for the modular arrangement of partitions are available:
7.1 At junctions, one of the two partitions continues past the other. There are no posts. Panels having widths $n \cdot M$ and $n \cdot M - d/2$ are needed, together with corner elements with dimensions $d \cdot d/2$.
7.2 The partitions are composed of panels with mitred edges so that a standard panel of width $n \cdot M$ can be used anywhere. Filler panels are needed, but there are no posts.
7.3 If posts are used which allow panels to be connected on all four sides, a standard panel of width $n \cdot M - d$ and a post with dimensions $d \cdot d$ will suffice.

8.1 The inner face of the external wall generally coincides with a grid line. The floor units correspond to full grid units. Partitions, as envisaged in any of the three above-mentioned systems, can be connected: in the case of **7.2** it is necessary to use filler panels, while **7.3** requires half posts.
The internal dimensions of rooms between two external walls are $n_1 \cdot M$; between an external and an internal wall they are $n_2 \cdot M - d/2$; and between two internal walls they are $n_3 \cdot M - d$, assuming all the internal walls to be of equal thickness d.
8.2 If the inner face of the external wall is inset a distance $d/2$ from the grid line, all the room dimensions are $n \cdot M - d$.

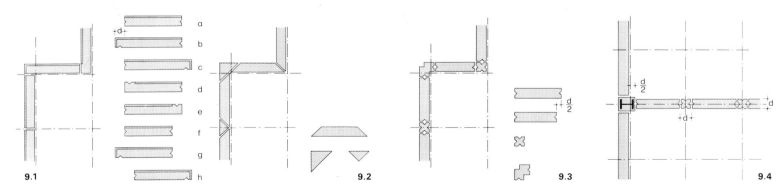

Modular external walls

For the modular subdivision of the external walls (Figure 9) there are — as in the case of the internal partitions — fundamentally three variants:
9.1 The external wall without posts, mullions or filler units requires, in addition to the standard unit a, basic elements for salient corners b, c, and for re-entrant corners d, e, as well as shortened standard units f, at the re-entrant corners $n \cdot M - d$. For wall indentations of reduced depth, shorter salient corner units g, h, of length $n \cdot M - d + d = n \cdot M$ may also be required.
9.2 The external wall, composed of elements with mitred edges, comprises only one basic unit and two types of filler unit.
9.3 The external wall with posts comprises two standard units and two types of post which are also provided with attachments for connecting the partitions.
9.4 The structural columns of a steel-framed structure can, when closely spaced, be combined with the external posts envisaged in **9.3**. If the columns are inset a distance $d/2$ into the room, they can be readily co-ordinated with a partition system with posts, **7.3**, or without posts, **7.1**.

Tolerances

Dimensional co-ordination

Modular vertical dimensions

With regard to the modular co-ordination of vertical dimensions, the Standard states:
'One of the basic vertical dimensions is the storey height. Generally its modular measurement is taken from the finished floor level of one storey to the finished floor level of the next, but it may alternatively be taken from one unfinished structural floor level to the next or from one staircase landing to the next.'
The height of steps is seldom a modular dimension, but groups of steps can be combined to have modular measurements, for example:

$$2 \times 100 = 200 \text{ mm} = 2 \text{ M}$$
$$2 \times 150 = 300 \text{ mm} = 3 \text{ M}$$
$$3 \times 167 = 500 \text{ mm} = 5 \text{ M}$$

Further particulars on the dimension coordination of steps are given on p. 291.

TOLERANCES

In a general way, tolerance can be defined as the acceptable range within which some dimension is allowed to vary, that is, it represents a degree of latitude within which the components of a structure can be suitably assembled and will fit together without difficulty.
In building construction a system of tolerances is unnecessary in circumstances where all the parts of a structure are made or shaped on site, so that each part can be adjusted and made to fit the others. But with prefabrication of components it becomes necessary to take due account of dimensional variations of individual parts and assemblies. More particularly, in the construction of buildings from factory-made standard components, is it essential to lay down definite tolerances for the manufacturing dimensions, in order to avoid any need for any cutting, shaping or other adjustment of the size of the components on site.
In Germany, a start has been made with the systematic introduction of tolerances in the building industry, as laid down in Standards, of which the following had been published by 1972:
DIN 18 201: *Terminology and principles for the application of tolerances in building.*
DIN 18 202: Sheet 1: *Tolerances for openings in buildings, including staircase wells*; this information is of interest only in connection with conventional masonry or concrete buildings.
Sheet 2: *Flatness tolerances for concrete or masonry surfaces.*
Sheet 3: *Tolerances for the flatness of floors.*
Sheet 4: *Allowable tolerances for structural dimensions.*
DIN 18 203: *Allowable tolerances for the dimensions of precast concrete components.*
These Standards are essentially geared to masonry and concrete structures and do not adequately meet the needs of prefabricated construction using components produced by different manufacturers, for example, for steel structures, for precast concrete units which are to be used in combination with steel structures, or for buildings with external metal cladding and with internal metal partitions.
Better guidance is given by Information Sheet M 2 issued by the Arbeitsgemeinschaft Industriebau (AGI). Structural tolerances can be legally specified and applied only if the physical conditions to which they relate are properly defined:

• The temperature, which may vary externally and internally.
• The loading, comprising permanent and imposed loads.
• The point in time, more particularly with regard to time-dependent deformations such as creep and shrinkage of concrete.
• The moisture content of the material, for example, in the case of timber components.

Steel components and the buildings constructed with them are fabricated and erected to much smaller tolerances than buildings constructed with other materials. The reasons for this are:

• Rolled steel sections are manufactured with relatively close tolerances.
• Fabrication of structural steel components is carried out in workshops using accurate machining and other equipment, including electronically controlled production lines operating with great and verifiable accuracy.
• Steel components are unaffected by time-dependent influences and moisture, which are liable to cause deformation in other materials.
• The elastic deformations of steel components due to loads and their volume changes due to temperature variations can be calculated with considerable accuracy.
• Precast concrete components which are assembled in combination with structural steel frames must themselves be made to close tolerances and may, thanks to the support provided by the steel frame, be very accurately positioned in the structure.

A steel frame therefore provides favourable conditions for the incorporation of other prefabricated components with relatively close tolerances.
In building construction, dimensional tolerances are necessary:

(1) for the efficient use of a building, for example, verticality of liftshafts and flatness of façades, or horizontality and evenness of floors;
(2) to ensure that components fit together properly without the need for cutting them down to size or carrying out other dimensional alterations on site. These tolerances relate only to those parts of the components where connections have to be formed, such as at joints, bolt-holes, etc.

In prefabricated construction with standardised units, the dimensions which have to conform to the specified tolerances must be identifiable on the drawings. In German steelwork detailing, it is normal practice to enclose these dimensions within a rectangle in which the tolerance limits are also written. A dimension may have bilateral tolerance, allowing variation in both directions from the design size, or unilateral tolerance, allowing variation only in one direction. In the former type of tolerance the limits may be equal, for example, 3000 ±10 (**1.1**) which means that the component in question is allowed to have a length not greater than 3010 mm nor less than 2990 mm. Bilateral tolerance may have limits that differ in the two directions, for example, 3000 +15/−40 (**1.2**), so that the length must not be greater than 3015 mm nor less than 2960 mm. A unilateral tolerance may be a plus tolerance 3000+20, (**1.3**) or a minus tolerance, 3000−20 (**1.4**).
In determining the tolerances, it should always be remembered that fabrication costs show a disproportionately steep rise as the tolerance specified becomes more exacting.
A distinction must be drawn between constructional tolerances and component tolerances.

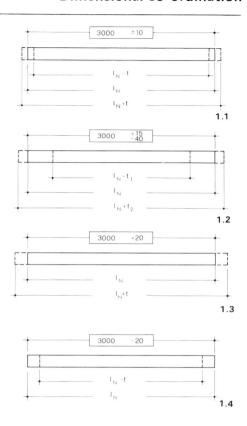

Constructional tolerances

These have to ensure that the structural components are indeed accommodated within the spaces assigned to them, the position of which is defined by a system of dimensions with certain fixed points at ground level. The following must be established:

• Permissible deviations for lengths and widths of structures and structural components, for reference lines on plan, for centre-to-centre distances and for grid dimensions.
• Permissible deviations for alignment on plan and for angles (more particularly: deviations from right angles).
• Permissible deviations for vertical distances measured from a fixed starting point, depending on the spacing of the reference points on plan and vertically.
• Permissible deviations for vertical alignment.

Component tolerances

The actual dimensions of components must be within the tolerance range, this being the range between the upper and the lower tolerance limits. Within the tolerance range the actual dimensions are allowed to deviate from the nominal dimension (or design size) for the particular dimensional quantity in question, such as length, angle, evenness, **2**. Joints are formed between adjacent components, **3**. The nominal dimension of the joint may be increased or decreased by an amount equal to the sum of the tolerances of the adjacent components. In the limiting case the joint may have zero width.
The jointing elements (fastenings or jointing material) between two components must be so adjustable or deformable that they can adapt themselves to the largest as well as to the smallest dimension of the joint. In addition, it

Dimensional co-ordination — Tolerances • Deformations

must be checked whether movements due to loads or to volume changes of the materials (temperature difference, shrinkage, etc.) are liable to occur in the joints whereby the jointing elements might suffer permanent elongation or shortening (p. 223). The following should be determined in advance:

- permissible deviations for the dimensions of components;
- permissible deviations for the surface condition of components.

To determine and specify suitable tolerances is a prerequisite for successful prefabricated construction with standardised units. Besides, it provides a legally acceptable basis for deciding who must bear the cost of any subsequent alterations necessitated by failure to comply with manufacturing tolerances.

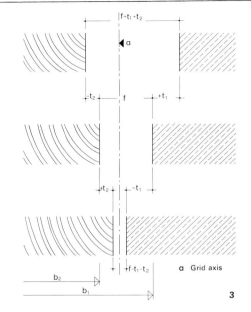

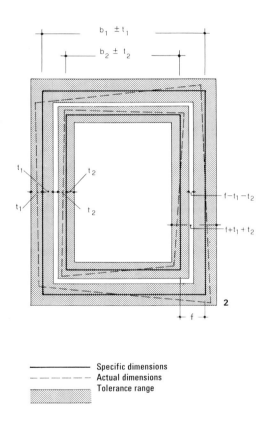

— Specific dimensions
--- Actual dimensions
▨ Tolerance range

DEFORMATIONS

Deformations in structures or structural members are governed by natural physical laws and are inevitable. However, in order to obviate objectionable effects, especially in the form of damage to buildings or finishes, the deformations must be investigated and their magnitude limited.

The deformation of steel members is somewhat larger than that of concrete members comparable in strength. With correct design and careful construction of the structural steel frames and their finishes, the detrimental effects of deformation can be eliminated.

First, the types of deformation that may occur, and their causes, will be considered. Structural members undergo deformation in consequence of volume changes in the materials from which they are made; structures as a whole deform as a result of displacement of the intersection points of the centre-lines of their component members.

Changes of shape

External loads or internal forces cause changes of shape in members. Conversely, it can be said

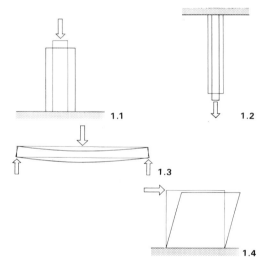

that no transmission of load is possible without associated strain in the loaded members involved. A column which has to carry load must shorten, a loaded floor must deflect. With this type of deformation the members undergo no change in volume. For example, a strut, on being loaded, becomes shorter but thicker (1.1); a tension member becomes longer but thinner (1.2). Other examples are the flexurally loaded member (1.3) and the member loaded in shear (1.4).

Deformations may be elastic, in which case the members will regain their original shape on removal of the load. If they remain deformed after load removal, the permanent deformations are known as plastic. In structural engineering, as a general rule, materials are subjected only to loads which remain within the elastic range of deformation behaviour. Some materials (for example, concrete), when subjected to long-term loads, undergo a certain amount of plastic deformation, known as creep, in addition to the normal elastic deformation. Creep is approximately proportional to the elastic deformation produced by the loads concerned. On removal of the load, the deformation due to creep gradually but not completely disappears.

Gradually applied loads cause static deformation; rapidly applied ones have a dynamic action; rhythmically repeated loads may give rise to oscillations in individual members, assemblies of members or the whole structure, for example, vibrations in floors when people walk on them or vibrations in high-rise buildings due to wind loading. Every object has its own particular natural frequency. Vibrations are produced when the frequency of the applied load is close to the natural frequency of the loaded body.

Changes of volume

Changes in the volume of a building material may be caused by:

- temperature changes;
- shrinkage or swelling due to chemical processes or to variations in the moisture content of the material.

In general, volume changes which cannot develop freely give rise to internal restraint forces which in turn cause deformation. When a simply-supported beam undergoes a rise in temperature, it can expand freely in all directions, that is, it can increase its length, width and depth without restraint (2.1). On the other hand, the beam fixed at both ends (2.2) is prevented from increasing its length when exposed to a rise in temperature; only its depth and width can increase. As a result of the longitudinal restraint a force Z develops. A similar effect occurs when a beam which has expanded due to a rise in temperature (2.1) is subjected to an axial compressive force Z which reduces it to its initial length. This shortening is accompanied by an increase in girth.

Unequal volume changes across the depth or width of a member cause bending (3). Thus a beam may be exposed to sunshine on its top face, while its underside is in the shade. A similar effect occurs in a timber beam sawn asymmetrically from a log: changes in atmospheric humidity cause differential changes in moisture content within the beam, with corresponding differences in shrinkage or swelling of the wood.

In steel-framed structures with their relatively slender members the restraint forces associated with volume changes generally do not produce any substantial increase in stress, whereas such forces may have quite important effects in concrete structures with members of relatively large section.

Displacements

A framed structure is composed of members whose axes are dimensionally associated with the overall structural geometry. The deformation of members may cause displacements of their joints and thus bring about a deformation of the whole building.

For example, a load acting on a floor (4.2) does

Deformations   Dimensional co-ordination

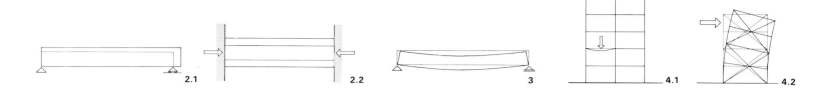

not appreciably alter the geometry of the structure, but wind forces (**4.2**), by producing changes in length in the lattice bracing members, cause the node points of the frame to shift laterally and vertically.

In general, the stability of a building depends only on the loads to which its structural members are subjected, not on their deformation. The permissible loading laid down in official building regulations ensures an adequate margin of safety against failure. On the other hand, the permissible amount of deformation is not usually stipulated in such regulations. It is, in each individual case, necessary to consider what effect the deformation of a member will have on utilisation and/or compatibility with other members of the structure and to decide how much deformation can be tolerated. Generally speaking, the more rigid a structure has to be, the more it will cost. It is therefore undesirable to impose unnecessarily strict limits on deformation.

Wherever two structural members intersect, there is a joint. The general principle to be applied is that

- either the deformation of the two members which meet at the joint must be able to develop freely and without restraint at the joint,
- or the restraint forces due to differences in deformation of those members are safely absorbed and transmitted by structural connections. In this latter case, elastic deformation is produced in the two interconnected members and in the connecting elements themselves.

This principle is applicable both to joints between individual structural members and to joints between parts of a building. Yet it is repeatedly ignored. This non-observance of so important a basic principle is one of the commonest underlying causes of structural damage.

Effect of dead load

Deformation should be considered with reference to the nominal condition of a building, this condition being attained:

- after all permanent loads have been applied;
- at a particular temperature (the erection temperature), usually taken as $+10°C$;
- at a time when the creep and shrinkage of the concrete have virtually ended;
- at a particular value of the atmospheric humidity, where applicable.

The structural members should be so designed and so erected that account is taken of these influencing factors.

For instance, the deformation due to dead load is calculated in advance, so that the eventual shape of the members in the completed building is known. In a steel structure, deformation affecting members or assemblies of members takes the following forms:

- Columns, struts and loadbearing concrete walls undergo a shortening, usually of negligible magnitude.
- Tension members elongate. This effect should be given due consideration in the erection of suspended structures, as the elongation of the hangers acts in opposition to the compressive shortening of the supporting structural core (p. 349).
- Vertical lattice systems are subjected to compressive forces due to the action of vertical loads. Such forces should be investigated and duly allowed for, more particularly with regard to the erection of the structure (p. 264).
- Floors deflect under the action of dead load. Steel beams can be provided with a certain camber, but as this involves an extra operation, and thus adds to the cost, only fairly long-span beams and girders (from about 10 m upwards) are normally provided with a camber (p. 242).

Effect of live load

Deformation due to live load is generally of little practical significance except for floors. It is important to limit the deflections of floors for:

- special use, for example, in laboratories containing delicate equipment;
- the attachment and waterproofing of external cladding;
- freedom from cracking (and subsequent loss of acoustic insulation) in partitions.

The calculation of the deflection of steel beams is dealt with on p. 245, while composite beams are considered on p. 259.

The limiting values frequently adopted for the deflection of floors under live load are:

for flexible floors: $f = \dfrac{1}{500} \cdots \dfrac{1}{300} \, l$

for rigid floors: $f = \dfrac{1}{800} \cdots \dfrac{1}{500} \, l$

Only live load deflections are relevant in this context and it is therefore only these that must be limited. Deflections due to dead load can be eliminated where necessary by cambering the floor beams prior to erection.

It should also be noted that not all live loads are moving loads. As a rule of thumb, it may be said that live loads are made up as follows: about one-third comprises fixtures such as light partitions and objects, such as furniture, which when once installed are rarely moved; about one-third is actually a variable load; and about one-third can be regarded as a reserve for exceptional circumstances, for example, people crowding together.

For all practical purposes, the first third can be added to the dead load. The last third rarely occurs and the resulting deflection generally disappears on removal of the load. Only the second third is truly a moving load.

Partitions should possess sufficient flexibility to be able to absorb deflection under variable loading without suffering damage (see p. 319; for fire-resisting walls, see p. 320; for concrete shear walls, see pp. 217, 267).

Effect of horizontal loads due to wind and earthquake

If seismic loads have to be taken into account, buildings must be designed for stability, not compliance with certain specific limits of deformation.

The horizontal deflection of buildings due to wind is of major importance, more particularly in tall multi-storey structures. Wind deflection not only has to be considered in terms of structural design, but may significantly affect the overall design, including the architectural aspects, of a high-rise building. Such buildings are liable to oscillate under the action of wind forces. These oscillations increase the static deflection due to wind pressure and may adversely affect the comfort or peace of mind of the occupants. For the bracing of high-rise buildings, see page 209. Specific problems relating to tall buildings are examined on p. 219.

Effect of temperature changes

Consider 5 (p. 178). To judge the effects of changes in temperature, it is necessary to consider the prevailing temperature conditions in any particular climate. In Central Europe, for example, the approximate range of air temperature is from $-30°$ to $+30°C$. However, individual members or parts of structures may reach temperatures as high as $+80°C$ as a result of direct exposure to sunshine.

Major temperature differences are liable to occur in any structural member during the construction of a building; later, in the completed building, such differences are confined to those members which are exposed to the external climatic conditions as opposed to members in the interior.

The deformation caused by temperature variations has to be absorbed in suitable joints between members and between parts of a structure. The correct arrangement and width of such expansion joints are important if structural damage is to be avoided.

Prefabricated structural members are generally subject to relatively minor temperature variations during their manufacture, since steelwork as well as precast concrete components are, as a rule, produced under cover in buildings. It can be assumed that such members will indeed have their theoretical design dimensions at the design temperature of $+10°C$.

During erection, structural steel members may be subjected to temperatures which differ greatly from the design temperature, as a result of which additional stresses may be developed in statically indeterminate systems. In-situ concrete and precast concrete members with joints filled in-situ are also subject to large temperature differences. The temperature of steel components which are protected from

Dimensional co-ordination — Deformations

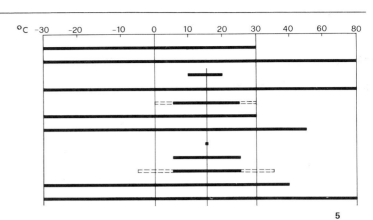

Ambient temperature in the open air
Over and above this: temperature of structural steel members exposed to direct sunshine
Fabrication temperature of members
Erection temperature of steelwork
Construction temperature of in-situ concrete members
Ambient temperature while the unheated building is under construction
Over and above this: temperature due to direct sunshine
Design temperature
Temperature of the floor slabs when building is in use
Temperature of roof structure, depending on insulation
Temperature of insulated external columns
Temperature of exposed steel columns

5

direct sunshine by concrete slabs, for example, in composite construction, will not rise above the ambient air temperature.

The unheated half-finished building, to which the external air has free access, is subjected to all the variations of the air temperature. High temperatures are liable to develop in members exposed to direct sunshine.

As soon as the cladding has been installed and the windows glazed, the range of temperature in the interior is greatly reduced. When the building is in use, with heating in cold weather, the temperature is normally between 15° and 25°C, though it may of course fall to a lower value in the event of a breakdown in the heating system. Somewhat larger variations, depending on the thermal insulation, may occur in the structural members of flat roofs.

The arrangement of expansion joints in buildings is discussed on pp. 223, 224.

External walls may undergo deformation due to the displacement of the structural frame of the building and to deformation of the cladding units themselves. Such deformation must be absorbed in the joints between the units. Problems associated with these effects are dealt with on p. 307.

Encased steel columns which are not in contact with the heated interior of a building participate in the external temperature variations. Although the heat-insulating casing will moderate the rise in temperature caused by direct sunshine, it cannot prevent cooling during a long period of cold weather.

During a fire, large temperature differences are liable to develop, which may cause considerable expansion in parts of the building, particularly floors. The possible effects must be given due consideration when designing the expansion joints, to limit the extent of the damage in the event of a fire (p. 223).

Effect of settlement of supports

For details of the consequences of such settlement on structures and the arrangement of joints, see p. 223.

Day nurseries	179	Department stores	183	
Schools	180	Residential buildings	184	
Universities	181	Hospitals	185	
Office Blocks	182	Multi-storey car parks	186	

Types of Multi-Storey Steel-Framed Buildings

The design of a building is governed by the purpose for which it is to be used, floor space requirements, internal layout, means of access and services, all of which determine its plan and its elevation. For buildings intended to serve similar purposes and thus having to fulfil the same requirements, certain typical designs exist which embody uniformity of plan and elevational treatment. In this chapter an attempt is made to classify the various types of building frequently constructed in steel. Of course, there are few clear-cut types: transitions from one type to another are gradual, and the requirements themselves are in constant evolution. The needs of the occupants and the conditions laid down by public authorities may also, in certain circumstances, call for different solutions.

DAY NURSERIES

Day nurseries are premises where children below school age spend a large proportion of the day under the care and supervision of staff appointed for the purpose. Such establishments must comprise facilities for children of different ages — rooms for playing, eating and resting, together with the necessary cloakrooms. As a rule, it must be possible to run them with no more than a minimum of supervisory staff. This sort of use requires:

Low buildings, with one or two storeys, with stairs easy to climb, an elaborately subdivided layout on plan with several means of access to the open-air, but partly roofed-in, play facilities, ample sanitary accommodation, kitchens, safe and robust equipment, together with a high degree of fire safety — not only of the load-bearing structure, but also of the finishes and fittings — and adequate fire escapes. Pavilion-type buildings with light structural steel frames are suitable.

Day nursery, steel-framed structure with highly subdivided layout: Motzstrasse, Berlin (architects: Hassenstein, Bratz).

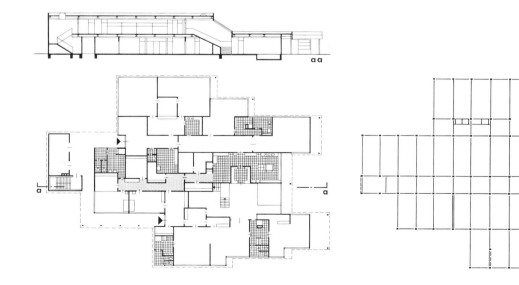

Scale 1:600

Types of multi-storey steel-framed buildings — Schools

As a result of rapid sociological and technical evolution, educational demands made upon school architecture are in a constant state of change. Yet, despite their different purposes, the various types of school (primary, secondary, comprehensive and special schools) show considerable similarity in the requirements that their buildings have to fulfil. Quite often, present-day schools have large numbers of pupils (running into several thousand in the case of comprehensive schools) with a wide age-range. Inside the buildings, there is continuous movement of people, including large groups. For this reason it is essential to provide wide corridors and staircases. Mechanical devices, lifts in particular, are not suitable, and that is why school buildings generally have only a limited number of storeys (up to four).

For general educational use it is desirable to provide classrooms whose size can be varied to meet changing requirements. Groups of between 20 and 30 pupils require rooms measuring approximately 8.40 × 8.40 m. Subdivision of available space to suit the needs of smaller groups or, alternatively, to accommodate larger numbers in one room should always be possible. Although they have not hitherto come into very extensive use, teaching machines are increasingly being installed in present-day schools, and the necessary arrangements for accommodating them should be made.

Technical and scientific education require machinery and equipment, which in turn means that certain parts of the school buildings must be provided with the appropriate services. Such parts accordingly possess little flexibility of layout, being largely committed to specific functions and purposes.

In general, the need for flexibility of internal layout calls for buildings with only a small number of internal columns and with long floor spans.

Where compact shapes on plan are adopted, higher demands are made on air-conditioning and ventilation, requiring more space above the ceilings in which to accommodate the ducts. The flexibility required in the layout is particularly well provided by floors comprising steel beams.

High standards of sound insulation are often required in school buildings and must be given due consideration in the design of floors and partitions.

On the other hand, the requirements of structural fire protection may be relatively few and simple, for the following reasons:

- the pupils are constantly under supervision;
- there is regular fire drill;
- the pupils are mobile and can move about quickly;
- they have plenty of experience in the rapid evacuation of the building;
- schools have wide corridors and many staircases, so that there are adequate escape routes;
- the fire load in schools is generally low.

Large expenditure on structural fire precautions is therefore hardly justifiable.

It is obviously desirable to insist on strict dimensional co-ordination for the numerous school construction projects in the next few decades, in order to promote industrialised building techniques and the use of system building and to reduce costs. Numerous school building systems are already on the market.

Steel is a very suitable building material for schools and is used for the majority of those constructed in Britain, France and West Germany at the present time.

Grammar school at Osterburken (architects: Bassenge, Puhan-Schulz, Schreck).

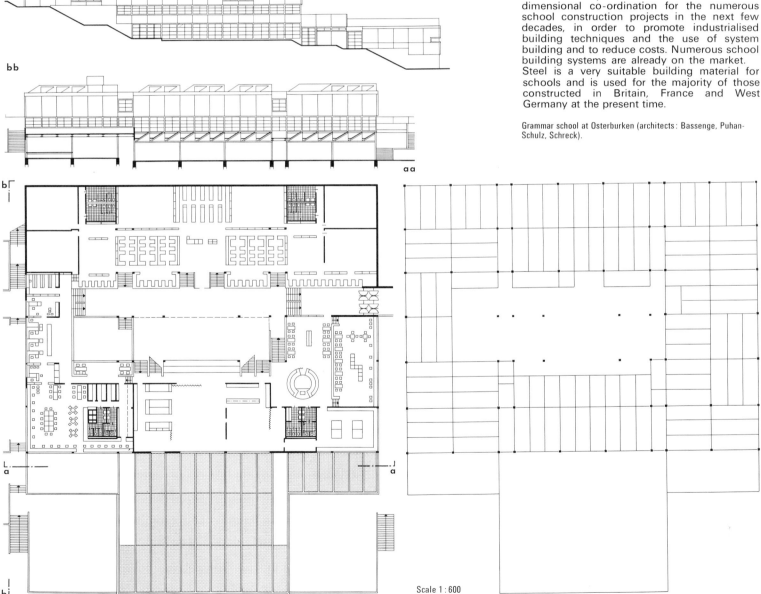

Scale 1:600

Universities
Types of multi-storey steel-framed buildings

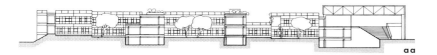

Scale 1:1500

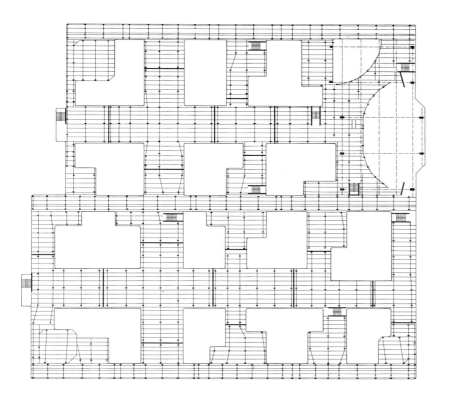

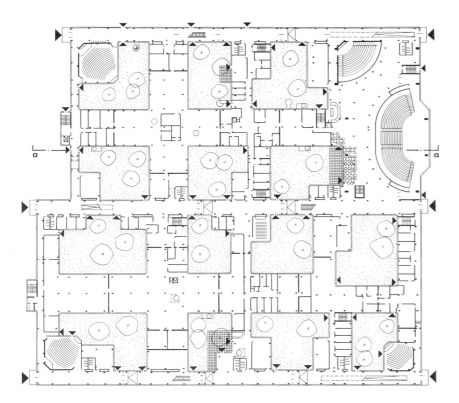

University buildings require large working areas. Even more than in schools, the rapid evolution of science and of teaching methods in universities demands flexibility in the subdivision of the available space. This calls for large interconnected working areas with widely spaced columns, uninterrupted by fixed structural features such as stairwells and service shafts. Such conditions can readily be satisfied by steel structures. The columns may be located at the perimeter of these areas and be merged into the external walls of the building or with the enclosing walls of the vertical circulation system. The latter may be concentrated in special cores or towers placed outside the working areas, while intensive horizontal circulation can be suitably routed at ground level. High-rise buildings in which a substantial proportion of the area on plan is occupied by corridors, staircases and lifts are not very suitable for universities. Low-rise construction facilitates communication, but it increases the length of the corridors and requires larger areas of land. Buildings with six to twelve storeys appear most economical.

Floor finishes and ceilings should be uniform over large areas to facilitate the rearrangement of internal layout. Despite having to fulfil exacting requirements for sound insulation, partitions should be readily demountable.

University buildings intended for predominantly theoretical activities require simple and relatively few technical services, whereas those intended for experimental work make high demands in this respect. However, the technical services must not be allowed to impair the flexibility of the layout. They must therefore be accommodated only in the floor spaces. If there is a possibility that a building will subsequently be converted from theoretical to experimental activities, sufficient space for the additional services should be provided from the outset.

The magnitude of university construction projects calls for industrialised building techniques. To this end, in the Federal Republic of Germany the Länderarbeitsgemeinschaft Hochbau (Interprovincial Working Group on Building Construction) issued in February 1972 a document entitled *Forderungskatalog zur Standardisierung im Hochschulbau* (*Catalogue of requirements for standardisation in university building construction*). The requirements for university building construction can be very suitably met by steel as the main structural material, more particularly in achieving widely spaced columns of relatively small cross-section and due compliance with specified tolerances, which is essential to ensure flexibility of layout and, indeed, the possible modification of the structure itself. For the convenient and readily accessible arrangement of services in scientific research buildings the two-layer arrangement of floor beams in structural steelwork is especially suitable (pp. 192–193). The structural danger arising from aggressive fumes in laboratories is greater in concrete than in corrosion-protected steel structures.

Social sciences block in the Free University of Berlin (architects: Candilis, Josic, Woods, Schiedhelm).
See pp. 211 (**12**), 262 (**10**), 258 (**5**), 296 (**5.2**) and 311 (**6**).

Types of multi-storey steel-framed buildings

Office blocks

Large office blocks are generally situated in the central areas of cities, in highly developed peripheral areas, or in the immediate vicinity of major industrial plants. Frequently, they take the form of high-rise buildings with elaborate mechanised facilities for vertical circulation. It is essential to have a minimum of internal supports within the areas destined for office accommodation, so as to have maximum flexibility of layout for open-plan or individual offices. Rectangular or triangular planning grids are commonly adopted but sometimes a circular shape on plan is preferred. A central ventilating or air-conditioning system is usually provided and is indeed indispensable in high-rise blocks and in buildings covering a very large area on plan. The low-pressure supply ducts for ventilating the inner zones are accommodated in the ceiling spaces, while the high-pressure ducts for air-conditioning are generally located alongside at the external walls: horizontally within the floor structure, vertically beside the columns. Often, too, elaborate electrical equipment is required, more particularly in buildings which contain electronic data processing terminals. If these services are installed in the ceiling spaces, they are accessible — for subsequent modifications or additions — only from the storey below. This is inconvenient and can give rise to difficulties in cases where a building is occupied by different tenants. To allow electric cables to be accessible from within the storey that they actually serve, they can be accommodated inside the cavities of metal floor decking or, in concrete floors, they can be housed in ducts resting on the structural floor and embedded in a 6 to 10 cm thick screed. The latter, of course, adds substantially to the dead weight of the structure.

A high level of sound insulation in office buildings, if it is required, is often confined to certain parts thereof, for example, executive floors, board rooms, etc.

For economic reasons, very short construction times, allowing no interruption of work in winter, are often specified for multi-storey office blocks. In solving such structural problems the following properties of steel frames are advantageous: long spans, small column sections, scope to install elaborate ducting within the depth of the floor structure and electric cables in metal floor decking, rapid erection unaffected by weather conditions.

Office building for Deckel, Munich (architect: Henn).

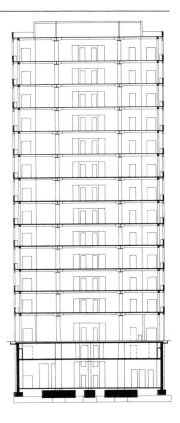

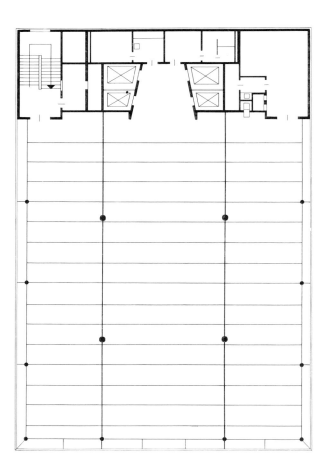

Scale 1 : 300

Department stores

Types of multi-storey steel-framed buildings

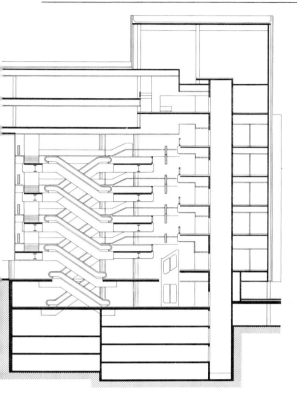

With the increasing concentration of retail trading, department stores are gaining in importance. Large interior spaces free from columns are needed. Modern department stores seldom have a large number of storeys. Indeed, in the form of supermarkets or shopping centres in the suburbs they are often built as single-storey structures. The long-span floors have considerable construction depths which are used to accommodate the elaborate services and ventilation ducts required to keep the air fresh in the inner zones of these extensive buildings.

To deal with the movement of large numbers of people and the internal transport of merchandise, multi-storey department stores have to be generously provided with mechanical appliances such as passenger lifts, escalators and travelators, as well as goods lifts, pneumatic despatch systems, etc. The need for wide passageways, numerous staircases and exits is, of course, self-evident.

Fire protection is a matter of the greatest importance in department stores, more particularly because the fittings and finishes as well as merchandise, such as textiles, constitute a heavy fire load. As a rule, sprinkler installations are provided as a first line of defence against the outbreak of fire. Wide and conveniently located escape routes, clearly signposted, both in the horizontal and in the vertical direction are a very important requirement. In addition, in German multi-storey buildings, the loadbearing structure must have a fire rating of 90 min.

Kudamm-Eck department store, Berlin (architect: Düttman).

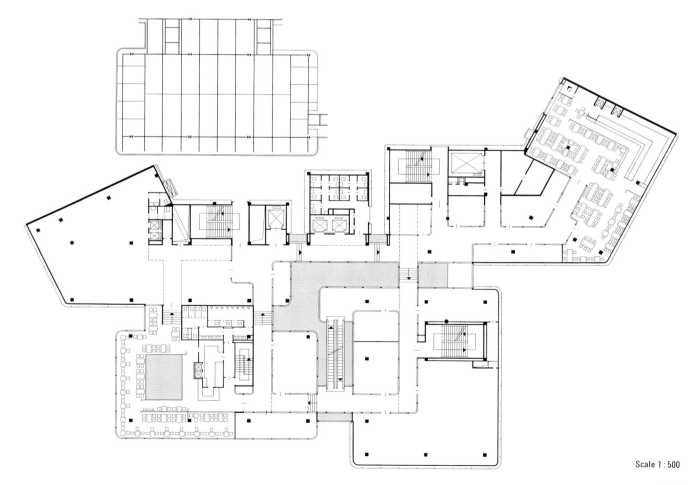

Scale 1 : 500

Types of multi-storey steel-framed buildings

Residential buildings

The major housing projects of the post-war period were mostly characterised by internal layouts with rigid subdivision into relatively small rooms, with no scope for rearrangement. Because of this approach, large-panel prefabricated construction systems gained predominance in residential building.

By using loadbearing partitions the designers produced an arrangement of cells whose dimensions could not subsequently be varied. The concrete slabs were supported on the walls and were given a smooth finish on the top and bottom surfaces to receive the floor finish and the ceiling coat respectively. This form of construction was acceptable in residential buildings because, in general, no horizontal services (other than electric wiring) had to be installed, the rooms requiring pipe connections (kitchen, bathroom, etc.), being installed one above the other in a multi-storey block of flats and served by shafts in which these services were accommodated.

With the growing demand for larger rooms and flexibility of internal layout, framed building construction, and therefore structural steelwork, is becoming an attractive proposition for housing projects. With a structural steel frame the following types of floor construction are possible:

1 The floor slab spans the full width of the room; the floor beams are concealed within the walls; the floor slab is relatively heavy and has a smooth finish on both faces. This is essentially the same solution as that adopted in large-panel construction, with rigidly planned internal layout.

2 The floor beams are closely spaced; the floor slab is thin and light; a separate ceiling installed below the structural floor provides fire protection for the steel beams, but adds to the cost of the building.

A residential building has too few services (other than electricity) to utilise the ceiling space economically. This solution is therefore generally not very suitable, except for very high-class blocks of flats.

3 The same floor construction principle as above, that is, with closely spaced beams, but these are exposed and are given a fireproof casing. The underside of the floor slab forms the ceiling of the storey below. This is an inexpensive solution, but if the arrangement of the exposed beams bears no relationship to the internal layout of the rooms, there may be difficulties in installing the partitions.

4 Long-span thick concrete floor slabs, without beams, provided with a smooth finish on the top and bottom surfaces, obviate the problems associated with ceilings, but they are heavy (examples: p. 277 **3**, p. 190).

Fire protection: Official regulations vary considerably from one country to another. However, they generally require the structural steelwork in multi-storey buildings to be encased. Also, party walls between adjacent flats are required to be of fire-resisting construction. Furthermore, good sound insulation is demanded in residential buildings.

On the other hand, only very moderate fire requirements are applicable to ordinary houses (as distinct from blocks of flats), so that in such buildings the structural steelwork can be left exposed and utilised as a design feature. There are numerous examples of this.

Hotel construction is in many respects similar to housing, but the need for flexibility of internal layout is not a very significant consideration, for the size of hotel bedrooms with their individual bathrooms is largely standardised. Floor structures of type **1**, as envisaged above, offer economical solutions. Sound insulation has to be of at least as high a standard as in blocks of flats. Similar considerations apply to hostels and homes for the elderly.

Parc Saint Clair residential project, La Baule, France (architect: Faivre-Rampant).

Scale 1 : 375

Hospitals — Types of multi-storey steel-framed buildings

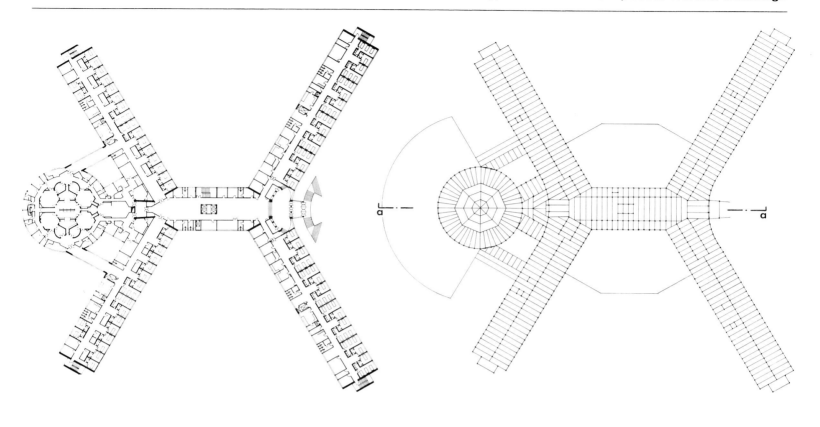

Scale 1:1300

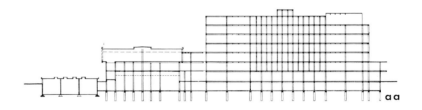

The great variety of functional requirements and the frequent difficulty of accommodating adequate internal circulation systems have led to widely differing solutions in present-day hospital architecture. Broadly speaking, a functional, and generally also a structural, separation into three zones may be distinguished:

1. *Medical treatment facilities*

In this sector a wide range of elaborate services and equipment has to be installed. These facilities comprise consulting rooms, operating theatres, sterilisation and preparation rooms, intensive care units, delivery rooms, X-ray department, therapy units, laboratories, dispensaries, etc. Large rooms free from internal supports alternate with relatively small units, but a variable internal layout may also be required. In view of the complex services and the possible need for subsequent additions and modifications, ample room within the ceiling space is desirable.

2. *Wards*

The varying size of the nursing units and their assignment to individual medical departments, together with the associated ancillaries such as nurses' and doctors' duty rooms, surgeries, examination rooms, laundry rooms, ward kitchens, etc., determine the shape and floor area of the storeys in the ward blocks.
Usually, double-zone layouts with central corridors are adopted. The wards or patients' rooms are oriented to face south to south-east. The ward blocks generally have a fairly rigid pattern of subdivision into one-, two- or four-bed rooms with their associated sanitary facilities. The remarks made on p. 184 with regard to hotels are applicable also to the ward blocks of hospitals, but technical equipment and services are more elaborate in the latter.

3. *Staff and service sector*

This comprises administrative offices, central laundry, patients' admission department, central kitchen, diet kitchen, dining rooms, nurses' residential quarters and training departments, etc. Some of these involve elaborate technical equipment and services, whereas others do not.

It is important to provide adequate means of communication and circulation within and among these various sectors of a hospital: for staff, visitors, patients and goods. Recent developments, such as the container principle, require a new planning approach. Most important among the requirements to be fulfilled is the attainment of clinical sterility, with proper separation of septic and aseptic zones. This must be given due consideration at the design stage. Hospitals must furthermore possess a high degree of safety against fire.

Hospital at Gonesse, France (architect: Rabaud).

Types of multi-storey steel-framed buildings

Multi-storey car parks

Multi-storey car parks have floors occupied partly by parking bays and partly by access roads. These features have standardised dimensions.
Parking bays generally measure 5.00 × 2.30 to 2.50 m, while roads range in width from 4.50 to 7.00 m, depending on whether the cars park at right angles to the access roads or obliquely to them (herring-bone system). These features in turn determine the overall width of a deck. To allow subsequent alteration of the subdivision of the available space and to facilitate the manoeuvring of vehicles, each deck, comprising a central access road and a row of parking bays on each side, should be free from internal supports. The long floor spans that this requires can be constructed economically in steel, with low construction depths, especially if composite construction is employed.
Low floor construction depths reduce the gradients of the ramps. The latter may be of the straight one-way type or, alternatively, two-way ramps may be used with split-level floors. Helical ramps in various arrangements are also extensively used. All these types of ramp structure can be constructed inexpensively from rolled steel sections and concrete slabs.
Since the dimensions of the main components of a car park are largely standardised, it is a logical choice to adopt a prefabricated construction system with units of standard size. Again, structural steelwork offers the most advantageous possibilities from this point of view.
The fire load in a car park is very low. Tests have shown that if a parked car catches fire, the risk of this spreading to an adjacent vehicle is very small, and there is negligible risk to human life, since only small numbers of people are present on the decks at any given time.
Moreover, since the decks are, as a rule, free from internal columns or other obstructions, there is ample opportunity for escape. Also, the decks are readily accessible to firefighting teams.
For these reasons, in many countries, including Great Britain and the Federal Republic of Germany, fire protection requirements are considerably eased in the case of car parks of the open type. For example, under German regulations issued in March 1972, unprotected structural steelwork is permitted for open multi-storey car parks in which the highest deck is not more than 16.50 m above ground level and the lowest deck is not more than 1.30 m below ground level, that is, with an overall height of 17.80 m, corresponding to seven or eight parking levels.
Louvre-type slatted external walls as protection against snow and driving rain are permitted for open-deck multi-storey car parks under the German regulations, but if these buildings are closed, fire-resisting structural members and artificial ventilation are required. Most car parks are unheated, so that there is no advantage to be gained, in cold weather, in using a closed rather than an open one. For this reason closed car parks are built only on sites where adjacent buildings prevent adequate natural ventilation of the decks. In such circumstances, a structural steel frame will have to be encased for fire protection. The same requirement applies to steelwork in underground parks.

Standardised car park, Krupp-Montex prefabricated system, with staggered floors. The traffic flows directly on the precast concrete slabs which act compositely with the floor beams.

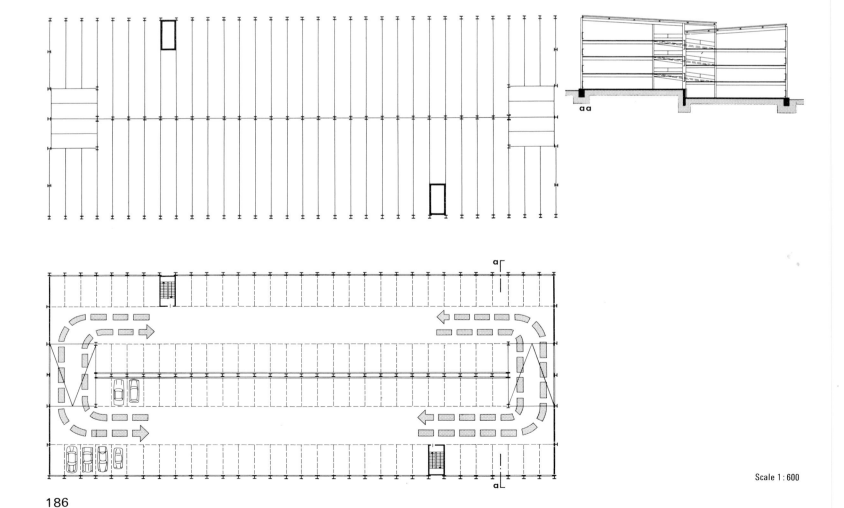

Scale 1:600

Types of Supporting Structure

Structural frameworks for multi-storey steel-framed buildings.

Arrangement of horizontal components
- Floor structure — 187
- Floors without steel beams — 190
- Floors with one-way beam systems — 191
- Floors with two-way beam systems — 192
- Triple beam systems — 195
- Combined floor systems — 196
- Triangular grid — 197
- Circular buildings — 198
- Widely-spaced external columns — 199
- Closely-spaced external columns — 201

Arrangement of vertical components
- Transmission of vertical loads — 202
- Interception of columns — 203
- Cantilevered structures — 204
- Suspended structures — 205
- Rigid frame buildings — 206
- Bridge-type buildings — 207
- Cellular construction — 208

Bracing for steel-framed buildings
- Principles — 209
- Rigid frames — 210
- Vertical bracing — 213
- Shear walls and cores — 217
- Bracing for high-rise buildings — 219
- Tubular buildings — 221

Joints in buildings
- Expansion joints — 223
- Settlement joints — 224

STRUCTURAL FRAMEWORKS FOR MULTI-STOREY STEEL-FRAMED BUILDINGS

A multi-storey building comprises a number of floors which are erected one above the other and which, in a framed structure, are supported at certain specific points. The vertical and horizontal forces acting upon the framework are transmitted to the foundations through the following assemblies of loadbearing members:

1. The floor structure directly carries the vertical loads and transmits them horizontally to the various points of support.
2. The vertical supporting system transmits the loads to the foundations of the building.
3. The horizontal and vertical stiffening systems also transmit the horizontal loads due to wind, earthquakes, earth pressure and other forces to the foundations.

In this chapter these three general loadbearing functions will be considered in a systematic manner for steel-framed buildings. Obviously, in different designs these functions may vary in their relative importance. On pp. 187–201, some buildings are presented in which the dominant design features are the arrangement on plan and the grid location of the points of support. The buildings discussed on pp. 202–208 derive their characteristic external appearance, that is, their elevational treatment, largely from the choice of the vertical loadbearing system. In the buildings on pp. 209–222, the conception of the system for the transmission of the horizontal forces, that is, the stiffening and bracing, is the dominant factor in the design.

ARRANGEMENT OF HORIZONTAL COMPONENTS

The floor structure

The load-carrying structure of an upper floor transmits the loads acting directly upon it to the supports. The arrangement, spacing and nature of the supports are determined by the design of the building on plan. The spacing of the supports, generally in the form of columns, determines the spans of the floor structure and therefore its construction depth.

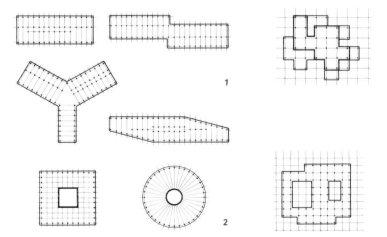

Shapes frequently adopted for buildings on plan

1 Buildings which are of narrow elongated shape on plan generally comprise two rows of columns, arranged along the external walls, and often one or two further rows in the interior. **2** Buildings which have a compact shape on plan are often designed with columns only at the perimeter, a central core being the only internal support. **3** Buildings covering a large area on plan, especially if they are of irregular or sprawling shape, comprise numerous columns, usually located at the intersection points of a co-ordinated planning grid.

Arrangement of horizontal components

The floor structure

Basic shapes of the floor structure

From the structural point of view, a floor in a framed building consists of a number of floor bays. The following shapes on plan may be distinguished:

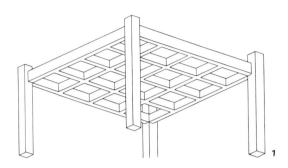
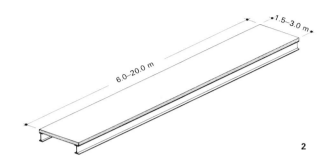

Square concrete slab

1 The simplest type of construction is the solid concrete slab. With long spans, such slabs tend to be very thick and heavy. To reduce weight, the slab may be formed with internal cavities or composed of a system of one-way or two-way ribs joined together by a relatively thin top layer of concrete, the latter being described as a waffle slab. The load transmission conditions being approximately the same in both directions, a planning grid with a square mesh is often adopted, this being typical of in-situ concrete construction.

Rectangular floor bay

2 The structural steel floor comprises the floor slab and the beams that support it. It is characteristic of steel-framed buildings that the floor beams transmit the loads in one direction only and that the floor slabs are generally supported, not on four sides, but on two. To construct a grillage for equal load transmission in both directions involves considerable effort in structural steelwork, as it is necessary to cut the intersecting beams at every junction and reconnect them in a manner suitable to transmit the loads.

The weight of the slab in such floors is often governed, not only by structural considerations, but also by other requirements, such as sound insulation and fire protection. In general, the weight of the slab should be as low as possible. To achieve this, the supporting beams must be fairly closely spaced, so that the actual slab spans only short distances. We thus arrive at the following fundamental rule for steel-framed buildings:

- The typical component of the floor structure of a steel-framed building is the narrow elongated floor bay.

The span of the floor beams generally ranges between 6.00 and 20.00 m, while the width of the bays varies from 1.50 to 3.00 m (**2**).

Column grid patterns for steel-framed buildings

Square grid

3 Buildings are often planned on the basis of a square grid, the columns of the structural frame being located at the intersections of the grid lines, so that square floor bays are obtained.

4 In steel-framed construction, these square bays are subdivided into elongated rectangular units. The floor is supported on secondary beams which in turn are carried by main beams. Both systems of beam have the same span, but this constructional arrangement is not typical of steel-framed buildings.

Rectangular grid

5 At little extra cost the floor beams can be given longer spans, so that the interior of the building can be kept largely free from columns. This approach can result in optimum structural solutions if the comparatively lightly loaded secondary beams have longer spans than the heavily loaded main beams.

- The distinctive feature of an economical steel-framed building is its rectangular column grid pattern.

6 This type of arrangement on plan is even more characteristic of steelwork. The floor beam arrangement is similar to that in **5**, but instead of widely spaced external columns interconnected by main beams, there are now closely spaced external columns corresponding to the positions of the secondary beams, which are connected directly to them, so that there are no external main beams. These external columns, which have to support the load from only one floor beam in each storey, can thus have much smaller cross-sectional areas than the columns in **5**, which have to support the loads from whole floor bays (shaded areas). Yet the total cross-sectional area of the columns in this arrangement is not much greater than in **5**, since this area is determined by the total load to be carried by all the columns, divided by the permissible stress. The only difference is that the slim closely-spaced external columns have a somewhat higher slenderness ratio, so that their total cross-sectional area is merely that much larger than that of the stouter columns spaced farther apart in the preceding solution.

Arrangement of external columns

Widely spaced columns

7 The external columns spaced relatively far apart, as in **5**, have to be offset some distance from the wall in order to provide space for the main beams interconnecting them. Thus such columns waste floor space and this in the vicinity of the windows, where space is most valuable. The loss of space is appreciable, more particularly in small rooms, especially if the partition does not coincide with the centre-line of the column. These considerations are particularly important in the design of high-rise buildings.

Closely spaced columns

On the other hand, columns closely spaced, as in **6**, can be substantially accommodated within the external wall structure itself or even placed on the outside. But even when such columns are set back some distance inward from the façade, they take up less space than is occupied by the radiators or convectors that are generally installed under the windows anyway.

8 A partition can be connected to any of these closely spaced columns. In addition, the external wall construction itself can be simplified because the cladding units can be suitably connected to the columns over the entire height of a storey and separate non-loadbearing mullions for the cladding and glazing are thus not required.

The floor structure | Arrangement of horizontal components

Types of floor with steel beams

In general, it may be said that the shorter the route taken by a load to its foundations and the fewer the elements involved in so doing, the greater the economy of the structure. Nevertheless, the space required for services plays a significant part in the selection of the arrangement of the main and secondary beams, their interrelationship and their depth.

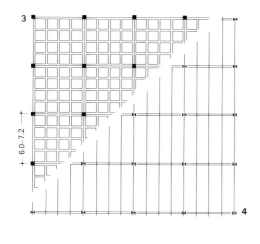

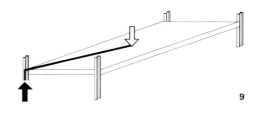

9 *Floor without steel beams*

The flat concrete slab, supported at its four corners, transmits the loads directly to the columns. With long spans the slab is very heavy. Therefore, this form of construction is appropriate only in combination with closely spaced columns.

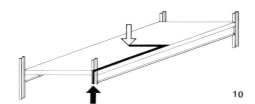

10 *Floors with one-way beam systems*

The floor beams are supported directly by the columns. The force transmission paths being short, this is an economical solution. The planning grid consists of elongated rectangles with columns widely spaced in one direction, closely spaced in the other.

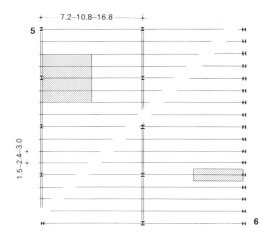

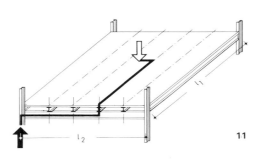

11 *Floors with beams in two directions*

If the columns are widely spaced in both grid directions, the loads from the secondary beams are transmitted to main beams which, in turn, transmit them to the columns. The forces thus have to travel farther than in the preceding examples. In general, it is advantageous to keep the span of the main beams l_2 shorter than that of the secondary beams l_1.

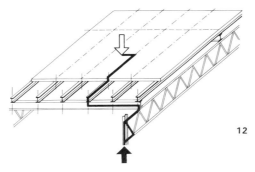

12 *Floors with triple beam systems*

For very long spans it may be necessary to support the main beams on a third set of members, such as lattice girders, which finally transmit the loads to the columns. In this case, the route taken by the forces is very long indeed.

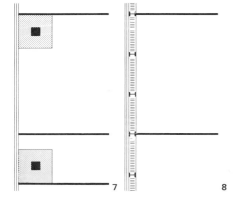

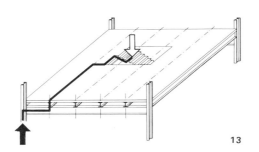

13 *Modifications to beam systems*

Openings in floors may have to be so large or so placed as to necessitate interruption of beams. The loads then have to be transmitted by detours, via trimmer beams, to the columns. Such arrangements add to the cost of construction, and it is therefore desirable to co-ordinate the design of the beam system carefully with the arrangements for vertical circulation.

Arrangement of horizontal components

Floors without steel beams

The concrete floor slab, with a smooth finish on both faces, is supported directly by the steel columns without the aid of steel beams. This form of construction can be used for short to medium spans (6 to 10 m), the floor slab ranging in thickness from 20 to 40 cm. No suspended ceiling is required, so the overall construction depth of the floor is not much more than the actual slab thickness. This inexpensive type of floor is well suited to residential buildings, hotels, student halls of residence and other buildings without elaborate services. Conduits for electric wiring can be incorporated in the loadbearing slab. To save weight, the latter may be formed with cavities or it may be a ribbed or a waffle slab.

1 This system is particularly suitable for use with the lift-slab method of construction. The floor slabs are concreted, one on top of the other, at ground level; the top face of each slab forms the soffit formwork for the next slab; a bond-breaking interlay prevents adhesion of the slabs to one another. Openings are left in the slabs to receive the steel columns, which are continuous over several storeys and are provided at the top with centrally-controlled lifting devices whereby the concrete slabs are raised one by one to their final positions (p. 349).

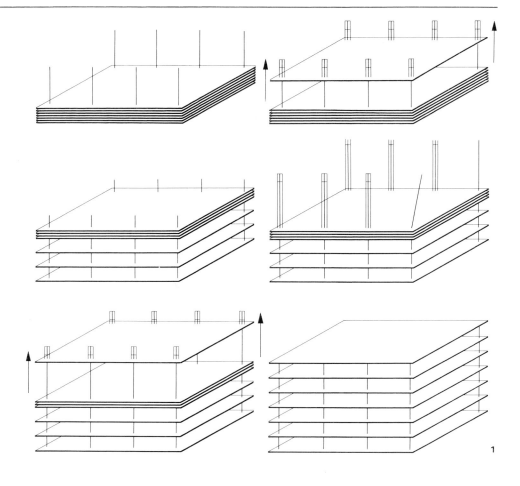

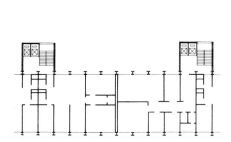

2 In relatively narrow buildings (up to about 10 m wide) the columns are arranged in two rows along the longitudinal edges of the floor slabs. There are no internal columns, and the floors thus offer great flexibility of internal layout.
Reference: *Wohnungssysteme in Stahl*, report prepared by Professor Ungers's department, Technological University of Berlin, 1968.

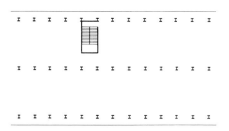

3 In buildings with relatively large dimensions in both directions the columns may be located on any suitable planning grid to support the floor slabs.

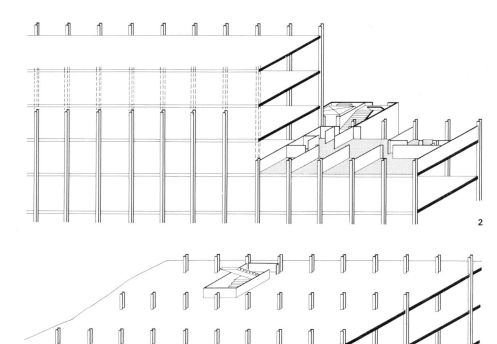

190

Floors with one-way beam systems

Arrangement of horizontal components

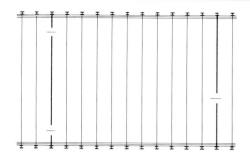

This type of column arrangement is characterised by long-span floors (up to 30.00 m) free from internal columns and flanked by closely spaced external columns, the latter generally being between 1.50 and 3.00 m apart.
The columns have small cross-sectional dimensions, so that they occupy little space or even merge with the walls. Separate mullions in the external walls are thus unnecessary. This is an economical form of construction, the floor slabs being supported only by the long-span floor beams which are connected at each end to a column, which is typical of structural steelwork: columns widely spaced in the direction of the floor beams and closely spaced in the direction transverse thereto.

1 This form of construction is suitable for relatively narrow buildings with external columns only, the column-free internal space offering great flexibility of layout. The absence of columns in the interior is an advantage in structures such as car parks, as freedom from obstruction makes for greater and safer manouevrability, besides allowing the layout of the parking spaces to be altered at any time.
The construction depth of the floor depends on the column spacing but it is generally between about 35 and 70 cm. Services are accommodated in the floor cavities, the longitudinal services passing through holes cut in the webs of the floor beams, while transverse distribution connections are accommodated in the ample spaces between the beams.
2 Buildings comprising several rows of columns can be constructed in a similar fashion. The closely spaced internal columns are located in the walls of the corridors. As the external columns are also of small section, they take up little space and may indeed be incorporated in the walls. There are no main beams interconnecting the columns, so that there is ample space for vertical services between the columns, for example, in buildings containing laboratories. Cross-walls or partitions can be connected to the external walls along any column centre-line.

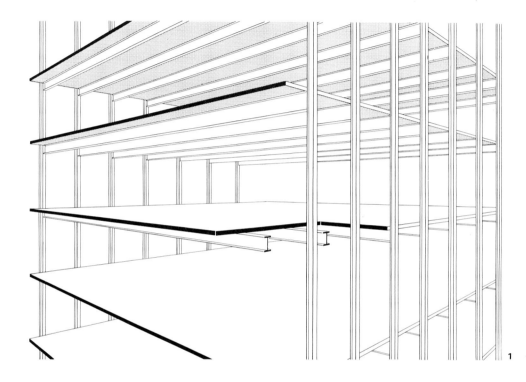

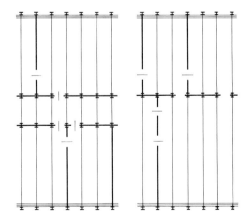

Because of the fixed positions of the internal columns, this form of construction is more particularly suitable for buildings in which no subsequent modification of the central corridor is likely to be required: for example, hotels and ward blocks of hospitals.
Alternatively, there may be only one row of internal columns, along one side of the corridor, an arrangement resulting in unequal floor spans but offering fresh possibilities for internal layout.

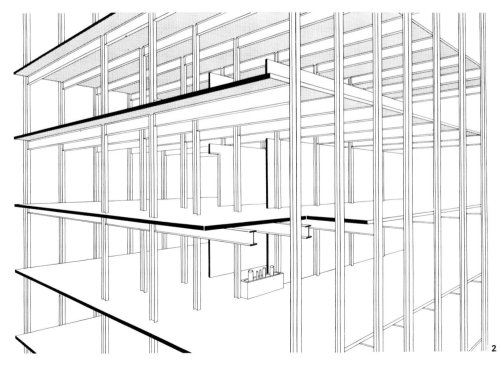

Arrangement of horizontal components — Floors with two-way beam systems

In framed building construction it is fairly common to have columns widely spaced in both directions. In such cases the floor structure comprises steel beams running in both directions, the secondary beams and the main beams which support them. For this arrangement, there is an optimum planning grid in which the relatively lightly loaded secondary beams have longer spans than the more heavily loaded main beams. In **1.2** the secondary beams have $1\frac{1}{2}$ times the span of the main beams, and in **1.1** they have twice the span. The most economical spans range from 6 to 12 m for the main beams and 7 to 20 m for the floor beams.

In this form of construction, the spacing of the floor beams need not correspond with the column spacing. In **1.1** every third floor beam coincides with a column centre-line, whereas in **1.2** the floor beams are staggered on both sides of the columns, an arrangement which has the advantage of leaving space for vertical services directly beside the columns, as well as somewhat simplifying the structural design of the steel frame.

The beams in either direction may be I-sections, castellated beams or lattice girders. They may be arranged in two distinct layers, one above the other, or they may interpenetrate. The nature and arrangement of the two families of beams, spanning in their respective directions, will depend on the available floor construction depth and on the services that will have to be accommodated within the floor structure (see also p. 285).

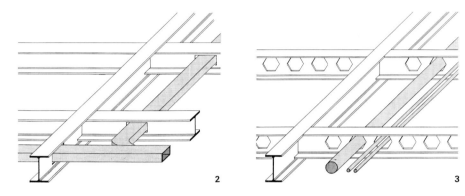

2 and **3** Floor beams and main beams installed in substantially the same plane, so that they interpenetrate, ensure low construction depth. Services may be passed through holes formed in the beam webs, castellated beams being especially advantageous in such circumstances. However, if elaborate services have to be installed, these must be accommodated in a special zone under the floor beam system.

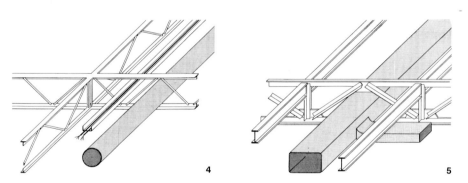

4 and **5** Lattice girders or bar joists are convenient because they present little obstruction to services, although of course the size of ducts is limited by the size of the triangular spaces between the web members (see tables, p. 288).

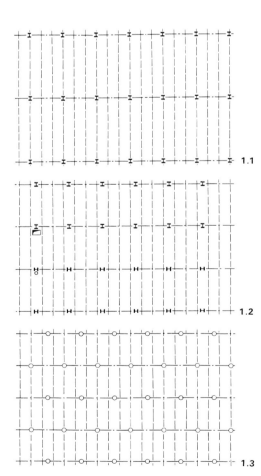

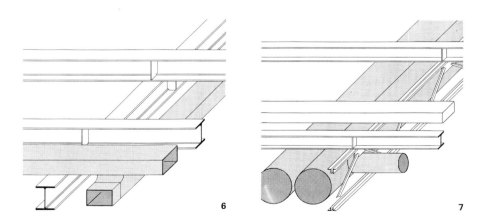

The columns do not necessarily have to be located at the grid points in both directions. On the grid lines corresponding to the centre-lines of the main beams the columns can be staggered in adjacent rows to produce a triangular column grid pattern (**1.3**).

6 and **7** A two-way beam system with the beams arranged in two layers, one above the other, requires considerable construction depth, but provides ample space for services. The latter, like the beams themselves, are arranged at two levels. Whenever any particular service changes its direction, it also changes to the other level. In structural steelwork, this arrangement of floor beams and services makes for economical design and very rapid erection.

Floors with two-way beam systems
Arrangement of horizontal components

The two-layer two-way beam system is especially suitable for buildings incorporating numerous and complex services.

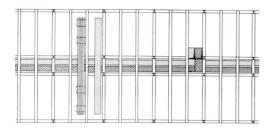

8 In a double-zone layout — central corridor flanked by rooms — this system leads to a solution where the principal services are installed at the level of the main beams, while the branch connections of the services are at the upper level, alongside the floor beams. The ceiling is suspended below the main beams. The branch services can extend across the full width of the building, passing under the corridor walls through the spaces left between the main beams and the floor slab. In the rooms beside the corridor the ceilings are suspended directly below the floor beams, i.e., at a higher level than in the corridor.

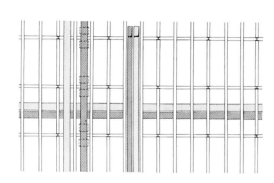

9 In a building covering a large area on plan and comprising one or more central service shafts, the principal horizontal services may, for example, be installed at the level of the floor beams, the principal branch connections at the level of the main beams, and the secondary branch connections again at the upper level.

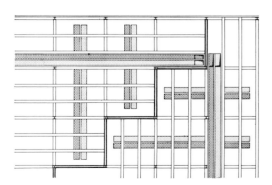

10 If the vertical service shafts in a building covering a large area on plan are located at one corner of the building, it is advantageous to interchange the direction of the main and secondary beams on each side of a diagonal of the structure, so that the upper beams run towards the service shafts and the principal services are installed at the upper level, while the main beams and branch connections extend at right angles to these principal services.

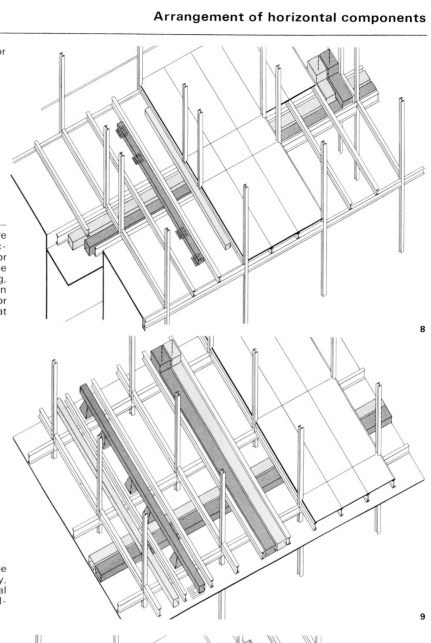

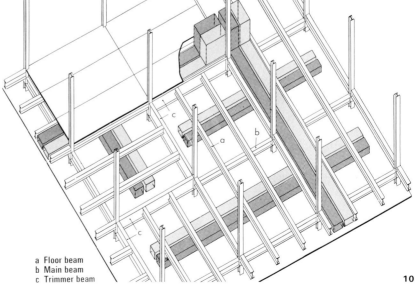

a Floor beam
b Main beam
c Trimmer beam

193

Arrangement of horizontal components | **Floors with two-way beam systems**

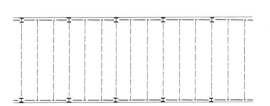

11 The only columns are external and widely spaced, being interconnected by main beams. The floor beams extend across the whole width of the building, the interior of which is free of columns and can be subdivided in any way desired.

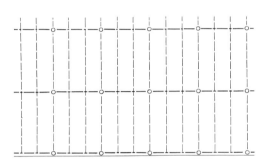

12 There are several rows of widely spaced columns, both external and internal. The external columns are interconnected by solid-web main beams, whereas the members interconnecting the internal columns are lattice girders. The secondary members are also lattice girders, with the same depth and installed at the same level. (Technical college at Brugg-Windisch, Switzerland. Architects: B. and F. Haller.)

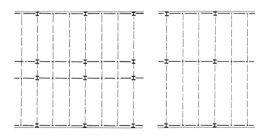

13 The central corridor need not necessarily be flanked by two rows of columns. Structurally, one row will suffice, in which case the floor beams have unequal spans.

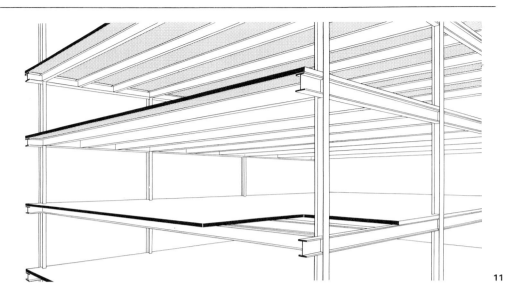

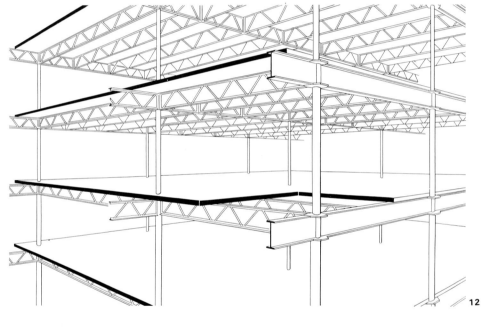

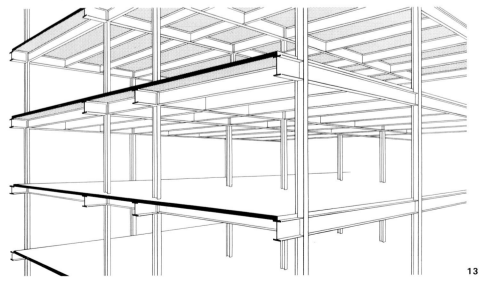

Floors with triple beam systems

Arrangement of horizontal components

Where large internal spaces unobstructed by supports are required, long-span floor structures are needed, which may necessitate a third set of horizontal members, such as plate or lattice girders, for supporting the main beams which, in turn, carry the floor beams.

1 In the arrangement envisaged here, the three sets of beams are installed at three levels, one above the other. The girders do not receive any direct loading from the floor.

2 Alternatively, the top chord of the lattice girder may be located at the same level as the floor beams (or the purlins, in a roof) and thus has to resist flexural loading in its own right, in addition to the force acting in it as the top chord of a lattice girder.

3 Another arrangement comprises two mutually perpendicular sets of lattice girders installed at the same level. The floor beams are positioned over the vertical web members of those girders which extend in the transverse direction. This is a space-saving solution for heavily loaded long-span floors.

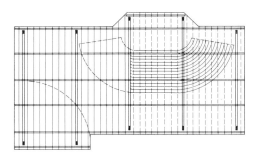

4 In this structural solution, the lattice girders are rigidly connected to the columns of a single-storey building to form portal frames. These girders are intersected by longitudinal lattice girders of smaller depth, which in turn support the roof beams. In this particular instance, the last-mentioned girders and beams are standardised elements of a building system. (Free University of Berlin, Audimax. Architects: Candilis, Josic, Woods.)

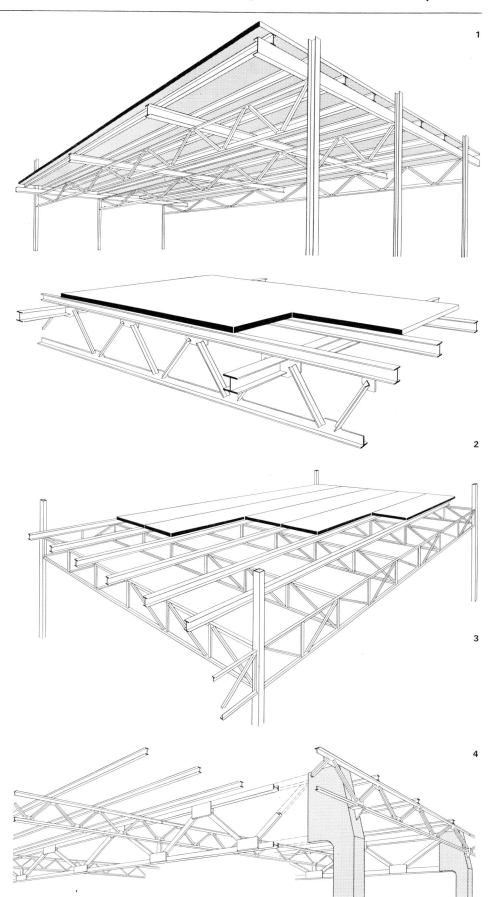

Arrangement of horizontal components

Combined floor systems

In many cases, a satisfactory solution can be obtained by a combination of some of the arrangements already described. Closely spaced external columns save space in the interior of a building and allow partitions to be conveniently installed on the centre-line of any particular floor beam. Internal columns are often set farther apart, the loads from the floor beams being transmitted to these columns via main beams.

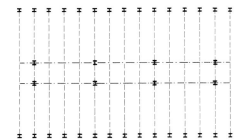

1 In this building the wide spacing of the internal columns allows large open-plan offices to be provided, but, alternatively, a subdivision into corridors and smaller rooms by means of partitions on the centre-lines of the external columns is possible. (Institute of Veterinary Medicine, Free University of Berlin. Architects: Luckhardt, Wandelt.)

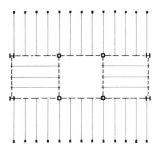

2 For this 100 m high building with beam spans of 7.00 m, the designers adopted external columns spaced at 1.80 m centres and having a cross-section of 16 × 16 cm. Each side wall of the building comprises only two loadbearing columns. If the difference in external column spacing is exploited as an architectural feature, the façade construction has to be so suited. Alternatively, to give all four façades a similar appearance, it is necessary to provide dummy columns in the side walls.

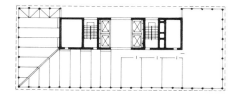

3 High-rise building with eccentrically positioned concrete core. The main beams and floor beams are installed at the same level and are connected to large diagonal girders extending between the core and the front corners of the building. The columns in the front and side façades are loadbearing columns, whereas the columns in the rear wall support the cladding only. (Administrative building for Mannesmann AG, Düsseldorf. Architects: Schneider-Esleben, Knothe.)

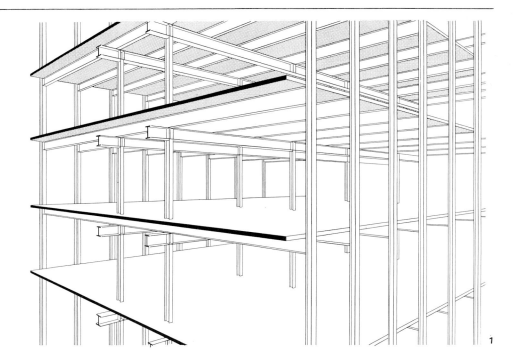

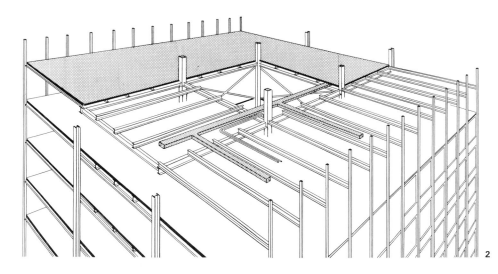

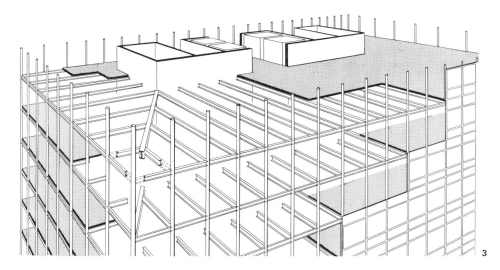

Triangular grid

Arrangement of horizontal components

Just as with the rectangular grid, the object is to divide up the floor bays by means of floor beams and main beams to obtain sufficiently small spans for the floor slab. Four possible arrangements for the beams will be described with reference to a specific example: the administrative building for Hamburg-Mannheimer Versicherung AG, Geschäftsstadt Nord, Hamburg (architects: Spengelin, Spengelin and Löwe).

1 A rectangular grid is superimposed on the entire ground-plan of the building. In the interior, this presents no special problem. Along the external walls, however, which are oblique to this beam grid, skew terminations to the floor structure arise. This is a grid which, though simple to achieve in structural steelwork, takes no account of the particular shape of this building on plan.

2 The main beams and the floor beams are placed at an angle of 60° to one another. The floor beams have longer spans than those in **1** and therefore require sections of greater depth. At the external walls the floor beams are either parallel to the wall or at an angle of 60° to it. Some of the floor bays have skew ends.

3 (Solution adopted in the actual building.) The floor beams are placed at right angles to the external walls, their spacing conforming to that of the external columns. The external walls are bolted to the webs of the beams, providing a simple method of fixing the cladding. The skew ends of the floor bays are in the interior of the building, instead of at the external walls, as in the preceding solution.

4 In this example the floor bays are triangular, the beams being arranged in star-shaped patterns which provide suitably short spans for the floor slab. Precast triangular concrete slabs can be used, the dimensions being such that they can be transported by road even when column spacings of 10 to 12 m are adopted. Columns of tubular section are used, with twelve beams connected to each, requiring some annular form of connection. This is an expensive solution which can justifiably be adopted only in situations where the steelwork can be left exposed, so that the aesthetic character of its pattern can be appreciated. However, as the structural system for the floor comprises only two types of beam, the cost can be reduced by standardised quantity production.

b	$h = 3b$	$s = \dfrac{2h}{\sqrt{3}}$
2.4	7.2	8.3
3.6	10.8	12.5

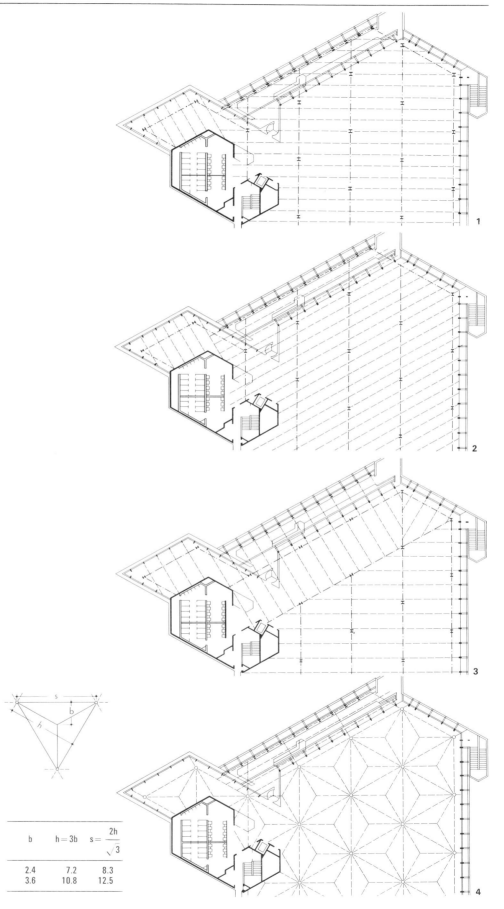

Arrangement of horizontal components

Circular buildings

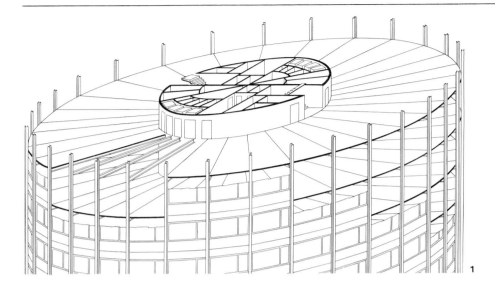

1 If the distance from core to external wall is not too large, a suitable arrangement for a circular building is to provide radial floor beams which are supported at their outer ends by external columns and at their inner ends by the concrete core. This is a particularly economical solution.

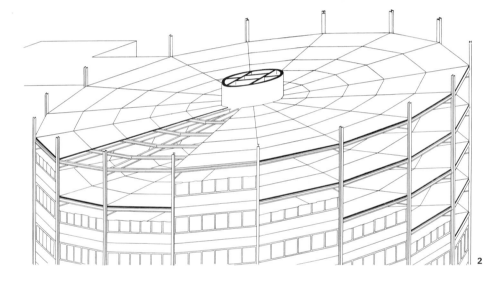

2 If the core-to-wall distance is large, it also becomes necessary to use tangential beams, which function as the floor beams, while the radial members now assume the role of main beams.

3 and **4** Helical ramps for multi-storey car parks are a special type of circular structure. The ramp may spiral through 180° (**3**) or 360° (**4**) per storey. Such structures can be built very elegantly with steel frames. All the radial beams terminate at an outer and an inner column. The floor bays are sections of a helical plane. Precast concrete slabs can be used, as all the floor units are identical, so that they can all be cast in the same mould.

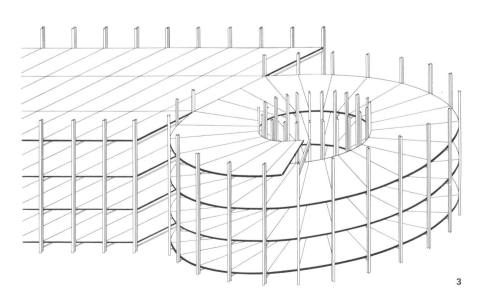

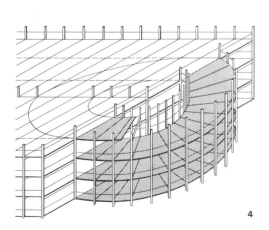

Widely-spaced external columns

Arrangement of horizontal components

In the design of a building, not only is the spacing of the external columns important, whether far apart or close together, but also their location in relation to the external wall.

1 If the columns are set a sufficient distance from the external wall — either inside (a) or outside (e) — they have no effect on the subdivision of the wall and its cladding features.
On the other hand, if the columns are set close to the external wall — inside (b) or outside it (d) — or are entirely merged into it (c), the design and construction of the wall and its cladding are considerably affected by the columns.
If the external columns are exposed, that is, without thermal insulation, they will undergo changes in length due to temperature variations caused by sunshine and by general changes in the outside air temperature. To avoid damage to external and internal walls, such changes in length must be taken into account in the design. Furthermore, such columns may form easy heat transmission paths between the exterior and interior of the building.
For multi-storey buildings, columns located as in (a) or (b) must generally have fire protection. On the other hand, no protective casing is, in many instances, required for externally located columns as in (d) or (e) nor for the outer parts of columns in the wall, such as (c). In such circumstances, the architect has the opportunity to exploit the steel columns as architectural features.

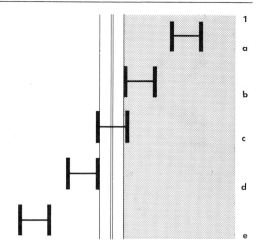

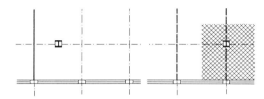

2 The column in this example is set back some distance into the interior of the building, so that the external wall is completely independent of the loadbearing frame. The subdivision of the windows is therefore unaffected by the column spacing. All the windows, installed between mullions all of the same type, are of equal width. A partition can be connected to any mullion. Vertical services and, within the floor structure, horizontal services can be installed in the space between the column and the external wall, which is an advantageous arrangement, more particularly for high-pressure air-conditioning ducts. If the column coincides in position with a mullion separating two windows, partitions must be connected to the column, necessitating the use of special partition units. This can be avoided by placing the columns between the mullions, in which case the windows at these columns may be blind windows.
A disadvantage is that the columns take up space within the building and restrict more particularly the valuable space in the vicinity of the windows. The sterile floor area (shown hatched in **2**) is substantially larger than the cross-sectional area of the column. The drawback is especially objectionable in small rooms, whose useful capacity can be substantially reduced by the presence of such a column.

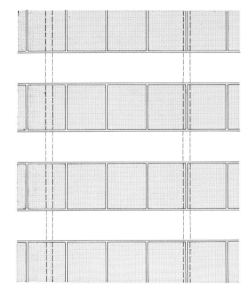
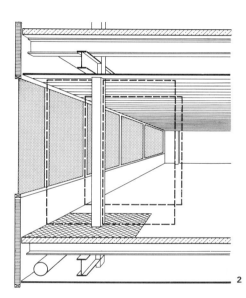

3 Columns located completely outside the building are connected to it solely by beams protruding through the façade. With this arrangement the elevational treatment and the internal layout of the building can be independent of the column spacing.

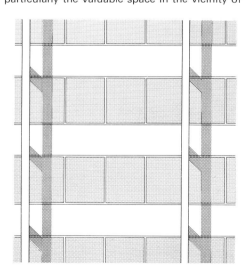
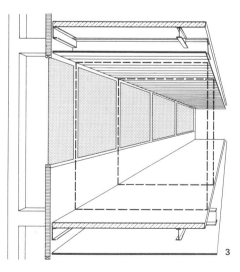

Arrangement of horizontal components   Widely-spaced external columns

Widely-spaced external columns which are in direct contact with the external wall will be of larger section and therefore wider than the mullions. If the window spacing is to conform with the planning grid dimension a, the window units b on each side of a column must be somewhat narrower than the normal window units between mullions. To give equal width to all the window units would result in irregularities in the internal grid. Depending on the location of the columns in relation to the external wall, there are therefore three possibilities, as follows:

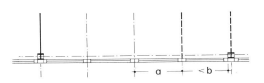

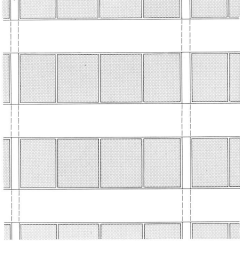

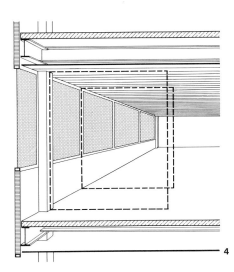

4 The column stands behind the external wall and is therefore not in contact with the outside air, since the wall continues past the front of the column. The column protrudes into the interior of the building. Where internal partitions consisting of standardised units are employed, this arrangement necessitates different units for connection to columns and mullions.

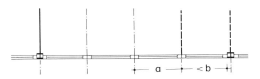

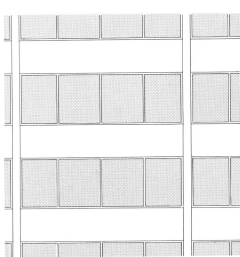

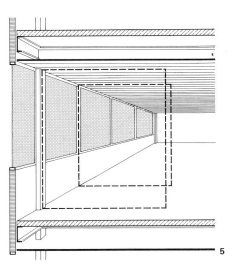

5 In this arrangement, the column is incorporated in the external wall and forms part of it, appearing as a feature of the façade, but not one that stands out prominently. Care must be taken to ensure that the joints between the cladding and the column are watertight. If the column is flush with the inside of the wall, special units for connecting a partition to a column (as distinct from a mullion) are not necessary.

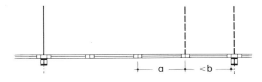

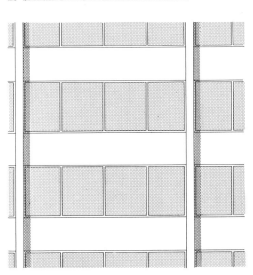

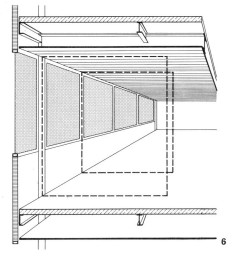

6 The column located completely in front of the external wall emphasises the verticality of the façade. Here again, no special partition units for connection to the columns are necessary.

Closely-spaced external columns
Arrangement of horizontal components

Close spacing of the external columns results in a more economical loadbearing structure. The columns are arranged at 1.50 to 3.00 m centres, corresponding to normal window spacings. They have to transmit relatively small loads and therefore have small sections, their width being equal to that of the mullions ordinarily employed in curtain wall construction. These external columns, to which the cladding can be directly attached, thus dispense with non-loadbearing mullions. As all the columns are identical, all the cladding units can be identical. What is more, in the interior of the building the conditions for connecting the partitions are the same for every column. Depending on the actual location of the columns in relation to the wall, the following possibilities exist:

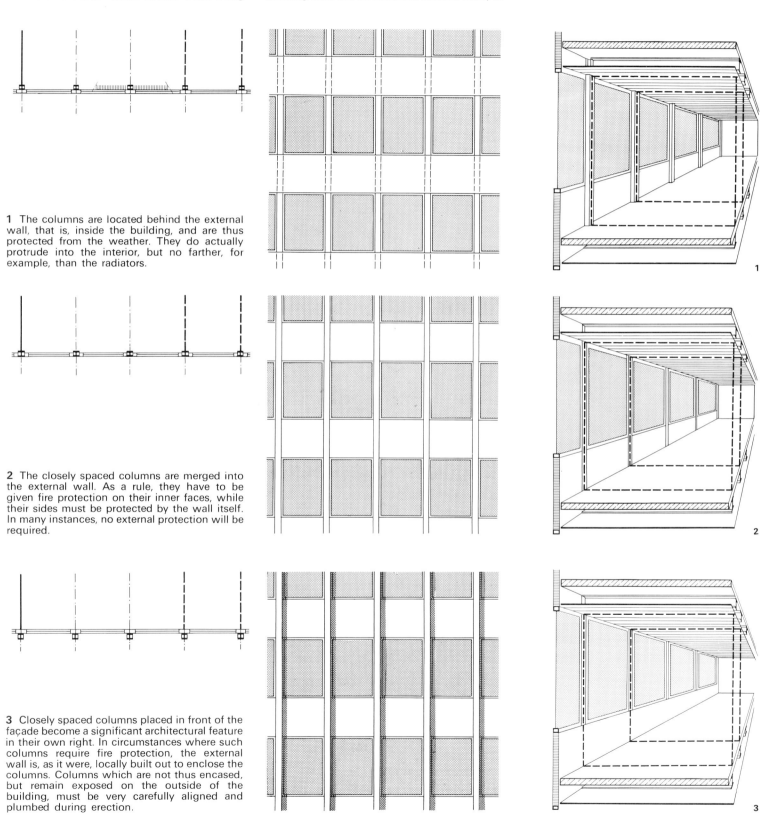

1 The columns are located behind the external wall, that is, inside the building, and are thus protected from the weather. They do actually protrude into the interior, but no farther, for example, than the radiators.

2 The closely spaced columns are merged into the external wall. As a rule, they have to be given fire protection on their inner faces, while their sides must be protected by the wall itself. In many instances, no external protection will be required.

3 Closely spaced columns placed in front of the façade become a significant architectural feature in their own right. In circumstances where such columns require fire protection, the external wall is, as it were, locally built out to enclose the columns. Columns which are not thus encased, but remain exposed on the outside of the building, must be very carefully aligned and plumbed during erection.

Arrangement of vertical components

Transmission of vertical loads

The shape of a building in elevation is closely related to the structural principles embodied in it. With present-day constructional resources — especially in steel — virtually any concept of design is technically practicable, limits being imposed solely by economic considerations.

The loads acting on the floors are transmitted to, and concentrated at, points of vertical support whence they are transmitted to the foundations. The spacing of the supports determines the spans of the floor beams. The floor slab may be supported at its edges or it may cantilever out. The vertical loadbearing members are the columns of the structural steel frame, possibly augmented by concrete walls and cores or shafts for vertical circulation and services.

If the points of support of the consecutive floors in a multi-storey building are exactly one over the other, continuous vertical columns are normally used. Sloping columns, inclined from the vertical, are also possible. These can transmit the loads either directly or by following a polygonal line in elevation, with a horizontal support at each change in direction.

In other types of multi-storey building the vertical components act as hangers or suspenders and are always connected to horizontal supporting members. The latter may thus carry the load of one storey, or of a group of storeys, or of all the storeys of the building. These various possibilities lead to the following forms of construction:

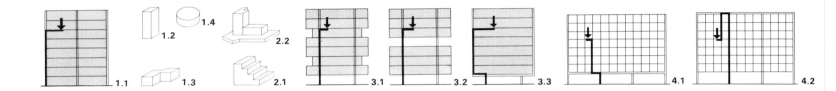

1 and **2** The columns extend in a continuous vertical line to the foundations, the load thus being transmitted by the shortest path. The various storeys may either be identical on plan, **1**, or differ, **2**. The building is composed of prismatic or cylindrical blocks.

3 While the external wall may be recessed in certain storeys, **3.1**, or even be entirely absent, **3.2**, the structural frame is essentially similar to that in **1**. If the external columns are set back from the perimeter of the building at ground floor level, the forces are transmitted in a detour through a horizontal framework, **3.3**.

4 Wide spacing of the external columns at ground floor level and close spacing of those in the upper storeys necessitates the use of some kind of intercepting or ring beam, **4.1**. If such beams are located at the top of the building, the secondary vertical components function as ties, **4.2**.

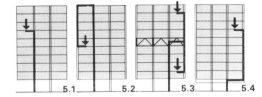 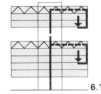 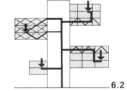 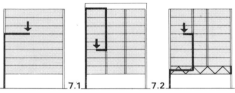

5 The loads are transmitted solely by the vertical, centrally located loadbearing components (core consisting of concrete walls or steel columns), the outer zones being free from supports at ground floor level. The loads from the upper floors may, for example, be transmitted to the core as follows:
5.1 Storey by storey, through the action of cantilevers at each floor.
5.2 Through suspenders connected to a cantilever frame at the top of the building and thence to the core (suspended buildings).
5.3 Through a rigid cantilever truss at mid-height of the building. The storeys below this truss are suspended from it, while those above are supported by it in the normal way.
5.4 Through a rigid cantilever beam near the base of the building.

6 The loadbearing core of the building becomes a dominant feature, both structurally and architecturally. The structure comprises several cantilever beams with groups of storeys suspended from them or supported by them from below. These cantilevers can conceivably be designed as trusses of great depth, several times the storey height.
6.1 The cantilever beams project symmetrically from both sides of the core.
6.2 The cantilevered groups of storeys are located in a haphazard manner with respect to the core, which is thus subjected to very large bending moments. The idea of subsequently adding further storeys, as and when required ('plug-in' construction) is not a realistic proposition.

7 The ground floor is free from internal columns, the loads being transmitted to the foundation by external columns only. More particularly, the following solutions are possible:
7.1 All the floor beams have very long spans.
7.2 The internal supports are suspended from girders spanning across the top of the building (portal frame buildings).
7.3 The internal columns stand on storey-height girders spanning over the first storey of the building.

Transmission of vertical loads

Arrangement of vertical components

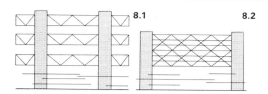 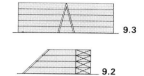

8 The structural loads are transmitted to two widely spaced main vertical loadbearing cores (bridge-type buildings).
8.1 Every alternate storey comprises a storey-height girder incorporating the adjacent floors as structural members thereof. The intermediate storeys thus remain free from supports.
8.2 The building is constructed as a long-span bridge, the full height of the building being utilised for the construction depth of the lattice girder.

9 Inclined columns require lateral support. The horizontal thrust increases as the slope of such columns departs from the vertical.
9.1 In symmetrical buildings, the horizontal forces induced by uniformly distributed dead loads cancel each other out. As a result, only the forces due to non-uniformly distributed live loads have to be resisted.
9.2 Asymmetrical buildings require strongly constructed lateral supporting structures to absorb the horizontal thrusts.

9.3 Inverted V-shaped columns effectively stiffen the building against wind forces.
9.4 With a V-shaped column system, unequal live loads give rise to horizontal thrusts.
9.5 Funnel-shaped buildings with inclined columns develop very large horizontal forces which can, in symmetrical buildings, be partly resisted by means of cable ties.

INDIRECT TRANSMISSION OF VERTICAL LOADS

In many buildings the bottom storey has to perform a function or serve a purpose different from that of the upper storeys. This frequently necessitates a different arrangement of the columns at ground floor level. Architectural requirements, such as raising the whole building on pilotis, may have similar consequences. In general, the ground floor columns are more widely spaced than the upper ones, so that the latter, instead of continuing straight down to the foundation, terminate on intercepting beams or girders spanning between the ground floor columns. Constructed in steel, such girders have low construction depth and can thus provide an economical solution. They can also serve as an architectural feature which strikingly reveals the path of force transmission.

1 The ground floor is recessed. The columns in the façade of the upper storeys have to be supported by heavy intercepting girders which often extend across the entire width of the building.
2 The external columns are closely spaced in the upper storeys, whereas the ground floor columns are spaced much farther apart. The upper columns terminate on an intercepting girder supported by the ground floor columns. Such girders are usually designed as plate girders.

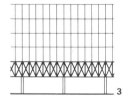 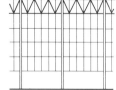

3 Alternatively, the intercepting girders may take the form of storey-height lattice girders, which are an economical means of constructing long spans and can appropriately be incorporated into storeys intended more particularly for the accommodation of plant.
4 If the storey-height intercepting girder can be installed only at or near the top of the building, the intermediate vertical components assume the role of hangers or ties, the whole loading of the building being transmitted through the main columns.

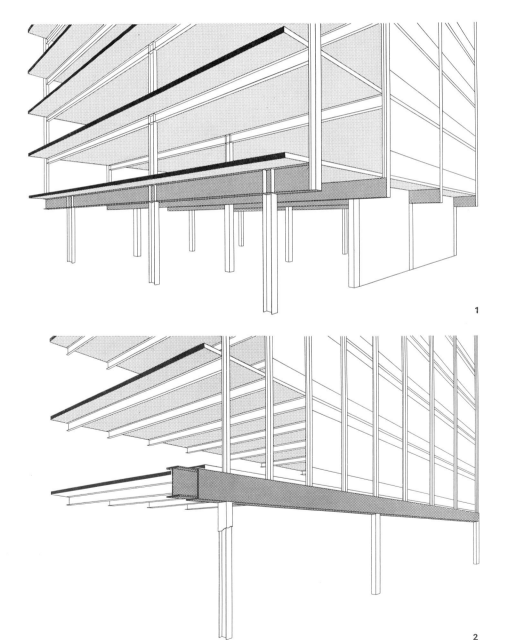

Arrangement of vertical components — Cantilevered structures

If a building is designed solely with internal columns from which the floors cantilever out, considerable freedom in the architectural treatment of the cladding elements and in the layout of the partitions is obtained. However, the deflection of the cantilevers and the resulting movement between the floors and the outer wall must be duly taken into account. Three possible solutions will now be considered.

1 The structure is composed of transverse frames, the beams of which cantilever far out from the columns. The external cladding of the building is suspended from deep, and therefore rigid, lattice girders installed at roof level, so that it does not participate in the deflections of the cantilevered floors under the action of varying loads. The connection between the cladding and the floors must therefore be carefully designed to leave the latter their freedom of movement and yet provide a soundproof joint. Stability is provided in the longitudinal direction of the building by rigid connections between the columns and the longitudinal beams. (Caisse Centrale d'Allocations Familiales, Paris. Architects: R. Lopez and M. Reby. Consulting Engineer: S. Pascaud.)

2 The building comprises two cores which serve to stiffen it in both directions. In every alternate storey a pair of Vierendeel girders extend longitudinally, cantilevering out from the cores. The top and bottom chords of these girders carry cross beams which, in their turn, cantilever out on both sides of the girders. Those storeys which contain the Vierendeel girders have no internal obstructions other than the vertical members of these girders, while the other storeys are completely unobstructed, except for the supporting cores, so that large open-plan layouts are possible. Because of their great depth, the Vierendeel girders are relatively stiff. Nevertheless, in the design of the cladding system, the effect of differing live loading on two consecutive Vierendeel girders must be taken into account.
This particular design has not yet been carried out. Although it offers some attractive possibilities from the point-of-view of flexibility of layout and utilisation, the structural system is likely to be very expensive.
The same principle is embodied in a design, by the present author, for a building of elongated shape on plan, in which the Vierendeel girders span between two pairs of cores. A bridge-type building is thus obtained (p. 203). (Design: Helmut Weber and Dieter Gans. Reference: 'Kernstützenbauweise' ('Construction system with core supports'), Bauen+Wohnen 1/1967.)

3 This example relates to a savings bank building at Ludwigshafen in Germany. It comprises four columns interconnected by heavy girders which cantilever out about 5 m in both directions. The floor beams are flush with the tops of these girders. The building is stiffened by an eccentrically positioned concrete core.

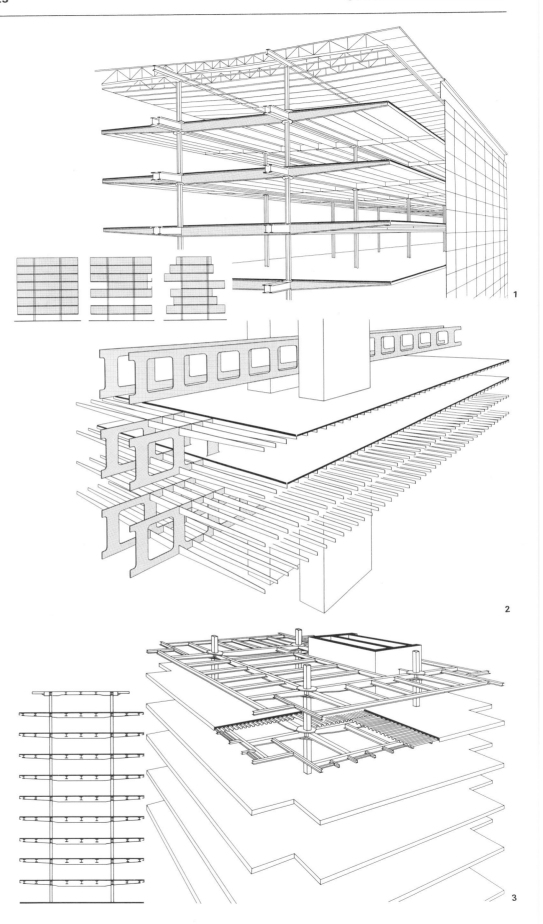

Suspended structures — Arrangement of vertical components

In a suspended building all the loads are transmitted through the central core, which may be constructed either in structural steelwork or in concrete, using sliding shuttering. The loads from the outer ends of the floor beams are supported by hangers suspended from robust cantilever girders projecting from the core which, in turn, transmits the loads to the foundations. At ground level, the area around the core is thus free of all supports. The hangers, being slender, can conveniently be incorporated in the framework of the outside wall. For particulars of erection, see p. 349.

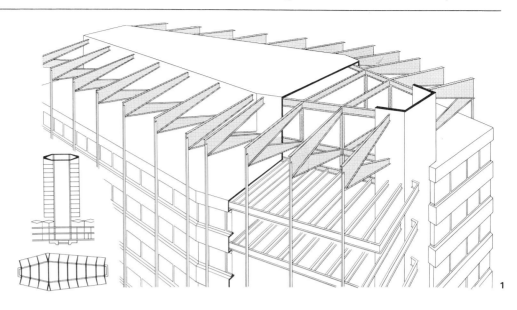

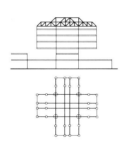

1 Core of structural steelwork with elongated shape on plan. The cantilevers are of plate girder construction. (Philips administrative building, Eindhoven, Holland. Architects: Luyt, de Jongh.)

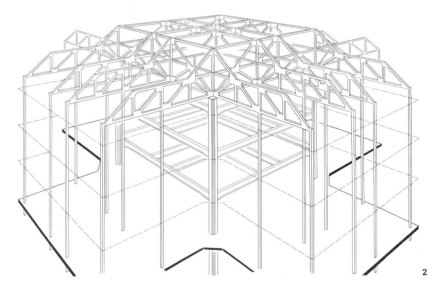

2 The core is square on plan, with heavy steel stanchions at the corners. Wind loads are transmitted through multi-storey rigid frames formed by the interconnection of these stanchions. Cantilevers are steel lattice girders. (Alpine Montan administrative building at Leoben, Austria. Architects: Huth, Domenik.)

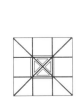

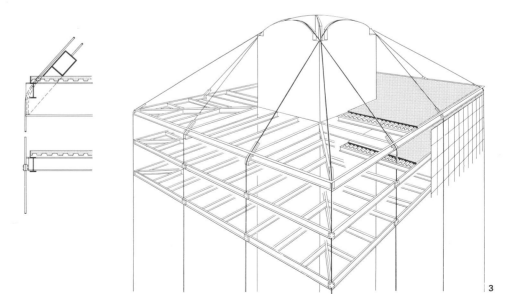

3 Suspended building, square on plan, with square concrete core, which is continued upwards beyond the top floor. The suspenders are cables which pass over the top of the core and rest on saddles. (Office building in Vancouver, Canada. Architects: Rhone, Iredale.)

Arrangement of vertical components — Rigid frame buildings

These buildings have no internal columns. All the loads are suspended from the cross girders of the rigid frames. The stiffness of the latter serves to absorb wind forces.

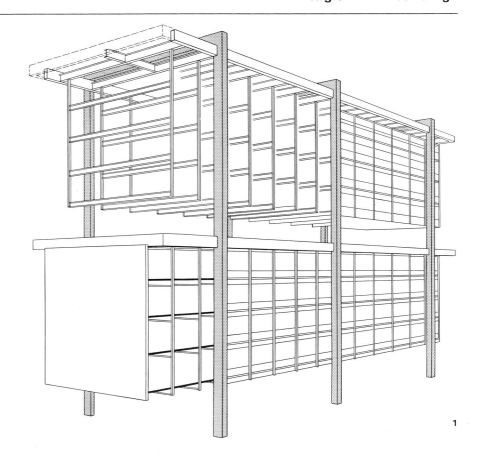

1 This building comprises three rigid frames, which are of box construction and which at the top and at mid-height support longitudinal box girders, between which there are transverse beams from which, in turn, four-storey blocks are suspended. (Iranian pavilion in the Cité Universitaire, Paris. Architects: Parent, Forough, Ghiaï.)

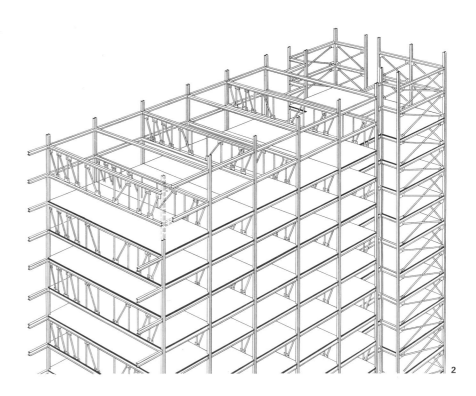

2 The zone between the vertical circulation cores comprises external columns, at 16.25 m centres, which are interconnected in the transverse direction by a framework based on the staggered truss system. The structural floor is thus of low construction depth and great rigidity. Some of the panels in the trusses are without diagonal members, to form unobstructed openings for doors and for the central corridor. In combination with the external columns, the trusses form multi-storey rigid frames for the transverse stiffening of the building. Additional stiffening is provided by lattice bracing in the vertical circulation cores. (Radisson South Hotel, Minneapolis, USA. Architects: Cerny Associates.)

Bridge-type buildings

Arrangement of vertical components

1 Multi-storey units spanning between vertical circulation towers allow organic extension of buildings on sites with limited space, through the incorporation and gradual replacement of existing older buildings. The long-spanning floors can be built elegantly and efficiently in steel. (First prize in architectural competition for Technological University in Berlin. Architects: Lambart, Özyar, Eisele.)

2 Bridge-type building with a span of 84 m. The loads from twelve storeys are transmitted through hangers to two 8.50 m deep lattice girders, assisted by two catenary members of I-section. The horizontal forces from the catenaries are resisted by the lattice girders. Four corner piers transmit the loads to the foundations. Additional storeys may subsequently be added, the loads from which will be carried by a steel arch structure which will surmount the lattice girders and the thrust from which will reduce the horizontal force acting in them. (Federal Reserve Bank, Minneapolis, USA. Architects: Birkerts, Skilling, Helle, Christiansen, Robertson.)

3 Two-storey bridge-type building supported by four columns and forming the upward extension of an existing building. The structure comprises two-storey-high Vierendeel girders, those in the external walls being exposed in the façades. The framework comprises the following Vierendeel girders: internal longitudinal girders a carried directly by the columns, external transverse girders b supported by those girders and in turn supporting external longitudinal girders c, and internal transverse girders d. (Extension of the Czechoslovakian Parliament Building in Prague.)

Arrangement of vertical components

Cellular construction

As a result of the increasing tendency to exploit industrialised building systems, a number of schemes embodying the use of modular cells have been developed. The object is to transport these modules to the site as completely prefabricated and fitted units, where their erection involves little more than merely stacking them one upon another and connecting the various services. Structurally, the following types of module are available:

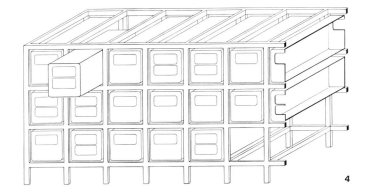

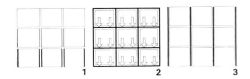

1 The modules are stacked directly one upon another, without any additional components. Those at the bottom of the structure have to carry the loads of those on top and must therefore be designed to withstand larger horizontal and vertical forces. Alternatively, all the modules may be of identical construction, in which case, of course, the upper ones will be structurally over-designed.

2 The modules are completely enclosed and self-contained boxes which are fitted into a structurally independent steel frame designed to take all the loads.

3 A compromise between the two above alternatives is to make all the modules identical and to insert flat steel bars – vertical, horizontal and, if necessary, diagonal – in spaces provided between the modules. These bars, in co-operation with the edge members of the modules themselves and restrained by these members against buckling, serve to absorb all the compressive and tensile forces in the system.

The essential parts of a module are the floors and the four corner columns. The longitudinal and end panels may be composed of prefabricated elements if they are needed as external or internal walls. In some designs, two possible methods of arranging the panels present themselves. Solution **1** is structurally unsatisfactory, but shares with solution **3** the advantage that the modules are packed sufficiently close together that walls and floors need not be of double-leaf construction. In a building based on solution **2**, the modules have to be completely weatherproof and insulated on all six faces, which of course involves extra expense.

So far, this form of construction based on box-shaped units has not been very extensively used. Some relatively minor projects have been carried out, as well as various experimental buildings and numerous designs which have not gone beyond the drawing board Some characteristic systems are described here:

4 The Doernach system (structurally corresponding to **2**) comprises a self-supporting framework of box-section columns and beams filled with water as fire protection. The actual modules are made of plastic and can subsequently be inserted into the framework. These modules, which are the property of the individual occupants of the building, are transportable by road and can accompany their owners if they decide to move house.

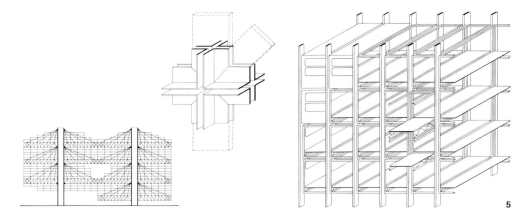

5 Entry for a design competition for the new Technological University in Berlin, 1968 (Architects: Von Sartory, Kohlmaier). The modules are constructed from welded angle sections with spaces left between them. Flat steel bars for the transmission of vertical loads can be inserted into these spaces; diagonal members, welded to these vertical bars and also installed in the spaces between the modules, provide wind-bracing. In addition, the diagonals can serve as web members for lattice systems extending the full height of the building.

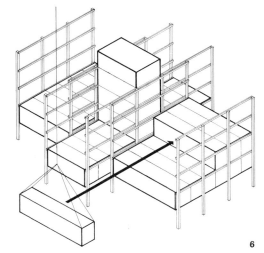

6 This system, called RSC (Reversible Steel Container), is intended for residential buildings. Each dwelling is composed of a group of three to six prefabricated modules (containers) which are suspended from the cross beams of multi-storey rigid frames. In appropriate positions, vertical circulation cores installed between pairs of frames serve also to transmit horizontal forces. (Architects: Meyer, Rinne.)

Principles

Bracing for steel-framed buildings

Horizontal forces are caused mainly by wind loads, which can be resolved into pressure and suction. In certain parts of the world, however, buildings have also to be designed to resist the possible effects of seismic forces. In tall buildings of relatively flexible construction the deflections due to horizontal forces may be quite substantial. When the columns lean over as a result of such deflections, they are additionally subjected to horizontal components of the vertical loads. Horizontal forces may furthermore be caused by earth pressure, friction in roller bearings or sliding bearings of bridge-type structures, pull in the belts of belt conveyors, or the dynamic action of machinery with oscillating masses.

Every building should be adequately stiffened against horizontal forces, considering their action in both directions.

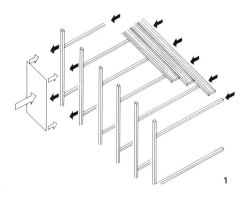
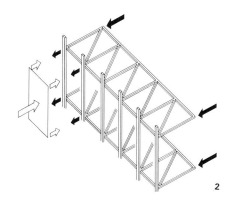
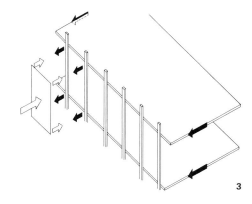

Horizontal bracing systems

Wind pressure acts on the external faces of the building, that is, on the façades, roof, etc., which in turn transmit these forces to the internal stiffening system comprising vertical and horizontal frameworks.

1 If the wind pressure acting on the façade is transmitted to the foundation at every row of columns, no horizontal load distributing systems are necessary.
2 If, on the other hand, the horizontal forces may be transmitted to the foundation only at a limited number of points, horizontal stiffening systems are required which, in steel-framed buildings, may be in the form of lattice or plate girders.
3 The function of transmitting the horizontal forces may be performed by the floors if these are of suitable construction to act as rigid diaphragms.

Arrangement of vertical bracing systems

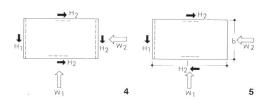

The vertical bracing systems or shear walls for transmitting the horizontal forces to the foundations must be arranged in at least two, non-parallel, directions and in at least three different planes in order to resist wind forces acting in two mutually perpendicular directions.

4 In a symmetrical building, shear walls arranged symmetrically absorb forces only from wind acting parallel to them.
5 If the shear walls are arranged asymmetrically, those perpendicular to the wind direction are subjected to additional forces due to moments set up in the structure.

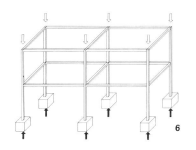
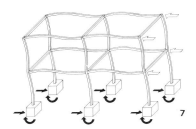
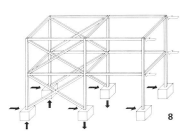
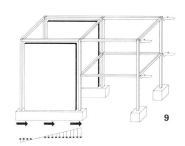

Types of vertical bracing system

6 The steelwork can be stiffened against horizontal forces in the following ways:
7 The structure is composed of rigid frames which may comprise some hinged joints but there must be sufficient rigid joints to ensure that none of the nodes of the frame is free to move sideways. The members may be straight or curved and a variety of shapes may be chosen for such framed structures.
8 The bracing system consists of a triangulated framework of rolled sections. The centre-lines of members converging at a joint should intersect at one point. The joints themselves are conventionally assumed to be hinged, so that the members are either ties or struts, loaded purely in tension or compression respectively. However, the overall bracing effect of a lattice system can be enhanced by constructing it with stiff members and rigid joints.
9 Shear walls in the form of more or less solid diaphragms, usually of concrete, transmit the wind forces by shear and bending.

Bracing for steel-framed buildings

Rigid frames

Rigid frames are assemblies of members comprising a sufficient number of rigid joints to ensure stability without the need for diagonal bracing, the advantage being that there are no obstructions in any of the bays between columns. Nevertheless, the use of rigid frames is one of the most expensive ways of bracing structural steelwork. They should therefore be used only as a last resort.

Long-span rigid frame structures are relatively flexible, and the horizontal deflection (sidesway) due to horizontal forces is greater than that in other vertical systems.
As a rule, the legs of the frames are also structural columns in the building, while the cross beams are an integral part of the floor structure.

Frame shapes

Frames may be of many different shapes but the following remarks apply mainly to rectangular frames, the components of which may be joined with hinged or rigid connections. The hinges may be introduced at either the feet or the tops of the columns.

1 A frame comprising two legs and a cross beam, conventionally called a portal frame, is able to resist horizontal forces if at least one of its four joints is rigid, in which case this sole rigid joint must resist the entire bending moment: three-pinned portal frame.

2 When the frame has two rigid joints, the moment applied to a single joint is reduced by about a half: two-pinned portal frame. Of the two frames shown, the one on the left with hinged feet is by far the most usual. In this case, under wind loading, only horizontal and vertical forces are transmitted to the foundations.

3 A frame with four rigid joints possesses the greatest stiffness, but substantial foundations are required to absorb the bending moments induced at the feet of the columns.

4 Any number of bays comprising only pinned-joints may be added to a portal frame.

5 If some or all the joints in such additional bays are also of rigid construction, a multi-bay rigid frame is formed and the bending moment to be resisted is shared among the various rigid joints.

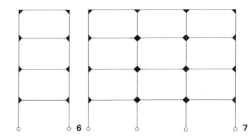

6 and **7** Single-bay or multi-bay portal frames can be placed one upon another to form multi-storey frames.

8 A long cross beam in a portal frame can be given intermediate support by means of rocker columns.

9 For shed structures, the rafter sections connecting the legs of the portal frame follow the pitch of the roof.

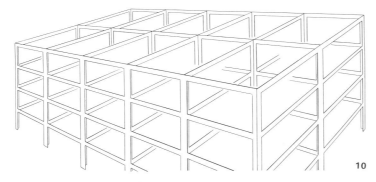
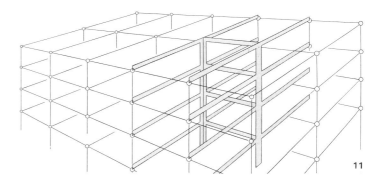

Optimum number of rigid frame members

The designer is often faced with the question whether all or only some of the columns should be incorporated into rigid frames. The considerations involved in answering this question can be illustrated with reference to some simple examples:

10 In this solution, all the transverse frames are designed as triple-column four-storey rigid frames, while all the longitudinal frames are similarly designed as rigid frames each comprising six columns. The horizontal forces are distributed among all the columns, each of which thus has to resist only a relatively small share. All the columns have to be rigid members and the foundations have to be designed accordingly. This is a possible solution for tall buildings.

11 In the second solution, only two of the transverse frames are designed as multi-storey rigid frames and in these only the centre columns are rigidly connected to the beams. The outer columns are designed as being pin-jointed at top and bottom. In the longitudinal direction of the building only one bay is provided with rigid joints and alone serves to stiffen the structure in that direction. By thus concentrating the stiffening function upon a small number of very rigidly constructed internal assemblies, the designer can use relatively light and slender columns for all the other vertical members, resulting in an overall saving, which is also often reflected in the foundations. Although pad footings generally have to be made larger, most other types of foundation can be of lighter design.

Rigid frames

Bracing for steel-framed buildings

12 In this example, a university building constructed with standard units, structural rigidity is obtained by means of multi-bay multi-storey rigid frames extending in both directions, there being no shear walls to interfere with the flexibility of internal layout. In the transverse direction the rigid frames are formed by the main beams, each comprising a pair of channel sections, and the structural columns; longitudinally the columns are interconnected by special wind-bracing which is independent of the floor structure. The main beams support the composite floor which comprises floor beams at 1.8 m centres, to which strictly standardised slabs, based on a 60 cm module, are bolted to provide shear connection. (New building for the Free University at Berlin-Dahlem: Krupp-Montex building system.)

13 In this 12-storey office building, square on plan, the columns are widely spaced at 11.00 m centres. The building is stiffened in both directions by two three-bay multi-storey rigid frames. There are only four internal columns and the interior is otherwise unobstructed, there being no shear walls or diagonal bracings, so that the core zone of the building is freely accessible on all four sides. (Administrative building of Warren Petroleum Corporation, Tulsa, Oklahoma, USA. Architects: Skidmore, Owens and Merrill.)

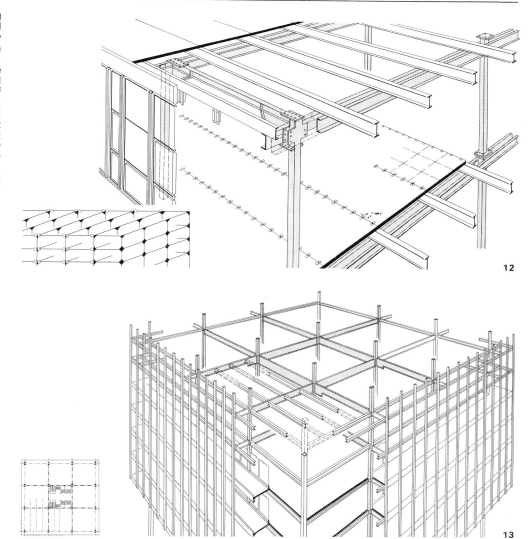

14 The two halves of this multi-storey car park have floors which are staggered half a storey in height in relation to one another and interconnected by ramps based on the d'Humy system. The ramp structure is utilised to provide structural stiffness in the transverse direction, the triangular ramp girders being rigidly connected to the centre columns of the transverse rigid frames. The vertical tensile and compressive forces developed at the ends of these girders by wind loads are transmitted through small secondary columns.

In the longitudinal direction the centre columns combine with the rigidly connected guard rails to form a close-mesh multi-storey rigid framework. From the purely structural point of view, the latter is highly indeterminate and considerably overdesigned, so that if part of the barrier is destroyed by a collision, there will be no danger to the structure as a whole.

This building illustrates how structural components required for other purposes can be economically utilised for stiffening. (Standardised car park construction based on the Krupp-Montex system.)

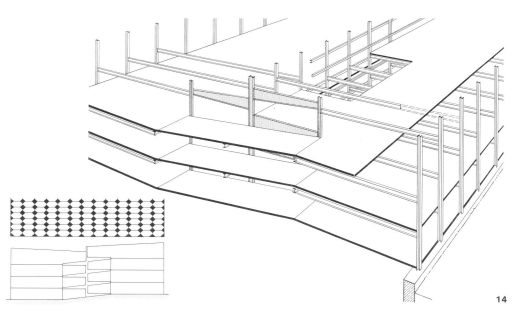

Bracing for steel-framed buildings

Façade frames

15 Behind the spandrel panels of the external walls there are rigid structures which, in combination with the closely spaced external columns, form a rigid framework within the plane of the façade (façade frame). In comparison with the columns, the spandrel-height façade girders possess very great stiffness, so that they undergo hardly any deformation under the action of horizontal loads. The columns are rigidly connected to these girders and can thus deflect only over the height of the windows. Thus, the unrestrained height of the columns is small and, since the columns are so numerous, each has to resist only a small proportion of the force. Therefore, despite the small cross-sectional dimensions of these external columns, their deformations remain small.
16 Plate girder.
17 Lattice girder.

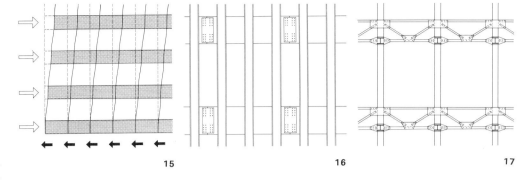

15 16 17

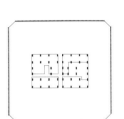

18 The twin 110-storey towers of the World Trade Center in New York, 411.50 m in height, are stiffened by external walls which together form a rigid tube (p. 221, **11**). The external walls are assembled from units comprising three columns connected to three spandrel plates.

18

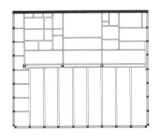

19 In this eight-storey building, three of the external walls are designed as façade frames with lattice girders rigidly connected to the columns, while the fourth is a fire-resisting party wall. (Union des Assurances de Paris (UAP), Marseilles. Architect: Jaume.)

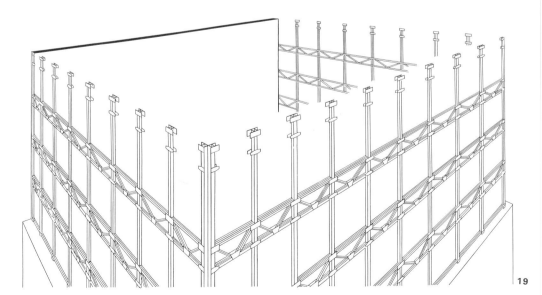

19

212

Vertical bracing

Bracing for steel-framed buildings

As a rule, the most economical and effective method of stiffening a steel-framed structure is by means of vertical lattice bracing.
1.1 From the structural point of view they are vertical cantilevers fixed at the base.
1.2 Slender vertical bracing systems develop large forces in their members which produce large strains. As the columns are closely spaced, relatively large horizontal deflections are induced.
1.3 As the forces in the members of wide bracing systems are smaller, there is a reduction in the horizontal deformation. Where practicable, the bracing should extend over the full width of the building.

 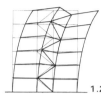 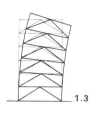 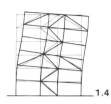 

1.4 The stiffness of a slender bracing system can be improved by providing additional diagonal members to connect it to the outer columns in selected storeys, so that these columns contribute to overall structural rigidity.
1.5 A deep horizontal girder (provided, for example, in the plant room of a high-rise building) has a similar effect. It reduces the tilting of the uppermost member of the frame and thus provides extra restraint against lateral deflection of the building.
The stiffness of a lattice girder can be increased by constructing it with rigid joints, so that a combination of lattice and rigid framework is obtained.

Arrangement of lattice bracing on plan

It is necessary to provide vertical bracing in at least two directions on plan. Such bracings in the interior are an obstruction, interfering with the freedom of internal layout. If internal bracings have to be provided, they should be incorporated in permanent walls pierced by few openings.
2.1 Narrow lattice bracings enclose a staircase.
2.2 Building with three transverse and one longitudinal bracing systems. If the cores are narrow and the building is high, additional stiffening as indicated in **1.4** or **1.5** may be needed.
2.3 Transverse bracings in unglazed end walls are economical and effective. Longitudinal bracing between two internal columns.

 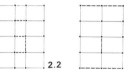 

2.4 Consistent application of the principle of making the bracings as wide as possible leads to their installation in the external walls. This has a decisive effect on the architectural appearance of the building.
2.5 A high-rise building square on plan and braced between the four internal columns often requires additional stiffening as indicated in **1.4** or **1.5**.
2.6 If the bracings are installed in the external walls of a building which is square or almost square on plan, particularly economical and architecturally effective solutions can be obtained.

Arrangement of lattice bracing systems in elevations

3.1 All the braced panels are located vertically one above the other.
3.2 The braced panels are staggered in relation to one another in successive storeys, the horizontal forces being transmitted through the floors from bracing to bracing. Each storey must be suitably stiffened to provide the necessary structural stability.

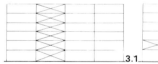  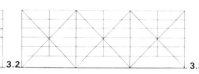

3.3 Lattice systems which comprise entire external walls also participate in transmitting the vertical loads and should be designed accordingly.

 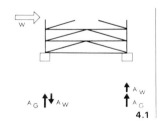 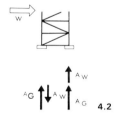 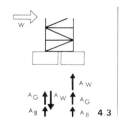 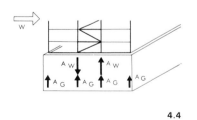

W = wind force A_W = reaction due to wind A_G = reaction due to vertical load A_B = reaction due to weight of foundation

Effect of vertical bracing on the foundations

As a rule the vertical chords of the lattice system are formed by the columns of the building. They are therefore subjected to vertical forces from both the floor and the wind loads. The latter produces alternate compression and tension in the columns functioning as lattice chords. The vertical loads always produce compression in the columns. Thus, in one chord the compressive forces are added together, while in the other the compressive force due to vertical load is reduced by the tensile force due to wind load. For structural stability there should always be a compressive force at the base of the columns. With unfavourable design conditions, however, a tensile force may develop there, in which case it is necessary to compensate for this by increasing the weight of the foundations.
4.1 The corner columns carry only relatively small vertical loads, but the forces induced by wind loads in the columns are also small because of the large width of the bracing. No extra ballasting of the foundation to increase its weight is necessary in this case. Most economical load transmission.
4.2 The internal columns have to carry large vertical loads and also, because the bracing is narrow, large vertical forces due to wind load.
4.3 The wind forces here are of the same magnitude as in **4.2**, but the vertical loads are small, as the columns in question are external columns. Heavier foundations are therefore necessary.
4.4 Ballasting is unnecessary if the external columns are held down by a deep basement wall which is able to compensate for the negative reaction.

Bracing for steel-framed buildings — Vertical bracing

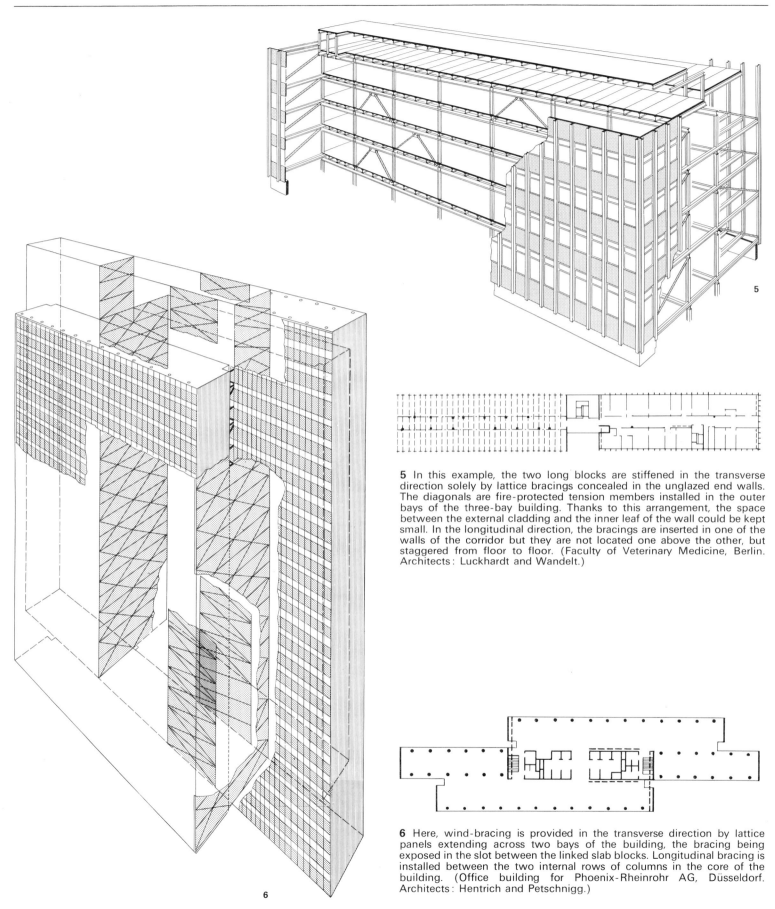

5 In this example, the two long blocks are stiffened in the transverse direction solely by lattice bracings concealed in the unglazed end walls. The diagonals are fire-protected tension members installed in the outer bays of the three-bay building. Thanks to this arrangement, the space between the external cladding and the inner leaf of the wall could be kept small. In the longitudinal direction, the bracings are inserted in one of the walls of the corridor but they are not located one above the other, but staggered from floor to floor. (Faculty of Veterinary Medicine, Berlin. Architects: Luckhardt and Wandelt.)

6 Here, wind-bracing is provided in the transverse direction by lattice panels extending across two bays of the building, the bracing being exposed in the slot between the linked slab blocks. Longitudinal bracing is installed between the two internal rows of columns in the core of the building. (Office building for Phoenix-Rheinrohr AG, Düsseldorf. Architects: Hentrich and Petschnigg.)

Vertical bracing

Bracing for steel-framed buildings

7 This is a triple-bay building with column spacings of 7.00, 3.50 and 7.00 m. Narrow transverse bracing is installed between four pairs of internal columns together with longitudinal bracing between two internal columns in one row. Because of the narrowness of the transverse bracing, the calculated horizontal deflections due to wind pressure were found to be too large. To reduce these movements, additional diagonal members, extending to the external columns, were installed in the four transverse bracing planes in both the second and the fifth storeys. These diagonals, functioning as stays, consist of flat steel bars on edge. After all the dead load had been applied, the diagonals were prestressed to such an extent that, with wind acting on one side of the building, the tension in the diagonal on that side was doubled, while the tension in the opposite diagonal was reduced almost to zero, so that, in every case, diagonals function in combination. (Head office building for Bewag, Berlin. Architect: Prof. Baumgarten.)

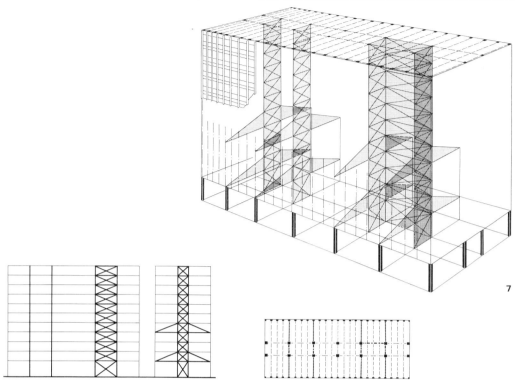

7

8 This building has external columns only. The floor beams span 12.50 m while the columns are spaced at 7.50 m centres. In the higher part of the building, the wind-bracing extends across the full width, a very economical solution because the width is larger. As the columns have to transmit very heavy loads, the resulting compressive forces compensate for the tensile forces induced by the wind load. The end wall of the higher part of the building projects 2.50 m beyond the columns, the bracing installed in this wall being continued between the two end columns below the first floor, in which horizontal bracing is provided to effect the transition. The resulting reactions are transmitted by a storey-height plate girder installed in a plant room. This girder is situated between the penultimate and the last column and cantilevers out to the end wall. (Free Berlin Television Centre. Architect: Tepez. Consulting Engineer: Treptow.)

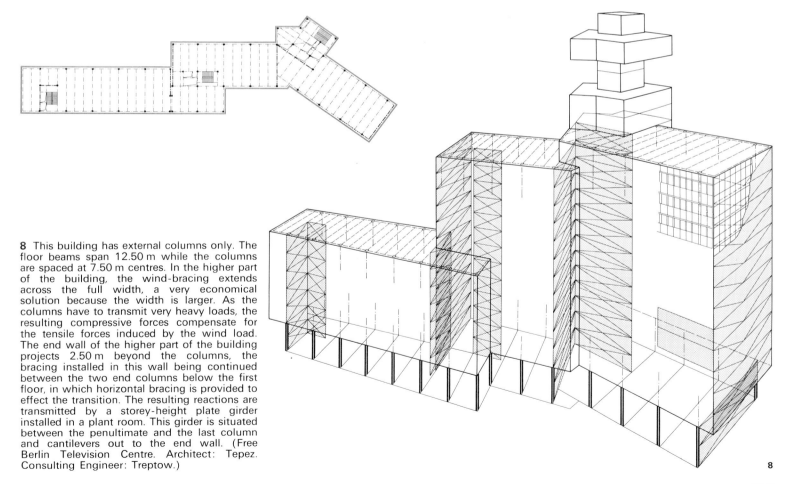

8

Bracing for steel-framed buildings

External bracing

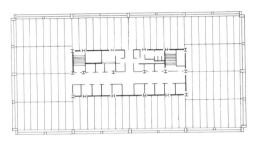

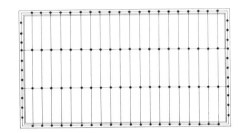

9 Stiffening of this building is obtained by means of heavy lattice bracings in front of the façades. These bracings also co-operate with the intermediate columns in transmitting vertical loads. For details, see p. 266, **1**. (Alcoa Building, San Francisco. Architects: Skidmore, Owings and Merrill.)

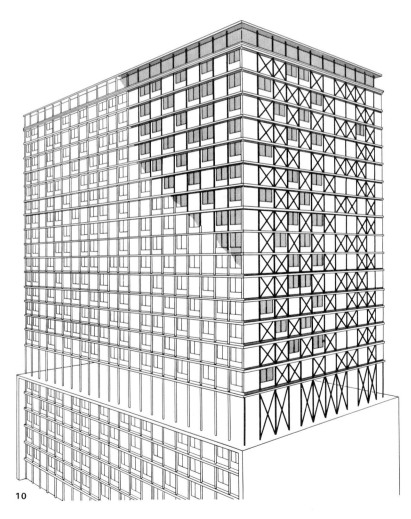

10 The lower part of this building is stiffened transversely by a heavy concrete shear wall and the upper part by tubular lattice bracings, exposed in front of the façades. The bracings are staggered from storey to storey, six of the twelve panels being braced in each storey. Stiffening in the longitudinal direction is provided by individual lattice systems in the two central rows of columns. For details, see p. 266, **2**. High-rise block of flats for Société Immobilière, Rue Croulebarbe, Paris. Architects: Albert-Boileau, Labourdette.

Shear walls and cores

When concrete walls are needed as fire-resisting compartment walls, stairwells and lift or vertical service shafts, it is obviously a good idea to utilise them for stiffening the building. However, since a steel-framed building can be stiffened quite simply by means of vertical steel bracing, it would be uneconomical to construct concrete shear walls solely for the purpose of stiffening.

Vertical circulation cores

The concept of solid cores, that is, tower-like shafts accommodating lifts, stairs, services and often cloakrooms, has been adopted from concrete construction.
This form of construction may also be appropriate to steel-framed buildings, though often other solutions are better, because in such buildings the vertical circulation facilities really require no more than appropriate openings in the successive floors. Flights of stairs and landings are attached to the supporting steelwork just as the floors are. To provide a fire-resisting enclosure around the staircases and lifts in a steel-framed building there is no need for heavy concrete walls. There are lighter and cheaper materials that satisfactorily perform this function (for example, gypsum, see p. 323). There is no compelling technical reason why cloakrooms should be accommodated in the concrete core. Doors and apertures for services sometimes necessitate such large openings in the walls of the shafts that they become difficult to design as wind-resisting shear walls. It is therefore often better to stiffen the walls of the building by means of wide-span lattice bracing. It is not possible, however, to lay down hard and fast rules.

Construction procedure

Shuttering systems

Concrete shear walls and cores may be constructed with in-situ concrete using individual panels of shuttering but, when the buildings are tall, cores are usually erected with sliding or climbing shuttering.

Construction schedule

In the construction schedule for the job, the concreting and the erection of the steelwork must be carefully correlated. Individual shear walls generally have to be constructed as the erection of the steelwork proceeds, as they usually do not possess the necessary stability to stand alone. On the other hand, concrete cores are in most cases completed to their full height in advance, so that they can be utilised as a platform for the cranes used in erecting the steelwork.

Tolerances

The differences in the tolerances attainable with in-situ concrete construction and structural steelwork, respectively, should be duly taken into consideration. Adequate scope to compensate for dimensional inaccuracies and discrepancies due to such differences must be provided. With tall buildings, allowance must be made for the differing strains, shrinkage and creep in the concrete and steel components under the progressively increasing loads developed in them during and after construction.

Structural connections

The beams of the steel frame must be suitably connected to the concrete cores or shear walls to transmit their vertical and horizontal loads to them. As the cores and shear walls usually have to perform their stiffening function during erection, effective connections should be established from the outset. Examples of such connections, allowing for the differences in dimensional tolerances, are given on p. 268.

Precast concrete components

A well-balanced construction procedure is obtained when the shear walls are constructed of precast concrete. Transmission of forces between the precast units, and connection to the structural steelwork, is ensured by means of steel sections embedded in the concrete. By such means, it is possible to establish sound and effective connections during the actual erection. This method of construction is, however, suitable only for buildings of limited height. For structural details, see p. 269.
To help the designer arrive at a decision, the arguments for and against the use of concrete walls to stiffen a building will now be reviewed:

Concrete shear walls or cores are advantageous:

• if the combined liftshaft and stairwell can adequately stiffen the building with no more than the wall thickness necessary for fire protection (10 cm of concrete for fire-resisting walls in general, 14 cm for fire compartment walls);
• if it is impracticable to provide the structural steel frame with the necessary lattice bracing;
• if the cores are located outside the main ground-plan as exposed external features, when the main body of the building may then be constructed with simple widely-spaced columns providing maximum flexibility of internal layout.

Lattice bracing is more appropriate:

• if it is practicable to provide light wide-span vertical lattice systems;
• if lifts and staircases are not located close together;
• if the staircases are not exactly one above the other, but are staggered in the successive storeys;
• if lift and staircase enclosures are planned as light glazed frameworks outside the actual building;
• if the construction time available is too short to allow cores to be constructed in advance of the steelwork;
• if the core walls have to be pierced by excessively large openings.

1 Fire-resisting compartment wall as transverse stiffening.
2 Two staircase walls as transverse stiffening.
3 Stairwell outside the actual buildings serves to stiffen two adjacent blocks.

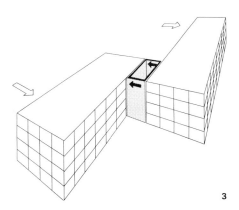

Stability of steel-framed buildings

Shear walls and cores

4 A 14-storey office block is surrounded by a five-storey podium block. The high-rise structure is stiffened by a vertical circulation core, two lower secondary cores, and fire-resisting wall. The cores were concreted with the aid of climbing forms before erection of the steelwork. The beams were temporarily supported in recesses in the concrete walls, the connections subsequently being made permanent by concreting in the recesses. During erection, the steelwork had to be stiffened with temporary bracing. (Administrative building of Allianz-Versicherung, Hamburg. Architect: Hermkes.)

5 This 20-storey office building is stiffened by its triangular concrete core, constructed by slipforming. Derricks for erecting the steelwork were mounted on the core. After completion, two more storeys were added to the building. (Unilever House, Hamburg. Architects: Hentrich and Petschnigg.)

6 In this 22-storey office building the concrete core accommodating the staircases, lifts and services also provides structural stability. It was constructed in advance by slipforming. Two derricks were then positioned on the core to assist in erecting the steelwork. The floor beams were connected to steel plates which were embedded in the concrete core and which were secured in position by a framework of angle sections similarly embedded during slipforming. (Europa Center, Berlin. Architects: Hentrich and Petschnigg.)

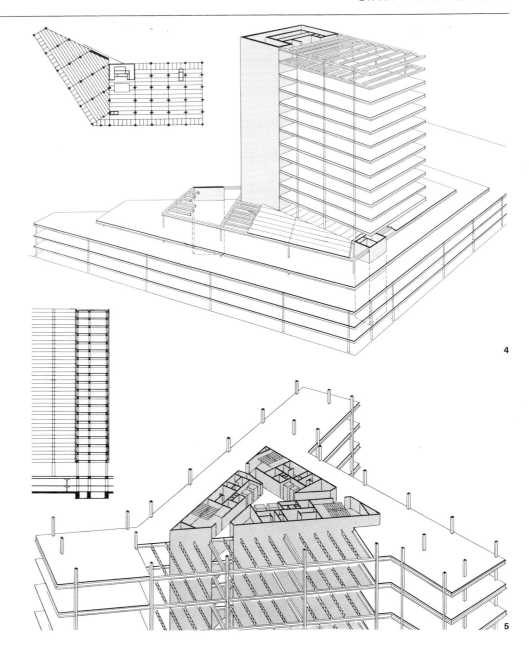

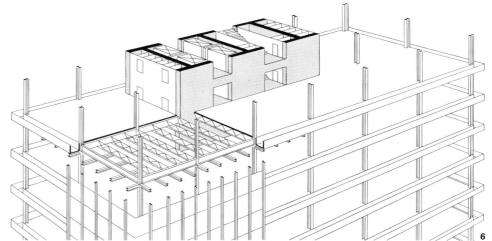

Bracing for high-rise buildings

Stability of steel-framed buildings

For high-rise buildings the choice of the correct stiffening system is of major importance, since a high proportion of the total cost of the structural frame in such tall buildings is taken up by the stiffening or bracing (up to 50% of the weight of the steel). The principal horizontal loads arise from wind and earthquakes. Since the latter generally do not arise in connection with the design of European buildings, only the effect of wind will be considered here. As a result of the static pressure it develops, wind causes deformations; furthermore, oscillations of the building are set up by eddy effects.

Wind pressure

Wind speed

The static wind pressure depends on the wind speed, the height of the building, its shape and the smoothness of its external surfaces.
In many countries the calculation of wind pressure is based on the highest measured wind speed, at a height $h_{10} = 10$ m above ground level, in a 50-year or 100-year period (design wind speed). In the Federal Republic of Germany this velocity is: $v_{10} = 30$ m/s.
The wind speed v_h increases with increasing height above ground level. In **1** the ratio v_h/v_{10} has been plotted for various site conditions. German Standard DIN 1055, sheet 4, envisages only one curve (a) for the Federal Republic generally and does not give any special rules for wind on high buildings. In the USA, according to Lippoth, the rate of wind velocity increase with increasing height in built-up areas is less pronounced than would correspond to curve d.

The dynamic wind pressure $q = v^2/16$.

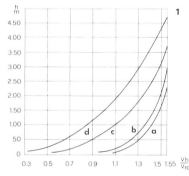

Curve a applies to DIN 1055, curve b to open flat landscapes, curve c to suburbs and curve d to city centres.

Shape coefficient

The shape of a building is of considerable effect on the magnitude of the wind load $w = c \cdot q$. In this expression c is a shape coefficient whose value is between 1.4 and 1.6 for high buildings. Precise information on the effect of shape can, however, be obtained only by means of wind tunnel tests. As a rule, symmetrical shapes on plan should be chosen, **2.1**, **2.2**. Asymmetrical or indeed rotationally symmetrical shapes (**2.3**) are not the most suitable for high-rise buildings because they give rise to torsional as well as flexural loading conditions, so that their stiffening involves considerable extra cost.
The smoothness of the façades and the architectural or structural details of a building can considerably affect the value of c. These, too, require wind tunnel tests in the case of high buildings.

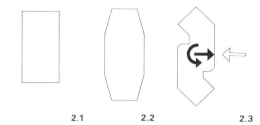

2.1   2.2   2.3

Horizontal deflection

Magnitude of the deformation

The magnitude of the horizontal deflection due to wind load normally increases with the height of the building, but the stiffer the building, the smaller the deflection. The permissible deflection is not subject to official regulations. According to American observations, the theoretical maximum deflection of high-rise buildings actually erected is between 1/800 and 1/200 of their height. Generally speaking, buildings which do not deflect more than 1/600 to 1/400 of their height are to be regarded as possessing adequate structural rigidity. In designing and detailing the connections in a building, it is of importance to know how much it is likely to deflect. External and internal walls which do not participate in the transmission of horizontal forces should be so connected that they can follow the expected deformations of the building without cracking or suffering other damage.

Vibration behaviour

The vibration behaviour of tall buildings should be investigated. The natural frequency of the structure should differ sufficiently from the eddy frequency. It is important that the acceleration due to wind-induced oscillations is below 0.5 m/s^2 to prevent objectionable physical discomfort to occupants of the building. The damping effect of the structural components not forming part of the actual stiffening or bracing system can be taken into account.
Figure **3** presents the relationships between the natural period of vibration λ of a building, the amplitude δ_d, and the maximum acceleration as a percentage of the acceleration due to gravity (g = 9.81 m/s^2). The significance is as follows:

Range	Acceleration as percentage of g	Degree of comfort
A	≤ 0.5	not perceptible
B	0.5–1.5	perceptible
C	1.5–5	objectionable
D	5–15	very objectionable
E	> 15	unbearable

To illustrate the point, the acceleration which is 15% of g = $0.15 \times 9.81 = 1.47$ m/s^2, which corresponds to that of a motor car which reaches a speed of 53 km/h from scratch in 10 s.

Total deflection

The above-mentioned deflections are in each case the sum (**4**) of the static deflection δ_s and the dynamic deflection δ_d measured at the top of the building, hence: $\delta = \delta_s \pm \delta_d$. Quite often the dynamic deflection is roughly half the static deflection, so that the maximum deflection is approximately 1.5 δ_s.

Design possibilities

The stiffening of high-rise buildings is usually achieved by a combination of two or more of the possibilities described in this chapter: rigid frames, lattice bracings, shear walls.
For example, in many instances all the beams and columns are additionally connected to lattice bracing systems, besides being provided with rigid joints, or concrete shear walls are introduced, especially in the bottom storeys.
As a general principle, the stiffening or bracing system should continue as far as possible straight down to the foundation. If the forces transmitted in these structural features have to be diverted, for example, in order to provide large column-free internal space in the bottom storeys of a building, such arrangements are liable to involve substantial extra cost.

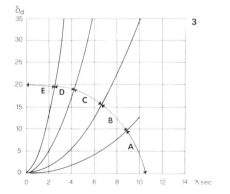

Reference: W. Lippoth, 'Windwirkung auf hohe Gebäude', *Der Bauingenieur*, 12/1968.

219

Stability of steel-framed buildings — Bracing for high-rise buildings

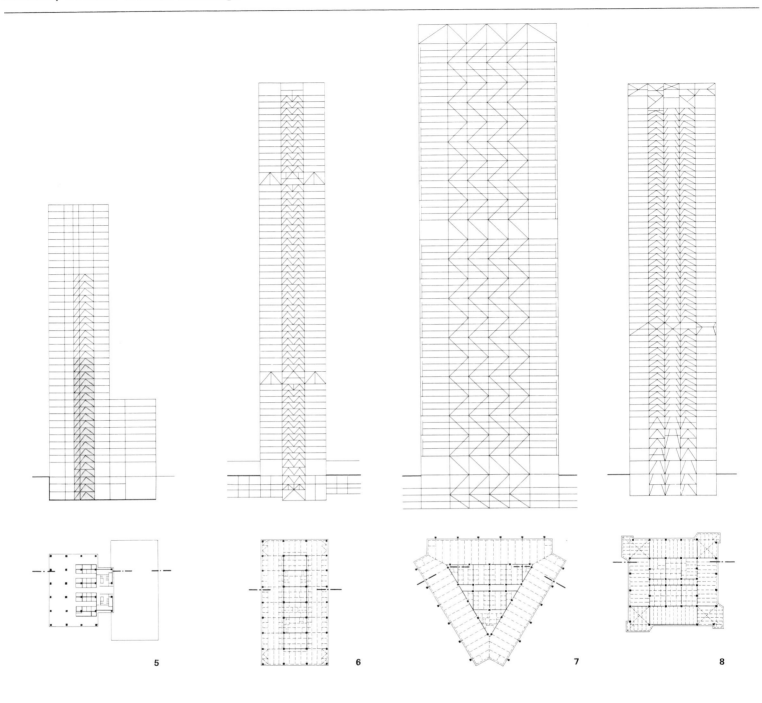

5 Seagram Building, New York (architect: Mies van der Rohe). Height 158.50 m. Upper part stiffened by rigid frame action only. Below the 29th floor there are K-bracings, augmented by 30 cm thick concrete walls from the 17th floor downwards.

6 Dominion Center, Toronto (architect: Sidney Bregman). Height 224.00 m. Stiffening by vertical bracing between the lift shafts. In addition, in the plant rooms there is transverse vertical bracing extending across the full width of the building, whereby the deflection is limited to 1/500 of the height.

7 Office building for United States Steel, Pittsburgh (architects: Harrison and Abramowitz and Abbe). Triangular on plan. Stiffening by vertical lattice bracing in the walls of the triangular core, with additional vertical bracing across the full width in the top storey, whereby lateral deflection is reduced by 30%. For the design wind velocity (statistical frequency once in fifty years) the horizontal deflection at 256 m above ground level is 1/500 of the height, half of which is due to dynamic action (see also p. 240).

8 'One Astor Plaza' office building, New York (architects: Kahn, Jacobs). Height 222 m. The building is stiffened by vertical lattice bracing in the four sides of the vertical circulation core and by four more bracing systems inside this core. Additional stiffening is provided by connecting the external columns to the main lattice bracing at the 17th and 54th floors.

| Tubular buildings | Stability of steel-framed buildings |

The high-rise building conceived as a rigid tube

If a high-rise building is of suitably compact shape on plan (circular, square, not too narrow a rectangle), the external columns can be structurally merged with the external lattice bracing or with spandrel girders so as to form a vast rigid tube. This stiffening system is particularly effective and economical. This is due not only to the optimum distribution of the bracing, but also more particularly to the co-operation of all the columns and bracing or spandrel girders in the external walls.

This can be clarified as follows: if a single vertical bracing system extending across the full width of a building is located in the interior (as in **9.1**) only those external columns a which are located within the plane of this bracing assist, as chord members, in resisting the wind load. The other external columns b, though sharing in the horizontal deflection of the building, contribute hardly anything to wind resistance. On the other hand, if vertical bracing is located in the external walls (**9.2**), wind load on wall a involves the vertical bracings in the walls c and d like the

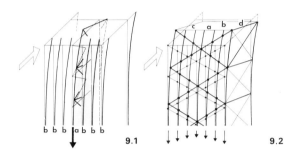

bracing in **9.1**. The corner columns now function as chord members. By means of the bracing in walls a and b, all the columns in these walls are connected to the corner columns and participate fully in resisting the wind load. Thus, all the columns in walls a and b function as chords for bracing systems c and d. Thus the overall chord cross-sections are very large and the strains developed in the external columns are small; so that the building has great structural rigidity, while the bracing is relatively inexpensive.

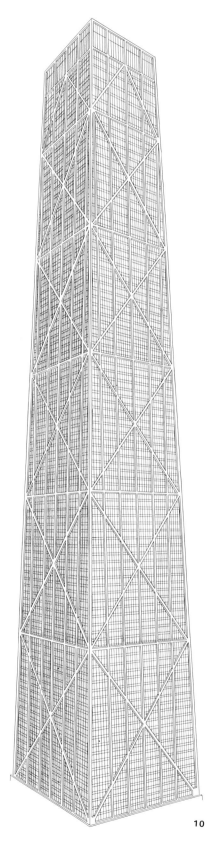

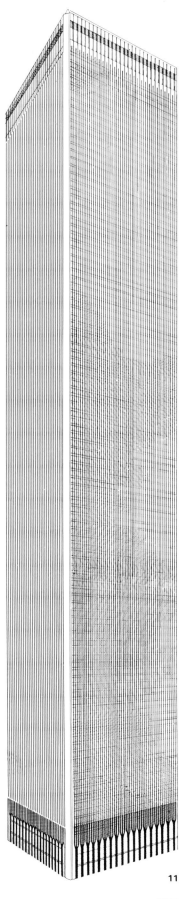

10 John Hancock Center, Chicago (architects: Graham and Skidmore, Owings and Merrill). In this 335 m high 100-storey building all the horizontal forces are transmitted through external bracing, whereby a substantial saving has been achieved in comparison with a bracing system located in the interior. The external lattice members form a distinctive architectural feature, but the diagonals, which are of considerable width, intersect the windows. For details, see p. 266 **4**.

11 World Trade Center, New York (architects: Yamasaki and Associates). The twin towers are each 411.50 m high and comprise 110 storeys. They are stiffened entirely by the rigid-frame gridwork of the external walls comprising closely spaced columns interconnected by deep horizontal members. For details, see p. 212 **18**, p. 349

Stability of steel-framed buildings

Tubular buildings

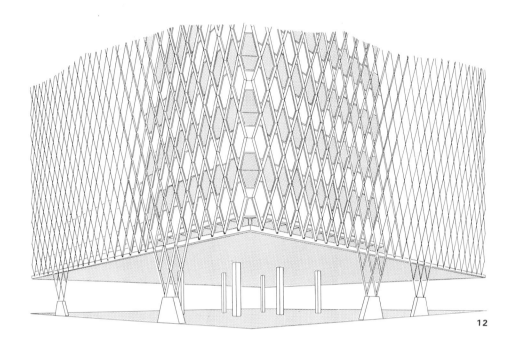

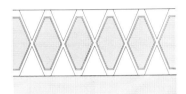

External columns and inclined bracing members can be merged to form a distinctive diamond-pattern latticework. Its members perform the dual function of columns and wind-bracing. The windows may be diamond-shaped or triangular.

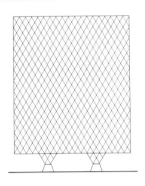

12 The external columns of this building are combined with bracing members to form a diamond latticework. The external walls are supported on triangular extensions of the latticework which, together with internal pilotis, raise the building off the ground. The lattice system is a determining feature both of the external and internal architecture of the building. The external columns are sheathed in stainless steel. Windows are lozenge-shaped with truncated points. For details, see p. 266 **3**. (IBM Building, Pittsburgh. Architects: Curtis and Davis.)

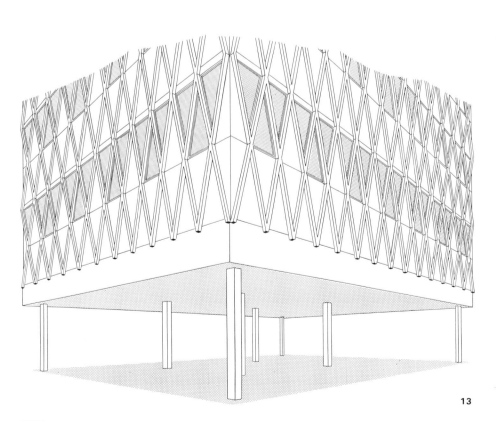

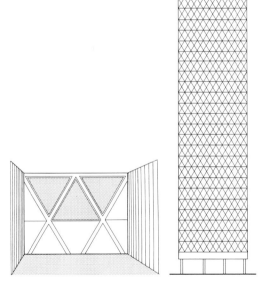

13 Design for a building 85 m in height. Diamond lattice of 12 cm wide members. In the interior of the building, which measures 30 × 30 m on plan, there are only four supports, designed as simple columns with cross-sectional dimensions not exceeding 30 × 30 cm. This form of construction provides a large amount of space. The triangular shape of the windows is dictated by structural requirements.

Expansion joints

Joints in buildings

Arrangement of joints

Joints between various parts of a building make it possible for those parts to undergo relative movements without setting up restraints in the structure. These joints mostly form vertical planes of separation. See also p. 176: deformations.

The principal causes of movement between different parts of a building are:

• changes in length of the floors and roof in consequence of temperature variations, including those caused by fire (expansion joints);
• vertical displacement between the parts of a building as a result of differential settlement (settlement joints).

The movements may occur during the construction period and/or during the subsequent service life of the building. Joints are undesirable features in the sense that they interfere with the homogeneity of a structure and involve additional expense. Hence, they should be spaced as far apart as possible.

EXPANSION JOINTS

The spacing of the joints should be so chosen that changes in length of the floors and roof will not cause damage and that, in particular, no excessive stresses are developed as a result of restraining effects in the structure.

Temperature variations

Temperatures during construction

The largest variations, in the case of a multi-storey building, normally occur during its construction and not, subsequently, when the building is in service. It may even be necessary to provide temporary additional expansion joints during construction, these being subsequently closed up and immobilised in the completed structure. For the temperatures liable to occur, see p. 178.

Shrinkage of concrete

Changes in the volume of concrete due to shrinkage are similar in their outward effect to changes in temperature.

Effects of fires

In the exceptional conditions of a fire, high temperatures may develop in the loadbearing structure. The protective casings for steelwork are so designed (p. 331) that the critical temperature of 450°–650°C will not be reached before the end of the specified fire resistance period. Such temperatures occurring in steel beams can be transformed into unrestrained linear changes only if:

• the floor slab is not composite with the steel beams; or
• the floor acquires the same temperature as the steel beams.

Because of the much larger cross-sectional area of the floor, the horizontal displacements, as a rule, do not depend on the temperature of the steel beams, but on the temperature of the floor. For concrete slabs, it can be assumed that the slab has reached a temperature of about 200°C when the steel beams reach their critical temperature. The movements in the floor which occur at that temperature should be taken into account — using higher stresses than the normally permissible values — in determining the spacing of the expansion joints and for analysing the stability of the vertical stiffening arrangements. These displacements cause:

• restraining forces, for example, as in the case envisaged in **4**; or
• horizontal thrusts, involving a tilting of the columns.

Spacing and arrangement of joints

Spacing of joints

The joint spacing in a steel-framed building depends to a great extent on the type of stiffening:

• Rigid-framed structures, on expansion of the floors, are subject to reactions which can be analysed. Joints can be spaced 30 to 50 m apart.
• Steel structures which are stiffened only at individual fixed points are to be regarded as otherwise pin-jointed. The cross-sections of the columns are relatively so small that, despite the high modulus of elasticity of steel, deformations usually do not give rise to appreciable stresses. Thus, in such structures, the expansion joints can be spaced far apart, sometimes even as much as 100 m.

Arrangement of joints

The location of the fixed points (cores, shear walls, bracing, etc.) is the determining factor in the arrangement of the joints in a building. In **1**, showing a building rectangular on plan, the bracing elements are substantially symmetrical, so that linear expansion will take place uniformly in both directions. In **2**, because of eccentric location of the core, expansion is virtually limited to one direction only.

Where several fixed points are provided in a building, the following considerations apply:

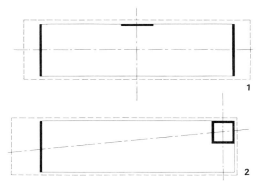

3 Where several elements are provided as transverse stiffening, each receives a certain proportion of the horizontal wind load in accordance with the principles of structural analysis, presupposing that suitably rigid floors (for example, concrete slabs) ensure the distribution of this load to the bracing elements. Thermal expansion does not give rise to forces in such a structure.

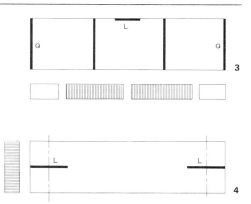

4 If bracing panels are arranged longitudinally, one behind the other, they also both participate in the transmission of wind load. However, thermal expansion induces restraining forces in this case. Such situations should be avoided. As large stresses are most likely to occur during the erection of the building, it is advisable — unless a permanent expansion joint is provided — not to fix one of the two shear panels until after the structure has been completed.

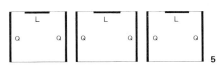

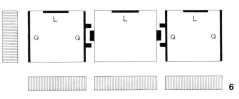

5 Subdivision of the building into three largely independent sections by two expansion joints will, in comparison with **3**, require two additional longitudinal bracing systems L, each of which must be made as strong as the single bracing in that example, because each section of the building must be self-contained and designed to resist the full wind load, even though the centre section is admittedly protected by the outer sections. Also, each section requires two transverse shear panels Q, so that there will be two more of these features than in **3**.

6 Alternatively, it is possible to interconnect adjacent blocks in such a manner that only shear forces, not longitudinal forces, are transmitted across the joints. For instance, if the middle block is connected to the shear walls of the two adjacent blocks, two walls are saved in comparison with **5**.

Joints in buildings | Settings joints

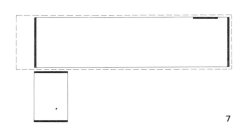

Joints in the roof

If a flat roof is of such construction that it is likely to be heated to higher temperatures than the floors, the deformation pattern shown in Figure 8 will develop. In such cases, it is advisable to provide some additional expansion joints in the top storey (Figure 9).

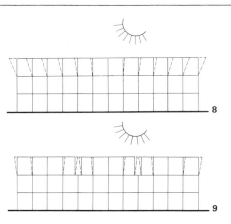

7 Buildings of relatively irregular shape on plan may require expansion joints to accommodate the movements of any projecting parts.

SETTLEMENT JOINTS

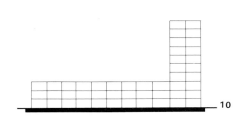

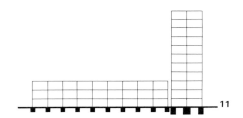

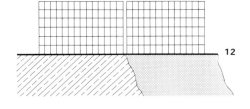

10 Different parts of a building may vary considerably in weight, resulting in differential loading on the foundations. If the latter comprises some kind of raft system extending under the whole building, there will be no differential settlement between the various parts. In such circumstances the low-rise block can be joined directly to the high-rise block without any special joint to separate them.

11 On the other hand, if individual foundations are used, differential settlement is a phenomenon to consider. In such circumstances, it is necessary to provide expansion joints between blocks of different weight.

12 Adjacent blocks may be of equal size and weight, but may be founded on different types of soil, more particularly differing in compressibility, so that differential settlement may occur.

Expansion joints are needed in such cases also. Buildings erected in mining subsidence areas should be subdivided into relatively small units by means of numerous expansion joints. Also, the structural systems employed should be statically determinate, that is, without continuous beams or rigid frames.

Columns	
Principles	226
Types of section	227
Design charts for steel columns	228
Base plates	232
Anchorage of column bases	233
Basement columns	234
Column splices	235
Beam-to-column connections	235
Fire protection of columns by encasement	238
Fire protection of external columns	239
Fire protection of columns by water cooling	240
Services in or beside steel columns	241

Beams and girders	
Structural function	242
Solid web beams and girders	243
Design charts for rolled steel beams	245
Beam-to-beam connections	248
Vierendeel girders	250
Castellated beams	250
Holes for services through webs of beams	252
Forms of lattice girders	253
Light lattice girders	253
Medium and heavy lattice girders	254
Tubular frames	255
Structural action of composite beams	256
Construction procedure and sections for composite beams	257
Precast concrete slabs for composite beams	258
Design chart for composite beams	259
Northlight frames	260

Stiffening elements	
Rigid frames	261
Hinged column bases	261
Joints in rigid frames	262
Horizontal wind-bracing	263
Vertical wind-bracing	264
External bracing	266
Transmission of forces to concrete walls, Tolerances	267
Connection of beams to concrete walls	268
Connection of floors to concrete walls	269
Connections for precast concrete shear wall units	269
Precast concrete basement walls	270

Structural Steel Frames

STRUCTURAL STEEL FRAMES

Clear pattern of forces

This chapter is concerned with the structural details of loadbearing steel frames. The forces acting on the frame are concentrated in relatively slender members with optimised cross-sections producing clear-cut, straightforward structures reflecting the true pattern of the forces.

Requisite properties

Some of the more usual solutions will be given for the structural components of multi-storey steel-framed buildings, to meet the following requirements:

Economy

- The work required in the fabrication shop can be minimised by labour-saving design and by the use of mechanically and electronically controlled fabrication plant.
- Ease of erection with the aid of mobile cranes and thanks to close dimensional tolerances.
- Choice of simple structural shapes which facilitate corrosion protection and which, particularly for structural components in the open air, prevent the accumulation of dust and moisture on them.

Compatibility with internal finishes and fittings

Any structural design should give due attention to:

- simplicity of connection of floors, claddings and partitions to the structural frame (see also pp. 271–284, as well as pp. 299–326);
- the requirements of structural fire protection (see also pp. 279–284, 320 and 327–336);
- the interaction and interconnection of the structural frame and the services (see also pp. 285–289).

Aesthetics

The architectural possibilities offered by steel components which remain exposed either on the inside or outside of buildings should be given due consideration.

Design charts

Design charts for certain important components of multi-storey steel-framed buildings are useful for preliminary designs. They cannot, however, take the place of more precise design calculations. The following items can be determined with reasonable accuracy.

- column sections;
- beam depths;
- the weight of the structural steel frame.

The members of the steel frame

The following components of the three structural systems which together constitute the steel frame of a building (see pp. 187–229) will be dealt with:

- The members for vertical load transmission. The design of steel columns with due attention to subsidiary problems: basement columns, fire protection, arrangement of services, etc. Concrete shear walls for stiffening (see below) may also perform a vertical load transmitting function.
- The members for horizontal load distribution. Only steel beams will be considered. For composite floor slabs, see pp. 276 and 278.
- The bracing members. Connection of lattice members and rigid frame members; discussion of problems of connecting steel components to concrete walls.

Columns

Principles

The columns of a framed building transmit the vertical forces to the foundations and thence to the subsoil. They are thus subject to direct forces. A distinction must be drawn between columns proper, on the one hand, and suspension members, on the other. These act essentially in compression and in tension, respectively. Intentional restraint or fixity of columns and eccentric application of vertical loads will additionally set up bending moments in columns. Casual unintended restraint effects and trivial eccentricities generally produce only minor secondary stresses which are usually neglected in steelwork design.

Columns

Columns must be designed for buckling. As this may occur in both directions, the direction of least stiffness should more particularly be considered. In general, the most economical column sections are those with almost equal moments of inertia in both directions. Very slender sections can be used for columns if they are suitably braced or stiffened against buckling between floors.

Types of section

A wide variety of sections are used for structural steel columns (p. 227). The loadbearing capacity can be closely adapted to the loads by appropriate adjustment of the external dimensions and the thickness and grade of the steel.
Steel columns are sometimes constructed as multiple members, as is frequently the case with stanchions in industrial buildings in order to facilitate the connection of other loadbearing members. Multiple construction may also be adopted for light and slender columns in which the overall buckling resistance is enhanced by spacing the components some distance apart. Suspension members, which are loaded in tension only, need not be stiffened to resist buckling.

Space occupied by steel columns

Steel columns are space-saving, especially those designed as box-section members with equal buckling resistance in all directions. Solid sections have the smallest dimensions.
1 This diagram compares the external dimensions of concrete and steel columns, the latter with solid and with box-sections, for loads of 100 t and 1000 t, for an effective length of 3.50 m. The external dimensions of the steel columns include a 25 mm thick fire encasement. In any particular column extending through the height of a building, the load to be transmitted progressively increases from top to bottom, and so does the requisite cross-section of the column. It is often desirable to give the column a constant external shape and size in all the storeys, in order to facilitate the installation of standardised internal fittings such as column casings, partitions and ceilings. With box or tubular sections this requirement can be fulfilled by appropriately varying the wall thickness of the sections and by using different grades of steel. The use of solid sections in the bottom storeys provides the least possible external dimensions.

Varying the column section in increments

The continental IPB sections are very suitable for columns, increments in section to suit differences in load being obtained by using the light, standard or reinforced ranges of sections and also by using steel of grade St 37 or St 52. Since the sections in the reinforced range have somewhat larger external dimensions than the others, it is often advantageous to combine the reinforced range of the next smaller size of section with the light and the standard ranges of the next larger size. In the bottom storeys the columns can, if necessary, be additionally strengthened, without increasing their external dimensions, by means of welded plates between the flanges.
2 Example of the possible variation of a column section.

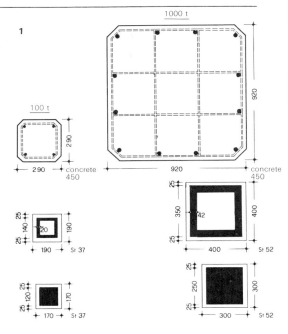

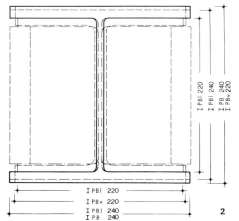

Internal column

Load t	Column section	Column load t	Floor load t
			10
		10	
60	IPBl 220 St 37		30
		40	
			30
		70	
116	IPB 240 St 37		30
		100	
			30
		130	
174	IPB 240 St 52		30
		160	
			30
		190	
219	IPBv 220 St 52		30
		220	
			30
		250	
290	IPBv 220 (box section) St 52		30
		280	

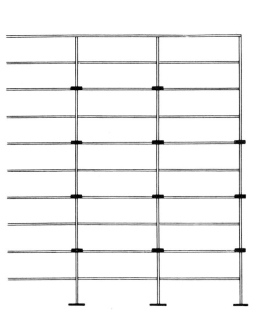

External column

Floor load t	Column load t	Column section	Load t
5			
	5		
15		IPBl 220 St 37	60
	20		
15			
	35		
15			
	50		
15			
	65		
15		IPBl 240 St 37	84
	80		
15			
	95		
15		IPB 240 St 37	116
	110		
15			
	125		
15		IPBv 220 St 37	140
	140		

Types of section

Columns

I-sections

This is the most frequently used shape for column sections. It is very suitable for connections to beams in both directions, as all parts of the section are accessible for forming bolted joints.
1 IPE sections for light loads.
2 IPB sections are eminently suitable for columns.
3 Rolled sections can be strengthened by means of plates welded on the outside of the flanges, when they are called compound sections.
4 For very heavy loads it may be necessary to use built-up columns composed of plates welded together. If suitably large plate thicknesses are used, it is possible to cope with practically any load encountered in structural engineering.

Rectangular hollow sections

These are suitable for columns which have to carry heavy loads or have to resist two-way bending or have long effective lengths in combination with relatively small cross-sectional dimensions. Because of their smooth external faces such columns present an aesthetically acceptable appearance and may be left uncased if fire regulations allow this.
5 IPB section transformed into a box-section by means of plates welded across the flanges.
6 Welded rectangular box-section.
Over the height of a building it is possible to achieve substantial changes in the cross-sectional area of a column by varying the thickness of the plate. Minimum thickness about 8 mm. The edges can be welded in a number of different ways.
7 Solid square or rectangular steel column, giving minimum dimensions, high fire resistance even with little protection, low cost of fabrication, ability to be accommodated in a partition, optimum utilisation of floor space.
8 Two channel sections welded toe-to-toe. Suitable for secondary columns, as the cross-sectional area cannot conveniently be varied (welding of internal plates rather awkward).

Cruciform columns

9 Column composed of four angle sections, sometimes adopted for aesthetic effect. Very suitable for columns at intersections of partitions into which they must disappear.
10 Same general type as **9**, but strengthened by interposed flat steel bars.
11 Heavy column composed of two rolled I-sections or built up from steel plates. Especially suitable to resist bending moments in both directions.

Structural hollow sections

12 Rectangular or **13** square hollow sections with rounded corners present a pleasingly smooth exterior. Transmission of forces into such sections requires special arrangements. Cross-sectional areas can be varied by using hollow sections with constant external dimensions but varying wall thickness.
14 Circular hollow sections are structurally very favourable because the moment of inertia is the same in all directions.
15 The cross-sectional area is variable to suit load requirements by using tubes varying in wall thickness but with the same outside diameter. Transmission of forces into thin-walled tubular sections requires special arrangements. For equal weight of steel, tubes cost about one and a half times as much as rolled sections, so that, despite lower fabrication costs, they can be more expensive than built-up box sections (such as **6**).

Battened or laced columns

Such column sections are frequently used in industrial buildings. They are suitable also for other buildings in circumstances where beams have to be fitted between the two shafts of such columns or where services must be readily accessible in the columns. The external dimensions are larger than for sections **5** and **6**. At intervals, the component members of the column have to be interconnected by batten plates or lacing to stiffen them against buckling.
16 Column composed of two channels. **17** Heavy column composed of two universal sections. **18** Four angles assembled to form a light column. The section can be easily varied to suit different loads by the appropriate choice of angles.

Suspension members

Suspenders or hangers, being subjected only to tensile loading, need not be stiffened against buckling.
19 Round steel bar. Forces are transmitted into the bar by means of screw threads and the bar is spliced with threaded sleeves.
20 Steel flat. **21** Two channels welded back-to-back. **22** Locked coil wire rope. Forces are transmitted into it through special fitted sockets.

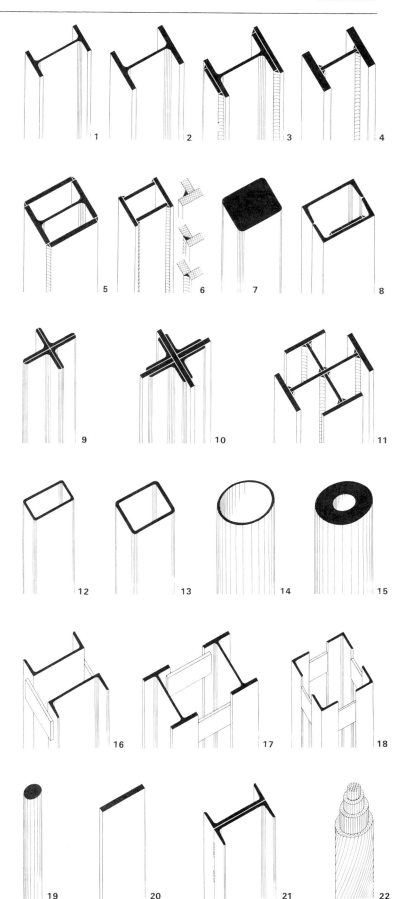

Columns

Design charts for steel columns

Loadbearing capacity of rolled steel columns

(Chart: p. 229)

The first approximation of the requisite section for a given load can be ascertained from this chart which is valid for a storey height (equal to the effective buckling length) h = 3.50 m. It can also be used for values of h differing a little from this average value. For h = 3.00 m, the loadbearing capacity increases by C% and for h = 4.00 m, it decreases by C% (C being the length coefficient). For storey heights significantly in excess of the latter value, the chart should not be used.

How to use the chart

Choice of section

Section A of the chart on the opposite page gives the loadbearing capacity of four series of rolled steel sections in two grades of steel. To choose a column section for a given load, first locate the value of the load on the vertical axis for P. Then proceed horizontally to the right until a suitable series and grade of steel is intercepted. The corresponding depth of section may be read off the horizontal axis for H vertically below, but when this axis is intercepted between two sections, the deeper section (to the right) must be selected. Farther below, at the foot of the chart, the appropriate value of the length coefficient C may also be ascertained, if required.

Weight of column

The weight of the column chosen is also ascertained by proceeding vertically downwards into section B of the chart until the curve for the appropriate series of sections is intercepted. The corresponding weight g in kg/m is read off the vertical axis horizontally to the left.

Economy coefficient

In section C of the chart, by proceeding horizontally from the column load P as far as the curve for the series chosen and then vertically downwards to the horizontal axis, it is possible to ascertain the economy coefficient W, which is expressed in terms of the ratio of the loadbearing capacity t of the selected section to its weight g in kg/m.

Examples

P, tonnes h = 3.50 m	St	Section	g kg/m	W t/kg/m	C %	P, tonnes, for h = 3.00 m	h = 4.00 m
300	52	IPBl 450	140*	2.2	4	312	288
	37	IPB 600	210	1.5	4	312	288
100	52	IPBv 160	80	1.4	14	114	86
	37	IPB 220	70	1.3	7	107	93

* rounded to nearest 10 kg/m

Loadbearing capacity of square and circular hollow sections

(Charts: pp. 230, 231)

With these charts, the column section can be quickly estimated and the appropriate grade of steel selected for given values of the column load. In each case, the right-hand section of the chart relates to the larger column sections with a breadth or diameter greater than 24 cm. The left-hand charts, drawn to a larger scale, are for small sections.

The loadbearing capacity is given for an average storey height (effective buckling length) h = 3.50 m. Small variations in storey height do not materially affect the loadbearing capacity. The values of C given at the bottom of the charts indicate the length coefficients, that is, the percentage increases or decreases in loadbearing capacity for columns down to 3.00 m in length or up to 4.00 m, respectively. The larger the column section, the smaller the difference in loadbearing capacity for different storey heights so that, for the stouter columns, the user has more latitude in applying the charts to storey heights other than those indicated. All columns are assumed to be of constant wall thickness. In the choice of section and grade of steel, the following general rules apply. For a given grade, moderate changes in the breadth or diameter of the section will result in only a small change in weight. The use of the higher grade St 52 steel, which is more expensive than the general purpose St 37 steel, will result in a substantial saving in steel only in the case of fairly massive sections. For more slender columns, the advantage gained in terms of reduced weight is lost because the slenderness is so much higher than for mild steel.

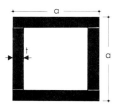

Examples of square hollow sections

P tonnes	St	a cm	t cm	g kg/m	C %
600	37	26.4	6.0	390	3
		33.0	4.0	380	3
	52	25.5	4.0	270	3
		29.5	3.0	240	3
300	37	18.8	5.0	210	7.5
		20.3	4.0	200	5
	52	17.6	4.0	160	7.5
		18.9	3.0	150	7.5

Examples of circular hollow sections

P tonnes	St	d cm	t cm	g kg/m	C %
300	37	39.0	6.0	490	2
	52	27.5	7.5	370	3
200	37	18.5	4.5	150	10
		21.8	3.0	130	5
	52	18.0	3.0	110	10

Loadbearing capacity of rolled steel sections — Columns

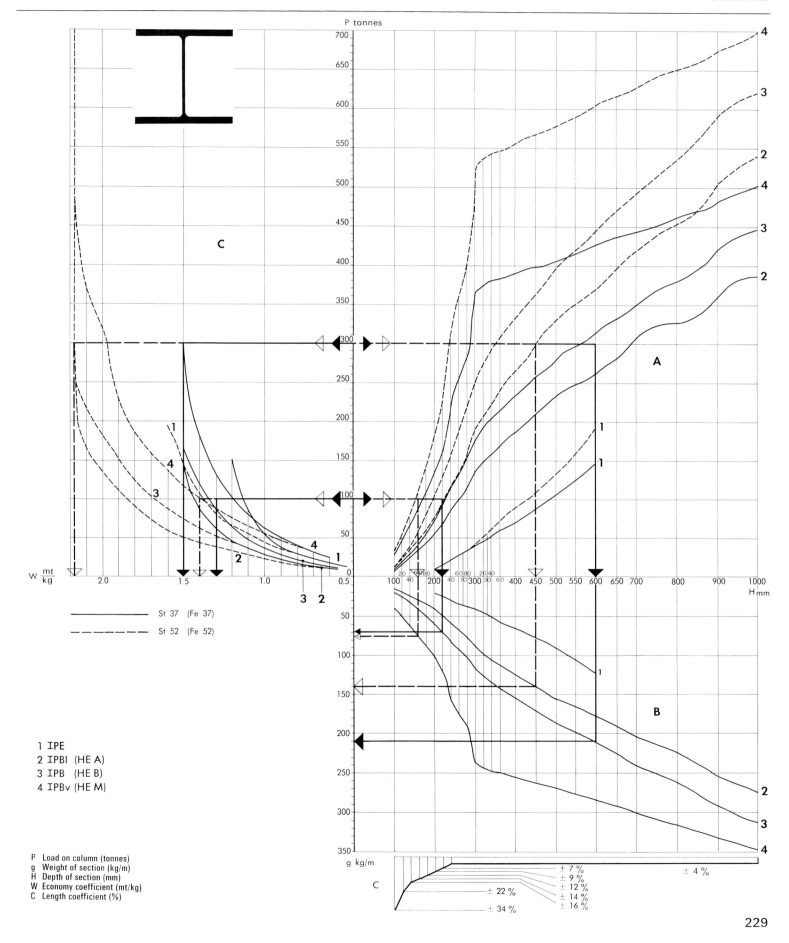

— St 37 (Fe 37)
--- St 52 (Fe 52)

1 IPE
2 IPBl (HE A)
3 IPB (HE B)
4 IPBv (HE M)

P Load on column (tonnes)
g Weight of section (kg/m)
H Depth of section (mm)
W Economy coefficient (mt/kg)
C Length coefficient (%)

Columns
Loadbearing capacity of square hollow sections

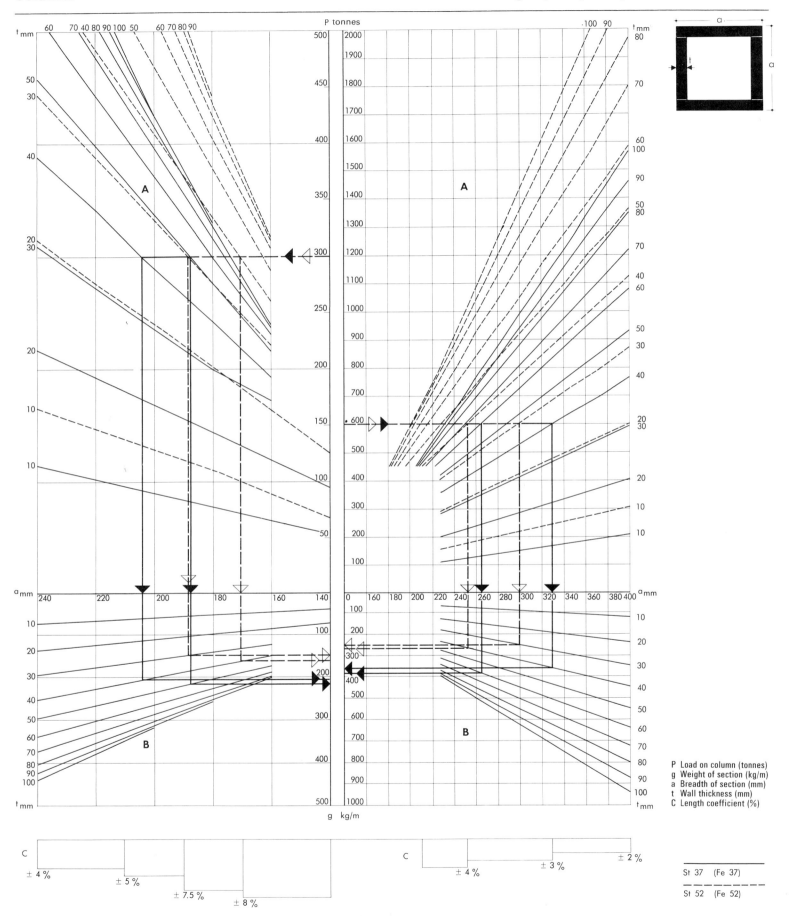

P Load on column (tonnes)
g Weight of section (kg/m)
a Breadth of section (mm)
t Wall thickness (mm)
C Length coefficient (%)

St 37 (Fe 37)
--- --- ---
St 52 (Fe 52)

Loadbearing capacity of circular hollow sections — Columns

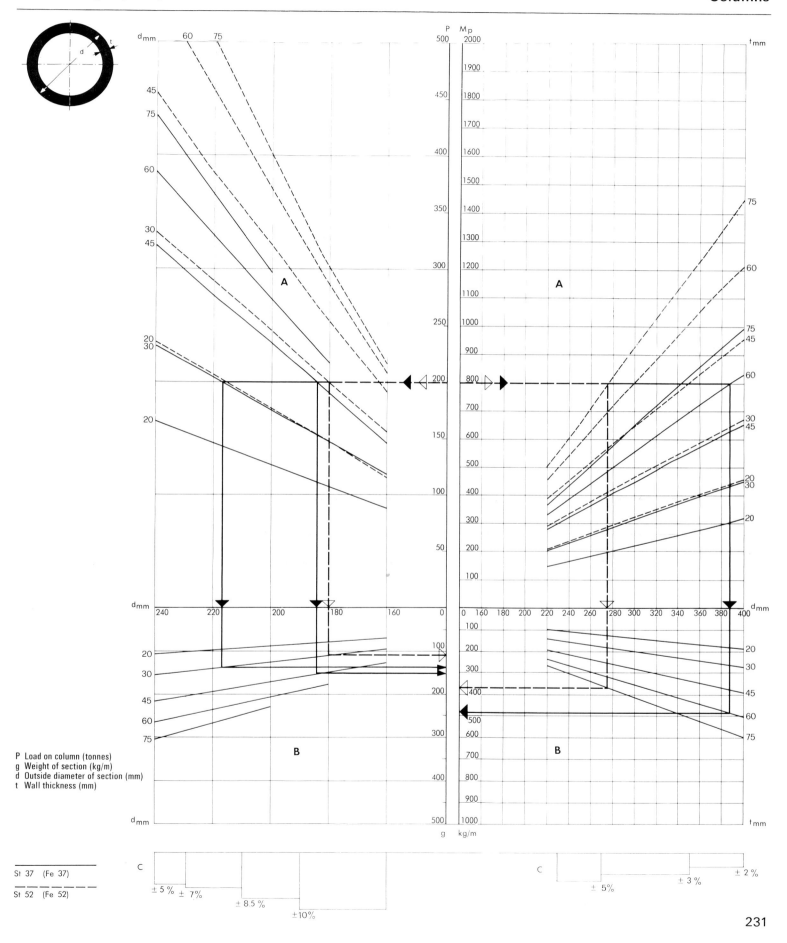

P Load on column (tonnes)
g Weight of section (kg/m)
d Outside diameter of section (mm)
t Wall thickness (mm)

St 37 (Fe 37) ———
St 52 (Fe 52) – – –

Columns — Base plates

Since the allowable stresses in structural steels are very much higher than those in concrete, the forces concentrated in the foot of a steel column must be distributed through a steel base plate to the concrete foundations to ensure that the allowable stress in the concrete is not exceeded.

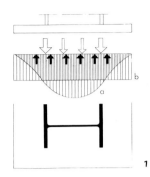
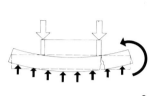
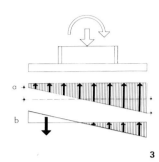

1 The actual pressure a on the concrete is least at the edges of the base plate and increases to a maximum directly under the column. For the purpose of structural calculation, however, a uniformly distributed pressure b on the concrete is generally assumed.

2 Bending stresses develop in the base plate, which must therefore be made sufficiently thick to resist these stresses alone or be provided with stiffening ribs.

3 This column, which is fixed at its base, is subjected to a bending moment as well as a vertical load.
(a) Compared with the vertical load, the bending moment is relatively small.
(b) In this case, the bending moment is sufficiently large to develop tension along one edge of the base plate, necessitating the use of holding-down bolts.

4 To allow for differences in the dimensional tolerances between steel and concrete construction, the top surface of the foundation is kept 3 to 5 cm below the underside of the column base plate. During erection, the latter is temporarily supported on steel packs and wedges, permitting adjustment of the column before grouting.
5 When adjustment has been completed, the space under the plate should be grouted or packed with high-strength concrete. When the grout or concrete has hardened, the steel packs and wedges are removed.

6 A thick base plate requires less fabrication than a thinner with stiffeners (as in **8**) and is, despite the larger quantity of material, generally the more economical solution. Also, it has a lower construction depth.
7 Tubular columns are also supported on thick base plates without stiffening ribs. The latter are necessary, in a radial arrangement, when a thin plate is used.

8 Simple column base comprising a thin base plate and stiffening ribs, the latter being welded to the flanges and as extensions to the web of the universal column. This arrangement is somewhat lighter than that in **6**, but has a greater construction depth and requires more fabrication.
9 Heavier base, but with thin plate, under a box-section column. The stiffening ribs form extensions to the walls of the column.

10 and **11** Horizontal forces at a column base can in many cases be transmitted through friction. Large horizontal forces may, however, require special arrangements, more particularly if the vertical load in the column is relatively small or if uplift due, for example, to wind loads is liable to develop. In such cases the plate must be provided with some form of projecting feature on the underside to develop dowel action, such as a welded-on steel block (**10**) or a short piece of rolled steel section (**11**). These projections engage with, and are grouted into, corresponding recesses in the foundation.

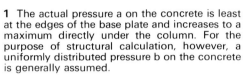
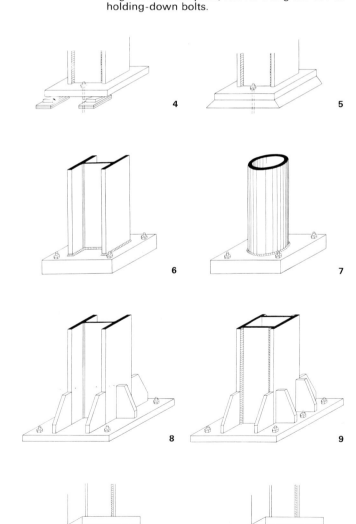

Anchorage of column bases

Columns

Simple columns which in theory transmit no bending moments to the foundation are generally secured at the base by means of light holding-down bolts which serve to locate the columns during erection. If the columns transmit bending moments or tensile forces to the foundation, provision must be made for them. The holding-down bolts are installed in preformed recesses in the concrete. Large forces require carefully and accurately constructed bases. If the holding-down bolts are to be concreted into the foundation in advance, they must be carefully set and held in position during concreting by means of templates, as the tolerance between the bolts and the holes in the base plates is seldom more than 1 or 2 mm.

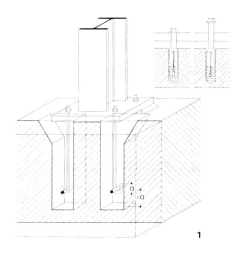
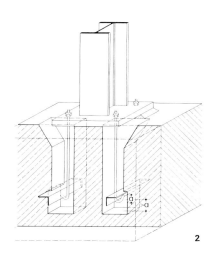
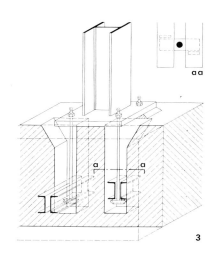

1 Simplest form of construction, used for small tensile forces. The holding-down bolts are hooked around reinforcing bars. During erection, the column can be temporarily secured by means of expansion bolts in holes drilled in the concrete.

2 To resist somewhat larger forces it may be necessary to install angle sections for anchoring the holding-down bolts. The recesses should be deep enough to allow the bolts to be hooked around the angles.

3 Hammer-head bolts can be used for transmitting even larger forces, in which case the anchor bars are usually composed of pairs of channel sections, these being provided with stops welded to the underside to prevent rotation of the bolts. To facilitate the insertion of the bolts and to check their position, rings are welded on to their upper end.

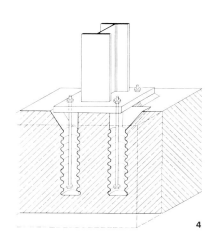
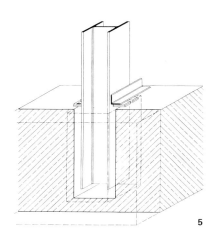
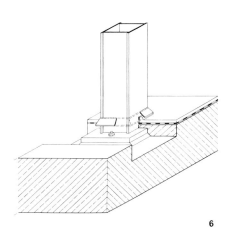

4 Anchor bars can be dispensed with if the recesses are in the form of sockets with corrugated walls formed with ribbed metal sheathing of the type used for the ducts of prestressing tendons. The holding-down bolts are provided with suitably enlarged heads and are grouted into these sockets.

5 Columns which have to transmit relatively small loads may be installed directly in to recesses (so-called pocket bases), no base plate being provided. For adjustment during erection such columns should have welded-on bars, such as angle sections, which can subsequently be removed, if desired.

6 Columns which have to be secured to footings in the open air require particular attention in the detailing of the base. Maintenance of the base is simplest if it remains entirely accessible above ground level. If the bottom end of the column itself has to be embedded in concrete, precautions must be taken to prevent any water trickling down the column from penetrating into the concrete at its contact face with the steel. Despite good anti-corrosive treatment, rusting is liable to start here. An effective barrier — such as a sealing compound which preserves its elasticity even in the long term — is therefore essential. If there is a waterproofing membrane below floor level, the column can be encircled by a welded steel collar, the membrane under this collar having its edge raised a little.

Columns

Basement columns

1 to 4 The foundations, external walls and floors of basements and cellars are usually constructed of concrete. Hence, it appears only logical to construct the columns in basements also of concrete and to begin the steel columns only above ground level, **1**. It must be borne in mind, however, that the column bases, including the holding-down bolts and their recesses, generally have a construction depth of 100 to 140 mm. Thus, if the steel columns stand on the ground floor, a thickness of 120 to 160 mm is needed to accommodate the column base structure, **2**.

In general, the base plates of the steel columns cannot be embedded in the structural concrete floor because here the main reinforcing bars of the two intersecting concrete beams are located at the top, **4**. Hence, it is advisable to continue the steel columns down through the basement to the foundations, **3**, and to construct the ground floor of the building with a steel floor structure, like the upper floors of the building.

5 and 6 If it is desired to continue the steel columns down through the basement to the foundations but, nevertheless, to construct the ground floor of reinforced concrete, the steel columns can be provided with steel shoes as bearings for the concrete beams. If the concrete is rigidly connected to the column, bending moments will be induced in the latter, which must be taken into account in its design. Alternatively, resilient bearing pads may be inserted to ensure that the concrete beams may undergo a certain amount of angular rotation without transmitting any appreciable bending moments to the column.

7 and 8 The external columns of a structural steel frame may extend down through the basement to the foundations or be supported on the basement walls. In the latter case, if the walls are narrow and the forces in the columns are large, the base plates have to be of elongated rectangular shape with triangular stiffeners. Corresponding pockets are left in the basement walls.

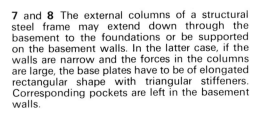
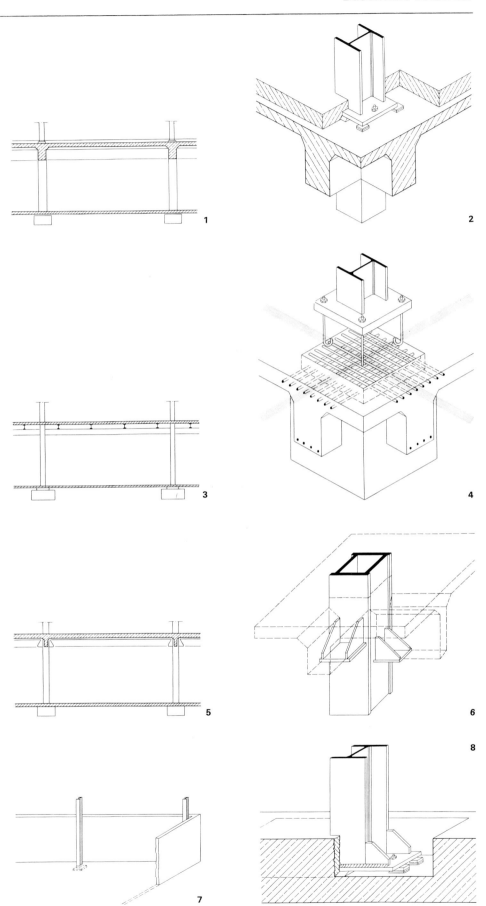

Column splices • Beam-to-column connections

To simplify erection, columns are delivered to the site in the greatest lengths possible, even up to about 20 m. Lengths consisting of a single rolled section are generally limited to 15 m, for reasons of cost, as the steel rolling mills make a surcharge for supplying greater single lengths. Very heavy columns often have to be delivered to the site in shorter lengths to keep their weight within the lifting capacity of the available cranes or derricks.

For various reasons, for example changes in column section, splices may have to be made within the units despatched from the fabricating shops. These are shop splices, as distinct from the erection splices formed between the units on the site.
To ensure efficient fit at the splices, the ends of the lengths of column must be finished square to the centre-line of the column. For lightly loaded columns a saw-cut end face suffices, but for heavier ones these faces must be machined. Under some codes of practice, part of the load at the splice can be transmitted by direct contact, the remainder being transmitted through welded or bolted connections. In any case, any tensile forces due to flexural loading conditions must be fully transmitted by such connections.

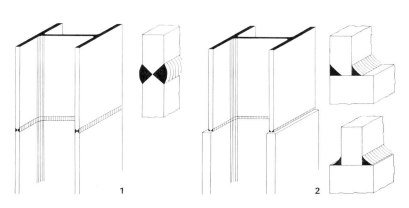
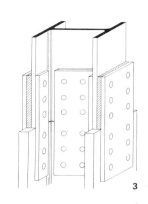
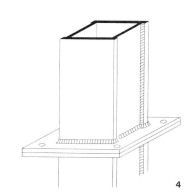

1 and **2** The simplest type of splice is the butt joint, **1**, which can also be used at changes in section, **2**. In the case of erection joints, temporary locating devices (lugs, butt straps, etc.) are needed until the welds are formed.
3 Bolted butt splice. The forces are transmitted through flange plates and web plates. If changes in section occur at such splices, packs (shown shaded) must be inserted; thin packs may also be needed to compensate for dimensional variations between nominally identical rolled steel sections. This type of splice dispenses with welding in the fabricating shops and on site, but is not always practicable because of the local increase in section.
4 One of the commonest types of column splice uses end plates. The cap and base plates welded to the upper and lower ends of the column must fit snugly together. Warping of such plates may occur as a result of welding stresses, in which case their faces may have to be machined by planing. Sometimes such splices are formed also in the fabricating shops in cases where large differences of section occur between the two parts of the column: only one welded butt plate is then used and, of course, no bolts.
5 Frequently, it is advantageous to let the beam continue at a column splice. The upper and lower parts of the column are provided with end plates similar to those in the preceding example. Stiffeners are welded between the flanges of the beam, so that the forces from all parts of the column section are effectively transmitted through the beam. These stiffeners must fit closely to the flanges and should therefore be accurately fitted, more particularly in rolled steel beams in which dimensional variations due to rolling tolerances occur.

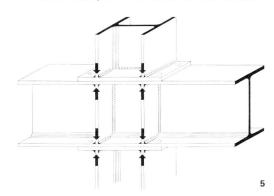

BEAM-TO-COLUMN CONNECTIONS

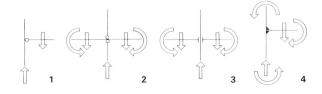

At these connections the vertical loads are transmitted from the beams to the columns. Such connections must:

- provide a positive connection during erection,
- be adjustable,
- be capable of execution in a simple manner without staging, if at all possible. The labour needed for fabrication and erection, and therefore the economy of the structure, is affected by the detailing of the beam-to-column connections.

1 Shear connection. The shear forces in the beam are transmitted as normal forces to the column. Such a connection, even if it is fully bolted, is conventionally treated as a pinned-joint, since the bolts develop a certain amount of flexibility.
2 Shear and moment connection, but without transmission of bending moment into the column. This connection ensures structural continuity of the beam, while the column is pin-jointed above and below the beam, so that no moment is transmitted.
3 Quite often, columns designed as being pin-jointed at top and bottom in each storey are so flexible in comparison with the beams that, despite rigid beam-to-column connections, it can safely be assumed that the columns receive no bending moment from the beams.
4 At corners of portal frames, and at external columns of rigid frames in general, shear forces and bending moments are transmitted into the columns. The connections should be appropriately designed.

Columns
Shear connections

1 Simple connection: cleat consisting of a single flat plate welded to the column. The bolts act in single shear and the connection is eccentric.

2 Connection by means of double angle cleats bolted to the column.

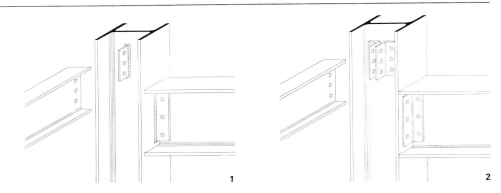

3 Connection by means of butt plates welded to the ends of the beams.

4 The connection comprises a pair of vertical angle cleats in conjunction with a seating cleat, also of angle section.

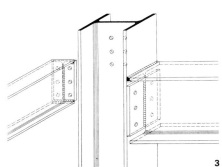
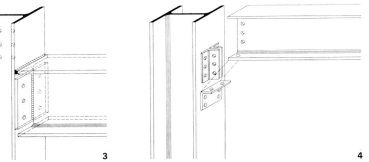

5 Beam connected to a box-section column by means of a single welded cleat (as in **1**) or with the aid of a butt plate welded to the beam, with additional support from a seating formed by a plate welded to the column. Stud bolts welded to the column engage with the holes in the butt plate of the beam and are secured with nuts. This connection cannot transmit bending moment, nor can any of the preceding ones. They are intended to transmit shear forces only.

6 Connection to a tubular column. Cleats may be welded to thick-walled round tubes, but thin-walled ones may have to be slotted to allow a connecting plate to pass through the column. Single shear or double shear joints can thus be formed, depending on whether one or two cleats are used.

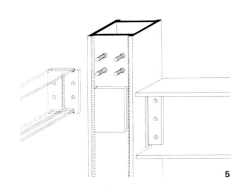
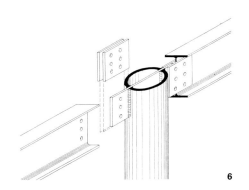

7 Beam is continuous over the column, the latter being provided with a capping plate to locate and secure the beam. Stiffeners are welded between the flanges of the beam; they need not be closely fitted to the top flanges, however.

8 Bearing for supporting a beam grillage on a cruciform column. The rocking ball bearing ensures a well-defined point of support. This particular detail is from the New National Gallery in Berlin (architect: Mies van der Rohe).

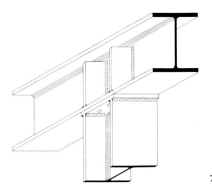
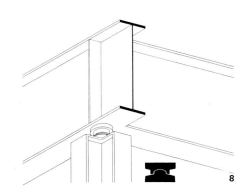

Shear and moment connections — Columns

1 The flexural tensile forces acting in the top flange of the beam are transmitted across the column by a continuous flange plate which also functions as the butt plate between the upper and lower parts of the column. The compression in the bottom flange of the beam is transmitted through packs and through fitted stiffeners welded inside the column (this type of connection is usually an erection joint).

2 Same principle as in **1**, but applied to a tubular column. In this case, transmission of the compressive forces in the bottom flange of the beam necessitates sawing through the column and inserting a compression plate.

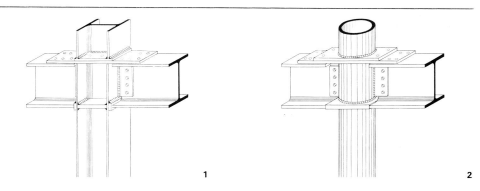

3 Transmission of bending moments from beam to column is effected through butt plates welded to the beams. These plates are connected to the column by means of high-strength friction-grip bolts. Compressive forces are transmitted by contact.

4 Welded connection. Flat cleats welded to the column are bolted to the beam during erection; later, welds are added. The top and the bottom flanges of the beam are fully welded to the column; during welding, a backing plate secured under the flange serves to prevent weld metal trickling through. In general, this form of construction is possible only with thick-walled column sections. It is extensively used in American high-rise building construction.

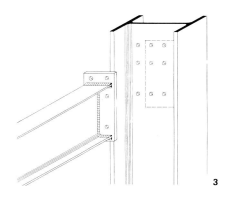
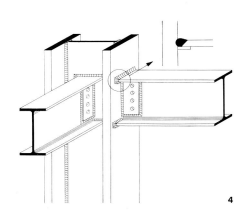

5 Shear force in the beam is transmitted by cleats welded to the column, which itself is continuous. The forces in the beam flanges are transmitted through side plates connected by double fillet welds.

6 The shear force in the beam is transmitted by cleats welded to the column, which itself is continuous. The forces in the beam flanges are transmitted through gussets bolted to the flanges. With this solution and the preceding one, no bending moment is transmitted to the column. A check should be made, however, that side plates or gussets do not obstruct vertical services installed beside the column.

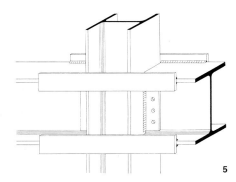
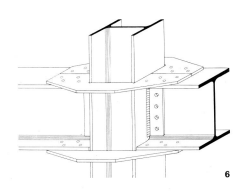

7 Same principle as in **6**, here applied to two intersecting beams. The gussets are flush with the beam flanges and are joined to them by butt welds. Shear forces are transmitted by ribs welded to the column under the gussets. No bending moment is transmitted to the column. (Solution adopted for the Jelmoli building at Zürich-Oerlikon.)

8 Square solid-section column with welded I-section brackets carrying a double I-section main beam which in turn supports secondary beams. Vertical services pass between the two main beam components. (Offenbach building system, Stahlbau Lavis.)

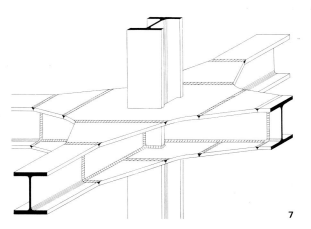
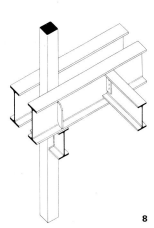

Columns

Fire protection of columns by encasement

Fire protection of internal columns

Encasement in concrete

1 to **3** The direct encasement of steel columns in concrete provides weatherproof and impact-resistant protection. Thin binders or wire fabric serve to secure the concrete casing and prevent spalling on exposure to fire. For a fire rating of 90 minutes, the concrete casing must be at least 40 mm thick, and the concrete should be at least of class 160. The concrete can be placed in vertical formwork around the column before or after its erection, or in horizontal form before erection. If a rolled steel column is to be surrounded with a square or rectangular casing (**1**), it is possible to fill the spaces between the flanges with a lightweight material in order to save weight.

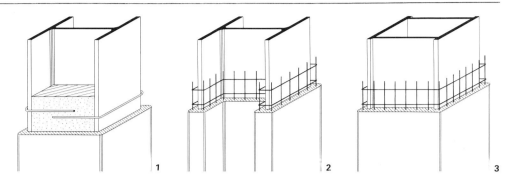

Plastering

4 Plaster applied to expanded metal lathing wrapped round the column, with light rods embedded in the casing, which is additionally provided with metal edge-protecting strips: various types:

35 mm pearlite plaster with cement or gypsum;
35 mm gypsum plaster;
35 mm vermiculite-cement plaster with 5 mm lime or lime-cement plaster finishing coat.

5 and **6** Sprayed asbestos or vermiculite casings with various binding agents are the most widely used and least expensive means of fire protection for steel members. Spraying is, however, a rather messy operation and should therefore be suitably timed to cause minimum disturbance. Sprayed casings have a rough surface and generally need additional treatment.

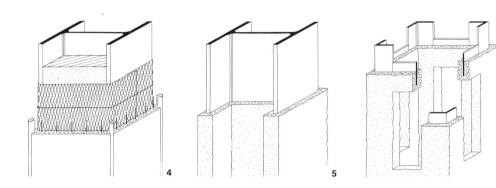

Slabs as cladding to columns

7 Vermiculite concrete slabs fixed with synthetic resin adhesive. At the joints between these slabs it is necessary to insert diaphragms of thicker material between the flanges of the column if the latter consists of an open section such as an I-section. A facing or cladding applied as a finish to the slabs is desirable; for this purpose glass-fibre fabric embedded in a suitable skimming coat and finished with a coat of paint will suffice to meet simple requirements.

8 Asbestos cement slabs are strong and can be worked like wood. They can be fixed by nailing or screwing. The horizontal joints in the slabs should be staggered by at least 50 cm when the height of the column exceeds the maximum length of the slabs.

9 Three-layer casing for fire rating F 90. There are three 15 mm thick layers of gypsum plasterboard, a, secured by steel bands, b, and a bonding agent, c.

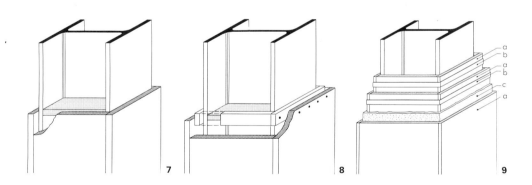

Jacketed columns

10 and **11** Steel column enclosed in mineral fibre mat with a sheet-steel jacket (0.75 mm thick). For fire rating F 90, the mat must be 50 mm thick for the tubular and 40 mm for the I-section column.

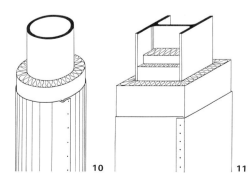

For general principles of fire protection, see the chapter 'Fire protection'. Further information on the protection of steelwork is given on p. 335.

Fire protection of external columns

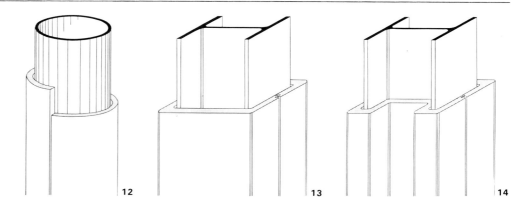

12 Asbestos cement jackets for round tubular columns.
13 and **14** Prefabricated shaped casings made of gypsum plaster or concrete with grouted or mortar joints. Rectangular casings, **13**, have the advantage of being independent of the actual cross-section of the column, so that the same casings can be used in all parts of a building, irrespective of the size of the columns they enclose. Smaller-section columns are padded out with fixing strips. Alternatively, the casings may be shaped to the column section, **14**. The prefabricated casings can be supplied as storey-height units.

Fire protection of external columns

External columns exposed outside the façades determine the architectural character of a building. For aesthetic reasons, and also to save expense, it is desirable to leave the columns uncased. An external column is threatened by fire only from the direction of the interior of the building, and if the column is actually outside the façade, the only danger will come from hot gases escaping through shattered windows.

1 If the columns are sufficiently far away from the building and/or if the windows on each side are sufficiently far from the columns, so that no flames can reach the latter, there is no technical need to encase the columns at all.

2 and **3** A 25 mm thick slab of a suitable fireproof material to serve as a flame barrier behind the column, and extending a sufficient distance on each side thereof, can prevent flames shooting out of a window from excessively heating the column. The recommended extra width of this barrier on each side of column, according to Bongard and Portmann in *Brandschutz im Stahlbau* (*Fire protection for structural steelwork*) is as follows:

Fire rating	IPB ≤ 260 IPE ≤ 220 □* ≤ 400 cm	≥ 280 ≤ 360 ≥ 240 ≤ 300 > 400 cm
F 30	7.0	9.0
F 60	10.0	12.0
F 90	12.0	15.0

* Box section with at least 8 mm wall thickness

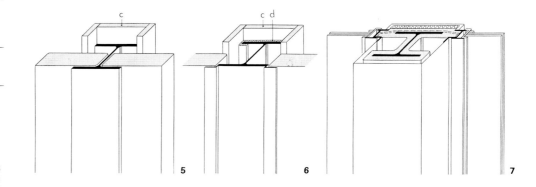

4 In this example the web of the column is protected by fireproof slabs (Vermitecta), a, in a sheet-steel jacket, b. The external flanges of the columns, which are an architectural feature of the building, are left exposed. (Institute of Veterinary Medicine, Free University of Berlin. Architects: Luckhardt, Wandelt.)

5 and **6** Column with exposed outer flange. Columns which are accommodated within the external wall and are adequately protected against fire on the inside, while only the outer flanges are exposed in the façade, are in little danger from fire, provided that there are no windows directly beside them: c internal fire protection; d thermal insulation.

7 The external column is totally encased and is provided with sheet-metal cladding on the outside, which is an architectural feature of the façade. (Institute for Mining and Metallurgical Engineering, Technological University of Berlin. Architect: Kreuer.)

Columns

Fire protection of columns by water cooling

Principles

The most effective method of preventing an excessive rise in temperature in a steel column exposed to fire is to design it as a hollow member and fill it with water. The column functions like a kettle of water on a stove in the sense that it will not become excessively hot so long as there is water in it; or it is comparable to the radiator of a water-cooled internal combustion engine.

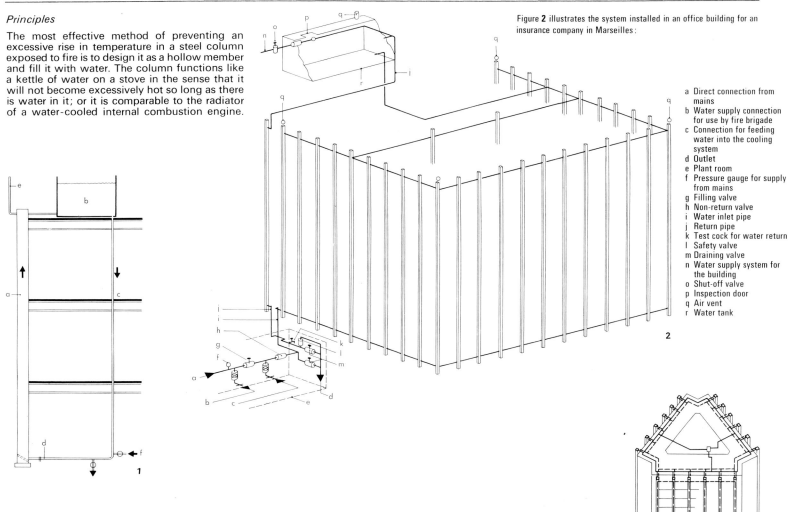

Figure 2 illustrates the system installed in an office building for an insurance company in Marseilles:

a Direct connection from mains
b Water supply connection for use by fire brigade
c Connection for feeding water into the cooling system
d Outlet
e Plant room
f Pressure gauge for supply from mains
g Filling valve
h Non-return valve
i Water inlet pipe
j Return pipe
k Test cock for water return
l Safety valve
m Draining valve
n Water supply system for the building
o Shut-off valve
p Inspection door
q Air vent
r Water tank

Equipment required

1 Water cooling of structural steel columns under fire conditions requires the following arrangements:

At the top the column, a, is connected to a supply tank, b, from which a pipe, c, also leads to the base, d, of the column.
In the event of fire, the water in the column is heated, so that circulation develops; the warm water rises and cold water flows down from the tank to replace it. The pipes through which the water circulates must be of sufficient capacity to carry away the heat generated. Steam escapes through the vent, e. There is a connection from the mains at f to keep the system suitably topped up at all times.

2 The diagram illustrates the application of the system to a building with water-cooled external columns, the latter being linked to one another at top and bottom. If only one column is heated, at first a movement of water develops between this column and its neighbours, that is, water rises in the hot column and descends in the cooler ones, so that this cooler water is fed to the base of the hot column. If the fire spreads, the circulation becomes more general, now also involving the water in the roof tank from which cooler fresh water flows into the system. In the event of a fire of long duration, additional water can be fed in from the mains or from a fire tender. Large buildings may be split up into several circulation systems. A tall building must be subdivided into independent water cooling systems, one above the other, to avoid high pressure at the bases of the columns.

3 In the United States Steel administrative building at Pittsburgh, USA, groups of about 16 storeys, each 4 m in height, are each combined into one cooling system. The water column in each system is thus about 64 m high, with a maximum hydrostatic pressure of 6.4 kg/cm^2. The columns in such buildings consist of circular or rectangular hollow sections or welded box-sections. The joints in the columns are welded and their capping and base plates are provided with suitably dimensioned holes. Each column can, if necessary, be completely drained from an outlet at the base. From the constructional steelwork point of view there are no special difficulties involved.

To prevent corrosion inside the columns, special additives are put into the water. In a similar manner, antifreeze agents are included to prevent freezing in winter. As only a small number of such water cooling projects have been actually carried out, it is not yet possible to give general information on the cost of this method of structural fire protection.

Services in or beside steel columns

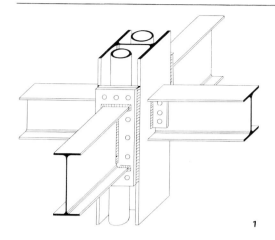
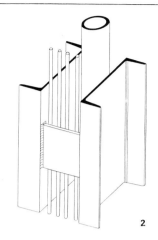
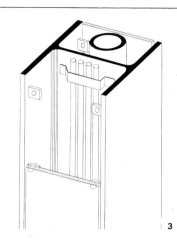

1 When thin-walled steel sections are used as columns, their recesses or cavities can be utilised to accommodate vertical services, bearing in mind of course that these must also continue through or past splices in the columns and, in the case of waste water discharge pipes, must also pass through the base plates. At points where beams are connected to the column on all four sides this necessitates special arrangements, **1**. In a tall building the columns are generally not of constant section over its entire height. In planning the services, it must therefore be borne in mind that in the top part of the building the space available to accommodate them is likely to be smaller and that in the bottom part it may be necessary to have welded columns instead of open I-sections.

2 Services installed between the two members forming a double column. This is a very favourable arrangement in circumstances where no fire protection is needed.

3 Services accommodated between the flanges of an I-section column in a school building. The cavities are closed by bolted timber panels. There is no fire protection. (School at Osterburken. Architects: Bassenge, Puhan-Schulz, Schreck.)

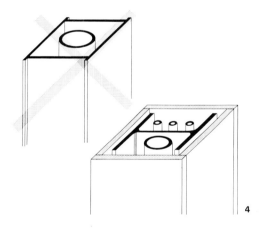
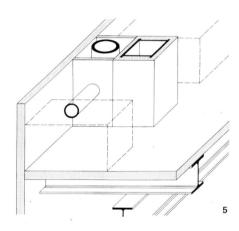
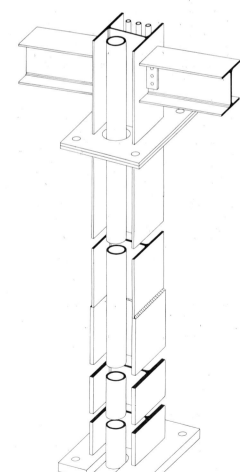

4 If the services are enclosed within the cladding for fire protection, parts of the latter should be detachable for access. It should be noted that leakages in water pipes may cause damage to the building.
The practice of installing pipelines in welded box-section columns is to be deprecated. To ensure reliable internal protection from corrosion, such columns should be completely sealed by welding. Ingress of moisture may cause undetected corrosion. Besides, in buildings with two or more storeys, the installation of such pipes inside columns is impracticable. This method is permissible only in exceptional cases in single-storey structures, such as canopies, with exposed structural steelwork.

5 In this example, the outermost columns are in fact set some distance inside the façade, leaving a space in which vertical high-pressure air-conditioning ducts are installed, with lateral connections to air-conditioning units on each side of the column. The external cladding is attached to brackets.

6 Rainwater downpipe accommodated between the flanges of an I-section column passes through top and base plates.

Beams and girders

Structural function

Function

Beams and girders are horizontal or, more particularly in roofs, slightly inclined structural members which carry vertical loads by flexural action and transmit them to their supports. The loads induce bending moments and shear forces in the members which, in turn, result in bending and shear stresses.

Shapes and types

1 A distinction can be drawn between solid-web beams or girders and open-web beams or lattice girders. A solid-web beam or girder, which may be either a rolled steel section or a plate girder, consists of two flanges and the web. A lattice girder comprises chords (or booms) interconnected by vertical and/or diagonal web members, which are either tension members or struts. Furthermore, a beam or girder can be designed to act in composite construction with the concrete slab which it supports.

Composition of beams and girders

Because of their slender cross-sectional shapes, steel beams are very economical, all parts of the section being effectively employed. The flanges or chords which resist the normal forces (bending, tensile and compressive forces) are concentrated at the extreme fibres of the section, where they are most effectively employed. In a simply supported composite beam, the concrete slab performs the function of a compression flange, while tension is largely resisted by the bottom flange of the steel component. Transmission of the vertical forces in a solid-web beam is effected through shear in the web, whereas in a lattice girder this transmission is effected through tensile and compressive forces in the web members.

In multi-storey buildings the flanges or chords of beams or girders are usually parallel, but other shapes are possible.

Up to a certain point, the greater the depth of a beam, the more economical it becomes, since the requisite cross-sectional dimensions of the flanges decrease. Above that limit, however, the additional cost of the web, or of the web members of a lattice girder, more than offset the saving on the flanges, or the chords.

In the case of composite beams it is necessary to pay due attention to the conditions of composite action of the two materials, concrete and steel, differing in strength and in elastic and plastic behaviour, as well as to the design of the shear connectors and the technique of constructing such beams.

Deflection

In the design of beams, it is necessary not only to remain within the allowable stresses, but also within the permissible deflections (p. 176). Deflections under dead load can be eliminated by giving the steel components an initial camber. Those due to imposed load should be compatible with the purpose for which the beams in question are used. In practice, therefore, the load-carrying capacity of a beam is often limited by considerations of deflection rather than strength. Such beams are therefore less economical than beams in which the strength of the material is fully utilised.

Limiting the deflection

The deflection of a beam is conventionally expressed as a certain fraction of its span 1, 2. If the designer wishes to specify a maximum permissible deflection, he must state whether this relates to the total load or to live load only. Usually, deflections are limited to some value between l/200 and l/800.

Frequently, the deflection under live load is limited to l/500 and the deflection under total load is limited to 1/300. A beam in which the deflection does not exceed l/800 is very stiff, which necessitates a correspondingly large depth and high cost, justified only in special circumstances. In laying down a limit for the deflection is should be ascertained what proportion of the live load actually varies. The magnitude of the deflection is particularly important in relation to the design of the partitions in a building. The absolute magnitude is of less significance than the angle of inclination ζ of the floor, **3**, which is greatest at the beam supports and thus induces the largest shear deformations δ in the partition. See also p. 319 for further details.

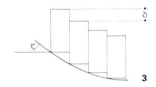

From the point of view of the occupants of the building the vibration behaviour of floors is also important.

It should be remembered that, especially in the case of heavy imposed loads, a large part is usually of a more or less permanent character, for example, bookshelves in libraries, while only a very small proportion, which could be more appropriately called live loads, can cause significant variations in deflection or give rise to objectionable vibrations. The large values specified for the live loads in buildings such as assembly halls are intended to cope with exceptional crowd conditions or emergency situations, in which the absolute magnitude of the structural deflection is not a significant consideration. The permissible deflection should therefore be decided after due deliberation, to avoid adding unnecessarily to the cost of the structure.

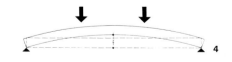

Camber

If a camber is specified, **4**, it should also be stated under what loading and at what stage the beam should be straight. Generally, this requirement has to be fulfilled when all the dead load has been applied to the structure. In many cases, however, such as buildings in which a large proportion of the imposed load is virtually of a permanent character (as in libraries), it is more appropriate to specify that the beam will become straight only after a certain proportion (for example, one-third) of the imposed load has been applied. With composite beams, the deformations due to the creep and shrinkage of the concrete must also be considered. As a rule, it is required that the beam should attain its desired final shape after these deformations have ceased.

Stiffening of beams

The top flanges of slender beams must be stiffened to prevent lateral buckling. The concrete slab over floor beams provides such stiffening. Alternatively, both floor and roof beams can be stiffened by steel-troughing units if these form rigid horizontal diaphragms which are adequately connected to the beams. In tubular structures the triple-chord girder is commonly employed: its tension chord consists of a single tube, while the compression chord comprises two tubes interconnected by bracing members. At bearings or at points of application of heavy concentrated loads, the girder should be restrained against lateral instability.

Simply-supported and continuous beams

In general, a beam which is continuous over several spans is more economical than a single-span beam, **5**, because in the continuous beam the bending moments and deflections are smaller. If the spans of a continuous beam are all

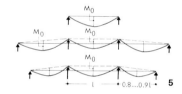

equal, the bending moments in the end spans will be larger than in the inner spans. It is therefore more economical to make the end spans shorter: about 80 to 90% of the inner spans.

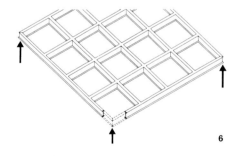

Beam grillage

A grillage or grid system, **6**, is obtained where two sets of beams are mutually perpendicular and the beams are rigidly interconnected and structurally continuous at their intersections. Thus, each of the two sets carries a certain proportion of the load, depending on the respective spans and relative stiffnesses of these sets. This type of construction can be adopted for long-span floors where the shape on plan is square or nearly square. It has the advantage of low construction depth.

Solid-web beams and girders

Beams and girders

The solid-web beam or girder is the type most widely employed in multi-storey buildings. A distinction must be drawn between rolled steel sections and welded plate girders. The rolled section is the more economical type and is generally preferable to the built-up section even if the weight involved is greater, as the latter section involves the cost of fabrication. Plate girders are used for long spans and/or heavy loads. Automatic welding equipment helps to reduce the cost of fabricating such girders.

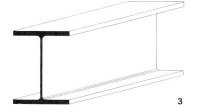
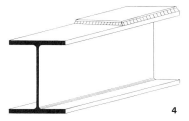

1 The slender rolled steel beam with tapered flanges has been largely superseded by sections of the IPE range, **2**, which have parallel flanges. Therefore, these sections are now used mainly for small beams and more particularly for purlins.

2 The IPE section, ranging in depth from 80 to 600 mm, is now the section most widely employed in multi-storey buildings. Because of the thin web, such beams are very economical, while the parallel flanges facilitate the structural connections. The supplementary IPEo and IPEv sections ensure a continuous range in the series.

3 The three ranges of universal sections (IPBl, IPB and IPBv), with depths from 100 to 1000 mm, are the ideal rolled steel members for supporting heavy loads. Because of their broad flanges they possess considerable stiffness in the transverse direction. They are suitable for use as both main and floor beams.

4 Where the construction depth available is restricted, use may be made of a compound beam, that is, a rolled steel section to which additional flange plates have been welded in the zones of maximum bending moment.

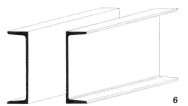
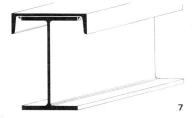
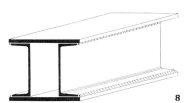

5 Channel sections can also suitably be used as beams, more particularly as edge beams.

6 Channel sections, employed in pairs, can serve as main beams under floors.

7 Rolled steel beam with channel section welded to its top flange to increase transverse stiffness. This type of compound section is often used for crane girders or for northlight valley girders.

8 Heavy welded beam, of low depth, comprising two channels and two flange plates. Suitable for carrying heavy loads in conditions of limited construction depth.

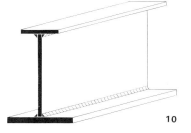
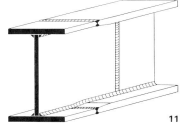
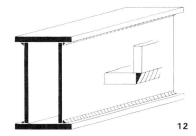

9 If standard rolled steel sections cannot meet the design requirements within the available construction depth, plate girders are used. Optimum choice of plate thicknesses ensures the most economical use of material.

10 Asymmetrical welded plate girder, designed to act compositely with a reinforced concrete slab.

11 For very long welded plate girders it may be economical to vary the thickness of the flanges and even of the webs along the span. This is quite normal practice in bridge construction.

12 Welded box-sections may be used as intercepting girders under tall buildings or as cantilevers, including more particularly those at the top of suspended buildings. If desired, the exterior of such girders can be finished smooth, without projecting flanges.

Beams and girders

Types of solid-web girder

Beams fabricated from rolled sections

13 Rafter sections may be produced by cutting a V-shaped notch extending up to the top flange of a rolled steel member and by welding the edges of the notch together, with the insertion of stiffeners, if required. Alternatively, the beam may be cut right through and the two parts welded together with the insertion of a ridge plate.

14 and **15** A rolled steel member may be slit obliquely in the longitudinal direction and the halves suitably rearranged and welded together to form girders of varying depth.

16 A haunch at the end of a beam can be formed by the insertion of a triangular plate welded into the web.

Types of welded girder

17 Similarly, a reduction in the depth at the end of a beam can be obtained by cutting away a triangular portion of the web. Or a built-up girder with a suitably shaped web plate can be used; in this way it is possible, in conjunction with welding, to produce girders of any desired size and proportions.

18 Another example of a plate girder with reduced depth at the ends.

19 Deep plate girders must be adequately protected against web buckling by means of welded stiffeners. Longitudinal, as well as transverse stiffeners, may be required. The web must also be stiffened over the bearings.

20 Example of a heavy cranked intercepting girder. The web plate is stiffened below the heavy point loads. Similarly, the flange plates are provided with ribs at changes of direction to absorb radial forces.

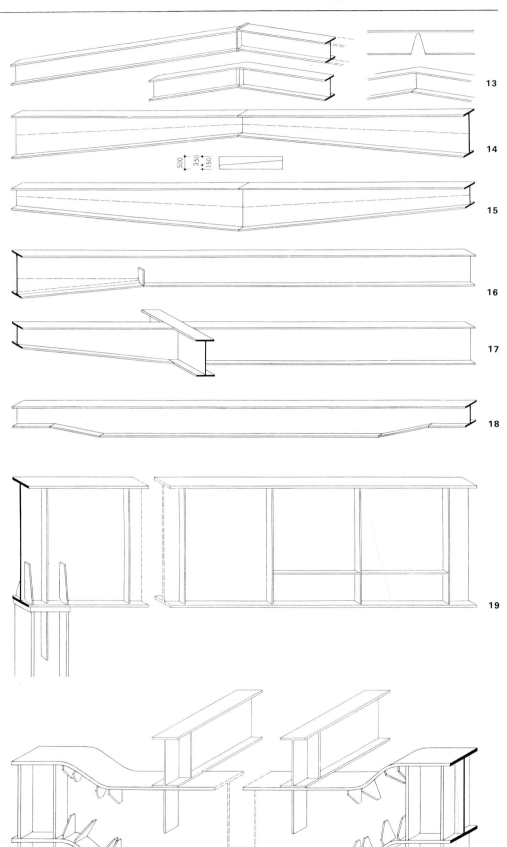

244

Design charts for rolled steel beams

Beams and girders

The charts on pp. 246 and 247 are intended for the preliminary design of beams consisting of rolled steel sections: p. 246 comprises shallow sections and p. 247 the deeper sections.

Loads

The uniformly distributed load q (t/m) is derived from the following loads per unit area:

dead load g (t/m²)
imposed load p (t/m²)
the breadth of application
of the load b (m)
Therefore, $q = b(g+p)$ (t/m)

Section A – bending moments

When a uniformly distributed load q is applied to a simply supported beam of span l, the bending moment

$$M = \frac{ql^2}{8} \ (t \cdot m)$$

The chart is entered by proceeding vertically upwards from the span l until the curve for a given load q is reached. The bending moment M is read off the ordinate to the right.

Section B – beam sections

The curves indicate the maximum bending moments which may be applied to the various sections without exceeding the following allowable stresses for the grade of steel quoted:

Steel St 37 $\sigma = 1.6$ t/cm²
Steel St 52 $\sigma = 2.4$ t/cm²

Deflection is not considered at this stage. To use section B for determining the required beam section, proceed to the right from the load curve in section A (ignoring the moment M) until the curve for the appropriate section is reached in section B. The depth of the section concerned is read off the horizontal scale vertically below the relevant point on the curve. If the value is intermediate between two sections, the deeper must be chosen.

Section C – weight of beam

Proceed vertically downwards from the section depth found on the horizontal scale until the appropriate range of sections is reached. The weight per metre is then read off the left-hand scale.

Section D – deflection

The deflection of a simply supported beam of span l under uniformly distributed load depends on the depth of beam h, the extreme fibre stress and the modulus of elasticity E. Then the ratio of deflection to span:

$$\phi = \frac{f}{l} = \frac{5}{24} \cdot \frac{\sigma}{E} \cdot \frac{l}{h}$$

or, for $E = 21$ kg/mm²,

$$\phi \approx \sigma \frac{l}{h} 10^{-4}$$

The required depth of beam for the full use of the allowable stresses with steels of the following grades is:

St 37 $\quad h = 1.6 \frac{l}{\phi} 10^{-4}$

St 52 $\quad h = 2.4 \frac{l}{\phi} 10^{-4}$

The radial lines in section D have been derived from the ratios

$$\phi = \frac{f}{l} = 1/200, \ 1/300 \text{ and } 1/500.$$

In most cases, the limit imposed on the ratio ϕ relates to the imposed load p. Let α denote the proportion of imposed load p in the total load $g+p$, then

$$\alpha = \frac{p}{p+g} < 1$$

The imposed load therefore induces only the proportion $\alpha \cdot \sigma$ of the total stress σ and the proportion $\alpha \cdot f$ of the total deflection f in the beam. The dead load deflection can be counter-balanced by cambering the beam. Therefore, only the deflection $\alpha \cdot f$ induced by the imposed loads need be limited. For equal fibre stresses, the deflection is proportional to the depth of the beam. To satisfy the condition that $\alpha \cdot f$ is not greater than the allowable deflection f, the minimum depth of beam required is therefore:

$$h_{min} = \alpha \cdot h$$

where h is the required depth obtained from section D for full utilisation of the allowable stress in the extreme fibres of the section. To facilitate the use of the chart, scales for $\alpha = 0.4$, 0.6 and 0.8 are given beside the scale for the beam depths, so that h_{min} can be read directly from them.

The procedure for determining the minimum depth of a beam is as follows: From the appropriate span on the horizontal scale for l, proceed vertically downwards to the radial line for the allowable deflection and the appropriate grade of steel. The corresponding minimum depth h_{min} is then read from the left-hand scale for the appropriate value of α. The sections printed in bold type are those considered in the charts on p. 246, while the deeper sections are on the next page. The section selected from section B must, of course, be at least as deep as that determined from section D.

Examples, p. 246

g t/m	p t/m	g+p t/m	$\alpha = \frac{p}{g+p}$	b m	q=b(g+p) t/m	l m	M tm	φ	St	h_{min}	Section	kg/m
0.450	0.350	0.800	0.43	2.4	1.92	6.0	8.65	1/500	37	200	IPB 200	61
											IPBl 220	50
											IPE 300	42[1]
0.500	0.750	1.250	0.6	3.0	3.75	9.0	3.80	1/300	37	270	IPBv 280	189
											IPB 360	142[2]
									52	380	IPBl 340	105

[1] Section chosen is IPE 300 St 37, being the lightest and stiffest.
[2] Section chosen is IPE 360 St 37: although IPBl 340 St 52 is somewhat lighter, it is more expensive than the section chosen.

Examples, p. 247

g t/m	p t/m	g+p t/m	$\alpha = \frac{p}{g+p}$	b m	q=b(g+p) t/m	l m	M tm	φ	St	h_{min}	Section	kg/m
0.500	0.500	1.000	0.5	8.0	8.0	16.0	256	1/300	52	577[3]	IPBv 800	317[4]
											IPB 900	291
											IPBl 1000	272
0.500	0.500	1.000	0.5	1.5	1.5	16.0	48	1/300	52	577	IPE 550	106[5]
									37	375	IPB 450	171[4]
											IPBl 500	155
											IPE 600	122

[3] bei $\alpha = 0.6 \ h_{min} = 690$
$\alpha = 0.4 \ h_{min} = 465$
$\alpha = 0.5 \ h_{min} = \frac{690+465}{2} = 577$

[4] Section chosen based on available construction depth.
[5] The IPE sections in grade St 52 are too shallow; therefore use St 37.

Beams and girders
Design charts for shallow rolled sections

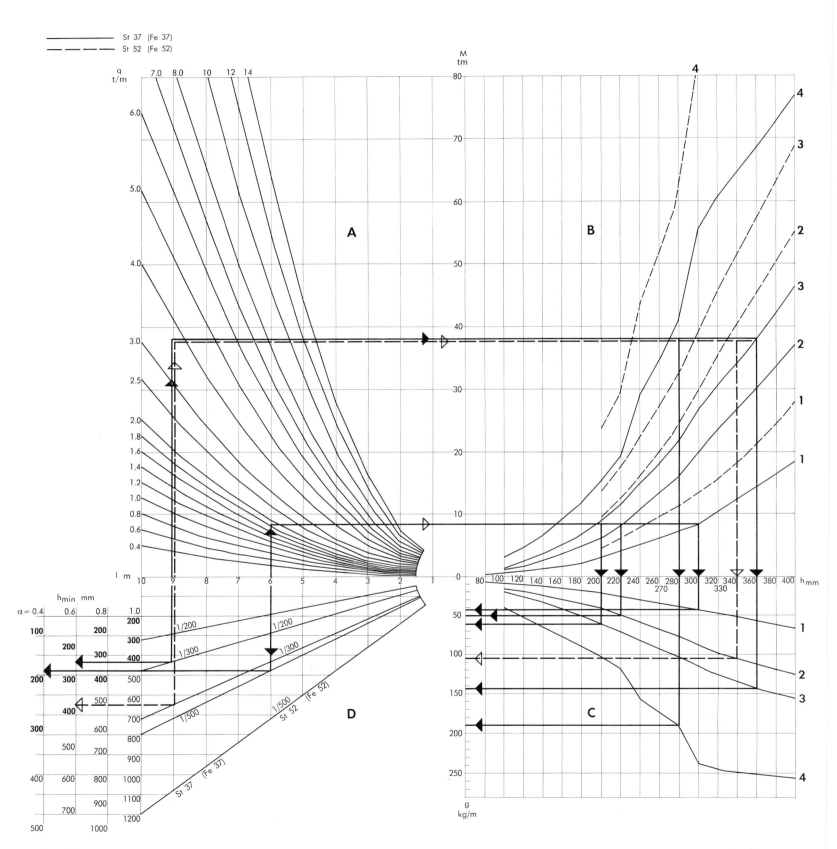

246

Design charts for deep rolled sections

Beams and girders

1 IPE
2 IPBl (HE A)
3 IPB (HE B)
4 IPBv (HE M)

St 37 (Fe 37)
St 52 (Fe 52)

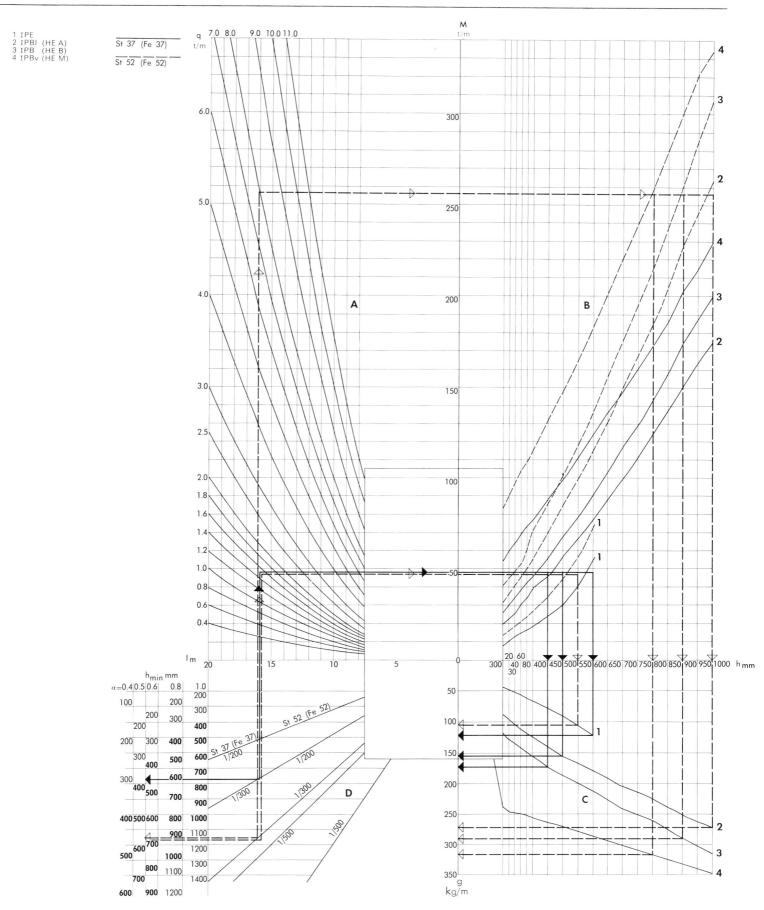

247

Beams and girders — Beam-to-beam connections

Beam-to-column connections have been described in the section on columns (pp. 235–237). In a beam-to-beam connection, the shear force from the secondary beam must be transmitted to the main beam. When it is required that the secondary beam should be continuous through the joint with the main beam, the bending moments also have to be transmitted, necessitating continuity in the flanges to transmit the tensile and compressive forces acting in them.
The following forms of construction are commonly employed for connections which have to transmit shear forces:

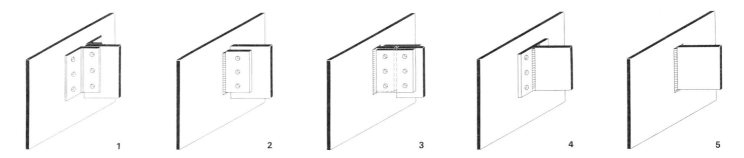

1 Connection by means of two angle cleats attached to the webs of the beams. **2** Single-shear bolted connection with welded vertical cleat. **3** Double-shear connection formed with welded vertical cleat and splice plates. **4** Welded butt plate on secondary beam. **5** Welded web connection. Welding must be continuous around the perimeter in the tensile zone.

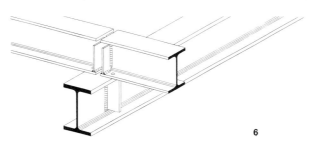

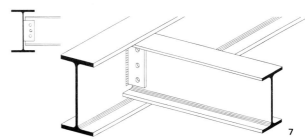

Connections for shear only

6 The secondary beams rest on top of the main beam and can be connected there. If necessary, the webs of the beams are provided with stiffeners at the point of support.

7 In this instance, the secondary beam is shallower than the main beam that supports it and can thus be continued right up to the web of the latter. This is the cheapest solution in terms of fabrication.

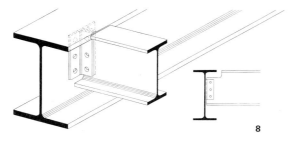

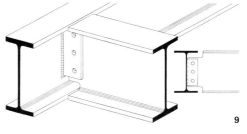

8 If the top flanges of the two beams, of unequal depth, have to be flush with each other, the top flange of the secondary beam must be cut away at the junction.

9 Where the two beams are of equal depth and at the same level, both the top and bottom flanges of the beam being supported must be cut away.

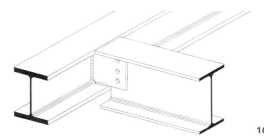

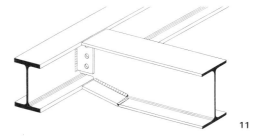

10 In this example, the supported beam is deeper than the supporting beam. The former has been cut off square, with no cutting away of flanges. This is an inexpensive solution, but bending moments are developed in the welded cleat, with the result that the supporting beam is subjected to torsional loading.

11 The solution in **10** is not aesthetically pleasing. If the junction of the two beams is visible, its appearance can be improved by raising the bottom flange of the supported beam.

Beam-to-beam connections — Beams and girders

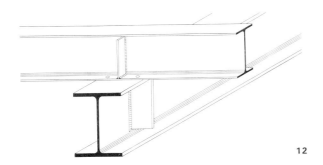

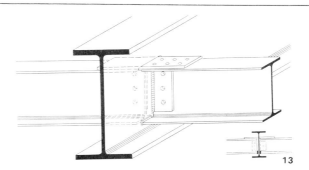

Connections for shear and bending moment

12 The secondary beam is continuous over the top of the main beam. This is the most economical type of connection, involving only a few bolts through the flanges to prevent the beams from shifting. Both beams should be provided with welded web stiffeners.

13 The secondary beam is shallow and its top surface is below that of the main beam. To transmit the tensile forces in the top flange, the secondary beam is provided with a splice plate which passes through a slot cut in the web of the main beam. The compressive forces in the bottom flange may be similarly transmitted by a splice plate or by contact through the web of the main beam. If necessary, steel filler plates should be inserted and secured in position by spot welds.

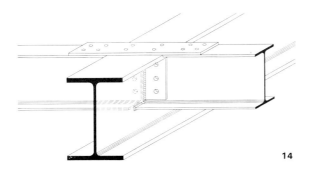

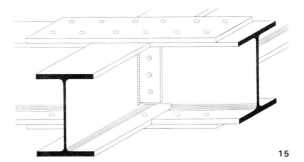

14 If the secondary beam is of lower depth than the main beam but has to be flush with the top thereof, the tensile splice plate may be laid over the top flange of the main beam. The compression in the bottom flange of the secondary beam is transmitted by contact, with inserted filler plates, as mentioned in the preceding example.

15 In this instance, the two beams are of equal depth and at the same level. The top and bottom flanges of the supported beam have to be cut away. Tensile and compressive forces are transmitted by splice plates passing over and under the supporting beam.

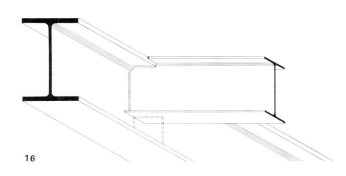

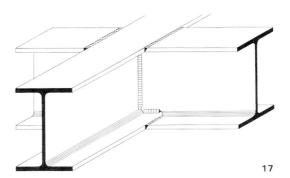

All-welded beam-to-beam connections

Welded connections are aesthetically more pleasing and give clear expression to the paths of force in a steel structure. If the welds are ground smooth, an even more monolithic effect is obtained.

However, the effort and cost involved in forming welded connections, especially during erection, are very much greater than in bolted connections. For this reason, all-welded beam-to-beam connections are generally employed for aesthetic reasons only. Prior to welding, the parts to be joined must be accurately fitted to each other and the weld faces suitably prepared. To hold the parts together during welding it is necessary, for larger members, to provide temporary fastenings allowing them to be bolted. These fastenings are subsequently removed by burning and the weld marks ground smooth. Smaller members can be temporarily secured by means of screw clamps.

16 Connection of a secondary beam to a deeper main beam. If the shear force to be transmitted is large, a stiffener must be provided under the secondary beam.

17 Welded connection of two beams of equal depth.

Beams and girders — Vierendeel girders • Castellated beams

The Vierendeel girder resembles a lattice girder, but comprises only chords and posts, no diagonals. The chords and posts are subjected to normal forces and bending and the efficient transmission of the forces at the joints must be ensured. Such girders are used as storey-height girders or as floor girders permitting the passage of bulky services through them, but they have the drawback of being expensive to fabricate.

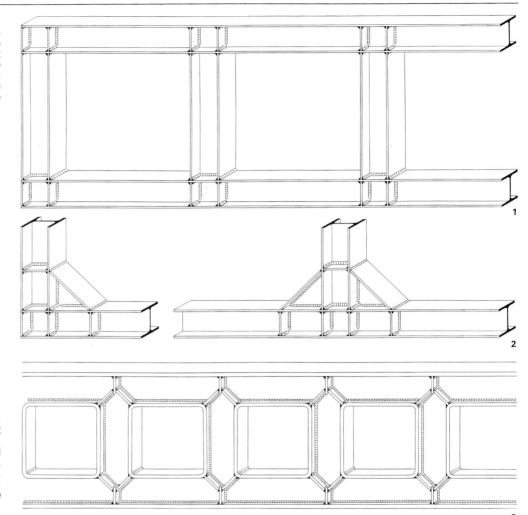

1 Welded Vierendeel girder composed of universal sections. It is a relatively light component for fairly small loads.
2 The load-carrying capacity can be increased by means of welded haunches at the joints, which can be conveniently cut from I-sections.
3 Heavy all-welded Vierendeel girder for carrying large loads. Fabricated from plates and wide flats.

CASTELLATED BEAMS

Open-web expanded beams, commonly known as castellated beams, are produced from rolled steel beams, **1**, by flame-cutting them along the web to a castellated zig-zag pattern, the two halves then being joined together at the crests of the castellations by welding, so that an open-web beam of greater depth than the original beam is obtained, **2**. Such beams are economical for the transmission of large bending moments in long spans, but their shear transmitting capacity is limited because of the weakening of the web. Near supports it may be necessary to weld in hexagonal plates to close one or two of the web apertures. The depth of a castellated beam can be further increased by the insertion of rectangular plates, **3**, so that a kind of Vierendeel girder is obtained. These particular components are known as Litzka beams. The web apertures are convenient for the passage of relatively small-diameter services, but are generally too small for air-conditioning ducts. For further details, see p. 289.
The fabrication of these beams requires special equipment, because the two halves of the beam tend to develop a curvature when the rolling stresses locked up in the original beam are released as a result of cutting the web. Fabrication of such beams becomes an economical proposition only when they are mass-produced.

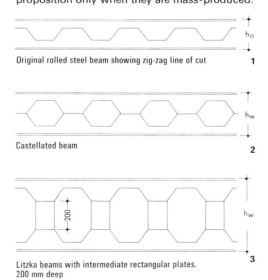

Original rolled steel beam showing zig-zag line of cut **1**

Castellated beam **2**

Litzka beams with intermediate rectangular plates, 200 mm deep **3**

Design charts

The charts on p. 251 are applicable both to the conventional castellated beams and Litzka beams.
Section A gives the bending moments that can be safely resisted by castellated beams (without intermediate plates), while Section B gives similar information for Litzka beams with 200 mm deep intermediate plates. The curves are for steel grades St 37 and St 52. Section C gives the weights g (in kg/m) for castellated beams, while Section D gives them for the Litzka beams.
The depths h_0 of the original sections are indicated along the horizontal axes of Sections A and B.
Sections A on pp. 246 and 247 can be used for determining the bending moment on a simply-supported beam due to uniformly distributed load. The deflection sections D on these pages can be used to find the minimum requisite beam depth to remain within a certain permitted deflection.

Castellated beams

Beams and girders

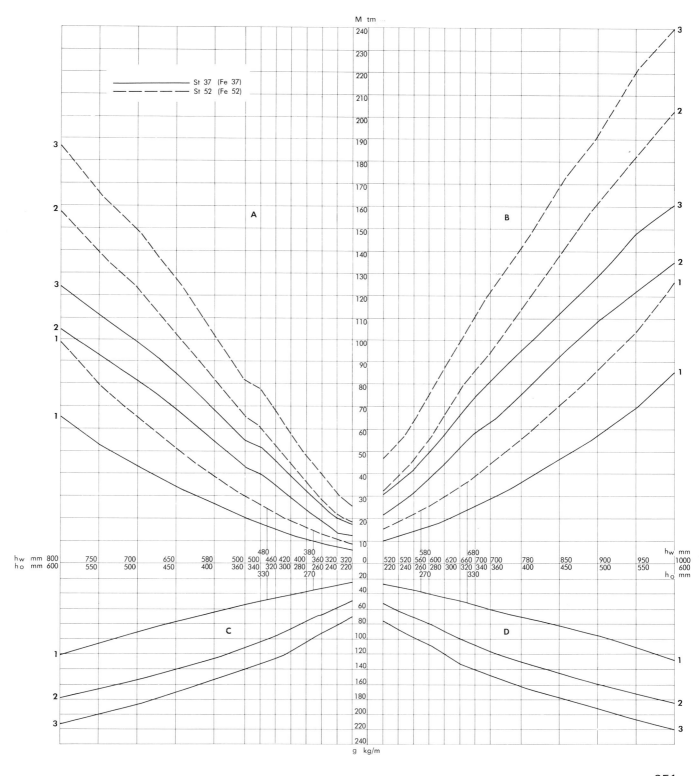

1 IPE
2 IPBl (HE A)
3 IPB (HE B)

Beams and girders — Holes for services through webs of beams

Rolled steel beams and plate girders offer the great advantage over comparable concrete members that services can easily be passed through openings formed in the relatively thin webs. Small openings may be formed by burning or, if they are circular, by drilling. This should, as a rule, be done in the fabricating shop, though of course in exceptional cases it can be done on site. Large openings may, however, weaken the web to such an extent that shear force cannot be safely transmitted unless suitable stiffening or strengthening arrangements are made. Therefore, large openings should preferably not be located in regions of large shear force near the supports, but in the mid-span region.

Where a large number of services are to be installed, it is preferable to adopt a two-way system of floor beams in which those in one direction pass over those in the other direction, so that the services can be arranged in two layers and can be passed over and under the beams.

Also, it is often advantageous to reduce the beam depth in certain parts and to let the services pass under the beams.

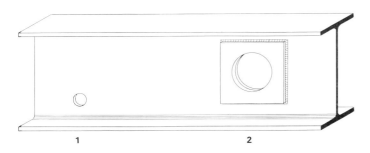
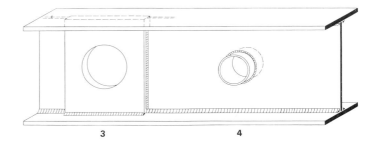

Circular openings

1 Small openings are burned or drilled in the web and require no local strengthening.
2 The web can be strengthened by means of plates welded to one or both sides.

3 For larger openings it may be necessary locally to replace the web by a thicker plate, but this is expensive.
4 Circular opening strengthened by welding in a short length of steel tube.

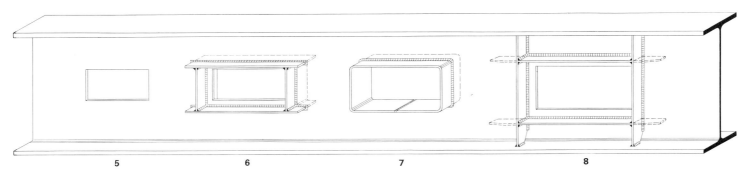

Rectangular openings

5 In general, a rectangular opening should be formed with its width greater than its depth, so as to leave sufficient depth of web above and below for adequate transmission of shear force.
6 Stiffening and strengthening elements should preferably be of labour-saving design, so that they can be welded directly to the beam web without having to be accurately fitted. They may be applied on one side of the web or, if large forces have to be transmitted, on both sides.

7 A fitted sleeve, more particularly in an opening with rounded corners, is laborious and expensive and should therefore be used only for aesthetic reasons where the beam is exposed to general view.
8 Example of strengthening and local stiffening around a relatively large opening.

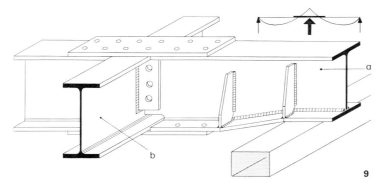
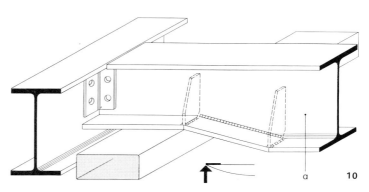

Reduction of beam depth

9 The continuous floor beam, a, has its full depth at its junction with the main beam, b, but is of reduced depth in the mid-span region, so that services can pass under it. At the changes of direction the bottom flange and the web must be provided with stiffeners. This method of providing space for services involves extra fabrication work and is used only in exceptional cases.

10 In this example the floor beam, a, is simply supported. As the bending moments in the floor beam are small at the joint, its depth can be locally reduced by raising the bottom flange. The shear stresses in the web should be checked; if they are too high, the web must be strengthened. Stiffeners at the changes of direction of the bottom flange are usually unnecessary because the forces are relatively small near the end of the beam.

Forms of lattice girders • Light lattice girders

Beams and girders

The lattice girder (or truss) is the form of construction for girders involving the least quantity of material. On the other hand, its fabrication is relatively laborious, so that in many instances a plate girder may, despite its greater weight, be a more economical alternative.

Lattice girders can be designed for any loading and for any length of span, ranging from very light ones with very slender lattice members to the heaviest bridge girders. In multi-storey buildings they are employed both as girders for supporting heavy vertical loads and as bracing systems for resisting wind load and providing structural rigidity. Lattice girders are employed as light secondary girders and main girders in floors and also as heavy (often storey-height) intercepting girders at the base or as cantilever girders at the top of buildings, the latter more particularly for suspended buildings. Lattice girders in floors have the advantage that services can penetrate them almost without restriction. For dimensions of the services that can pass through the lattice openings in relation to girder type and depth, see p. 289.

A lattice girder comprises the top chord, the bottom chord and the web members, the latter being composed of vertical and diagonal members or diagonal members only. To resist buckling, compression members must have larger cross-sections than tension members carrying equal force. It is therefore generally advantageous to use a lattice system in which the longer web members are loaded in tension.

The joints in lattice girders are usually designed as pin-joints, so that the members have to transmit normal forces (tension or compression) only. Of course, if the chord members have loads applied to them between the joints, then such members will, in addition to the normal forces acting in them, be loaded locally in bending, in which case they will require a suitably rigid cross-section.

Various types of parallel chord lattice girder

1 The web members consist of vertical struts and diagonal ties inclined in the same direction. So-called Pratt or N-girder.
2 In this system, the diagonal members are alternately ties and struts, acting in tension and compression respectively. The vertical posts serve essentially to transmit the applied loads to the bottom joints.
3 Lattice girder comprising inclined web members only, conventionally known as a Warren girder. The slope of the web members is approximately 60°. There is an aesthetic advantage in that, when two or more such girders are viewed one behind the other, the absence of vertical members produces a more restful overall impression.
4 Long-span lattice girders of the type envisaged in **2** may be provided with secondary lattice systems for load application points between the main joints. Tension in a member is indicated by a positive, compression by a negative sign; the signs in parentheses refer to the forces due to the local loads.

Forms of lattice girder

5 and **6** Trapezoidal girders.
7 Parabolic girder.
8 Fish-belly girder.
9 and **10** Triangular roof trusses.

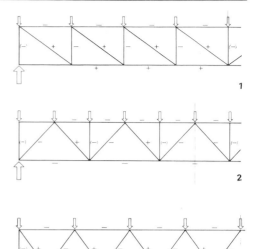

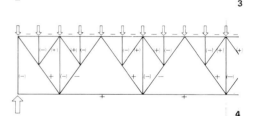

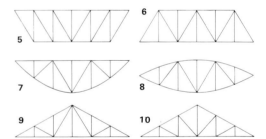

LIGHT LATTICE GIRDERS

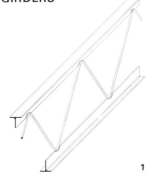

1 Bar joists. The chords consist of T-sections; the web members are individual round steel bars or a continuous zig-zag bar, as illustrated. Such girders are economical to produce, but they can carry only small shear forces. They may be used, for example, for roof girders or lightly loaded floor girders.

2 Girder composed of light-gauge cold-formed sections. As the thickness of steel is small, good corrosion protection is essential, preferably in the form of hot-dip galvanising. Such girders can be mass-produced by automated methods, such as the one illustrated (manufacturer: Keller KG, Munich), which is galvanised and can be used as a purlin under steel trough roof cladding.

Loadbearing capacity

Allowable shear forces Q_x and bending moments M_x

Allowable stress (kg/cm^2)	Q_x (kg)	M_x (kgm)
H = 1400	2200	6288
HZ = 1600	2520	7200
H = 2100	2980	8976
HZ = 2400	3410	10272

Beams and girders — Medium and heavy lattice girders

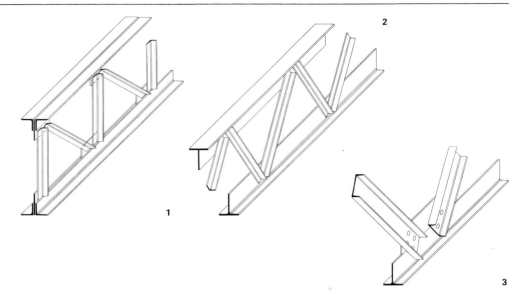

1 Lattice girder composed of angle sections and gusset plates. The chords comprise equal or unequal angles in pairs; with unequal angles the longer legs of the latter are placed horizontally in the top chord to increase its buckling resistance. The web members are single or double angles. Girders of this general type may be of bolted or welded construction. With regard to corrosion protection, see **4** below.

2 The chords of this girder are T-sections or halved I-sections. It is a very economical and frequently used form of construction. The web members are single or double angles or channels. If the forces to be transmitted are not large and the chord members are of sufficient depth, the web members can be welded directly to the latter.

3 For the transmission of large forces and with chord members of limited depth, it may become necessary to weld gusset plates to the chords. The girders are welded or bolted.

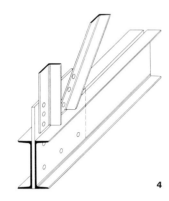 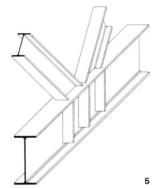 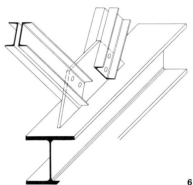

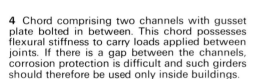

4 Chord comprising two channels with gusset plate bolted in between. This chord possesses flexural stiffness to carry loads applied between joints. If there is a gap between the channels, corrosion protection is difficult and such girders should therefore be used only inside buildings.

5 Chord consisting of a universal section. For relatively light girders the web members may be butt-welded to the chords. The latter should be provided with one or more stiffeners at each joint. This form of construction is rather expensive in fabrication.

6 Chord consisting of a universal section with welded gusset plate for connecting the web members. This is an economical and extensively used form of construction for heavily loaded lattice girders.

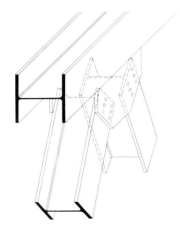  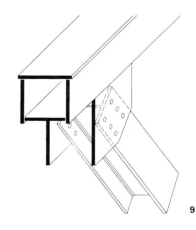

7 Chord consisting of a universal section placed with its web in the horizontal position. Gusset plate welded to each flange.

8 Open box-section with double gusset plates for very heavy lattice girders.

9 Closed box-section of welded construction with double gusset plates.

Tubular frames

Beams and girders

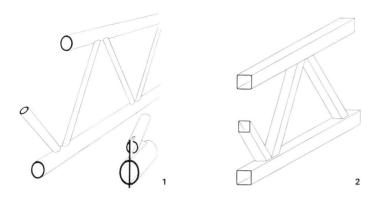

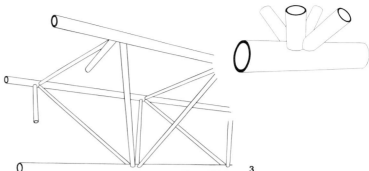

1 Lattice girder composed of tubular members. All the web members may be given equal outside diameter, but may have different wall thicknesses appropriate to the forces they have to transmit. Joints are formed by welding the appropriately prepared ends of the tubes, this being done by flame cutting or by machining with the aid of special equipment. Gusset plates are sometimes employed, in which case the tubes are slotted; but for reasons of corrosion protection this type of joint should be used only in the interior of buildings.

2 Square or rectangular hollow sections are generally connected by butt welding. Elegant, clean-looking structures are obtained with such sections and fabrication is easier than with circular sections.

3 Triple-chord girder composed of tubular members. The double top chord possesses high resistance to buckling and therefore need not be laterally braced. This is a typical tubular structure.

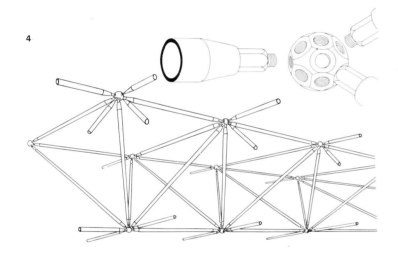

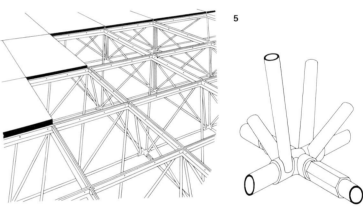

4 'Mero' tubular truss with screwed polyhedral joints. This is a versatile construction method suitable for roof trusses, space frames, etc., whether plane or curved, for permanent or temporary structures of all kinds, such as grandstands, exhibition halls, etc.

5 'Space-Deck' system. The space framework is composed of pyramidal prefabricated units which are interconnected by tie-rods with screwed sleeve couplers. The upper surface consists of square frames, composed of angle-sections, and covered with aerated concrete slabs.

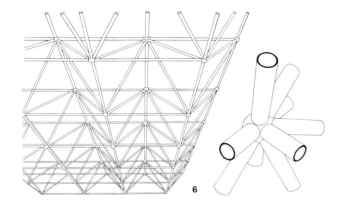

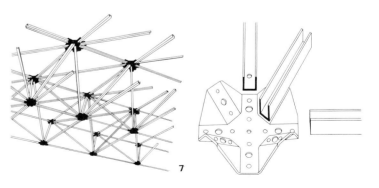

6 'Oktaplatte' system. Space framework consists of prefabricated units with joints comprising a spherical element to which the tubular members are welded.

7 'Unistrut' system. Space framework composed of channel sections, all of equal length, with bolted connections comprising standard gusset units.

Beams and girders — Structural action of composite beams

Definition

In composite construction, materials with different properties are so joined together that they co-operate structurally with each other. In structural steelwork, a composite beam more particularly denotes a steel beam with its top flange connected to a concrete slab. In the region of positive bending moment the slab functions as the compression flange, while the steel beam, particularly its bottom flange, resists the tensile force. Where negative bending moments occur the concrete slab may be regarded as acting compositely with the steel only if the tensile forces in the concrete are balanced by means of a prestress. The use of prestressing in combination with composite construction is economically justifiable only for heavy loads in long-span structures. It is therefore extensively used in bridge decks, but seldom in building construction.

Shear connectors

To achieve efficient composite action between the steel and the concrete it is necessary to establish a mechanical connection between the two materials to absorb the horizontal shear forces at their interface, various types of shear connector being available for the purpose. If the two components are not thus connected together, they will certainly contribute their respective shares to the loadbearing capacity, but the total load they will carry will be much less than if they were so connected. This can be illustrated by the simple example of a 'beam' composed of planks stacked one upon the other (**1.1**, **1.2**). If there is no mechanical connection between the planks, the overall loadbearing capacity is the same as though the planks were laid side by side. But if they are joined together with nails, analogous to shear connectors, they become a composite beam and their combined loadbearing capacity is greatly increased (**1.3**).

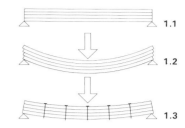

Advantages of composite construction

With composite action between the concrete floor slab and the steel beam, the following advantages are obtained:
1. saving in the weight of steel (up to 50%);
2. a stiffer floor and therefore a reduction of deflection.

Obviously, a part of the saving made by the reduction in weight of the steel beam is lost through the provision of shear connectors.

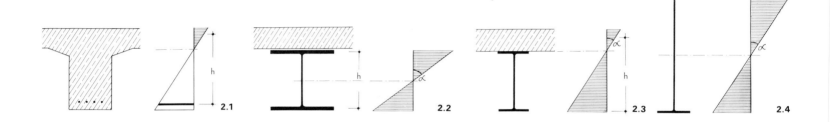

Stresses and deflections

2.1 In a concrete T-beam, the slab functions as its compression flange. There is equilibrium between the compressive force in the compression zone of the concrete and the tensile force in the reinforcement at the bottom of the beam.

2.2 In a composite beam, the rib of the T-beam is replaced by a steel section the main function of which is to resist the tensile force. If there were no mechanical connection between the two materials, the steel beam would have to carry the whole load. In that case, its effective depth would merely be the distance h between the centres of its flanges, and the neutral axis would be at mid-depth of the beam. The stress distribution diagram has a large angle α, the deflection is large and the beam has low stiffness.

2.3 If composite action is established by mechanical connection between the steel and the concrete, the effective depth h is increased, now extending from the centre of compression in the concrete to the bottom flange of the steel beam. The latter, and more particularly its bottom flange, can therefore have a smaller cross-section. The neutral axis is higher and almost coincides with the top flange of the steel beam, so that this flange too can be small. For the same allowable steel stress in the bottom flange, the angle α of the stress-distribution diagram is smaller, compared with the non-composite section shown in **2.2**, and the deflection of the composite beam is reduced.

2.4 To reduce the deflection to the same extent by using solely a deeper steel beam, the depth below the neutral axis in **2.3** would also have to be available above that axis, so as to produce a stress-distribution diagram with the same angle α.

Shrinkage and creep

Fresh concrete shrinks as it dries out. Also, it undergoes a gradually developing permanent shortening under the action of compressive stress, this being known as creep. In a composite beam, only the concrete slab is subject to these changes, not the steel. As a result of shrinkage and creep, the share of the loadbearing assignable to the slab will diminish in course of time and the steel beam will have to take a larger proportion of the load. Thus, there is a redistribution of forces. For large composite structures it is necessary to carry out an accurate analysis of these phenomena. With the shorter spans generally met in building construction, however, the margin of safety implicit in the permissible stresses is sufficient to cope with these effects and, therefore, no special analysis is generally required.

Composite action established in-situ

Steel-framed buildings have the advantage that the erection of the steel components can be carried out rapidly, unhampered by adverse weather conditions. This advantage cannot be fully exploited if the floor slabs have to be concreted in-situ. The overall construction time can be reduced by precasting the concrete slabs, in which case the composite action with the steel has to be ensured by suitable arrangements for shear connection. This method of construction calls for close dimensional tolerances and careful scheduling of the job.

Shear connectors

Two methods are extensively used for transmitting the shear forces between the concrete slabs and the steel beams in buildings:

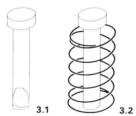

1. Headed studs (**3.1**). These are attached to the steel beams by means of a semi-automatic welding technique. The shanks of the studs transmit the horizontal shear forces, while the heads prevent the concrete slab from lifting off the steel beam. A wire helix placed around a stud will increase its load capacity by 20 to 25% (**3.2**). Care must be taken to ensure that the space between the helix and the stud is properly filled with concrete. The studs may be welded to the beam in the fabricating shop or on site. If necessary, the studs can be welded through steel decking laid on the steel beams in advance, prior to concreting the slab in-situ.
2. Friction connection. The precast concrete slabs are pressed on to the steel beams by high-strength friction-grip bolts in order to produce friction forces which resist the shear.

The slabs are laid on the steel beams either dry or on a bed of mortar or, by far the most effectively, on adhesive mortar.

Construction procedure and sections for composite beams

Beams and girders

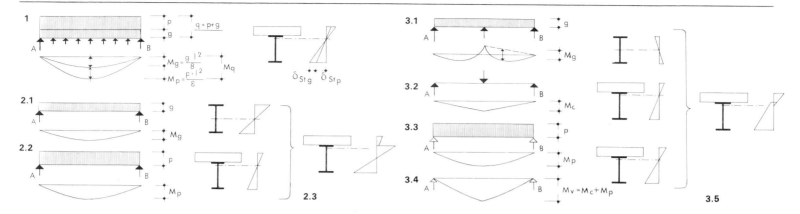

Composite action is not effective until the construction of the section has been completed. Until that stage has been reached, the following methods of supporting the components of the composite section are available:

1 The beam may be continuously propped along its entire length until the composite action has become effective. Then both the dead weight, g, and the subsequently applied loading, p, will be supported by the composite section.

2 Erection without propping. If the steel beam is connected to its permanent supports without temporary propping along the span and is then loaded with the dead weight of the concrete slab, the bending moment M_g due to this dead weight must be carried by the steel beam alone, without the benefit of composite action, **2.1**. Only the moment M_p due to the subsequently applied loading will be resisted by the composite section, **2.2**, the combined stress distribution diagram being as shown in **2.3**. This method of erection has the advantage of dispensing with propping, but requires a heavier section for the steel beam than in **1** or **3**.

3 Erection with propping. If the steel beam is provided with temporary supports at one or more points along its span before the concrete slab is added, the dead weight of the latter will produce only small bending moments M_g in the steel beam, **3.1**. This situation represents a practical approximation to the continuous temporary support envisaged in **1**. The props may be installed at the same level as the permanent supports A and B or they may be a little higher, so that the steel beam is given a certain amount of preliminary camber, whereby its subsequent stress conditions are further improved.

On subsequent removal of the props after the concrete has matured and the composite action with the steel has been established, the reactions hitherto developed by the props will have the effect of concentrated loads, producing a moment M_c, applied to the composite beam at this stage (**3.2**). The subsequently applied loads will produce a moment M_p which is resisted by the composite beam, **3.3**. On superimposing the stress-distribution diagrams associated with these three successive stages, we finally obtain the combined overall diagram, **3.5**, corresponding to the moment M_v acting in the composite beam, **3.4**.

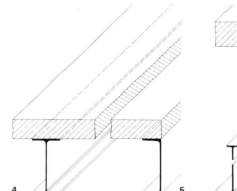 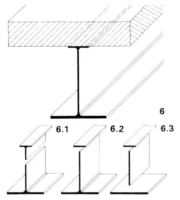 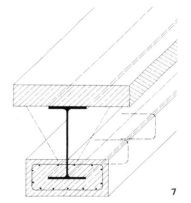 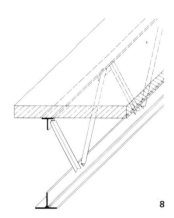

Beam cross-sections

4 The steel beams most commonly employed for composite construction intended for relatively small spans and loads are light joists such as those in the IPE range.
Heavier sections, such as those in the IPB range, are less frequently used because the top flange is not efficiently employed.

5 Channel sections are very convenient as edge beams. If the back of the channel is placed flush with the edge of the concrete slab, a clean smooth finish for exposed beams is obtained. Spandrel units, cladding panels or railings can be easily attached to the channel.

6 For longer spans and heavier loads, it is advantageous to use a built-up welded section comprising a small top flange, instead of a heavy rolled section. Such an asymmetrical section can be produced from two different sizes of rolled steel beam slit longitudinally, **6.1**, or by welding a narrow top flange on to the web of half a heavy rolled section, **6.2**. It can, of course be completely built up in the form of a plate girder, as in **6.3**.

7 The 'Preflex' beam is a special type of composite beam, developed and marketed by the company of that name. In the factory, the beam is subjected to a flexural load and in this condition a concrete casing is cast round its bottom flange, welded reinforcement being used to increase the bond. When this concrete has hardened, the flexural load is removed, causing the concrete to be precompressed. In this way a kind of composite prestressed concrete beam is obtained, which can be combined with a concrete top slab in the usual way. The advantage of the concrete bottom flange is that, although it does not greatly add to the load-bearing capacity of the beam, it greatly increases its stiffness and thus reduces deflections, which makes it especially suitable for long spans.

8 The top chord of a lattice girder to be combined with a concrete slab for composite construction need be no wider than is strictly necessary for proper erection and for welding the shear connectors. In some cases, the latter may be attached directly to the upper gusset plates, so that the steel top chord can be entirely eliminated.

257

Beams and girders   Precast concrete slabs for composite beams

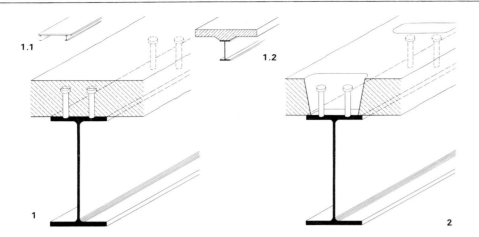
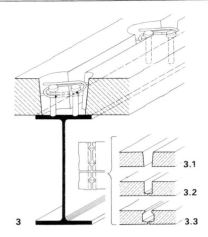

1 Composite beam with in-situ concrete slab and stud shear connectors. An alternative technique is to prefabricate the slab in strips with pairs of steel beams attached, **1.1**, but this is economical only for units of limited length; greater length and weight makes the cost of transport and erection excessive. Haunches cast in the concrete slab over the beam, **1.2**, increase the effective depth of the composite section and thus effect some saving in the steel beam, but they involve extra expense in shuttering the slab.

2 Precast concrete slabs with preformed cavities for shear connectors — single studs or pairs — may be laid dry on the steel beams. A mortar bed provides better corrosion protection for the top flanges of the latter, but involves an extra operation. The cavities should be somewhat wider at the top than at the bottom, to obtain a wedge action in the in-situ concrete or mortar filling to prevent the slabs from lifting.

3 Krupp-Montex system of composite construction with precast concrete slabs, the longitudinal joints of which are placed over the steel beams. The studs engage with semi-circular recesses in the edges of the slabs and are enclosed in reinforcing loops. For beams of longer span, the slabs also have transverse joints, **3.1**. The transverse joints may be provided with bottom nibs to obviate the need for formwork, **3.2**, and may be shaped to produce a keying effect across the joint to prevent unequal deflections of adjacent slabs under heavy loads, **3.3**. See also p. 278 **9**.

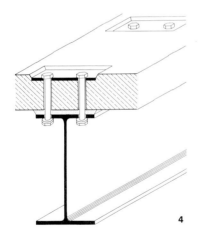
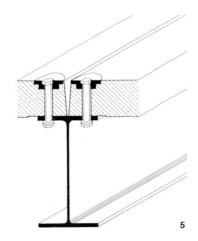
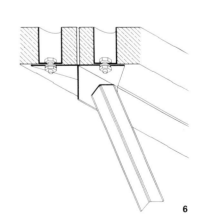

4 Composite beam with transmission of shear force by friction developed by high-strength bolts. Suitable both for in-situ and precast concrete slabs. The latter are usually laid on a mortar bed for better contact. Alternatively, an adhesive may be used; with in-situ concrete this should be a special water-resistant adhesive. The bolts exert their pressure on the concrete through steel bearing plates.

5 Krupp-Montex system with friction connection. The precast slabs are bolted dry to the steel beams. The bearing plates under the bolt heads are of forged steel. The slabs are cast in precision-made steel moulds and have smooth and flat soffits ensuring good contact with the steel beams.

6 Lattice girder with precast concrete slabs forming a composite member (Rüterbau system). The steel girder has no top chord, its diagonal members being welded to horizontal gusset plates which are bolted with high-strength friction-grip bolts to steel sockets cast in the slab. See also p. 278 **10**.

Design chart for composite beams

Beams and girders

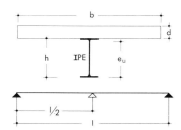

Section B – Beam section

To determine the beam section we continue horizontally to the right, thus entering Section B, and find the intersection with the curve for the appropriate slab thickness. The required IPE section is then read on the horizontal scale vertically below the intersection. If a value intermediate between two sections is found, the larger section should be adopted.

Deflection

The depth necessary to remain within a certain permissible deflection can be obtained from Section D on p. 246. The depth e_u of the neutral axis of the composite beam is equal to half the depth h_{min} found for an all-steel beam, that is: $e_u = \frac{1}{2} h_{min}$.

In a composite beam, the neutral axis is approximately at the level of the top flange of the steel section. Therefore, the deflection of a composite beam is about half of that of a steel beam of the same depth. In other words, the composite beam is about twice as stiff as the steel beam alone.

Examples

Loads								Composite action	Section			Deflection	Comparison with/without composite action	
g	p	$\alpha = \frac{p}{g+p}$	b	q	l	M	d		H	kg/m	St	f/l	Weight	Deflection
kg/m	kg/m²		m	t/m	m	tm	cm						%	%
300	500	0.625	2.4	1.92	10	24	10	with without	IPE 400 IPBl 320	66 98	37 37	1/800 1/330	67	41
400	750	0.650	3.0	3.45	12	62	12	with without without	IPE 500 IPBl 550 IPBl 450	90 166 140	52 37 52	1/570 1/320 1/260	54 too weak	56

St 37 (Fe 37)
St 52 (Fe 52)

Assumptions

The assumptions made in preparing the chart for the initial design of composite beams are as follows:

- Steel beams: IPE sections.
- Steel: St 37 and St 52.
- Concrete: B_n 450. If B_n 300 is used, the allowable bending moments will be somewhat smaller.
- Beam spacing: the chart is valid for beams with the usual spacing b varying from 1.80 to 3.60 m.
- Depth d of concrete slab:
for beam spacing $b \leqslant 2.40$ m, $d = 10$ cm.
for beam spacing $b > 2.40$ m, $d = 12$ cm.
- Span: up to $l = 20$ m.
- Creep and shrinkage:
creep factor $\phi_n = 4.0$.
Shrinkage strain $\varepsilon_s = 3 \times 10^{-4}$.
- Erection: It is assumed that the beams are temporarily supported, with at least one prop at mid-span, until the concrete has matured.

Design loads

To derive the required beam section, it is necessary first to determine the distributed load per unit area $g + p$ (t/m²) and then to find the uniformly distributed load per beam:
$q = b(g + p)$ (t/m).

Section A – Bending moments

The bending moments are obtained from Section A. From the appropriate span on the horizontal scale we proceed vertically upwards until we intersect the curve corresponding to the load q; the bending moment (in tm) is then read on the vertical scale.

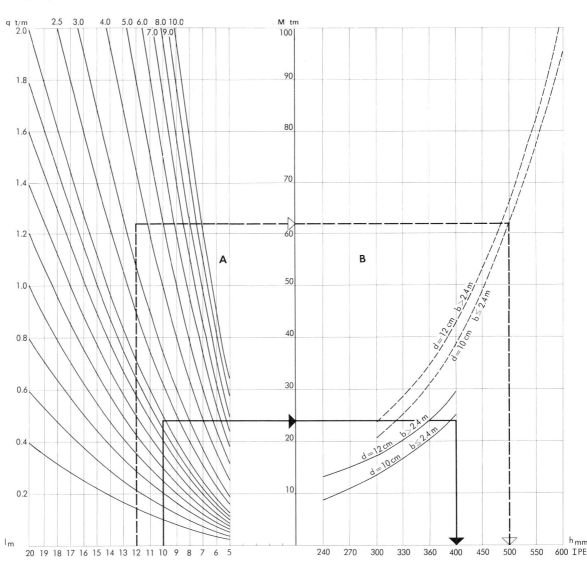

Beams and girders — Northlight frames

General

The northlight (or saw-tooth) roof is extensively used in industrial buildings. The glazed surface faces more or less north, its height, length and slope depending on the intensity of illumination required at the working surface inside the building; it may be sloped or vertical, while the unglazed surface may be flat or curved.

At the base of each glazed surface a generously dimensioned valley gutter must be provided. The roof covering is supported on purlins or rests directly on the main frames.

The northlight roof structure has to resist vertical and horizontal forces. In structures which have their horizontal bracing in the plane of the roof, the edge beam under the first glazed surface should have special horizontal bracing and erection should commence from this end of the structure. Otherwise, in the event of the collapse of one northlight bay, its neighbour would be deprived of horizontal stiffening and the whole structure would collapse by a kind of chain reaction. This hazard exists especially with sloping glazed surfaces.

Structural systems

1.1 Three-pinned frame. The simplest system for northlight roof structures. With sloping glazing, horizontal thrusts are developed under vertical loading. In addition, horizontal forces due to wind also have to be absorbed at the supports.

1.2 Two-pinned frame. If the frame is sufficiently stiff, the thrust due to vertical loads is substantially reduced. It can then be treated as a cranked beam on two supports. Wind forces need be resisted at only one bearing, the other being designed as a roller sliding bearing.

1.3 Three-pinned frame with tie-rod. As horizontal thrust is eliminated, only vertical reactions are produced by vertical loads. Wind forces are resisted at one end of the frame.

1.4 Northlight truss. The most appropriate form of construction when the frames are widely spaced and the roof covering is supported by purlins.

2 Parts of a northlight roof:

a ridge purlin
b purlin
c bottom purlin
d rafter ⎫ northlight frame
e strut ⎭
f intermediate strut
g valley beam
h main beam
B width of bay
L length of bay

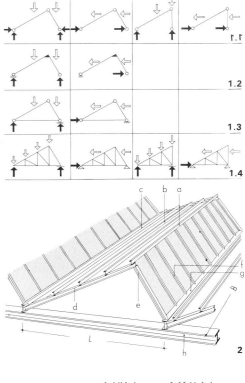

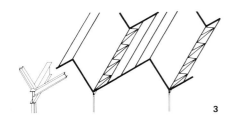

3 The roof load is transmitted through the purlins to the roof frames which in turn transmit it directly to the columns. A strong bottom purlin is needed under the glazing to carry the weight of the latter and to support the gutter. Horizontal bracing by lattice girder.

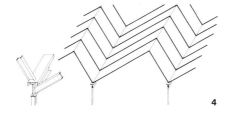

4 Vertical forces are carried by universal valley beams, while horizontal forces are resisted by reclining universal beams.

5 Truss over full depth of glazing. Advantages: very light, rigid girder. Disadvantages: diagonal members of the lattice girder intersect the glazing. Any arrangement of web members is possible. Horizontal stiffening is provided by the bracing in the plane of the roof.

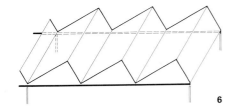

6 Where a number of northlight frames occur in succession, as is often the case, the repetition of individual columns may be an objectionable feature. The number of columns can be reduced by supporting the valley beams on long-span main beams. If these are of I-section, they are likely to be of substantial depth and weight.

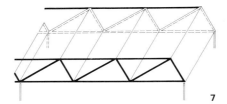

7 The most economical solution is to combine the successive northlight frames into a lattice girder spanning between widely spaced columns. As the bottom chord has to take tensile forces only, it can be made quite slender and will thus be relatively unobtrusive. The top chord must penetrate the roof covering and the glazing. These penetrations should be carefully insulated and sealed against the ingress of water, as otherwise damage is liable to occur.

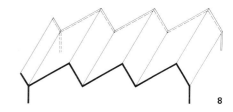

8 In this solution, the successive northlight frames are rigid members joined continuously to form one long zig-zag beam. In theory this is an elegant solution, but as heavy structural sections are necessary to ensure the requisite stiffness, it can be adopted for not more than three frames in succession.

Stiffening elements

Rigid frames

In multi-storey buildings the rigid frames which serve to stiffen the building are composed of beams and columns which usually also perform the normal loadbearing functions. In the top storey, sloping, cranked or curved beams may be used to suit the shape of the roof structure. Thus the members of a rigid framework do not differ significantly, in so far as their structural components are concerned, from the members of a pin-jointed framework. The essential difference is in the construction of the joints.
If external columns are rigidly connected to beams located in the same plane, façade frames are formed.

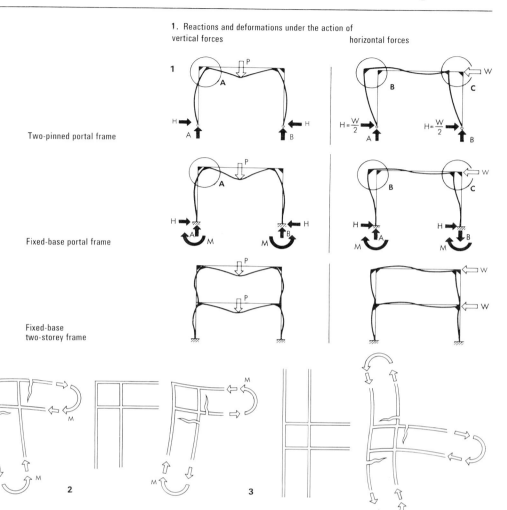

1. Reactions and deformations under the action of vertical forces / horizontal forces

Two-pinned portal frame

Fixed-base portal frame

Fixed-base two-storey frame

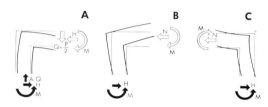

Deformation of knee of portal frame

Deformations and forces at knees

When a knee of a rigid frame is subjected to bending moments from the beam and the column, the angle between these members tends to decrease, **2**, or to increase, **3**. Figure **4** illustrates the deformation due to bending moments at a joint in a multi-storey rigid frame.
As appears from these drawings, the forces acting in the flanges of the members give rise to shear stresses in the knee webs. To transmit these forces effectively, the inner flange of each member should continue to the outer flange of the other member. In some cases the knee web must be strengthened, for example, by increasing the plate thickness or by means of diagonal stiffeners.

Hinged column bases

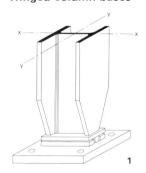

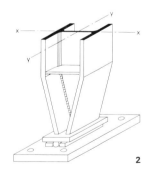

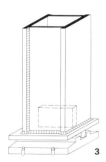

1 Hinged bearing permitting rotation about an axis parallel to the x-axis of the IPB section.

2 Hinge parallel to the y-axis of the IPB section. At the base the flanges of the column taper inwards to concentrate the forces at the hinge.

3 Linear rocker bearing for box-section column transmitting large forces. The base plate of the bearing is secured against displacement by means of studs embedded in the concrete footing.

4 Point rocker bearing. The convex part should always be underneath, and the ring always on top, so that no water can collect in the latter.

Stiffening elements | Joints in rigid frames

Knee joints for universal sections

Beam-to-column connection in a multi-storey frame comprising universal sections

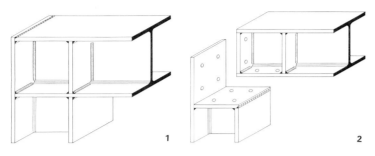
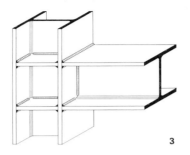
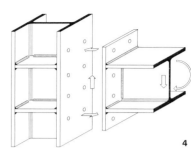

1 All-welded connection. **2** bolted connection. **3** All-welded connection. **4** Site-bolted end-plate connection.

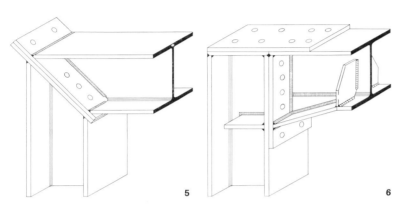
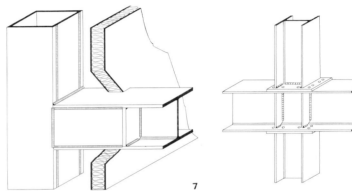

5 Mitred joint for portal frame knee. The butt plates welded to the ends of the members are connected by means of high-strength friction-grip bolts.

6 When the knee is subjected to large bending moments, the depth of the beam at the knee can be increased by haunching, for example, by slitting the web and inserting a triangular plate. In this example, the connection is made by bolting with the aid of an end plate welded to the beam and a splice plate over the top of the joint. The column web is stiffened at the junction with the bottom flange of the beam.

7 Welded connection between a beam and a box-column exposed in front of the façade of a building. The beam is a universal section locally built up to form a box-section at the joint. This type of connection can also be used in cases where external columns are filled with water for fire protection (p. 240).

8 This joint is similar to that shown in **5**, p. 235, but specifically designed to transmit bending moments to the column.

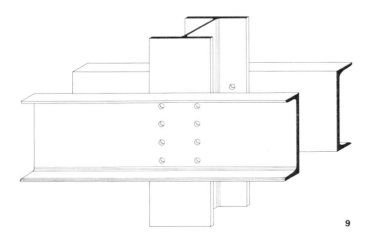
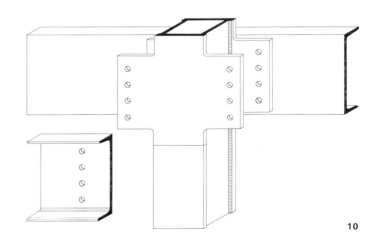

9 Here, the beam consists of a pair of channel sections passing on each side of the column, to which they are bolted. This type of joint has only limited capacity to transmit bending moments.

10 A box-section column provided with gusset plates on both sides can absorb large bending moments from the double channel section beam (this drawing is a detail of **12**, p. 211).

Horizontal wind-bracing

Stiffening elements

The external walls of a building transmit the wind forces to the roof and floors, which in turn transmit them to the cores, shear walls, rigid frames, etc. In most cases, the roof and floor structures form rigid horizontal diaphragms which can perform this function. Otherwise, it is necessary to provide horizontal bracing. In a multi-storey steel-framed building it is generally sufficient to insert a horizontal bracing system in every second or third floor, the stiffness of the columns being adequate to transmit the wind forces over such intervals of height.

If the floor structure can perform the wind-bracing function alone, but not until completion of the building, it is necessary to install temporary horizontal bracing during erection, which is subsequently removed.

Wind-bracing is not needed in all the bays of a floor or roof, but only in enough to ensure that every joint in the horizontal framework is maintained in its correct position.

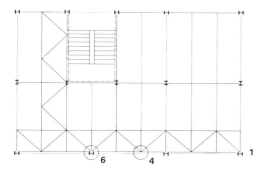

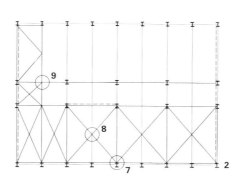

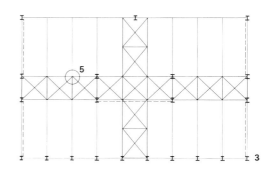

1 The staircase is enclosed by three vertical bracing systems. The horizontal bracing in the longitudinal direction of the building comprises a lattice girder, one chord of which is formed by the external main beams of the floor, while the other is a separate member running parallel to it. The transverse horizontal bracing is inserted between two floor beams.

2 In this example, the vertical bracing is inserted in the two end walls and also between a pair of internal columns. The horizontal bracing in the longitudinal direction is installed between one internal row of main beams and the external columns, which are interconnected by beams forming the outer chord of the lattice girder. The transverse horizontal bracing is inserted between two floor beams.

3 Here, also, the vertical bracing is inserted in the two end walls and between a pair of internal columns. The longitudinal horizontal bracing is installed between the internal columns (a favourable solution if a central corridor is to be provided). The transverse horizontal bracing is inserted between the two central floor beams.

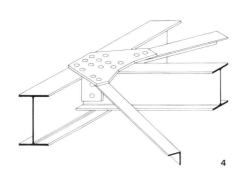

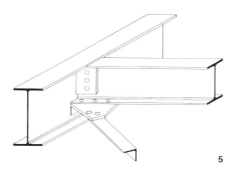

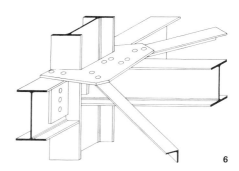

4 The web members of lattice bracing can conveniently consist of angle sections. Here, they are installed at top flange level of the floor beams. When profiled sheet-steel floor decking is used, however, the gusset plate and bolt heads may be a nuisance.

5 An alternative is to insert the wind-bracing at bottom flange level of the floor beams.

6 Connection of simple angle bracing at the junction between an external column and main and floor beams.

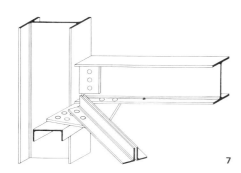

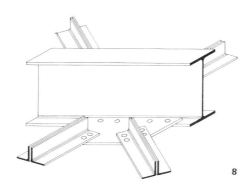

7 If the floor beams are all connected directly to columns, without main beams, a separate member (in this instance, a channel section) must be provided to form the chord of the lattice bracing (as in **2**).

8 Intersection of the wind-bracing members with a floor beam.

9 If the floor beams are supported on main beams, structurally the best solution is to attach the bracing members at the level of the bottom flanges of the floor beams.

Stiffening elements — Vertical wind-bracing

Vertical bracing elements

Lattice girder systems are also used for the vertical bracing of multi-storey buildings, the 'chords' of which are often adapted from the columns of the building and the 'posts' from the floor beams. The diagonal web members are usually purpose-made, but sometimes it is possible to adapt stair stringers or ramp beams. The bracing can be so designed that the diagonals are either loaded only in tension or are alternately loaded in tension and compression. Diagonals designed to resist tension only can have smaller cross-sectional dimensions than those which have to resist compression or those which, depending on the direction of the wind, function as ties or as struts. The latter must be able to withstand buckling, whereas ties can be of any cross-sectional shape, such as flat or round bars, which take up very little room in the construction of the walls. When ties are used as diagonal cross-bracings, for a given wind direction one tie in each lattice panel will be in tension, while the other will be unstressed.

Vertical loads on lattice bracing systems

It must be borne in mind that the lattice diagonals also participate in the transmission of vertical loads. The columns undergo compressive strain during the erection of the building, with increasing vertical load, and also after completion, when the live loads are added. The diagonal members are inevitably involved in this shortening and are thus loaded in compression. Hence:

- Diagonal members intended to function as struts in the lattice system must be designed to take these additional compressive forces, as must also their connections.
- The compressive forces in the diagonal members can be reduced by deferring the completion of the permanent connections until the greater part of the permanent load has been applied.
- In the case of diagonal members designed to function only as ties, the permanent connections should not be completed until the vertical loads have been applied to the structure. To ensure that the diagonals will not go slack under compressive live loads, it is advisable to prestress them with a load equivalent to this compression.

Types of vertical wind bracing

The centre-lines of the diagonal members should pass through the intersections of the centre-lines of the beams and columns. Eccentric connections produce secondary moments in the lattice members.

The diagonals should, as far as possible, form angles not differing greatly from 45° with the other members. If the lattice panels are of elongated rectangular shape, a V-shaped or A-shaped arrangement of diagonals may be adopted. In wide and high buildings, individual cross-bracings may extend over several storeys.
1 It should be borne in mind that vertical bracing is usually located in or close to planes that have to be closed by walls. To make provision for door and window openings, diagonal members may be kinked, in which case the point of change of direction must be secured by an extra member.

The diagonal members are usually connected by bolting. Narrow vertical bracing systems, including the diagonals, may, however, be welded in the works and transported to the site as prefabricated units.

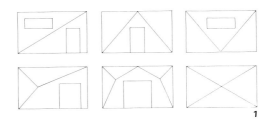

1

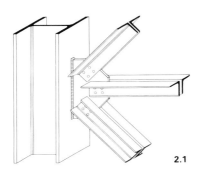

2.1

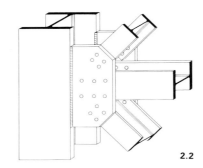

2.2

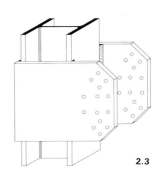

2.3

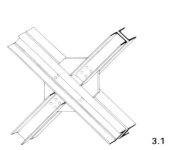

3.1

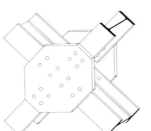

3.2

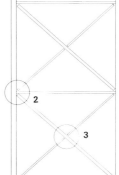

2.1 and **3.1** Crossed diagonals as light bracing between a pair of structural columns, to which they are connected by bolting to welded gusset plates. At the intersection, one diagonal continues uninterrupted, while the other is spliced with the aid of a gusset plate. Channels or angles can be suitably employed as diagonals.
2.2, **2.3** and **3.2** Crossed diagonals as heavy bracing between two columns. The diagonals consist of universal sections connected to the columns with pairs of gusset plates at each joint. If the gusset plates are in the plane of the flanges, they are connected with butt welds, **2.2**; if they are placed transversely to the flanges, fillet welds are used, **2.3**. At the intersection, **3.2**, one diagonal is continued uninterrupted, while the other is spliced with the aid of gusset plates.

Vertical wind-bracing
Stiffening elements

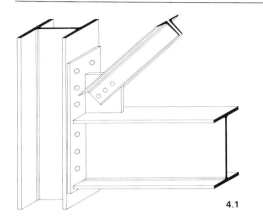
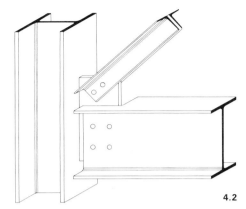
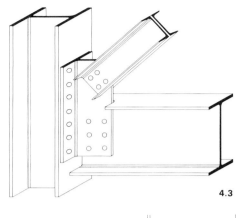
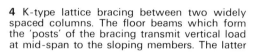

4 K-type lattice bracing between two widely spaced columns. The floor beams which form the 'posts' of the bracing transmit vertical load at mid-span to the sloping members. The latter are thus loaded in compression. They are connected by means of gusset plates, which may either be welded to the floor beam, **4.1**, or be welded (**4.2**) or bolted (**4.3**) to the column.

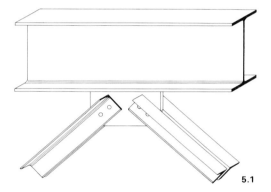
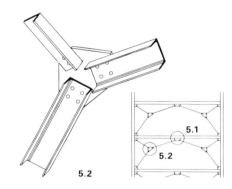
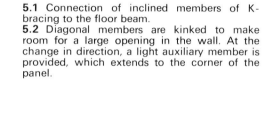

5.1 Connection of inclined members of K-bracing to the floor beam.
5.2 Diagonal members are kinked to make room for a large opening in the wall. At the change in direction, a light auxiliary member is provided, which extends to the corner of the panel.

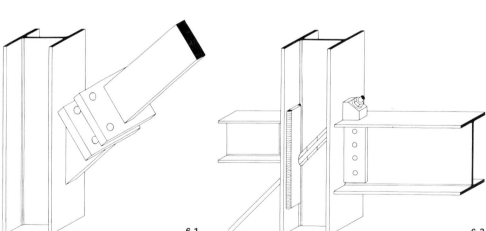
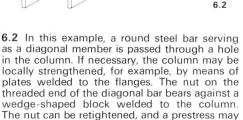
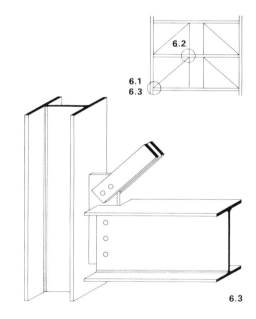

6.1 In this example the diagonal, consisting of a flat steel bar, has a welded end plate which is bolted to a corresponding cleat welded to the column. High-strength bolts are used to form the connection, by means of which it is possible to prestress the diagonal. Thus, under wind load, both diagonals participate: in one of them the tensile stress due to this load is additional to the prestressing force, while in the other the compressive stress correspondingly reduces it.

6.2 In this example, a round steel bar serving as a diagonal member is passed through a hole in the column. If necessary, the column may be locally strengthened, for example, by means of plates welded to the flanges. The nut on the threaded end of the diagonal bar bears against a wedge-shaped block welded to the column. The nut can be retightened, and a prestress may be applied to the diagonal, if required.

6.3 Double flat bars as tension diagonals. The connecting bolt holes are not drilled or reamed until all the loads have been applied to the structure. During erection, temporary attachments are used to ensure stability. Under wind load one pair of diagonals in each panel participates, while the other is inactive. With wind in the opposite direction, the roles are reversed.

Stiffening elements **External bracing**

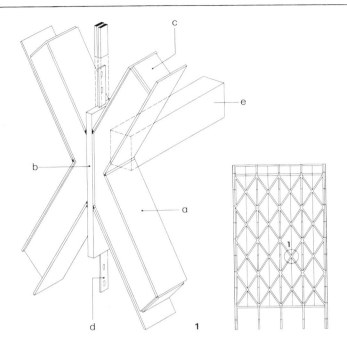

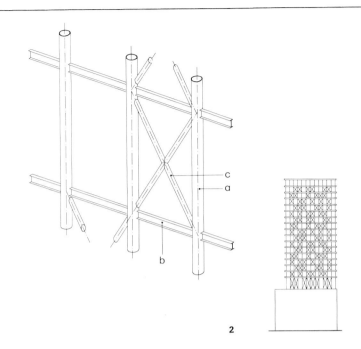

1 This vertical bracing is exposed in front of the façade. The diagonal stubs, a, are box-sections which are welded to the gusset plate, b, and butt-jointed to the main diagonals at c. The hangars, d, provide intermediate support for the floor structures while the stub, e, connects the floor structure directly to the joint in the diagonals. (Alcoa Building, San Francisco, USA: p. 216 **9**.)

2 External vertical lattice bracing, comprising tubular structural columns, a, floor edge beams, b, and welded tubular diagonals, c. The steelwork is exposed in front of the infilled façade. (High-rise residential building in Paris: p. 216 **10**.)

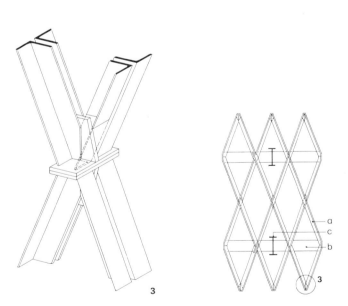

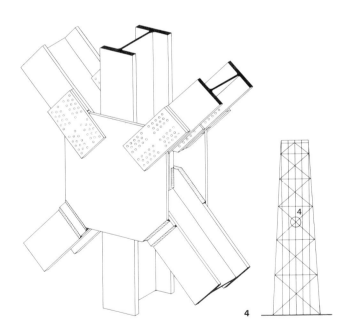

3 Diamond-mesh lattice system located externally. The inclined members perform the dual functions of the structural columns and of the lattice diagonals. The members, a, are double angles between which the main edge beams, b, are welded and to which the floor beams c are bolted. The lattice system was prefabricated in units 7.60 m high and 4.10 m wide in special jigs and assembled on site by means of high-strength friction-grip bolts. The lattice members have fireproof casings and are faced with stainless steel sheet. (IBM Building, Pittsburgh, USA: p. 222 **12**.)

4 Simplified detail of a main joint in the external bracing system for the John Hancock Center, Chicago, USA (see p. 221 **10**). It comprises heavy, almost storey-height gusset plates, with welded and bolted connections. The columns and diagonals are heavy welded I-sections, all of which have fireproof casings and sheet-metal cladding.

Transmission of forces to concrete walls

Stiffening elements

Structural action

Concrete walls are very suitable for the transmission of horizontal forces because, on account of their high shear strength, such walls undergo very little deformation. As shear walls they transmit only the forces acting within their plane, **1**, that is, each wall resists wind load in one direction only. To stiffen a building it is therefore necessary to have at least two walls more or less at right angles to each other or joined to each other at a corner, **2**. A box formed by four shear walls functions as a core which can also resist torsion, **3**, provided that the core is adequately stiffened by internal diaphragms or by rigid floor slabs to preserve its cross-sectional shape under torsional loading, which may arise in consequence of the eccentric location of the core within the ground-plan of the building.

Construction procedure

Shear walls and cores are, as a rule, concreted in situ, precast concrete construction, though possible, being infrequently employed. They are usually built before the steelwork is erected, since they must give support to the latter. Exceptionally, shear walls or cores may be built after steelwork erection, in which case temporary bracing must be used for the steel. Because of their low stiffness perpendicular to their plane, individual shear walls cannot be constructed more than a few storeys ahead of steelwork erection. On the other hand, cores, which are stable in themselves, can be completed in advance or be constructed in a timed repetitive sequence of operations well ahead of the steelwork. In the former case, cranes for handling the steelwork can be attached to, or erected on top of, the completed core. The following construction techniques are used for the core.

Concreting procedures

Individual formwork

A method in which the formwork is fixed and the walls are concreted separately for each storey is only suitable for relatively low buildings. The formwork, however, can be constructed to close tolerances and various recessed or projecting elements for connecting the floors can quite easily be incorporated.

Climbing formwork

For high-rise buildings the formwork, supported by the concrete wall itself, may be raised storey by storey. Projecting elements are, however, very difficult to incorporate in this system.

Slipforming

This is generally the most economical method. Nevertheless, it does presuppose that the core is virtually a closed box-section with no projecting elements. As the height of the slipformed concrete core increases, dimensional inaccuracies are also liable to increase; on the other hand, the correct cross-sectional shape of the core is very well maintained with this technique. The wall thickness can be reduced stepwise, but each change requires appropriate alterations and adjustments to the formwork.

Precast concrete components

Shear walls can be constructed to very close dimensional tolerances from precast concrete units, especially if these are accurately cast in steel moulds. The units, including those which project, can be accurately placed in position but the connections whereby the units are assembled should be designed and executed with considerable care. Unfortunately, walls assembled from precast units are not very suitable for transmitting large horizontal forces. Heavier units would improve their performance in this respect, but would be impracticable from the point of view of transport and handling. The precast concrete units are erected together with the steelwork, which is a time-saving procedure, as it dispenses with the need to wait for the advance construction of at least a substantial part of the core.

Transmission of forces

The concrete walls are designed and constructed in accordance with the reinforced concrete regulations. Arrangements to connect the floor structures to these walls call for special attention. As horizontal forces are transmitted by the floors, it is necessary to ensure effective connection to the concrete walls which usually also fulfil the other requirements such as fire protection and sound insulation. Of course, the floor beams must also have suitable bearings in the walls for the transmission of vertical forces, **4**.

Forces to be transmitted

The following forces have to be transmitted from the floor structure into the concrete wall or core, **5**:

• Vertical force V due to dead and imposed loads.
• Horizontal force H in longitudinal direction of beam, due to wind pressure and suction on the external wall parallel to the concrete wall and restraint forces due to temperature variation in the same direction.
• Horizontal force T due to wind force on the external wall perpendicular to the concrete shear wall, and restraint forces due to temperature variation in the same direction.
• The fixing moment M should be avoided by suitable design of the connections.

To stabilise the structural frame, a proportion of these forces must be resisted during erection. This may present a major difficulty arising from the lower degree of dimensional accuracy with which the concrete walls can be built compared with steelwork. To ensure effective connections between the steelwork and the concrete, while the construction of the building is still in progress, special arrangements are necessary.

Tolerances

The dimensional tolerances attainable with in-situ concrete construction, with good workmanship and using individual formwork, are approximately 2 to 3 cm. In the case of a high core concreted by slipforming, a deviation can occur of up to 10 cm from its theoretical centre-line over a height of 100 m. With precast concrete it is possible to work to closer tolerances.
Such dimensional inaccuracies must be allowed for in the design of the connections between the steelwork and the concrete. In theory there are six degrees of freedom and therefore six possible movements at a connection, **6**.

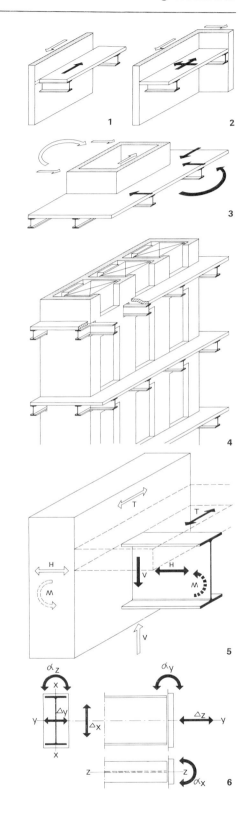

Δ_x height tolerance
Δ_y tolerance for lateral displacement
Δ_z length tolerance in longitudinal direction of beam
α_x rotation of connecting devices about vertical axis
α_y rotation of connecting devices about horizontal axis
α_z rotation of connecting devices about longitudinal axis of beam

Stiffening elements — Connection of beams to concrete walls

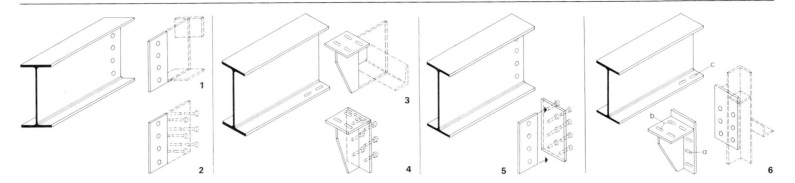

Cleats embedded in the concrete

This type of connection offers only a limited possibility, determined by the clearance of the bolts in their holes, of compensation for dimensional inaccuracy, namely, Δ_z, Δ_y and α_z. It is therefore only suitable for use with identical precast concrete units produced in steel moulds.
1 Transmission of forces H and V through anchor plates welded to the cleat.
2 Cleat with shear studs.

Brackets embedded in the concrete

This connection can be used for walls concreted in situ in purpose-made formwork, the latter being provided with re-entrant blocks or boxes to form the recesses. Tolerances Δ_z, Δ_y and α_x are accommodated by slots in the bracket and in the bottom flange of the beam. Compensation for height tolerance Δ_x is provided by shims. Tapered shims compensate for the angular tolerances α_y and α_z.

3 Absorption of forces V and H by an angle section welded to the back of the cleat.
4 The force V is transmitted through the shanks of the welded studs, while the force H is transmitted in compression by the end plate and in tension by the heads of the studs.

Connection to slipformed concrete

In general, with slipformed construction, no embedded steelwork can be allowed to project from the surface of the concrete.
5 The connecting plate is embedded flush and secured by studs. A vertical cleat is subsequently welded to it. Allowance for tolerances Δ_x, Δ_y, α_x and α_z is obtained by welding the cleat in the correct position, and compensation for Δ_z and α_y is obtained by cutting this plate, supplied somewhat oversize, to the precise dimension. As accurate measurement and positioning are necessary, the in-situ welding operation is relatively time-consuming and expensive.

6 The connecting plate is welded to an embedded vertical angle which in turn is anchored by a horizontal angle. This method requires a system of angle sections embedded in the concrete, but ensures accurate location of the connecting plate. The latter is provided with holes and with welded-on cap nuts at the rear. A bracket provided with slotted holes is bolted to the plate to support the beam, the bottom flange of which is also provided with slotted holes.
Allowance for the various tolerances is achieved as follows:

Δ_x by intermediate shims
Δ_y by slots a and b in the bracket
Δ_z by slots c in the beam
α_x by slots b and c
α_y by tapered shims
α_z by slots in the bracket

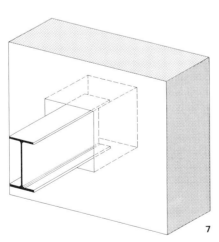
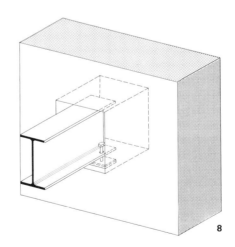
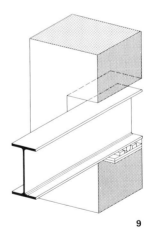

Bearings in pockets

Bearings in pockets are by far the most logical and economical method. They can be provided in any type of formwork. However, they have the drawback that, in addition to the dimensional inaccuracies of the in-situ concrete, there are often quite considerable inaccuracies in the location of the bearing surfaces. During erection, the steelwork must be secured by temporary bracing or other means, as positive connection to the concrete is achieved only after the pockets have been filled with concrete. When the beam has been set on skims or spacers in the pocket, mortar is packed under it. The final concreting of the pockets should be deferred until as much as possible of the dead load has been applied to the structure, otherwise substantial restraint moments are liable to develop in the beams.
7 The beam may be supported in the pocket with the aid of bearing plates or packings to allow for adjustment. It is best to install the plates accurately and secure them with mortar before the beams are placed in position.
8 During erection, the beam may be supported by two adjustable bolts, the space under it being subsequently packed with mortar to form the permanent bearing.
9 Here, the beam is supported by a knife-edge bearing. The base plate of the bearing should be installed, accurately positioned and grouted before the erection of the steelwork. Adjustment of the base plate is facilitated by providing it with three adjustable bolts engaging with cap nuts welded beneath holes in the plate.

Connection of floors to concrete walls

Stiffening elements

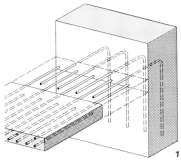 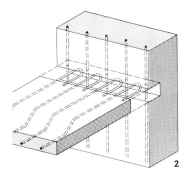 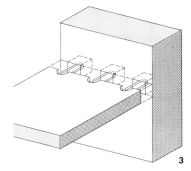 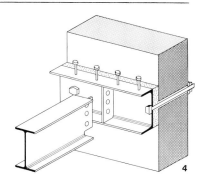

1 The horizontal compressive and tensile forces H and the shear forces T are transmitted by the floor slab to the wall. Starter bars projecting from the wall overlap the bars in the slab. If the latter is composed of precast units, an in-situ concrete joint can be formed at the junction with the wall which can also make allowance for dimensional inaccuracies. The width of the strip thus concreted in situ is determined by the required lap length of the reinforcement. The latter must be suitably designed to cope with the fixing moment if a rigid connection between the floor and the wall is formed.

2 Projecting reinforcement is impracticable if the wall is slipformed. In that case, a continuous horizontal recess may be formed, traversed by the vertical bars in the concrete wall. The bars protruding from the slab, bent to form hooks or loops, project into this recess. This form of construction can also be employed with precast concrete floor slabs, the joint then being formed by a strip of in-situ concrete placed around the projecting bars and extending into the recess.

3 More effective transmission of shear force is obtained by providing the wall with a series of recesses or pockets into which the loops projecting from the floor slab are inserted. If the floor is composed of precast slabs, these can be provided with corresponding recesses or indentations along the edge adjacent to the wall to obtain efficient interlocking action with the in-situ concrete forming the joint.

4 This bearing is formed by means of a continuous channel section secured to the concrete wall by means of bolts extending through tubes embedded in the latter. The clearance between these tubes and the bolts makes allowance for tolerances. The tubes are grouted, and grout or mortar is placed between the channel section and the wall and also between the bearing plate for the nuts at the rear and the wall. This form of construction may advantageously be used in circumstances where numerous steel components have to be connected to a concrete core and a bearing for the floor structure is also required. The concrete floor rests on the top flange of the channel, which is provided with studs for the transmission of shear.

CONNECTIONS FOR PRECAST CONCRETE SHEAR WALL UNITS

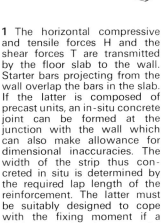 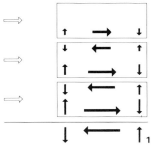 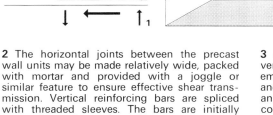 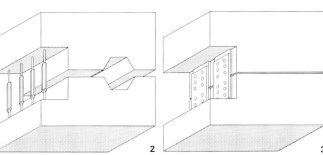

With precast concrete construction, the advantages of shear walls can be combined with a rapid method of erection which is independent of weather conditions.

1 The resultant wind force from the successive storeys has to be transmitted to the foundations by the bottom precast wall unit. At each horizontal joint the shear force has to be transmitted, while in addition the vertical forces due to wind and dead load must also find a suitable path to the foundations. This solution is practicable only for low-rise buildings (up to six storeys). As the concrete units have to be cast in precision-made steel moulds, to ensure efficient fit of the units being joined together, precasting becomes an economic proposition only when relatively large numbers of units have to be produced. The connection of floor beams to such shear walls may be made as indicated in p. 268 **1, 2**.

2 The horizontal joints between the precast wall units may be made relatively wide, packed with mortar and provided with a joggle or similar feature to ensure effective shear transmission. Vertical reinforcing bars are spliced with threaded sleeves. The bars are initially screwed into the steel moulds to ensure correct alignment on subsequent erection of the units.

3 Alternatively, a positive connection across the vertical joint may be formed with steel plates embedded flush with the face of the concrete and provided with welded studs at the rear to anchor them. After erection of the units, the connections are formed by means of bolted splice plates. The embedded plates are provided with holes and welded-on nuts to receive the bolts. During casting, these plates are bolted to the steel mould to ensure exact positioning. High-strength friction-grip bolts may be used at these joints, and the same form of construction may be used also for shear transmission.

Stiffening elements

Precast concrete basement walls

Basement walls may be constructed with precast concrete units and erected simultaneously with the steelwork. Construction time is thus reduced.

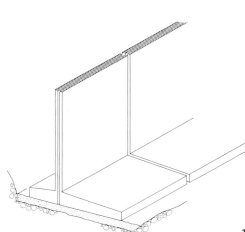

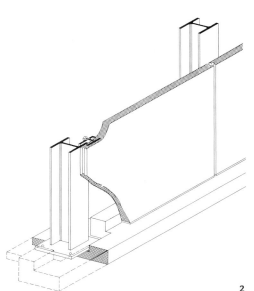

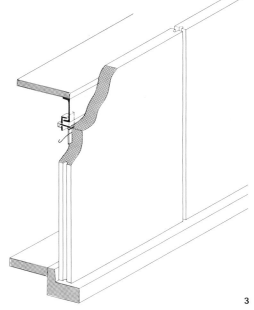

1 Concrete units, each comprising a wall and a foot portion, are bedded on a prepared base. As an angular retaining wall the structure is self-supporting when the soil is backfilled on the outside.

2 In this example, the retaining wall slabs resist the thrust by deriving support from the steel columns and from a ridge or nib formed on the strip footing. The thrust exerted by the backfill may be transmitted through floor slabs to shear walls, cores or other stiffening systems, or it may be resisted by fixed-base columns directly. The latter may be the structural columns for the building. The concrete basement wall slabs may be additionally supported by the basement floor structures, more particularly by the edges of the concrete floor slabs. The wall slabs used for basement construction range in thickness from about 10 to 14 cm.

3 The precast wall slabs are supported at the base by the strip footing and at the top by the floor slab. As these wall slabs are supported on two sides only and have a relatively deep span, they have to be made correspondingly thick and heavy. If desired, their thickness may be reduced from about 20 cm at the bottom to about 10 cm at the top.

4 The joints must be carefully detailed to ensure watertightness and yet allow some degree of movement between adjacent units. In simple cases it may be sufficient to employ concrete-filled or grouted grooves, **4.1**. A better joint is obtained by installing a water-bar, **4.2**. The water-bar may be precast into one unit and subsequently grouted into a groove formed in the adjacent unit, as in **4.3**. It is also advisable to add an external flexible waterproof strip, bedded in bitumen.

Floors and Staircases

Floors
- Composition and function of floors — 271
- Structural floor — 273

Steel decks
- Floors with loadbearing steel decking — 275
- Floors with loadbearing concrete slabs — 276

Concrete floors
- Long-span concrete floors — 277
- Short-span concrete floors — 278

Fire protection of floors
- Basic principles — 279
- Forms of protection — 280

Suspended ceilings
- Functions of the suspended ceiling Components — 281
- Plastered ceilings — 283
- Prefabricated panels — 283
- Demountable prefabricated panels — 284

Horizontal services
- Services installed in the floor — 285
- Heating installations in the floor cavity — 286
- Air ducts in the floor cavity — 287
- Passage of services through steel beams — 288
- Electrical installations in the floor — 288

Floor finishes — 289

Staircases
- Function and types — 290
- Staircase geometry — 291
- Structural aspects — 293
- Staircase enclosures — 293
- Arrangement of staircases in steel-framed buildings — 294
- Stairs in concrete shafts or cores — 294
- Concrete stair details — 295
- Steel stair details — 297
- Various forms of steel stairs — 298

COMPOSITION AND FUNCTION OF FLOORS

In a multi-storey building the floors form the horizontal separations between the storeys. As well as the transmission of the vertical and horizontal loads acting on it, a typical floor also has to provide some degree of insulation against sound, heat and moisture, besides performing a structural protection function. The horizontal services are usually installed in the floor, which, when taken to include the ceiling on the underside and the floor finish on top, forms the upper and lower visual boundary of the rooms below and above it.

To fulfil these requirements it is often necessary to employ a multi-layer form of construction. Thus, the thickness or depth of the floor structure as a whole depends on this construction and on the thicknesses of the constituent layers and, in turn, has some effect on the overall height of the building.

As a rule, three functional zones may be identified in a floor of a modern multi-storey building:

• The loadbearing, or structural, floor comprises the floor slab and the floor beams.
• The ceiling forms the soffit of the floor structure. It may simply be plaster applied to the underside of the floor slab and the sides of the beams or it may consist of a suspended ceiling, often of multi-layer construction.
• The floor surfacing, comprising one or more layers, lies on the structural floor and has to perform load-distributing and insulating functions of various kinds. It also forms the base for the floor covering. The various layers composing a floor structure are indicated in Figure **1** and the accompanying table.

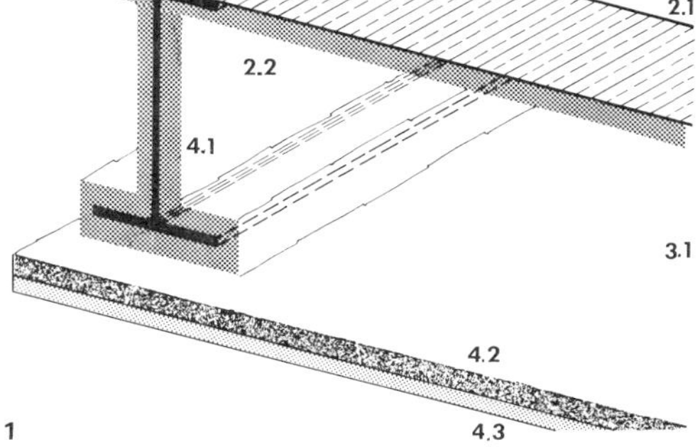

		Load transmission	Sound insulation	Fire protection	Heat insulation	Protection against moisture and condensation	Services	Visual boundary
1. Floor surfacing	1.1 Floor covering		▮			▮		▮
	1.2 Floor finish						▮	▮
	1.3 Waterproof membrane					▮		
	1.4 Insulation		▮		▮			
	1.5 Screed							
2. Structural floor	2.1 Floor slab	▮	▮	▮		▮		▮
	2.2 Floor beams	▮		▮				
3. Underfloor space	3.1 Cavity						▮	
4. Ceiling	4.1 Cladding to slab and/or beam			▮				▮
	4.2 Upper layer							
	4.3 Suspended ceiling		▮	▮			▮	▮

Floors

Composition and function

Load transmission

The floor takes the vertical dead and imposed loads and transmits them to its points of support in the structural steel frame. Furthermore, the floor transmits the horizontal forces due to wind or earthquakes to the bracing systems in the building.

Sound insulation

In modern high-rise buildings, involving large numbers of people concentrated in residential or working space in close proximity to one another, acoustic insulation is a very important consideration. The floor must provide the necessary insulation against airborne noise and footstep sound. In respect of this acoustic function, the floor comprises one or more layers:

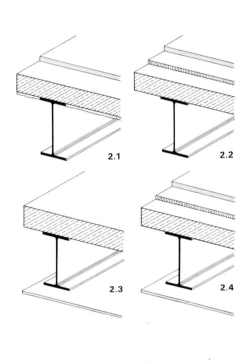

2.1 The single-layer floor comprises only the structural floor. Acoustically it rates as a single-layer system even if it is provided with an adhering top surfacing or finish and a plaster coat on the underside. Such a floor can provide adequate insulation against airborne noise if its weight is great enough, but it cannot ensure effective footstep sound insulation.
2.2 The two-layer floor comprises the structural floor and a surfacing separated from it by a resilient insulating material such as glass-wool (so-called floating floor).
2.3 Alternatively, a two-layer floor may consist of the structural floor with a suspended ceiling fitted underneath.
2.4 The three-layer floor is a combination of the two foregoing forms of construction, that is a floating floor with a suspended ceiling.
Evaluation of the acoustic properties of a floor should comprise the floor structure as a whole. The numerous possibilities in terms of materials and methods of construction cannot be discussed within the scope of this book. Reference should be made to specialised literature, for example: Moll, *Bauakustik* (*Architectural acoustics*), Wilhelm Ernst & Sohn, Berlin.

Fire protection

The requisite fire resistance of a floor can be achieved in various ways (p. 279):

- for the floor slab itself: choice of a fire-resistant form of construction or a protective coating or finish;
- for the steel beams: sprayed protection or cladding;
- as an alternative, the floor structure as a whole may be encased.

Heat insulation

If a floor separates a heated from an unheated storey, it should comprise a heat-insulating layer, which is not needed if the storeys above and below the floor are both heated. Terraces should be heat-insulated on the same principles as for flat roofs (p. 300). Floors over external passageways, cantilevered floors or floors over unheated basements require special insulation: the structural floor may be provided with a cladding on its underside, or a sub-floor may be provided on which a layer of insulating material is laid.

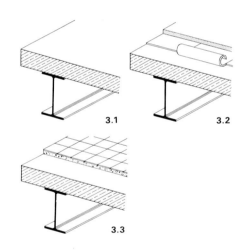

Protection against moisture

3.1 For waterproofing a floor which is directly trafficked or walked on, for example, a multi-storey car park, see p. 285.
3.2 Waterproofing of open-air floors, such as terraces, floors over basements in courtyards, or top floors of car parks. An asphalt surfacing is not sufficient by itself, as asphalt is liable to crack. For modest requirements a 1 cm thick mastic asphalt waterproofing course under the surfacing suffices. Over rooms or similar spaces, a waterproof layer of aluminium or copper sheet laid in bitumen, or plastic sheet bonded with an adhesive, should be used. This should be covered by a protective layer, on which an appropriate surfacing for vehicular or pedestrian traffic should be laid.
3.3 'Wet rooms' with high humidity levels, such as kitchens and bathrooms, should have waterproof floor coverings such as ceramic tiles, glued plastic tiles or plastic strips.

Condensation water may form on the underside of a floor supported on steel beams over open spaces (for example, ice skating rinks) or over enclosed spaces with high humidity levels (shower cubicles, laundries, swimming baths) and cause objectionable dripping. If there is relatively little condensation, dripping can be prevented by coating the floor slab and the steel beams with a porous plaster, for example, vermiculite or asbestos plaster. Such a coating is able to absorb a certain amount of water and to part with it later, by evaporation. If humidity is continually at a high level, it is necessary to extract the moist air and supply dry air. With regard to corrosion protection for structural steelwork, see pp. 357–358.

The floor as a support for services

This function of the floor becomes very important in large modern buildings with their increasingly sophisticated service installations. To ensure flexibility of internal layout, the vertical services are generally concentrated in a limited number of shafts, keeping the internal walls and partitions as much as possible free from services, while horizontal distribution of the services within each storey is accommodated within the floor structure. The latter group may comprise various units for lighting, heating and ventilation, along with sprinkler equipment, pneumatic despatch lines and many other items, some of which may additionally be fitted with switching and control devices, all accommodated in the floor–ceiling system. For further particulars, see p. 330.

Depth of floors

The floor must be designed with due regard to all the functions it will be required to perform. Its overall depth, which is an important feature in the design of the building as a whole, depends on:

- the span, the loading and the allowable deflection of the beams;
- the arrangement of the beams (in one or in two layers);
- the thickness of the floor slab;
- the space required by the various services to be installed in the cavity, especially the air ducts;
- the depth of the false ceiling construction;
- the thickness of the floor finishes.

The following are some approximate figures for guidance:

- Floor surfacing:

floor covering	5 mm
floating finish	35–45 mm
noise prevention and heat insulation	15–25 mm
screed with electric wiring	30–100 mm
screed	15–25 mm

- Floor slab:

reinforced concrete slab	100–150 mm
concrete slab on steel troughing	120–160 mm
flat slab or ribbed construction	150–400 mm

- Beams:

floor beams	240–400 mm
main beams	360–600 mm

- Ceiling:

false ceiling providing fire protection	40–80 mm

Composition and function

Floors

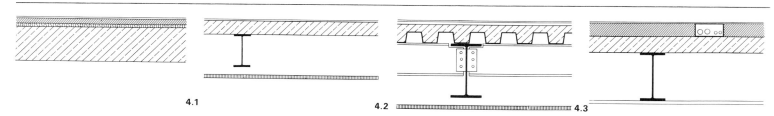

Examples of floor construction

4.1 Residential building: imposed load 200 kg/m²; span 4.50 m

floor covering	5 mm
floating finish	40 mm
insulation	20 mm
concrete slab (without beams)	250 mm
	315 mm

4.2 Office building: imposed load 200 kg/m²; span 6.00 m; without elaborate services

floor covering	5 mm
concrete slab	100 mm
floor beams	240 mm
false ceiling	80 mm
	425 mm

4.3 School or institutional building: imposed load 350 kg/m²; spans 7.20 × 7.20 m; without elaborate services

floor covering	5 mm
screed	15 mm
concrete slab on steel troughing	140 mm
main beams	400 mm
false ceiling	80 mm
	640 mm

4.4 Institutional building: imposed load 500 kg/m²; spans 9.00 × 12.00 m; elaborate services

floor covering	5 mm
screed with electric wiring	100 mm
concrete slab	120 mm
floor beams	360 mm
main beams	450 mm
false ceiling	80 mm
	1115 mm

STRUCTURAL FLOOR

Structural action

The structural floor in a steel-framed building comprises the floor slab with or without steel beams. For details of steel beams, see p. 242.
The floor slab performs various structural functions. In transmitting vertical load to the steel beams it acts as a slab in the sense that it is loaded transversely to its own plane and functions as a flexural member, **1**. It acts as a diaphragm, that is, a member loaded only within its own plane, in the transmission of horizontal loads to the bracings, cores, etc., of

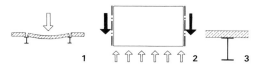

the building, **2**. In composite structures the floor slab acts as the top flange of the composite section, **3**. In steel-framed buildings only two materials are of practical significance for the construction of structural floors: steel and concrete.

Usual forms of construction:

1. Normal or lightweight concrete slab, usually of constant thickness, more rarely with ribs in one or in two directions or voided.
2. Steel-sheet decking, profiled to increase its rigidity and load bearing capacity, always provided with concrete filling for load distribution and to provide fire protection and sound insulation.

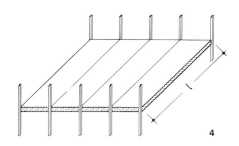

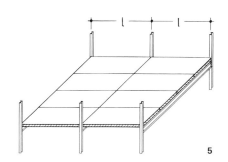

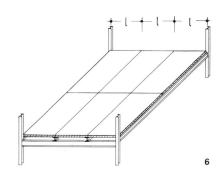

Range of spans

4 Very long-span concrete floor slabs (spans l ranging from 5.00 to 10.00 m, exceptionally even more) are supported directly by the steel columns, so that floor beams and main beams are dispensed with.

5 Long-span slabs (spans l from 3.00 to 7.00 m) which rest on main beams, dispensing with floor beams. These are generally constructed either as solid concrete slabs (ranging in thickness from 15 to 25 cm) or as ribbed concrete slabs (overall depth up to 40 cm).

6 In most steel-framed buildings the floor slabs have short spans, l from 1.20 to 4.00 m, the most economical range being between 2.40 and 3.00 m. These are solid concrete slabs of constant thickness or supported by steel decking.

Floors

The structural floor

The following considerations are relevant in determining the span to be adopted for the floor slab:

- The span of the slab should fit in with the dimensions of the planning grid for the building.
- The thickness d of the concrete slab depends in many instances not only on structural considerations, but also on such requirements as sound insulation and fire protection.

For optimum economy, the span b of the slab should be so chosen that, for this thickness, the quantity of reinforcing steel g_{st} remains within suitable limits. The following are some approximate figures for guidance:

$b = 2.40$ m; $d = 10$ cm;
imposed load $p = 500$ kg/m^2:
$g_{st} = 4-8$ kg/m^2
$b = 3.00$ m; $d = 12$ cm;
imposed load $p = 750$ kg/m^2:
$g_{st} = 6-10$ kg/m^2

- The spacing of the floor beams is determined by the maximum permissible transportable width of the precase concrete units, the main joints of which are parallel to these floor beams.

Lightweight concrete, used either as a filler material or as a structural material in its own right, reduces the weight of the floor. However, it may involve some loss of sound insulation and fire protection. Furthermore, with long spans and heavy loads, the weight advantage attainable with lightweight concrete may be reduced because deeper sections are necessary.

Surface finish of loadbearing floors

The degree of evenness of the surface of the structural floor depends on the method of construction. The dimensional tolerances inherent in the construction of an in-situ concrete floor require the application of a screed as a preliminary finishing layer over the slab. Smaller tolerances are obtained when in-situ concrete is placed on permanent formwork, such as precast concrete units or steel troughing. With careful workmanship, a screed may then be dispensed with. A floor composed of precast concrete units generally requires only a very thin screed. If these units are precast in standardised steel moulds, a surface of such evenness can be obtained that the floor covering may be laid directly upon the structural floor. Small irregularities at the joints can be made good with a filler.

Protection against surface water

Normally, a steel floor cannot be regarded as watertight. To achieve watertightness without applying a special waterproofing membrane, certain precautions should be taken.
An in-situ concrete floor should be laid in one continuous operation, without construction joints, and the concrete itself should be of good quality and well compacted. If the structural floor is composed of factory-made concrete units, the latter are themselves generally watertight. The joints, on the other hand, are not, unless special precautions are taken such as those for the floors of open-type multi-storey car parks. The joints may be sealed by bonding the precast concrete floor units together with a special adhesive mortar which should attain at least the tensile strength of the concrete, while the structure is so designed that this strength is not exceeded under actual service conditions.

STEEL DECKS

Steel decks are composed of profiled steel sheets from 1 to 2 mm thick. Normally, a single sheet is used but cellular decks require two sheets. They carry at least a 5 cm thickness of in-situ concrete or some other load-distributing layer.
If the decking is used merely as permanent shuttering, the thickness of the sheet may be below 1 mm.
The panel-type decking is produced from mild steel sheet, continuously galvanised on both sides and cold-formed, usually with a trapezoidal section. The flanges and/or the webs of the section are frequently stiffened by rolling in indented ribs or dimples which increase the bond and therefore the composite action with the in-situ concrete. Panel units are obtainable from the producers in breadths from 500 to 1000 mm and depths from 35 to 150 mm. In addition to the 14 to 25 μ layer of zinc, the underside of the sheet may be further protected against corrosion with a plastic coat or paint.

Types of section (Figure 1)

Shapes commonly used:

1.1 and **1.2** are individual trough sections.
1.3 to **1.7** are profiles used for panel units.
1.8 and **1.9** are cellular sections.

Advantages of steel decks:

- light weight;
- rapid erection;
- the concrete slab needs no shuttering;
- a working platform is immediately provided;
- protection for following trades.

Disadvantages of steel decks:

- the decking either serves merely as permanent shuttering, or,
- if it performs a loadbearing function, it may require fire protection underneath;
- compared with 'dry' construction methods, it

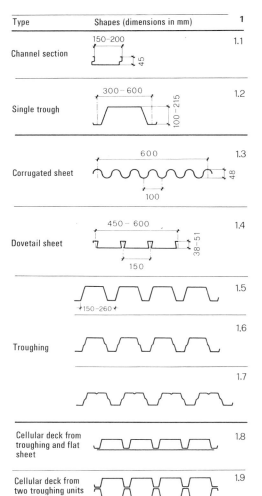

has the drawback of introducing moisture with the use of in-situ concrete.

Laying the steel deck

The panels are cut in advance to the size required, bundled and delivered to the site as and when needed for laying in accordance with the erection programme. Thus, floor construction can proceed smoothly, keeping up with the erection of the steel beams. The plain sheets are usually cut to length prior to cold forming but the decking can be cut with special shears suited to the particular profile. Oblique cuts are made with manual tools.

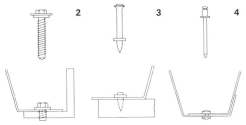

Fastenings

The following methods may be used to fasten the decks to the supporting beams:

- welding, in accordance with the manufacturers' instructions;
- self-tapping screws, in accordance with DIN 7513, or bolts, **2**;
- blind rivets, **3**.

The deck units themselves may be joined together in the following ways:

- pop rivets, **4**, which are installed mechanically from one side only, without being held up on the other side;
- screws;
- pressing or joggling the edges together, in accordance with the manufacturers' instructions.

Floors with loadbearing steel decking

Steel decks

Floors with loadbearing steel decking

With this form of construction the steel decking alone carries the load. In-situ concrete filling, precast concrete slabs or some kind of surfacing laid on an insulating mat distributes the load and provides fire protection and sound insulation on the top of the floor. Underneath, however, some kind of fire protection must be provided. This particular floor can only act as a horizontal diaphragm to resist wind loading if each trough of the decking can be fixed to the supporting beams. If these conditions cannot be fulfilled, the same function can be performed by in-situ concrete with sufficient depth and adequate reinforcement. Otherwise, wind-bracing must be inserted below the steel deck.

Design chart for steel troughing

5 The spacing of the floor beams varies from 1.50 to 4.00 m, the most economical range being from 2.00 to 3.00 m. For the usual loads between 600 and 2000 kg/m², many types of proprietary deck are available, differing from one another only in their profile. The usual depths of these sections vary between 35 and 120 mm, while the weight varies between 10 and 27 kg/m².

The chart is intended to be used to give a rough estimate of the loadbearing capacity of decking. For more precise design purposes, the safe load tables published by the individual manufacturers should be consulted, more particularly because with heavy loading the critical condition is not the allowable bending moment but the shear forces at the supports.

To use the chart, proceed from the total load (which includes the dead load) on the vertical scale and, for spans between 2.00 and 3.00 m, select an appropriate deck for a given thickness of sheet and depth of profile. The values are valid for decks continuous over three spans.

6 The weight of the deck is found under the depth of section. The weights are only approximations because of the different shapes of profile;

l Span of decking (m)
q Load on decking (kg/m²)
h Depth of section (mm)
s Thickness of sheet (mm)
g Weight of decking (kg/m²)

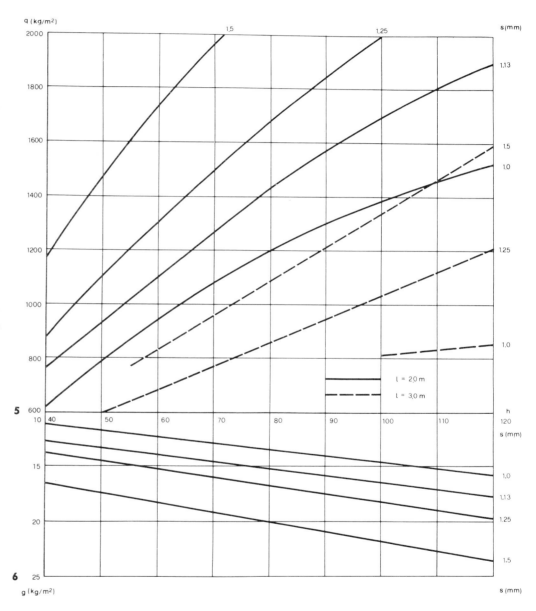

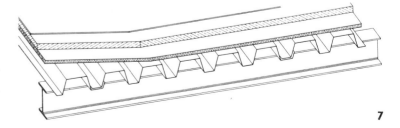

7 Self-supporting galvanised steel troughing, with a finishing course or precast concrete slabs on an insulation layer. This form of construction is considered to have adequate fire protection on top if the finish is thicker than 5 cm and/or the insulating layer consists of mineral fibre or mineral wool boards.

Steel decks

Floors with loadbearing concrete slabs

8 Robertson Q-Floor. The troughing can be closed, a, by introducing a flat sheet beneath it. By adding this sheet, the loadbearing capacity is increased while cells are produced which provide ducts for services, especially cables. The decks are supplied in depths of 40 and 80 mm. An especially wide cell Q-F5 can be inserted in the deeper type which can be used directly as an air-conditioning duct, b.

9 A floor with a much greater loadbearing capacity is produced by welding together pairs of the deeper trough sections. This double profile will carry 2000 kg/m^2 over a span of 4.00 m.

10 Units are joined along their edges by interlocking seams, made with special pliers, while connections are made to the supporting beams with spot welds, in accordance with specific instructions, or with the aid of various kinds of self-tapping screws, etc. Safe-load tables are available which give the allowable shear forces. A concrete topping of at least 5 cm provides the necessary fire protection above the floor. See pp. 279 and 280 for protection of the soffit. For further particulars about services, see pp. 285–288.

The depth of the main beams can be so chosen that the top flanges of these beams are flush with the top of the decking. With this arrangement very simple beam connections are possible. See p. 248 **7**.

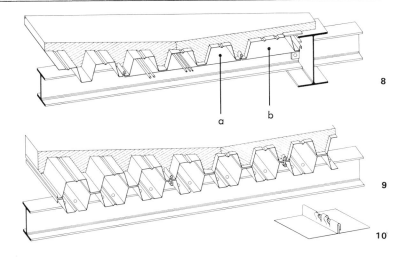

Floors with loadbearing concrete slabs

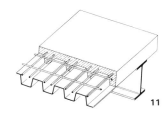

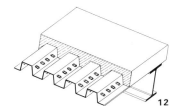

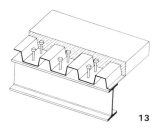

11 The decking serves merely as permanent shuttering for the concrete, the advantage being that rapid construction is possible. Round bars are provided for reinforcement and the resulting slab acts as a ribbed concrete floor. If the reinforcement has adequate cover, the floor slab rates as fire-resistant. It also functions as a horizontal diaphragm capable of transmitting wind forces.

12 Composite action between decking and concrete slab. The decking provides the reinforcement for the concrete and shear connection is produced by ribs or dimples rolled into the sheet. The soffit may require fire protection.

13 Composite action between the concrete slab and the floor beams is achieved by means of stud welded shear connectors. This is a very economical form of construction. The studs are welded on the site in accordance with the manufacturers' instructions.

14 Hoesch and Holorib decks are made by rolling dovetailed ribs into continuously galvanised steel sheets. They may be used either as permanent shuttering or as combined shuttering and reinforcement. In the second case the sheets must be anchored at their ends, say with stud welded shear connectors, to ensure that the sheet replaces the normal reinforcement. The range of spans for this type of decking lies between 2.40 and 8.40 m. As the profile is undercut, this decking has a fire rating of at least 90 min. The dovetail rib has the advantage that it facilitates the fastening of false ceilings and services.

In the diagram stud welded shear connectors are shown which provide shear connection between the concrete slab and the floor beam. The design of composite beams is carried out in accordance with the relevant national standards, without reducing the shear forces.

References: Merkblatt 203, *Stahltrapezbleche für Dach und Wände* (*Steel troughing for roofs and walls*), Beratungsstelle für Stahlverwendung (German Steel Information Centre), Düsseldorf.
IFBS – Verbindungsmittelkatalog (catalogue of fasteners).

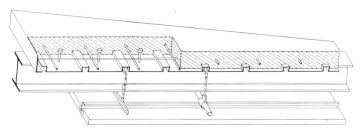

Long-span concrete floors

For structural reasons, long-span concrete floor slabs have to be between 20 and 40 cm thick. As solid slabs they become very heavy, and for this reason they are generally formed with voids or with concrete beams or ribs extending in one or in two directions. These types of floor may be concreted in situ or be precast. The latter alternative presupposes that the floor can be assembled from transportable prefabricated components the width of which for transport on public roads should not exceed 2.50 m, or 3.00 m in special cases, or up to 4.50 m with police escort. Units of greater width are produced in temporary sheds set up on the building site itself but, on account of the financial outlay involved, this is economically justifiable only for very large projects.

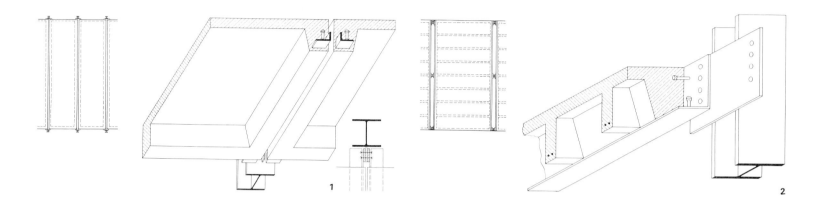

1 Large precast concrete slab units of transportable width, with integral concrete beams on both sides. These beams act compositely with embedded unequal angle sections, the wide horizontal leg of which is provided with stud shear connectors and forms the tensile reinforcement for the beam. The vertical legs of the angle sections in adjacent units are bolted together and also provide a means of connection to the steel columns in the building. The concrete slab, which is about 10 cm thick, can also comprise an edge beam to support external walls. The angle sections embedded in the main beams must have a sufficient amount of concrete cover to ensure their fire protection.

2 Large concrete slab which can be precast on site or in situ in the actual structure. The steel columns are interconnected by special steel sections comprising web and bottom flange and provided with shear studs for composite action with the concrete edge beam. The two edge beams are connected to each other by concrete ribs. The actual slab can be quite thin.

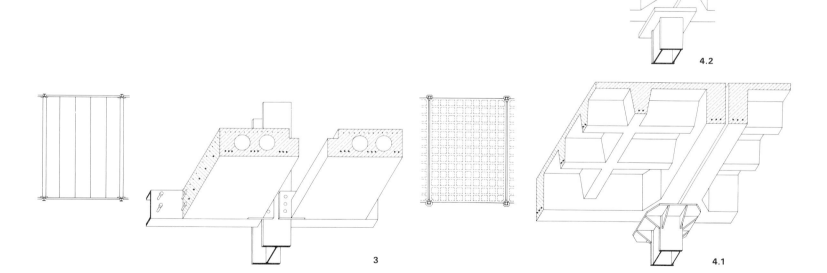

3 Large concrete slab for precasting on site or for concreting in situ. The steel beams are connected by means of studs to a concrete slab flush with their top and bottom flanges and provided with longitudinal voids (preformed with embedded cardboard tubes in accordance with the Röhbau system or with inflatable rubber tubes). The slab, with its flat top and bottom surfaces, is particularly convenient for residential buildings or for other buildings with relatively small floor loads and without services installed in the floor. (Reference: *Beton-Kalender*, 1970, p. 180).

4 Large concrete slab, almost square, comprising concrete edge beams and a slab ribbed in both directions. The unit as a whole is supported on bracket bearings welded to the steel columns. When the slabs are cast on the site in panels of great size, the edge beams are constructed separately. The brackets on the columns are provided with stiffening ribs, **4.1**, or, if construction depth is limited, are formed with thick plain plates, **4.2**.

Concrete floors — Short-span concrete floors

Nearly all types of concrete floor constructed in situ or commercially available as precast concrete units can be used in combination with steel beams.

The following types function also as horizontal diaphragms possessing the necessary rigidity to transmit wind load:

- in-situ concrete floors (**1** to **4**);
- precast concrete soffit planks serving as permanent formwork for in-situ reinforced concrete topping (at least 5 cm thick) (**5** to **7**);
- composite floors.

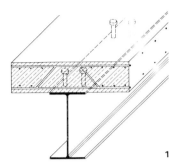
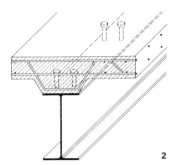
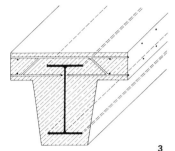
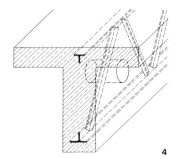

In-situ concrete slabs

1 Concrete slab on steel beams, with or without composite action (p. 258). Most frequently employed form of construction.

2 Haunched concrete slab. With composite construction this makes for greater construction depth, thus effecting a saving on steel in the supporting beam, but shuttering for the slab is more expensive.

3 Steel beam embedded in the concrete beam without composite action. As the top reinforcement is carried over the steel beam the concrete slab is continuous. The web of the steel beam must be adequately covered with concrete for fire protection.

4 Lattice girder completely embedded in concrete.

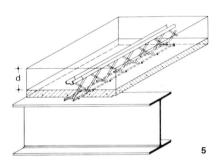
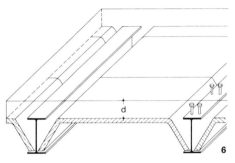
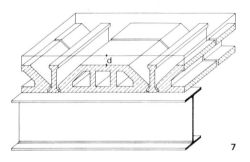

Precast concrete units with in-situ topping

5 Flat precast slabs, approximately 4 cm thick, incorporate the bottom reinforcement of the floor, while the shear and top reinforcement project from the precast concrete, to be subsequently embedded in in-situ concrete. A number of floor construction systems based on this method are commercially available.

6 Thin precast concrete soffit units, approximately 4 cm thick, are fitted between the webs of the steel floor beams and remain in position as permanent formwork for the in-situ concrete slab. The webs of the beams thus receive fire protection. The in-situ slab is reinforced and concreted in the normal way. The steel beams may be provided with studs to provide composite action (as shown on the right in the drawing).

7 Floor comprising light precast concrete beams (conventionally reinforced or prestressed) supporting lightweight concrete filler blocks. The in-situ concrete topping interconnects all the components to form a rigid diaphragm. This system is, however, only suitable for relatively light floor loads. There are numerous proprietary systems on the market; the system illustrated here is that of Arsen Schweizer, Berlin.

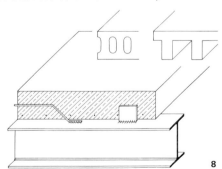
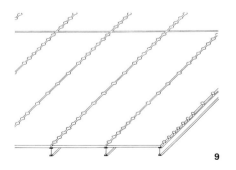
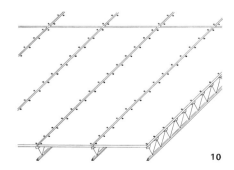

Precast concrete slabs

8 Precast slabs, flat on both sides, consisting of normal, lightweight or aerated concrete, resting on steel beams, but without composite action. To prevent displacement of the slabs, stops or locating devices should be welded to the top flanges of the beams, such as small plates, dowels or lengths of concrete reinforcing bar.

9 Precast concrete slabs in composite construction with rolled steel beams: Krupp-Montex system (for details, see p. 258, **3**). Slabs 1.80 to 3.00 m wide, 10 to 16 cm thick, with top surface ready for floor coverings. Length of slabs up to 8.40 m.

10 Precast concrete slabs in composite construction with lattice girders: Rüterbau system; the girders have no top chords, this function being performed by the concrete (for details, see p. 258, **6**).

Basic principles

Fire protection of floors

The floor is an important loadbearing component. Building regulations and bye-laws require the floor to possess a certain fire resistance rating. The following information is in accordance with the current regulations in the Federal Republic of Germany (see also pp. 330–333). In addition, as a rule, the floor forms a horizontal boundary for a fire compartment, just as fire-resisting walls form the vertical boundaries of such compartments. Thus, the building shown in **1** is divided into fire compartments

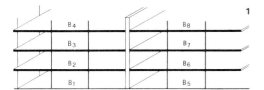

numbered from B1 to B8. To perform this function the floor should have a certain well-defined fire resistance rating to prevent the spread of fire from one storey to another. To fulfil this latter requirement, openings in a floor which exceed a certain permitted size, such as openings for staircases, liftshafts or services, should have casings or enclosures of similar fire resistance. In order to prevent the spread of fire from storey to storey at the external wall of a building, it is often specified that this wall should have a certain fire resistance rating and that the path along which fire can spread on the face of the building, more particularly the distance from the top of the windows of one floor to the top of the spandrel panel of the floor above, should not be less than a certain value (see p. 305, **3**).

The requisite fire resistance of a floor can be achieved by ensuring that the individual components, for example, the floor slab and the floor beams, possess the necessary fire resistance, or by combining the floor slab, beams, ceiling and lateral closures into a structure which as a whole can meet the requirements in this respect. Thus, it is not correct to speak of a fire-resisting false ceiling as such; it can only mean that the ceiling will, in combination with the floor that supports it, constitute an assembly possessing the desired resistance. The floor should fulfil its fire resistance requirements from above, below and the side.

If services presenting a fire hazard in themselves are accommodated in the floor cavity, this potential source of fire must also be given due consideration. The following combinations may occur:

2 If no definite fire resistance of the floor is required, no protective measures need be applied to the floor or its beams.

3 A definite fire resistance for the floor can be obtained by ensuring that both the floor slab, as a solid concrete slab, and the floor beams, as encased steel beams, independently possess this degree of fire resistance.

4 Alternatively, the floor cavity may be enclosed by a ceiling and lateral closures. To obtain a certain fire resistance for the floor as a whole, the floor slab must withstand fire attack from above, the ceiling must withstand attack from below, and the lateral closures must withstand attack from outside. This is an economical solution if a ceiling is to be installed anyway.

5 If the ceiling has no fire resistance of its own, the floor slab and floor beams must themselves possess the necessary resistance from above and below.

6 If the floor cavity as illustrated in **4** contains an air duct which could, in the event of a fire, convey hot gases which it would not be able to resist, then the floor slab and beams (at least in the immediate vicinity of the duct) should be designed to the requisite fire resistance.

7 Here, the situation is similar to that in the preceding example, but now the duct can retain the hot gases, preventing excessive emission of heat to the surroundings. In this case, the floor slab and beams need not be given any special fire protection within the floor cavity.

8 If a space of substantial depth, permitting access to service personnel, is provided between the floor and the ceiling, the structural floor will be in a similar situation to that envisaged in **3**. The ceiling, although acting as a gangway, is generally not required to conform to any particular resistance rating, as it is not classed as a loadbearing structure.

9 This drawing illustrates the various arrangements for ensuring a certain resistance and for forming the boundary to a fire compartment:
1. A ceiling of particular construction and quality, suspended at a certain distance a below the floor beams. For further details of ceilings, see p. 281.
2. Lateral closure of the floor cavity at stairs and other openings in the floor. For details, see p. 293.
3. Staircase or shaft enclosure with specified fire resistance. For details, see pp. 290–298 and 320.
4. External wall with a certain fire resistance, so designed that the flame-spreading distance is not more than $f_1 = s + b + h$, where s is the depth of the lintel and b is the spandrel height.
5. The flame-spreading distance can be increased to $f_2 = s + b + h + w$ by providing a horizontal baffle (see p. 305).

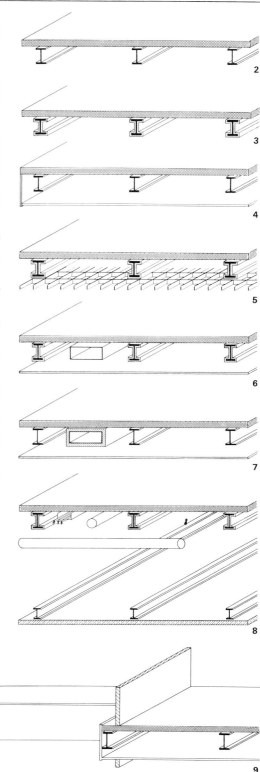

Fire protection of floors

If the slab is required to have a particular fire resistance, it is necessary to determine whether this must apply to fire acting from above or from below.

If the top of the floor is exposed to fire, the structural floor slab itself should possess the specified resistance. For a fire rating of 90 min a 5 cm thick layer of concrete is sufficient, without any special requirements for reinforcement.

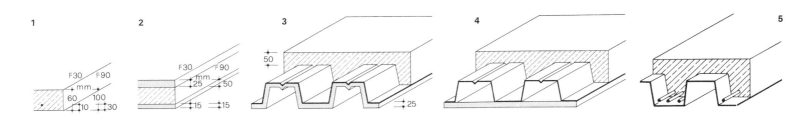

1 According to German Standard DIN 4102, Sheet 4, a flat concrete slab will attain a fire resistance period of:

• 30 min, for a thickness of 60 mm, for reinforced concrete or reinforced aerated concrete, with 10 mm cover to the reinforcing steel;
• 90 min for a single-span reinforced concrete or reinforced aerated concrete slab at least 100 mm thick, with at least 30 mm cover (concrete or concrete+plaster) to the bottom main reinforcement, or for a reinforced concrete continuous slab at least 100 mm thick, with 10 mm cover to the bottom reinforcement and with continuous top reinforcement equivalent to not less than one-third of the mid-span main reinforcement.

2 Floors as shown in **1**, but not by themselves possessing the required fire resistance:

• for a fire resistance of 30 min, for fire attack from above: 25 mm top screed; for attack from below: plaster 15 mm thick applied to lathing (wire netting, expanded metal);
• for a fire resistance of 90 min, for fire attack from above: 50 mm concrete layer; for attack from below: 15 mm plaster applied to lathing.

3 Concrete slab on steel troughing. For 90 min fire resistance, attack from above: 50 mm concrete; attack from below: 25 mm vermiculite, pearlite or asbestos plaster (applied in the usual way or sprayed).

4 Similar to preceding example. For 180 min fire resistance, attack from above: 50 mm concrete; from below: 30 mm thick Vermitecta bonded to underside.

5 Floor as in **3**. Under certain conditions load-bearing steel decking, either alone or in composite construction with the slab, is accepted for a fire rating of 90 min without protection on the soffit.

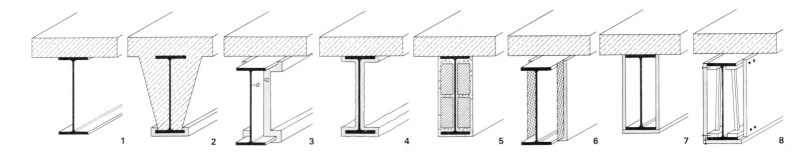

Fire protection of beams

1 30 min fire resistance is provided by several coats of intumescent paint. The beam thus remains visibly exposed as a structural member, which may be an advantage. For subsequent repairs or touching-up, the availability of the same paint must be ensured.

2 A beam embedded in concrete (as in p. 278, **3**, **6**) is adequately protected except for its bottom flange. The latter may be cased in 25 mm vermiculite plaster or 40 mm cement plaster applied to wire lathing or 25 mm Vermitecta board bonded on. Embedding the steel beam in this manner is effective, but expensive.

3 The beam is encased in concrete before erection. Wire netting must be embedded in the concrete to secure it in position. In addition, for deep beams, studs must be welded to the upper part of the web or apertures must be cut in the latter to bond the concrete on both sides together. For a fire resistance of 90 min the concrete casing must be 60 mm thick. This method is suitable only if large numbers of beams have to be treated in this way, as expensive formwork is needed.

4 Sprayed asbestos or vermiculite plaster, applied in several coatings, to a total thickness of at least 25 mm to give 90 min fire resistance. The beams must be given a preparatory treatment as instructed by the suppliers of the fire-resisting plaster. Rust and mill-scale must be removed, but firmly adhering light rust is harmless. In some cases, a thin priming coat of paint (15 μ) applied in the fabricating shops is permissible as a base for the sprayed casing. Relatively deep beams need wire mesh, secured to welded pins, embedded in the casing. The bottom flanges of broad-flange beams should have wire mesh wrapped round them.

Spraying can only be done in suitable weather and is messy. If fastenings for pipes or cables are subsequently attached to the beams, the plaster casing must be locally removed. This must subsequently be made good, preferably with insulating material applied 'dry', fixed with clips or similar devices.

5 The spaces between the flanges may be filled in with bricks or lightweight concrete blocks encased in plaster of at least 15 mm thickness. Wire mesh should be wrapped round the bottom flange.

6 The beam is wrapped in lathing, such as expanded metal, and a 25 mm thick casing of vermiculite plaster is applied.

7 The beam is encased in bonded-on slabs or boards, for example, 25 mm thick Vermitecta. The adhesive used for the purpose must be shown not to lose its bonding power during the fire resistance period envisaged. The advantage of this method is that it is independent of weather conditions and causes little mess.

8 Bolted or nailed slabs or boards, based on asbestos, asbestos cement or gypsum products. Before they are fixed, suitable supporting strips of fireproof material must be clipped to the beam, in accordance with the suppliers' instructions. The joints should be caulked with gypsum or some other suitable material. Advantages: 'dry' method, independent of weather and causing little mess. The form of construction illustrated will provide 90 min resistance, which can be increased to 120 min by insertion of fire-resisting insulation in the cavities.

Functions · Construction

Suspended ceilings

The term 'ceiling' in the present context comprises all constructional features situated underneath the structural floor, that is, in structural steelwork it designates those components of the floor structure which are installed below the slab and beams. In steel-framed buildings, the ceilings are usually of the suspended type.

Functions of the suspended ceiling

Space-enclosing function

The ceiling forms the upper boundary of the room, concealing the loadbearing components of the floor above as well as any services installed in the space between them.

Sound insulation

- Control of acoustic reverberation time in the room.
- Sound insulation between storeys and also between adjoining rooms in the same storey if partitions stop below the ceiling.

Ventilation

- In buildings with mechanically ventilated or air-conditioned rooms the floor cavity contains the ducts and air delivery and extraction appliances.
- Sometimes the ceiling is perforated for the direct distribution of fresh air from the space above, which serves as a large air duct (plenum chamber).

Lighting

- Light fittings may be suspended from the ceiling.
- Alternatively, light fittings may be incorporated in the ceiling itself and form part of it. The air-conditioning delivery and extraction appliances may be combined with these fittings.
- The ceiling may consist of ribs or baffles for light control and diffusion which also contribute to the architectural appearance of the room.

Heating

Radiant heating panels may be installed as ceiling components.

Sprinkler equipment

Fire detection elements and sprinklers are fitted in the ceiling.

Laboratory services

If laboratory benches are served by pipes or wires installed above the ceiling, special outlet units may be incorporated in the ceiling.

Fire protection of the structural floor

In combination with the structural floor above, a suitable suspended ceiling can form a vital part of an overall floor system possessing a certain specified fire rating. Even a perforated plenum chamber ceiling may perform this fire protection function under special conditions. Built-in light fittings and ventilation appliances should be suitably encased to ensure that they do not form vulnerable spots in the protection provided by the ceiling.

The forms of construction of suspended ceilings described below conform to the current regulations in the Federal Republic of Germany in so far as fire protection is concerned.

Construction of the suspended ceiling

In general terms, a suspended ceiling comprises the actual ceiling, a system of beams or bearers to support it, and the hangers whereby these in turn are suspended from the structural floor.

Many different materials and methods of construction are used for suspended ceilings, traditional ones being used as well as new standardised systems. New types are constantly being developed, often for special purposes. Within the scope of this book only those aspects of this extensive subject can be considered which are typical of steel-framed buildings and which should be taken into account at the design stage. Of particular importance are suspended ceiling systems which provide fire protection of the steel beams from below and also perform one or more of the other functions listed above. A typical ceiling of this type is illustrated in **1**. Such ceilings are not significantly more expensive than those which do not provide fire protection. For economical construction in steel it is therefore very important to choose suspended ceilings which, besides performing other functions, can also – and at no appreciable extra expense – provide fire protection. In buildings with such floors the cost of further protective measures against fire are not very significant, since the floors comprise by far the greatest proportion of components to be protected. In the cases envisaged in **5**, **6** and **8** on p. 279, despite the presence of a ceiling, direct fire protection of the structural floor components, that is, the slab and beams, is nevertheless necessary.

Three types of suspended ceiling may be distinguished:

Plastered ceilings

These comprise a plaster coat applied to lathing. They have the disadvantage that the space above the ceiling is not subsequently accessible without locally destroying the ceiling. Where services have to be installed in this space and may need attention or may have to be altered later on, this type of construction is therefore inappropriate. A further drawback is that such ceilings introduce moisture and dirt into the building at the time of construction, which are undesirable, more particularly in prefabricated structures.

Prefabricated panels with jointless finish

The panels are attached to a supporting framework and the joints are filled in and finished so as to produce a jointless surface when viewed from below. Like the plastered ceiling, this type of construction has the disadvantage that the space above is not accessible without local damage to the ceiling. On the other hand, it shares with the following type the advantage of largely dry and clean construction.

Demountable prefabricated panels

This method has the general advantage of dry construction, and the same applies to the arrangements, if any, for closing or covering the joints. Individual panels can subsequently be removed without damaging adjacent ones. In some systems, the panels thus removed will themselves be damaged and have to be replaced by new ones; in others, panels can be removed and refitted without any damage, so that there is access in any position to the space above the ceiling.

For a suspended ceiling to provide adequate fire protection for the floor as a whole, the conditions in an actual fire should correspond to those in the fire tests on which the rating of the ceiling was based, more particularly as regards:

- the material and thickness of the actual ceiling;
- the material, form of construction and dimensions of the supporting framework;
- the material, form of construction and dimensions of the suspension system.

In this respect, the following dimensions are critical, **1**:

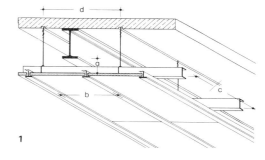

1

a Clear distance from the top of the ceiling to the underside of the structural steel members to be protected.
b Spacing of the battens or runners which directly support the ceiling.
c Spacing of the bearers and hangers.
d Spacing of the hangers in the direction of the bearers.

If different forms of construction have to be adopted, it is usually not necessary to carry out the whole fire test again. Quite often, additional tests of limited scope, or experts' reports based on experience in the testing of other comparable systems, provides sufficient information to predict the fire behaviour of such alternative materials.

Suspended ceilings — Components

Ceiling materials for plastered construction:

- cement plaster
- gypsum plaster
- anhydrite plaster
- vermiculite cement plaster
- pearlite cement plaster
- asbestos cement plaster

Materials for ceilings with sealed prefabricated panels:

- wood-wool panels, when fire rating requirement is small
- gypsum
- gypsum plasterboard
- asbestos cement
- mineral fibre
- combinations of materials, sometimes with mineral fibre blankets laid over them

Ceilings comprising demountable prefabricated panels:

- special impregnated chipboard, when fire rating requirement is small
- gypsum
- mineral fibre
- gypsum-pearlite
- vermiculite
- asbestos cement
- combinations of materials, sometimes with mineral fibre blankets laid over the panels

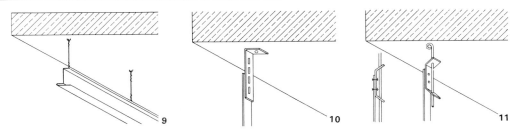

Hangers for ceilings

8 Wire hangers are used in simple types of construction. **10** Flat steel hangers with series of holes or slots for length adjustment at the bolted connections. **11** Round or flat bars secured by spring steel clips permitting 'infinitely variable' adjustment of the length.

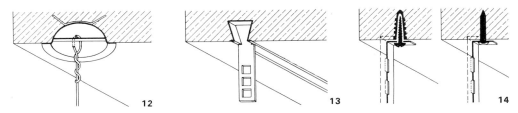

Attachments to concrete floor slabs

12 Plastic cup embedded in the concrete. **13** Embedded anchor rail. **14** Attachment by means of drilled plug and screw (Rawlplug) or cartridge-fired stud.

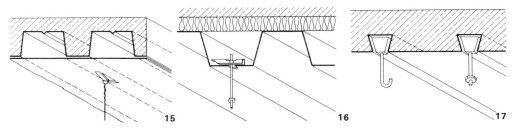

Attachments to steel decking

15 Robertson floor. In the type of construction with flat soffit sheets the latter are provided with stamped preformed loops. **16** Toggle fasteners. If the trough section is filled with concrete, a method as shown in **14** may be employed. **17** Holorib floor with dovetail ribs and fixings.

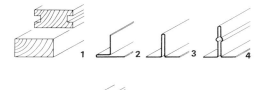

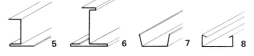

Members for the supporting frameworks of ceilings

These members may consist of wooden battens or light-gauge sheet-steel sections, often of proprietary types. Wooden supporting members, **1**, are restricted to ceilings which have to fulfil only low fire resistance requirements, generally not more than 30 min rating. They are easily workable and are especially suitable for plastered ceilings and also for ceilings constructed with tiles. T-section runners, **2, 3, 4**, are suited for panels nailed or screwed to the underside. In some systems the ribs are provided with preformed slots. The systems illustrated in **5** and **6** are intended for use with slide-in inserted panels or panels supported on the bottom flange, with exposed runners, which may be given a painted finish or faced with metal sheet. Channel-section runners are used for some systems, **7, 8**.

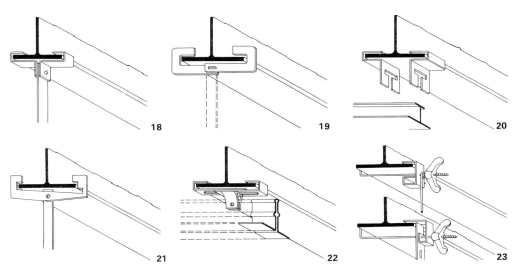

Attachments to beam flanges

18 Flange clip. **19** Plate in two parts with slotted hole to fit various widths of flange. **20** Double clips with T-shaped apertures for insertion of runners. **21** Quick-action fastener. **22** Two-piece clip with slots for adjustability. **23** Hook bolt with wing nut and clamping plates.

Plastered ceilings   Suspended ceilings

Plastered ceilings have only very limited sound absorption capacity. This may be increased by means of underslung acoustic panels or tiles made of soft sound-absorbent material, which may be combustible. Combination with built-in light fittings is possible, but not to be recommended, as the electrical connections are not subsequently accessible. Such light fittings should in any case be cased for fire protection. Combination of this kind of ceiling with air delivery and extraction openings is also unsuitable on account of inaccessibility of the space above the ceiling.

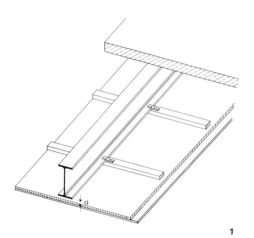

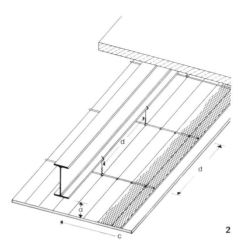

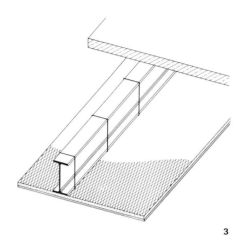

1 15 mm lime-cement plaster applied to lightweight building board or other lathing. A cheap form of ceiling construction for a fire rating of 30 min. The supporting framework consists of wooden battens of thickness a clipped to the bottom flanges of the floor beams.

2 Vermiculite or pearlite cement or gypsum plaster, 25 mm thick, applied to expanded metal supported by 5 to 8 mm dia steel rods at 200 to 600 mm centres d and suspended at 300 to 1000 mm centres c. Distance a between the ceiling and the floor beam = 10 to 60 mm. Fire resistance from 120 to 180 min is possible.

3 Vermiculite gypsum plaster, 20 mm thick, with finishing coat, applied to expanded metal lathing without battens. Spacing of the attachments 200 mm in longitudinal direction of beams, 700 mm transversely. 180 min fire resistance.

CEILINGS COMPRISING PREFABRICATED PANELS

The use of jointless prefabricated panels attached to an inexpensive supporting framework, is always to be recommended in cases where the space above the ceiling need not be subsequently accessible and the acoustic properties of such ceilings are adequate.

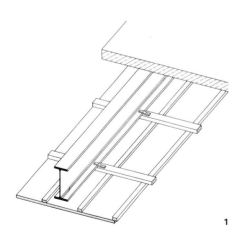

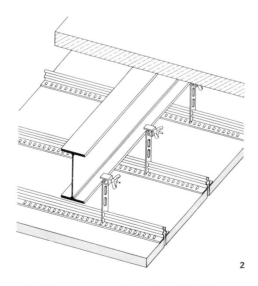

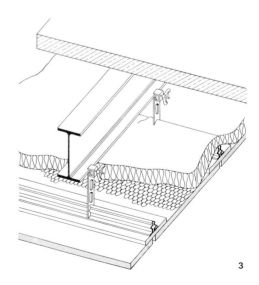

1 Cheap ceiling comprising gypsum or plasterboard panels fixed to timber battens. 30 min fire resistance. (Manufacturers: Knauf.)

2 15 mm gypsum plasterboard fixed by screwing to protected sheet-metal rails. 90 min fire resistance. (Manufacturers: Rigips.)

3 10 mm thick Isoeternit asbestos cement panels (manufacturers: Eternit AG, Berlin). Silan SMO 40 mineral fibre blanket stitched to wire netting. Joints between panels are provided with 100/10 mm Isoeternit cover strips.

Suspended ceilings — Demountable prefabricated panels

Ceilings of this general type may be erected on concealed or on exposed runners. In the latter case, a distinct grid pattern appears. The panels or tiles of which the ceiling is composed can readily be removed. Hence, such ceilings are more particularly suitable if easy subsequent access to the space above the ceiling is required. Mineral fibre panels, which also possess good acoustic properties, are suitable for these ceilings. Additional sound insulation is obtainable by means of mineral fibre blankets placed over the panels, whereby increased fire resistance is also achieved.

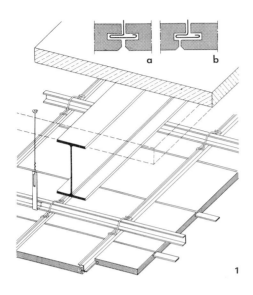

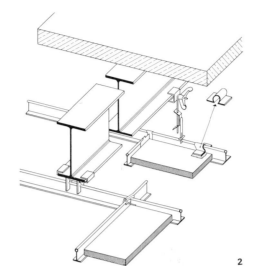

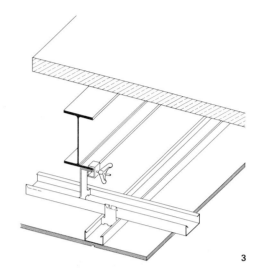

1 The edges of the ceiling panels have grooves into which the concealed runners are slid. The latter are in turn supported by transverse bearers. Fire resistance: 90 min. Individual panels can be removed by slitting around the edges, but only those around the sides can be removed intact; easier demountability is obtained with asymmetrical grooves b than with symmetrical grooves a.

2 Exposed runners with panels supported on flanges. Fire resistance: 90 min. Clips secure the panels against lifting. In the event of a fire the runners will expand, so they must not terminate in close contact with walls.

3 Ceiling panels screwed to channel-section runners can be individually removed and refitted without damage.

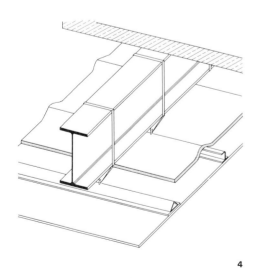

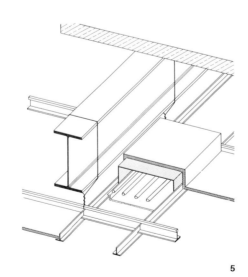

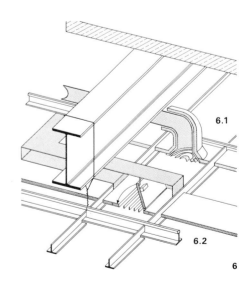

4 Fire resistance and acoustic properties can be improved by mineral fibre blanket. Here: 180 min resistance obtained with 10 mm Nobranda asbestos cement panels and 37 mm Basalan blanket. Manufacturers: Harpenfeld Bauelemente GmbH, Osnabrück.

5 Mineral fibre panels with built-in light fitting provided with a fire-resisting casing.
6.1 Air inlet and extraction ducts are encased (for example, with mineral fibre blankets) to ensure that hot gases rising in a fire will not come into direct contact with the steel beams.

6.2 A similar result is achieved with trapdoor-type flaps which are held open by devices which fuse at elevated temperature. Thus, in a fire, the flaps will close and form a fire-resisting seal to the openings.

Services installed in the floor

Arrangement of horizontal services

The more numerous and elaborate the services that are needed in a building, the greater is the care necessary in planning their layout. Detailed design calculations for heating, air-conditioning, lighting and power requirements have become a standard feature of modern building construction, just as much as the careful planning of the central installations for heating and cooling a building, controlling the humidity of the air, transformation of the voltage of the electricity supplied, switchgear operation, etc. Frequently, however, there is a lack of proper co-ordination of the layout of the services with the structural features of the building. Quite often, too, lack of time compels the structural frame to be designed, put out to contract and perhaps even built while the elaborate and time-consuming operations of planning the services lag behind. In such cases, the greatest possible amount of space should be made available for accommodating the services, to have some latitude to cope with services not foreseen in the earlier stages of planning and also to have extra space for subsequent additions.

More efficient and economical solutions are obtained by starting to plan the services early and by co-ordinating their layout with the structural design of the building. With an efficient arrangement of beams, columns and floors it is often possible to effect substantial savings in the cost of the services, so that, despite the sometimes higher cost of the floor, there is nevertheless an overall saving.

An optimum total cost (p. 171) is obtained by rigorously arranging the loadbearing members and the services on two levels, so that, for example, all those members and services extending in the north–south direction are on the upper level, and all those extending in the east–west direction are on the lower. In this way, the loadbearing members (more particularly the main and floor beams) and the services do not interfere with one another and the services can also cross one another, the full depth of the floor cavity being available for this. In addition, the services can be arranged much more systematically and conveniently than in any other system. Indeed, the designer of the service layout of the building is virtually compelled to be systematic. The only services that sometimes cannot completely fit in with such arrangements are waste water pipes which have to be laid to a suitable fall and may have to pass through beams.

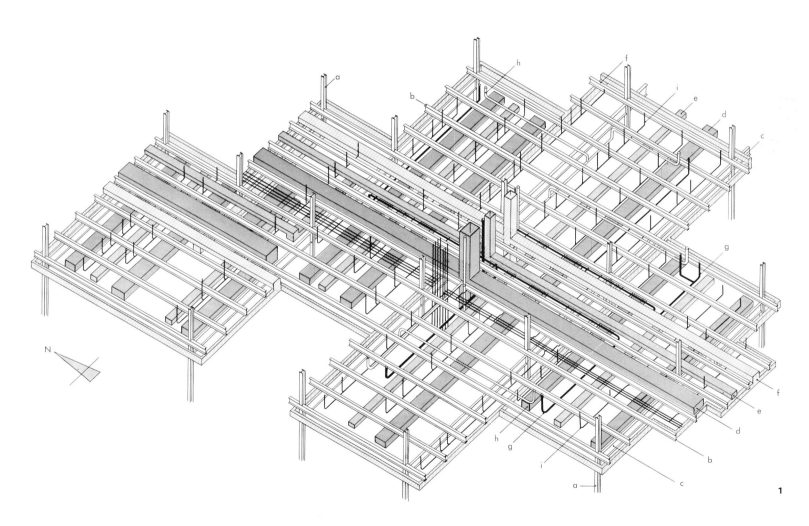

a column
b floor beam
c main beam
d exhaust air duct
e low-pressure air supply duct
f high-pressure air supply duct
g water feed pipe
h waste water pipe
i cables and wiring

1 The building, which has a complex shape on plan, comprises a central service shaft. The main ducts, pipes or cables extend from the shaft in the north–south direction on the upper level, that is, the level on which the floor beams are installed. The distribution ducts, etc., are arranged on the lower level, where they run in the east–west direction. The north–west part of the building requires main services which are additionally fed by supply ducts and pipes also on the lower level, that is, the level of the main beams. The high-pressure air ducts extend along the external walls and enter the interior at suitable points.

285

Arrangement of horizontal services — Heating installations in the floor cavity

The radiators for central heating are usually installed below the windows. If the external columns of a steel-framed building are placed directly behind the façade, the hot water pipes must pass round the columns or through them. There are the following possibilities:

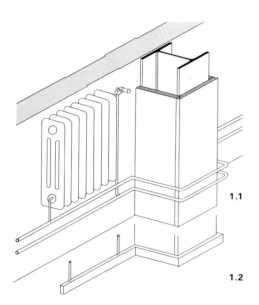

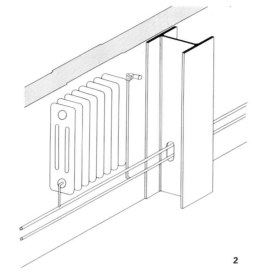

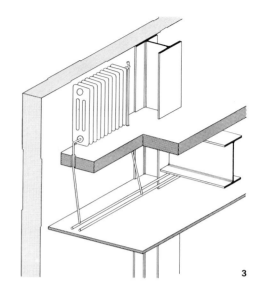

1.1 The two pipes of the two-pipe system are made to go round the column, a rather unsightly solution, hardly suitable for rooms with any claim to decorative merit.
1.2 In the case of a one-pipe central heating system, it may be acceptable to conceal the pipe in a plinth.

2 It is often possible to pass the hot water pipes through holes in the webs of the columns. They must of course also penetrate any fire protection casing provided.
3 A neat solution can be obtained by installing the pipes in the floor cavity, under the floor beams. The holes in the concrete floor slab

may be either drilled or preformed at the time of concreting, in which case they may be more accurately positioned with precast than with in-situ concrete construction.

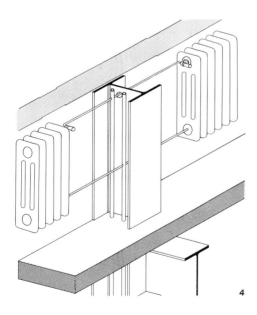

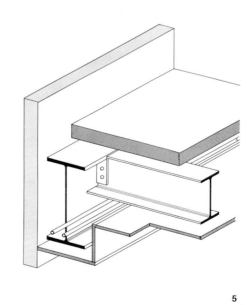

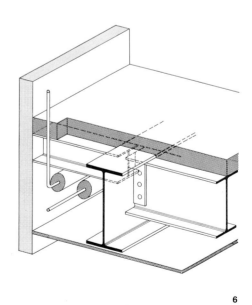

4 In this solution the flow and return pipes, installed vertically between the column flanges, supply only the radiators directly to the left and right of the column. Any casing around the latter must of course also be penetrated (see p. 238).
5 Where the main beams are close up against the external wall the space between the latter

and the beams can hardly be utilised for the accommodation of heating pipes or any other services, as it becomes inaccessible after the cladding has been installed. Hence, the pipes must be placed on the inside of these beams. The suspended ceiling may be raised to a little below the underside of the secondary beams, while the casing of the main beam at the external

wall may be combined with the lintel or window head features.
6 Here, the columns and main beam are set back so far from the external wall that the floor slab has to be additionally supported by short cantilever beams of limited depth. The pipes can conveniently be installed under these cantilevers.

Air ducts in the floor cavity

Arrangement of horizontal services

High-pressure ducts supplying air to units installed under the windows present further problems. Some possible solutions are illustrated in the following examples:

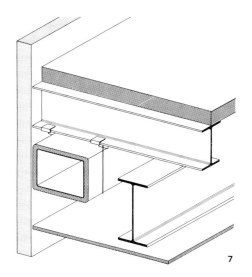
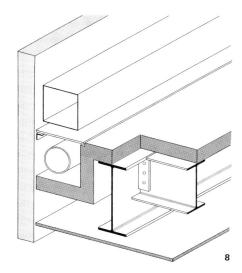
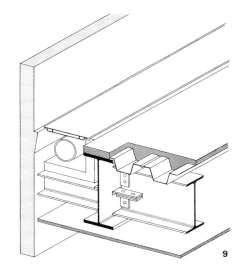

7 The floor beams rest on the main beams and project some distance beyond them. There is sufficient space for the ducts between the main beam and the external wall.

8 A good solution is to install the high-pressure pipes in an underfloor pipeway directly behind the façade, but in front of the columns. The bottom of the pipeway should be of the same general construction as the floor slab for reasons of fire protection and acoustic insulation. Constructing a stepped-down edge to the concrete floor slab is relatively expensive, but allows the air-conditioning units to be connected to the pipe at any point along the pipeway. The latter can also accommodate other services such as the hot water supply. The solution illustrated in **7** is cheaper, but involves drilling or pre-forming the necessary holes in the concrete floor slab. In deciding which solution to adopt, the total cost of floor plus services should be considered.

9 A pipeway which provides access from above can alternatively be constructed with prefabricated units supported on small low-lying cantilevers. The prefabricated units should of course comply with the requirements for fire protection and sound insulation.

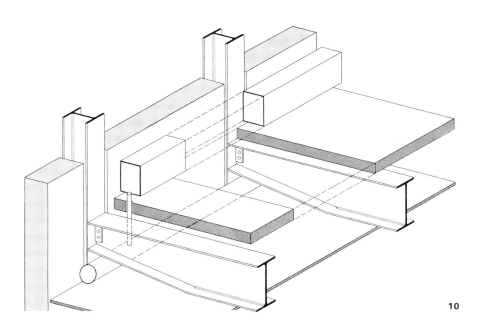
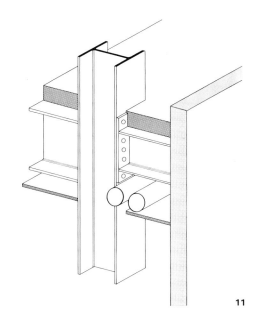

10 With closely spaced external columns, each having a floor beam connected to it, the high-pressure air pipe should be installed in the floor cavity. To reduce the construction depth of the floor, the end of the beam may be tapered, which is structurally quite practicable, but involves extra cost.

11 Air pipes installed in the floor cavity between the columns and the external wall. The pipes go under small cantilevers which support the floor and the façade.

287

Arrangement of horizontal services

Passage of services through steel beams

The tables on this page give some information on the dimensions of pipes and ducts that can be made to pass through steel beams and girders of various types. In the case of lattice girders it has been presupposed that the chords and the web members are of equal width, namely, 15% of the depth of the girder. The intermediate plates in the Litzka beams, **5**, are 20 cm in depth.

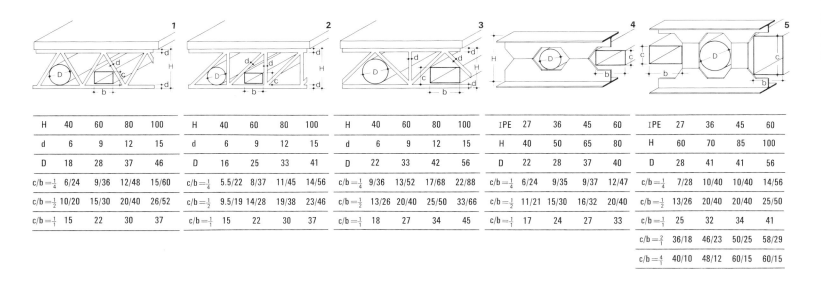

	1					2					3					4					5			
H	40	60	80	100	H	40	60	80	100	H	40	60	80	100	IPE	27	36	45	60	IPE	27	36	45	60
d	6	9	12	15	d	6	9	12	15	d	6	9	12	15	H	40	50	65	80	H	60	70	85	100
D	18	28	37	46	D	16	25	33	41	D	22	33	42	56	D	22	28	37	40	D	28	41	41	56
$c/b = \tfrac{1}{4}$	6/24	9/36	12/48	15/60	$c/b = \tfrac{1}{4}$	5.5/22	8/37	11/45	14/56	$c/b = \tfrac{1}{4}$	9/36	13/52	17/68	22/88	$c/b = \tfrac{1}{4}$	6/24	9/35	9/37	12/47	$c/b = \tfrac{1}{4}$	7/28	10/40	10/40	14/56
$c/b = \tfrac{1}{2}$	10/20	15/30	20/40	26/52	$c/b = \tfrac{1}{2}$	9.5/19	14/28	19/38	23/46	$c/b = \tfrac{1}{2}$	13/26	20/40	25/50	33/66	$c/b = \tfrac{1}{2}$	11/21	15/30	16/32	20/40	$c/b = \tfrac{1}{2}$	13/26	20/40	20/40	25/50
$c/b = \tfrac{1}{1}$	15	22	30	37	$c/b = \tfrac{1}{1}$	15	22	30	37	$c/b = \tfrac{1}{1}$	18	27	34	45	$c/b = \tfrac{1}{1}$	17	24	27	33	$c/b = \tfrac{1}{1}$	25	32	34	41
																				$c/b = \tfrac{2}{1}$	36/18	46/23	50/25	58/29
																				$c/b = \tfrac{4}{1}$	40/10	48/12	60/15	60/15

ELECTRICAL INSTALLATIONS IN THE FLOOR

With increasing mechanisation of modern offices, more and more attention has to be paid to the proper accommodation of electric cables and wires and their accessibility at all times. In office buildings it should be possible to install a good deal of additional electrical services at a later date. With the increasingly extensive use of data-processing equipment in offices, more data terminals are being provided, requiring cables for their connection to the data communications system. In schools and university buildings, too, the development of electronic teaching aids and other equipment will require more elaborate cable services in the future. The following considerations should be examined:

- Should it subsequently be possible to make branch connections to existing cables?
- Should it subsequently be possible to insert additional cables into cable conduits?
- Should it be possible to re-lay the cables:
- Should the cables be accessible from the storey which they actually serve or can subsequent alterations, additions, etc., be carried out from the storey below after removal of the ceiling?

Electric cables which are laid on the structural floor must be protected by a screed or surfacing applied to the latter. The thickness of the surfacing depends on the type of cable:

3 to 5 cm for thin cables directly embedded;
4 to 8 cm for cables in conduits;
8 to 10 cm for cables in ducts.

The surfacing adds to the weight of the floor. The advantages associated with the laying of cables on the structural floor must be weighed against the extra cost arising from this increased load. It should be borne in mind that a loadbearing concrete floor slab is generally not more than 10 cm thick and that the concrete over a steel deck may be as little as 5 cm. The surfacing may therefore double the dead weight of the floor, so that the cost of the loadbearing structure as a whole is substantially increased.

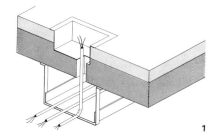

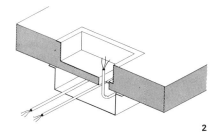

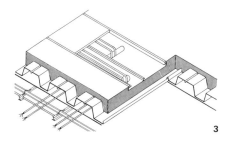

1 The cables are supported in a cable rack in the floor cavity and are accessible from the storey below. The structural floor has recesses in which floor junction boxes can be installed when the floor surfacing is being laid. This is a simple solution permitting the subsequent installation of additional cables, but each time a connection is made it involves local dismantling of the ceiling below.

2 An improvement on the preceding solution in that now the cables are laid in troughs attached directly to the floor. Access to them is provided by openings in the floor. If the floor is composed of precast slabs with a very smooth surface finish, plug sockets may be concreted in. At the openings giving access to the cables there will be a gap in the fire protection provided by the floor, but such openings can be closed with concrete covers or other suitable means.

3 Steel deck with cable ducts. Cellular deck comprising troughing unit welded to flat sheet.

Electrical installations in the floor

4 Steel or plastic cable conduits laid on the structural floor in a surfacing 8 to 10 cm thick. Though adding substantially to the weight of the floor, this arrangement offers considerable freedom of layout for elaborate wiring.

5 Floor mounted on short columns or pedestals, for buildings with elaborate and bulky services, such as control and switchgear rooms. The pedestals are adjustable for height to compensate for inaccuracies in the evenness of the structural floor on which they stand.

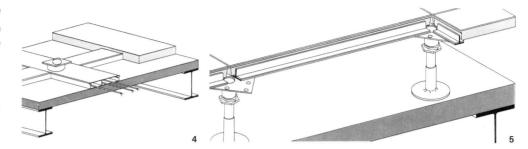

Floor finishes

FLOOR FINISHES

The structural floor of a steel-framed building can be provided with any desired type of surfacing or finish. The following considerations apply:

- The tolerances of the structural floor are often smaller than with other methods of construction.
- The floor surfacing should weigh as little as possible, to avoid losing the benefit resulting from the use of a structural steel frame.
- Services can be installed in the floor cavity, rather than in the surfacing.
- A floating floor should not be used if demountable partitions are to be installed. In that case, sound insulation should be ensured by other means (p. 321).

Structural floor with screed

A screed improves the evenness of the structural floor. A cement–sand mortar screed should be 4 to 6 cm thick and should, for relatively heavy loads, contain a light wire mesh reinforcement. If the screed is too thin, it is liable to break up or, in the event of fairly large deflections of the floor, to become detached. Thin screeds can, however, be used if a synthetic resin binding agent is employed instead of cement.

Since the dimensional accuracy of a steel-framed structure is greater than that of other forms of construction, the surface of the structural floor is generally also more accurately finished. This in turn has the following effects on the various floor construction methods:

- In-situ concrete floors on steel beams in general achieve the least dimensional accuracy. For these a levelling screed of 4 to 6 cm should be allowed for.
- Greater accuracy is achieved with precast concrete floor slabs or with in-situ concrete on steel decking. With careful workmanship, a thin synthetic resin based screed should suffice on such floors.

Floor covering direct on structural floor

With precast concrete slabs on steel beams it is possible to obtain such accuracy that the floor covering may be laid directly on the structural floor. The following conditions must be satisfied:

- The slabs should have a particularly flat and even surface, obtained by casting them face-downwards in steel moulds or, if they are cast face-upwards, by very careful and precise screeding and smoothing.
- The bearing surfaces of the slabs on the steel beams should be constructed to close dimensional tolerances.
- The slabs should be laid carefully, ensuring that no foreign bodies get in between the beams and the slabs to cause inaccuracies in the levels of the latter.

This method of construction has the following advantages:

- Dispensing with the screed effects a saving in weight.
- The whole floor can be very accurately adjusted and aligned.
- Various service items can be embedded in the concrete at the time of casting the slabs, such as junction boxes for electrical connections, heating and water pipe connections, fastenings and screwed sockets for items of equipment (for example, in classrooms or laboratories).

This constitutes an important step towards the industrialisation of building construction.

Structural floor without covering

In some cases, the structural floor may be left without any finish or covering, for example, in basements, on floors carrying technical equipment, or on floors which are in the open and serve to carry vehicular and/or pedestrian traffic (such as decks in multi-storey car parks). Here, there are two main aspects to consider:

- Resistance to wear, to avoid abrasion and dust.
- Watertightness.

Resistance to wear

The better the quality of the concrete, the better will be its resistance to wear. Trafficked concrete slabs are given a good flat surface which is finished with a coat of synthetic resin paint with sand or corundum particles scattered over it to produce a non-slip walking surface. If vehicular traffic is admitted to the floor, it is advisable to produce a ridged surface by brooming the fresh concrete or treating it with a profiled roller.

Watertightness

A distinction must be drawn between the watertightness of the concrete slabs themselves and that of the joints between them. Slabs made of good quality concrete are generally watertight but in some cases a waterproofing agent may be used as an admixture. Watertightness at the joints, including construction joints, can be achieved as follows:

1. A permanently elastic sealing compound may be used to close the joint. Assuming this compound to have a deformation capacity of ± 20 to 25%, the width of the joint should be equal to four or five times the expected range of movement of the two slab edges in relation to each other.
2. Alternatively, the joint may be closed with a sealing strip, such as a glass-fibre reinforced synthetic resin compound, which possesses the necessary deformation capacity.

Floating floors

A surfacing or screed laid on a soft supporting layer and not in direct contact with external or internal walls provides a very good method of insulation against the sound of footsteps (an improvement of 18 to 28 dB can thus be achieved). However, it is essential that partition walls should stand directly on the structural floor and that at such walls the surfacing must not have direct contact with them. This problem is considered in more detail on p. 321. A floating floor is appropriate in buildings with fixed partitions, but not in circumstances where frequent rearrangement of internal layout by varying the positions of the partitions is envisaged and such partitions have to meet acoustic requirements. Since steel-framed construction is particularly suitable for obtaining large floor spaces unobstructed by supports and thus providing flexibility of internal layout, the use of floating floor systems should be carefully considered, especially as an equally good effect may often be achieved by the use of soft resilient floor coverings which do not impose the same restrictions.

Floor covering

The choice of floor covering is of major importance from the point of view of sound insulation. The degree of protection against the sound of footsteps can be much improved by means of the following coverings:

Soft resilient linoleum or plastic coverings laid on a felt or cork base may effect an improvement of between 10 and 20 dB. Textile fabrics on foamed plastic or felt achieve a sound reduction of 18 to 31 dB and are thus superior in this respect even to a good floating floor (see Moll, *Bauakustik*, p. 238).

Staircases

Function and types of staircases

Function	Service stairs	Basement-roof emergency stairs	Main stairs	Outside stairs
Frequency of use	By service personnel	Irregular use	Infrequent	Frequent
Architectural importance	Slight		Great	
Building regulations	—	Essential stairs		
Rise	≤ 75°	≤ 35°	30°	20°
Width (cm)	≥ 60	≥ 90	≥ 110	≤ 250
Range of application	—	Residential buildings		
	Business premises	—	Business premises	
	Communal buildings	—	Communal buildings	

1.1

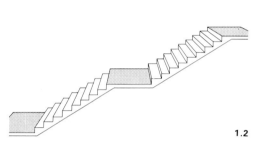

1.2

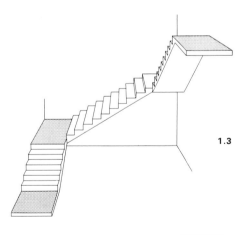

1.3

Function of staircases

Stairs are used for vertical circulation in buildings. A distinction must be drawn between fixed stairs and escalators. Ramps and travelators, as well as lifts, may perform similar functions. In high buildings, lifts are normally used almost exclusively as means of transport from floor to floor, though stairs must always be available for use in emergencies.

Since stairs are an important and essential feature of escape routes in a fire or other emergency, most countries have regulations specifying the dimensions, fire resistance, access, enclosure, smoke extraction arrangements, number and spacing of staircases incorporated in escape routes.

As staircases exposed in rooms or halls may constitute an important architectural feature, they should be designed with great care.

The relationship between the type of building, the functional purpose of the stairs, the frequency of their use, their importance in terms of architectural design and their dimensional characteristics, are outlined in the above table.

Further requirements are often imposed to suit particular conditions of use.

Types of staircases

Depending on the space available, or the architectural requirements, stairs may be straight, curved or spiral on plan. The most frequent type, especially in buildings designed for construction by industrialised techniques, is the staircase with straight flights. Stairs with curved flights or tapered heads are used only for the sake of architectural effect or, for example, where space restrictions necessitate them. They are not considered in this book.

Spiral staircases usually serve only as emergency stairs or to connect individual storeys. Sometimes, however, large spiral staircases are built as major architectural features in entrance halls. Because of the relatively minor importance of spiral staircases, only three examples are given (p. 298).

In the design of stairs and their incorporation into a steel-framed building, the following aspects should be given particular attention:

- the geometry of the staircase;
- the construction of its components;
- the vertical enclosure of the staircase in relation to the storeys and to the rest of the building generally.

Many different materials and forms of construction can be used for stairs. From the numerous possibilities, solutions more particularly suited for steel-framed buildings and especially for those embodying simple and economical industrialised methods of construction are presented and discussed here.

Straight staircases

The fixed straight staircase comprises flights and landings. A series of more than three steps is referred to as a flight. For reasons of safety and general convenience of use the length of a flight should be limited to about 18 steps. It is seldom that two floors in a multi-storey building are connected by a single flight of stairs, **1.1**. If, as is more usually the case, two or more flights are installed in each storey, one or more intermediate landings, as well as top and bottom landings, will be needed. Figure **1.2** shows a straight staircase with one intermediate landing. If the successive flights are at right angles to one another the landing at each change of direction is known as a quarter-space landing, **1.3**. In the arrangement shown in **1.4** the flights rise alternately in opposite directions, each landing being called a half-space landing. With two flights per storey this is the commonest form of construction and takes up the least space on plan. For greater storey heights the staircase may comprise three flights, when the top and bottom landings will be in relatively different positions.

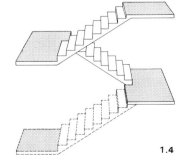

1.4

Staircase geometry

Width of flight, depth of landing

A distinction must be drawn between the structural width of a flight of stairs and the effective width, measured between handrails, usually laid down in building regulations. The effective widths are approximately:

- for one person   0.75–1.00 m
- for two persons   1.10–1.30 m
- for three persons 1.80–1.90 m

The overall height of a staircase within a storey comprises the clear height and the structural height. The clear height measured from the front edge of the step to the underside of the flight above should be not less than 2.00 m. The width of the space between two flights of stairs rising in opposite directions should not be less than a certain minimum, so as to give sufficient clearance between handrails. On the other hand, some building regulations require this interflight space, called the stair well, not to exceed a certain maximum, as a safeguard against accidents.

The depth of the landing should not be less than the effective width of the flight. Some regulations lay down a certain minimum depth. Doors are not allowed to swing open into the landing space, as this might involve a risk to persons using the stairs.

Construction of the steps

The convenience and safety of stairs very much depend on the relationship between the tread and the riser. Within each flight and, furthermore, within all similar flights of stairs in one and the same building this relation should not be varied. This not only makes for greater safety in using the stairs, but also for more economical construction, especially when precast concrete stairs are used. Furthermore, standardised tread and riser dimensions facilitate planning and enable the associated features such as balustrades and hand-rails to be produced to uniform dimensions.

For buildings whose design is based on the international system of modular co-ordination the depth of the landing and the length of the flight should be a dimension divisible by 60 or, if this is not practicable, by 30 or, in an extreme case, by 10. In any case, the overall length of the staircase, as well as its width, should fit in with this modular co-ordination.

Under this system of dimensions, the storey height is subdivided into modular units of 10 cm. Even if the storey heights in a building are not all the same, though always a multiple of 10 cm, they obviously have to be a multiple of the riser of the steps. Therefore, riser dimensions of $16\frac{2}{3}$ cm or 15 cm can suitably be used, since three steps of $16\frac{2}{3}$ cm give a height of 50 cm, and two steps of 15 cm give a height of 30 cm.

There is no true mathematical relationship between tread and riser, but there are various empirical formulas. One such formula, which gives acceptable results in terms of comfort and safety to stair users, is based on the consideration that the average length of a pace on a horizontal surface is 63 cm and that the average distance between the rungs of a vertical ladder is 31.5 cm. It is considered that the proportions of a step in a flight of stairs should fit in somewhere between these extremes. Hence, the relationship $a + 2s = 63$ is adopted, where a denotes the tread and s is the riser. If the number of risers is n, then the length of the flight, including the tread at the top landing, will be $L = n \cdot a$. The height of the flight will be $H = n \cdot s$. Therefore: $L + 2 \cdot H = n \times 63$, so that the number of steps in the flight is:

$$n = \frac{L + 2 \cdot H}{63} \text{ (dimensions in cm)}.$$

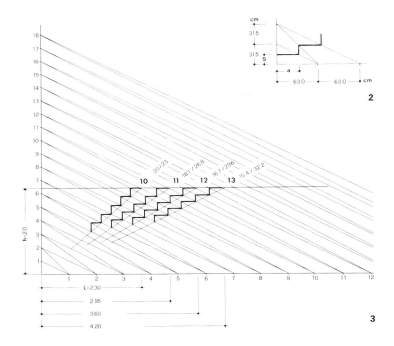

H	h	n	7	8	9	10	11	12	13	14
2.70	1.35		19.3/24.4	16.8/27.0						
80	40		20.0/23.0	17.5/28.0						
90	45		20.8/23.4	18.1/26.8	16.1/30.8					
3.00	1.50			18.8/25.4	16.7/29.6	15.0/33.0				
10	55			19.4/24.2	17.2/28.6	15.5/32.0				
20	60			20.0/23.0	17.8/27.4	16.0/31.0				
30	65			20.6/21.8	18.3/26.4	16.5/30.0				
40	70			21.2/20.6	18.9/25.2	17.0/29.0				
50	75				19.4/24.2	17.5/28.0	15.9/31.2	14.6/33.8		
60	80				20.0/23.0	18.0/27.0	16.4/30.2	15.0/33.0		
70	85				20.5/22.0	18.5/26.0	16.8/27.0	15.4/32.2		
80	90				21.1/20.8	19.0/25.0	17.3/28.4	15.8/31.4		
90	95				21.7/19.6	19.5/24.0	17.7/27.6	16.2/30.6		
4.00	2.00					20.0/23.0	18.1/26.8	16.7/29.6	15.4/32.2	
10	05					20.5/22.0	18.6/25.8	17.1/28.8	15.8/31.4	
20	10					21.0/21.0	19.1/24.8	17.5/28.0	16.2/30.6	

This relationship can be expressed in a graph in which the horizontal scale is in units of 63 cm and the vertical scale in units of 31.5 cm, the corresponding points being joined together, as indicated in **2**. If the diagram presented in **3** is drawn on tracing paper to the commonly used scales of 1:50 and 1:100 it can be used to select suitable riser:tread ratios (s:a) or, alternatively, if s:a is predetermined, to determine the flight length for a given height of flight.

The riser:tread ratios obtained from this formula for various heights of flight and numbers (n) of risers in a flight are given in Table **4** (source: Schuster, *Treppen aus Stein, Holz und Metall* (*Staircases in masonry, timber and metal*)).

Staircases

Staircase geometry

Arrangements at landings

At an intermediate landing, more particularly a half-space landing, the location of the top step of the lower flight in relation to the bottom step of the upper flight is important. Various possible positions are indicated in **5**.

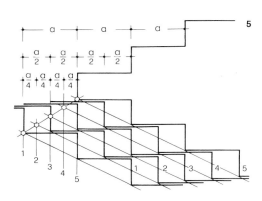

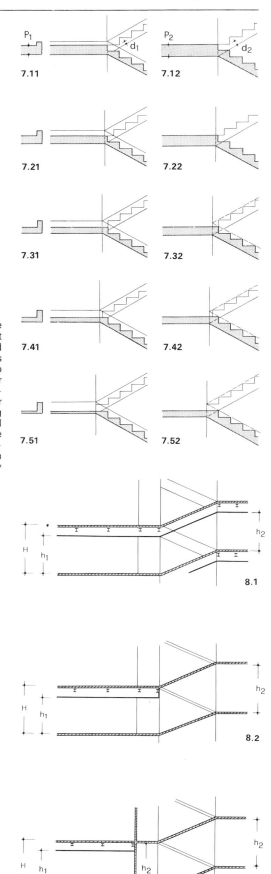

7 The arrangements of flights shown in **5** are set out separately in **7**, the flights on the left and the right being respectively with and without stringers. For equal flight thicknesses d_1 and d_2, the thicknesses of the landing slab p_1 and p_2 diminish as the first riser of the upper flight is displaced (to the left in these diagrams) with respect to the last riser of the lower flight. The minimum thickness of the landing slab depends, of course, on the structural requirements. These diagrams should therefore be helpful in choosing an appropriate arrangement at the landing where particular minimum values for the thicknesses d and p are structurally necessary.

Underside of stairs

8.1 Suspended ceilings are commonly employed in steel-framed buildings to conceal the floor beams and also to provide space in which various services can be accommodated. Such ceilings can be continued under the flights and landings of staircases. This is usually done in circumstances where the staircase is not accommodated in its own enclosure, but is a visibly exposed feature of a room or hall and has to conform to fire resistance requirements.

8.2 If flights and landings are constructed of concrete without visibly exposed steel beams, a separate ceiling is not needed, provided that there is sufficient cover to the reinforcement. The soffit of the concrete may either be left exposed or it may be plastered. With this form of construction there is a substantial reduction of depth in comparison with a staircase comprising steel beams and a suspended ceiling. The latter involves a sudden change in height at the junction between the normal ceiling and the soffit under the stairs. The solution illustrated in **8.2** should be avoided. If the ceiling is required under both the top and bottom landings, it should preferably be continued under the flights and the intermediate landing, as in **8.1**. All-steel staircases can be installed without any constraints, provided they are given fire protection.

8.3 If the stairs are in a separate enclosure, as is frequently required under building regulations, it is generally possible to effect a neat transition from the construction depth of a steel beam floor to the reduced construction depth of a concrete staircase.

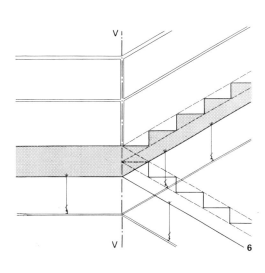

6 If the two flights converging at a landing have similar geometrical features, which is usually the case, it should be ensured that all corresponding lines of the two flights intersect at the same vertical line VV. This is a mathematical necessity and observance of this rule will help the designer to find an elegant solution, ensuring proper alignment of the break lines of the 'up' and 'down' landings as viewed from below.

Structural aspects   Staircases

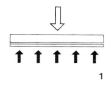 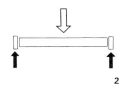 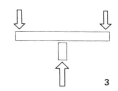 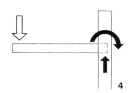

The step as a structural component

1 The step is supported by a slab along its whole length and is therefore not subject to flexural load.

2 The step is supported at each end by beams or stringers and is therefore loaded in bending in the manner of a beam on two supports.

3 The step is supported centrally and thus functions structurally as a double cantilever, the cantilevering length being equal to half the length of the step.

4 The step is rigidly fixed at one end to a wall or stringer and thus functions as a cantilever, the length of which is equal to the length of the step.

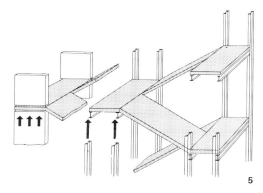 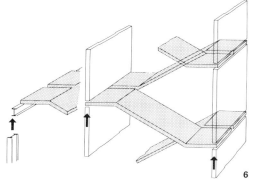 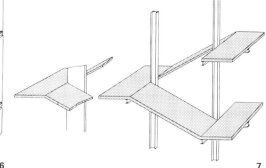

Transmission of loads from flights and landings

5 The flights span as beams from landing to landing. In this example the half space landings span transversely, transmitting their loads to columns or walls.

6 Each flight together with its top and bottom landing forms a doubly cranked beam supported on walls or beams.

7 In the arrangement illustrated here the flights and landings are supported by columns installed in the stair well and provided with cantilever beams. Alternatively, the steps may cantilever out from internal stringers or from the wall around the staircase.

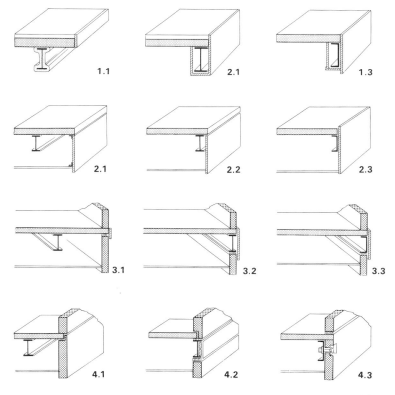

STAIRCASE ENCLOSURES

1 If the floors of the building have exposed beams with fire encasement, the stair flights usually also have exposed soffits. In **1.1**, the structural floor slab has a top surfacing or finish and is provided with a fascia strip at the edge. In **1.2** and **1.3**, the fascia at the stair opening is combined with the beam casing.

2 If a staircase passes through a floor with a suspended ceiling, the floor cavity must be closed in for aesthetic reasons and also for fire protection, where the ceiling has this latter function to perform. If the suspended ceiling is continued under the stair flights and landings, similar edge trimmings should be used for these parts too. Proprietary asbestos cement panels can be conveniently used for fire encasement.

3 The staircase may be enclosed within non-loadbearing fire-resisting walls. These walls stand on the floors and may be located directly over floor beams. There must be no rigid wall-to-floor connection, as the floor must be able to deflect under load without involving the staircase enclosure walls, which are not designed for this. If the floor beams are under the edge of the slab, they should be protected by thin fire-resisting panels.

4 The solution in these examples is similar in principle to the preceding ones, but here the enclosure has a flush face at the edge of the floor. In **4.1**, the wall stands on the edge of the floor slab and conceals the face thereof. In **4.2**, the wall stands on the beam, which is protected by a casing on its outer face. The wall panel may also be bolted to the floor beam, as in **4.3**, in this instance a channel section.

Staircases

Arrangement of staircases in steel-framed buildings

1 To accommodate a staircase in a steel-framed building, the most obvious solution consists in leaving an appropriate bay in each floor open and installing the stairs in the space thus provided. This is an especially appropriate solution if the stairs are to be left open, that is, not provided with an enclosure, so that the interior spatial continuity is maintained. The top and bottom landings are connected to the floor beams, whereas the intermediate landings have to be suspended by thin hangers from the floor above or be carried on struts from the floor below. The width and length of the flights should as far as possible fit in with the grid of floor beams, otherwise the floor openings will involve the use of trimmer beams. Escalators are usually installed as open features in the framework, as they are not rated as essential stairs and therefore do not require fire-resisting enclosures.

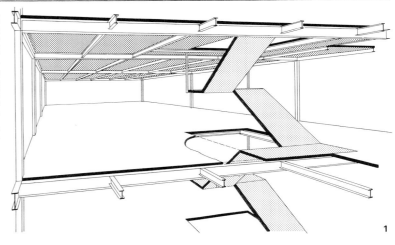

2 If structural conditions rule out an arrangement as envisaged in the preceding example, it is often a good idea to install light columns at all four corners of the top, bottom and intermediate landings. These columns facilitate the fixing of staircase-enclosing walls and, as the enclosure is structurally connected to the rest of the frame, the columns can also carry a share of the floor loads.

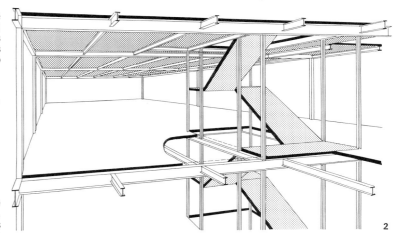

3 If the columns of the staircase enclosure are interconnected by lattice bracings, they can also be utilised in the overall stiffening of the building. With a staircase comprising two flights per storey this arrangement is nearly always possible, the two longitudinal sides and one end face of the enclosure being braced, while the fourth face remains free from bracings to provide unrestricted access at top and bottom. Lift shafts and service shafts are often located directly beside staircases, and the bracings may comprise these shafts as well.

STAIRCASES IN CONCRETE SHAFTS OR CORES

Frequently, the walls forming the staircase enclosure are required to perform a structural as well as a space-enclosing and fire-resisting function. Questions of economy have been considered on p. 167; for structural details see pp. 267–269. The core thus formed always carries the vertical loads due to the staircase itself and, in addition, usually takes a certain amount of load from the floors of the building. Depending on the extent to which the walls of the core could perform a further function in the transmission of horizontal loads, the following solutions are possible (see also p. 320):
1. The enclosure is unable to take horizontal loads. It leans against the structural steelwork, which is stiffened in some other manner. The walls of the staircase enclosure may be precast concrete panels which are erected at the same time as the steelwork.
2. The staircase enclosure is located between two parts of the building which are each stiffened to resist horizontal forces. The enclosure itself should possess adequate stability to ensure that, in the event of collapse of one part of the building, the stairs will still be available as an escape route for the other part. Hence, the enclosure should be designed to resist wind load acting on its surfaces. Precast concrete components should be effectively interconnected with some kind of steel device or bars embedded in the concrete (for details, see p. 269).
3. The core participates in the bracing of the whole building. It may be located within or outside the building; in the former case it may be in the interior or against an external wall. Frequently, the staircase enclosure is combined with liftshafts and/or service shafts and the core thus formed may additionally contain cloakrooms in each storey. By thus enlarging the horizontal dimensions of the core, its stiffness is of course increased. See p. 217.

Concrete stair details

Staircases

In a steel-framed building the staircases, if they have to be fire-resistant and soundproof, are generally made of concrete. They may be concreted in situ or, as is increasingly the case, precast either as individual steps or as complete flights.

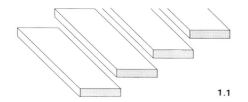

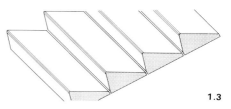

Individual steps of concrete or artificial stone

1.1 The steps are slabs forming the treads, while the risers are open gaps. Not permitted for 'essential stairs'.

1.2 The steps have upstands forming closed risers.

1.3 Each step is a prismatic block. A smooth soffit is obtained in this way.

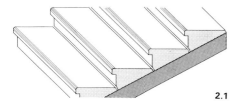
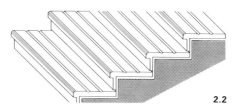
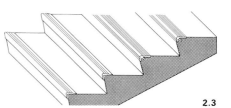

Stair flights

2.1 The steps are laid on a smooth slab forming the soffit. They are generally similar to the steps in **1.3** and may or may not be provided with a special tread covering.

2.2 In this example, the soffit slab and the steps are cast in one piece. Separate prefabricated units serve as facings to treads and risers. This form of construction has the advantage that the stairs can be used immediately after they have been installed, the covering units being added later, so that they will not be damaged during construction. Also, there is considerable dimensional latitude.

2.3 The finished steps may be cast monolithic with the soffit slab. Close tolerances are essential, necessitating accurate formwork, preferably made of steel. Tread coverings are fixed directly to the precast stair flight. On stairs of secondary importance the covering may be omitted; in such cases a synthetic resin coating, with corundum grit to provide a non-slip surface, may suffice. Plastic strips to protect the edges of the steps may be incorporated at the time of precasting or be stuck on later. During construction of the building, the steps should be suitably protected from damage.

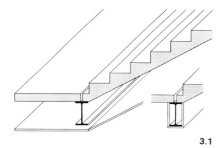
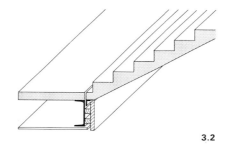

Methods of supporting the stair flights

3.1 The landing and flight rest on steel beams, the connection being formed by means of shear studs. Fire resistance is provided by a suspended ceiling or by encasing the beams.

3.2 The landing and flight are supported by a channel-section with an angle-section attached, there being fire encasement on the outside face.

3.3 If the landing slab is more than about 20 cm thick, it can be designed to support the flights without the assistance of a separate beam. The edge of the slab is provided with a 10 cm deep ledge as a bearing for the flights.

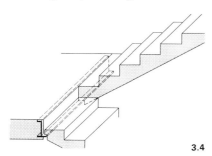
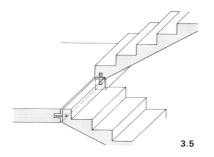
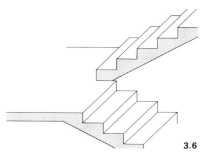

3.4 A steel section may be embedded in the edge of the landing, its bottom flange serving as a bearing for the flights. The ends of the embedded beam are bolted to the structural steel frame of the building.

3.5 The upper flight rests on the top of the landing slab, while the lower flight bears against it.

3.6 Another possibility is to cast the flights in one piece with the landings. Each flight is thus monolithic with half a landing slab at its top and bottom ends.

Staircases
Concrete stair details

Flights with stringers
In this form of construction the steps are supported laterally by stringers, which in turn are supported at the top and bottom landings. The stringers, which function as beams, may be of concrete or steel.

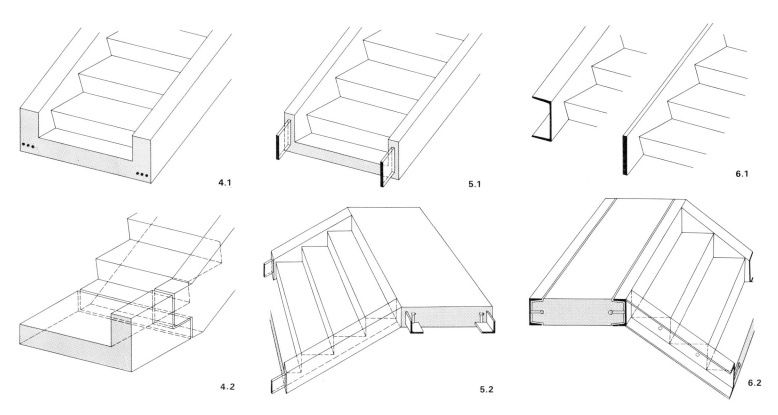

4 The stringers are relatively massive reinforced concrete members which are supported at the landing slabs by means of halved joints.
5.1 A narrower and more slender stringer is obtained by embedding a flat steel bar in it. The bar is the main structural feature, while the concrete serves as fire protection; for example, a 2 cm thick bar is provided with 3 cm of concrete cover on each side, so that the stringer has a total width of 8 cm. The flat steel bar may be provided with flanges or other fixing devices at the upper and lower ends of the stair flight and may thus be secured to the structural steelwork.
5.2 Angle sections are embedded in the landing slab along its two longitudinal edges. The horizontal legs of these angles are provided with shear studs for composite action with the slab. The stair flights are connected to the vertical legs at the ends of the slab. The angles protrude from the concrete, so that the landings can be fixed to the steelwork (Krupp-Montex system).
6.1 Concrete stairs with steel stringers. The latter consist of flat bars or channels, provided with welded-on reinforcing bars or studs for connection to the concrete steps. The stringers are installed in the moulds before the concrete is cast. In this way composite prefabricated flights are produced which can be conveniently fixed to the steelwork.
6.2 The landing slab is cast between channel sections to which the steel stringers can be connected. This form of construction does not rate as fire-resisting.

STEEL STAIR DETAILS

Steel as a constructional material for stairs offers a wide range of design possibilities. Because of its great strength it can produce very elegant solutions, such as stairs supported on slender struts or from thin hangers or carried by straight or curved beams. The steel staircase may be designed as some form of shell-type or stressed-skin structure, but generally it is more conventionally subdivided into well recognised parts: supporting struts and beams, steps and landings. The steps may be of wood, concrete or sheet steel. The steps may be supported on one central beam or two side beams. These beams may consist of flat bars or channel sections, sometimes welded to form box-sections, or various I-sections. Very elegant solutions are obtainable with rectangular hollow sections. A steel staircase will be fire-resistant if its steps are made of fire-resistant material and the supporting beams or the entire soffit of the staircase are suitably encased.

Steel stair details

Staircases

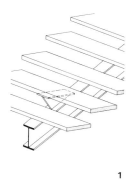 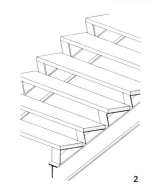  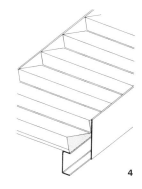 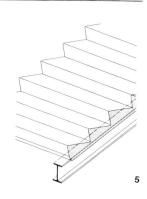

1 Wood or concrete steps carried on steel brackets with horizontal flanges. In this example, the brackets are welded to a central supporting I-section beam.

2 Wood or concrete steps on bearings of bent flat steel bars which may be of various shapes.

3 The steps consist of cold-formed steel sheet fitted between flat steel bars as stringers. Wood or concrete facings may be applied to the steps.

4 Individual precast concrete steps supported on channel sections welded to flat steel stringers. The steps are fixed to the steel with the aid of studs.

5 Flight of concrete steps cast directly on to steel side beams, with welded studs for connection.

 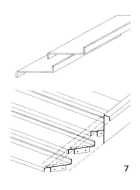 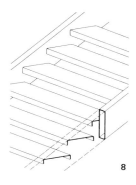 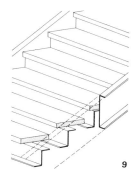 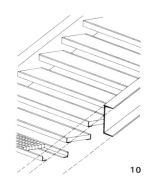

6 Cold-formed or welded steel steps between flat steel stringers.

7 Individual steel steps, of Z-shaped cold-formed section, between cold-formed stringers.

8 Cold-formed steel steps between rectangular hollow stringers.

9 Risers consisting of Z-shaped cold-formed units, treads of wood or grating, fitted between cold-formed stringers.

10 Treads comprising steel grating between angle sections, fitted between channel-section stringers.

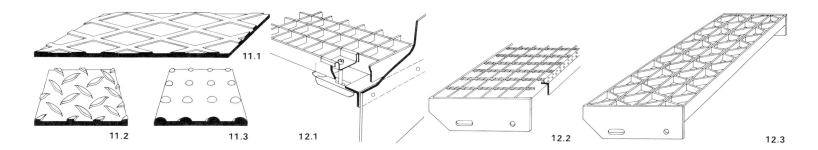

11 Steel treads without facings are used only in industrial buildings and for service and emergency stairs as they have the disadvantage of being noisy. Plain steel steps should be provided with a non-slip surfacing such as synthetic resin with corundum grit, textile fabric, etc.
Alternatively, the treads may be made from non-slip patterned plates, 4–5 mm thick, of which there are various types. The type illustrated in **11.1**, and known in Britain as diamond plate, has the disadvantage that dirt, water and ice are liable to collect between the raised ribs on the surface, while cleaning is difficult. The other patterns shown do not have these drawbacks. The type shown in **11.2** is known in Britain as Durbar plate. The plate in **11.3** has raised circular bosses.

12 Gratings are used for the treads of stairs in industrial buildings or for service stairs. There are many types, differing in pattern, strength and methods of connection. The grating may be supported on angle sections, **12.1** or be fabricated as complete step units ready for bolting to the steel stringers, **12.2**, **12.3**.

Staircases

Various forms of steel stairs

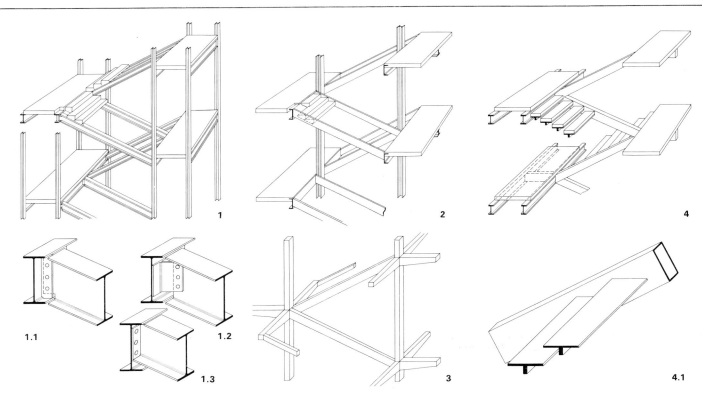

1 Simple construction with four columns per landing. Flights and landings are carried by steel beams. Flight beams can be connected to the landing beams by means of cleat plates, **1.1**, or angle cleats, **1.2**, welded to the latter or by end plates welded to the flight beams, **1.3**.

2 The stair system may be supported on just two columns in the stair well. These are provided with cantilevers which carry the landings and flights. Columns may be universal sections, while the landings are supported on channel sections and the flight beams consist of flat steel bars.

3 Similar to preceding example, but here the columns and beams are hollow sections. Thus, the columns consist of square sections, the flight beams of rectangular sections and the cantilevers of welded tapered sections.

4 The central stringer cantilevers from the top and bottom landings. It supports the intermediate landing and the steps, which cantilever out on each side of the stringer.

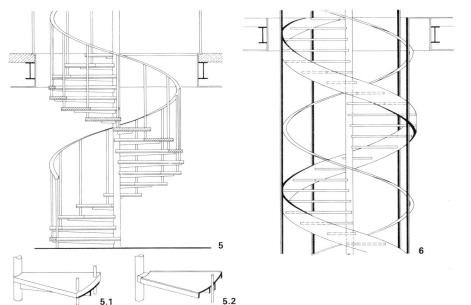

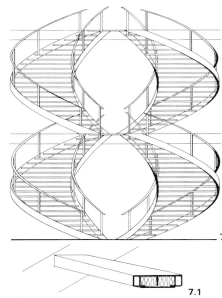

5 Spiral stairs. The loadbearing member is a vertical tube to which the steps are welded and from which they cantilever out.

6 The steps of the spiral stairs are welded between the central tube and an edge beam or stringer which is fixed to four suspension bars.

7 Spiral stairs with large-radius open well. The steps are supported between inner and outer spiral stringers which span between the floors of the building.

Roofs, external walls, internal walls

Roofs
  Function of the roof 299
  Flat roofs 300
  Flat roof details 302
  Pitched roofs 303

External walls
  Features of external walls for steel-framed buildings ... 304
  Functions of the external wall 304
  External wall construction 305
  Materials and components for external walls 306
  Deformations 307
  Attachment of the external wall to the loadbearing structure ... 308
  Façades in horizontal bands 310
  Mullion façades 312
  Panel elements 316
  Cladding 316
  Means of escape, galleries, rails for external maintenance equipment ... 317

Internal walls
  Requirements applicable to internal walls 318
  Tolerances 319
  Compartment walls 320
  Types of partitioning, connections 321
  Top connection of partitions 322
  Examples of partitions 323

FUNCTION OF THE ROOF

Multi-storey steel-framed buildings almost invariably have flat roofs. Pitched roofs are rare except for industrial buildings. Hence this chapter will be concerned more particularly with the flat roof. The range of constructional types to be considered is further reduced in that flat roofs are for the most part externally insulated and that for multi-storey buildings the downpipes for rainwater discharge are installed internally. Thus, the form of construction for the edges of the roof is limited to a few well-defined types.

The roof is subject to the following loads:

• Dead load: comprising the dead weight of the roof structure itself and, where the roof is accessible, paving slabs, plant pots or other roof garden features, rails and appliances for external maintenance.
• Imposed load: loading due to people moving about on roofs accessible to the public, otherwise occasional loading by maintenance personnel; moving loads due to external maintenance appliances.
• Snow.
• Wind, acting as pressure and as suction.
• Temperature differences.

The roof should provide protection against:

- surface water from outside;
- water vapour diffusion from inside;
- temperature influences;
- noise.

Roofs

Flat roofs

1 Construction of a typical externally insulated flat roof:

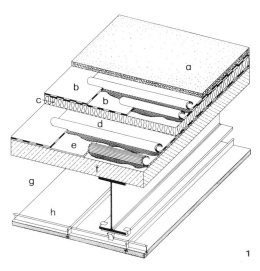

a gravel
b waterproof layers
c thermal insulation
d vapour barrier
e venting layer
f loadbearing roof structure
g cavity
h ceiling

Waterproof layer
Venting layer
Heat insulating layer
Vapour barrier
Venting layer

Such panels can be laid 'dry' and they must be carefully sealed at the joints under dry conditions, for example, under temporary shelters in winter. In this way, a temporary roof covering is obtained, which can be completed later under better weather conditions, by the addition of the finishing layers.

The following materials are used for vapour barriers:

- aluminium or copper foil, laid in bitumen or glued;
- plastic sheet, laid dry or in bitumen, with welded or glued joints.

The venting layer may consist of:

- felt coated with coarse granules;
- corrugated paper;
- glass-fibre tiles or other fibrous materials;
- corrugated asbestos panels.

To illustrate the physical behaviour of the roof, the following variants in roof coverings may be considered:

The requirements that the roof has to fulfil will significantly influence the type of construction to be adopted. For comprehensive information on the physical behaviour and on particulars of construction for flat roofs, reference should be made to the technical literature on the subject, for example, Henn, *Das flache Dach* (*The flat roof*), Munich.

The waterproofing systems employed on flat roofs are of various kinds:

- two or more layers of bituminous felt, with or without a covering of gravel or chippings, laid on and bonded together with bitumen;
- other materials (such as jute fabric) suitably combined with bitumen;
- plastic sheeting, sometimes with interlaminated metal foil, laid in bitumen or placed loose, with welded or glued joints;
- roof covering compounds, more particularly mastic asphalt.

Thermal insulation usually consists of material applied in the form of boards or panels, such as:

- natural insulating materials (for example, cork 200 kg/m^3, $\lambda = 0.04$ kcal/mh°C);
- rigid foamed plastics (many types in densities ranging from 30 to 50 kg/m^3, $\lambda = 0.025$ to 0.035 kcal/mh°C);
- foamed glass (200 kg/m^3, $\lambda = 0.047$ kcal/mh°C), laid in bitumen and serving also as a vapour diffusion layer.

In order to prevent the escape of warmth, the insulating panels are often laid in several layers with staggered joints. In some proprietary systems, the insulating panels are supplied with various other layers already bonded to them, for example, with the following composition, from top to bottom:

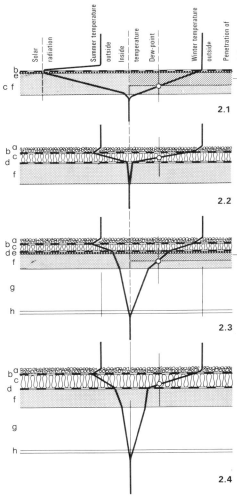

2.1 The structural roof slab (for example, aerated concrete) also provides thermal insulation and is provided with only a waterproof layer. The temperature gradient curves show that in cold weather the dew-point is located within the structural slab. As it is impracticable to apply an effective vapour barrier under the slab, moisture-bearing air from inside the building penetrates into the slab and condensation occurs in the upper part, so that it becomes saturated with moisture. As a result, its insulating effect is reduced, and drops of water due to further condensation form on the underside. Sunshine on the top of the thinly covered roof slab causes heating, so that water vapour makes its way out of the top of the slab thus producing blisters unless a venting layer, comprising air passages in communication with the external air, is provided.

The structural slab itself undergoes considerable temperature differences which may cause substantial variations in length and even result in hogging of the slab, especially on exposure to strong sunshine.

2.2 This represents a typical roof, comprising the following features:

- the layer of gravel prevents excessive heating by sunshine;
- separation of the loadbearing and the heat insulating functions prevents the development of excessive temperature differences in the structural slab;
- the heat insulating layer should be so thick as to ensure that the dew-point is always within it;
- the vapour barrier prevents moisture-laden air from the interior of the building from penetrating into roof layers in which it may be cooled to below the dew-point, causing condensation of water vapour;

Under normal temperature conditions a venting layer under the vapour barrier is superfluous. However, it may be necessary when a structural concrete slab has not had sufficient time to dry out, so that it may subsequently give off moisture. Nonetheless, it is better to connect a venting layer with the inner wall, so that the vapour barrier layer on the roof can be joined to that of the outer wall with a waterproof joint. If it is accessible to the open air, there is a danger of condensation and subsequent dampness at the outlets.

A venting layer over the thermal insulation is necessary if there is a possibility of the escape of moisture from the thermal insulation.

2.3 The space between the roof structure and the ceiling may be a factor to consider in connection with the roof insulation. If this space is not in effective communication with the air inside the room, the upper layer of stagnant air in the cavity may eventually acquire a temperature close to that of the outside air. Although this layer of air enhances the insulating effect, it may have the result of shifting the dewpoint to below the vapour barrier, so that the structural slab will become saturated with moisture. Small slots or holes in the ceiling hardly prevent this effect. On the other hand, it does not occur if the ceiling space is in really effective communication with the interior of the room, for example, if the ceiling is formed merely by slats and large apertures. However, such a ceiling cannot provide fire protection for the roof structure (should such protection be required). Alternatively, the cavity may be used as a plenum chamber for the circulation and distribution of conditioned and temperature-controlled air.

2.4 These disadvantages can be avoided by increasing the thickness of the insulating layer, ensuring that the dew-point will always be within it.

Flat roofs

The structure of a flat roof is generally similar to that of a floor, but lighter. In steel-framed buildings the roof materials most widely employed are steel and concrete.

Metal roofing

Metal roofing and cladding are composed of profiled sheets:

- corrugated sheet;
- ribbed sheet;
- steel or aluminium troughing.

Galvanised steel sheet is extensively used for externally insulated roofs. The methods of fixing and jointing are as described on p. 274. The requisite depth of the section, h, and the weight of the sheeting as a function of the total roof load, q, the span, l, and the support conditions, single or two spans, can be estimated with the aid of the accompanying chart **3**. If the upper surface of the roof has to be constructed to a certain fall to assist rainwater run-off, this should be achieved by appropriately sloping the roof beams, not by means of a sloping concrete topping or screed.

The steel sheet is in itself impervious to water vapour, but this is not necessarily true of the longitudinal and transverse joints. Under normal physical conditions (internal relative humidity 65%, internal temperature +20°C, outside temperature −15°C), however, a vapour barrier is not required if full air-conditioning is not envisaged or if, in schools and dwellings, thermal insulation in accordance with DIN 4108 is provided.

The openings which exist around the edges of the roof from the profiling of the sheet must be sealed to prevent deposits forming inside and to provide insulation against external weather conditions. Hence, foamed plastic closures are available for all the usual types of profiled decking.

The capacity for resistance to flying sparks and radiated heat in accordance with DIN 4102 is valid for all the roof decking and waterproofing complying with DIN 4102, Sheet 4, paragraph 7.6, together with steel troughing covered with thermal insulation complying with the regulations for the use of combustible materials in multi-storey buildings.

Concrete roof slabs

Concrete roof slabs, laid in situ or composed of precast concrete units and with or without composite action, are generally similar in construction to floor slabs. Such roofs are more particularly suitable in circumstances where relatively large loads have to be supported by the roof structure. If the building is to be subsequently raised by the addition of one or more storeys, the roof should be designed as a normal floor.

Lightweight concrete slabs, though possessing good thermal insulation capacity, must nevertheless be provided with additional insulating layers of adequate thickness to ensure that the temperature in the concrete always remains above the dew-point (see p. 300, **2.1**). Aerated concrete is very suitable as a lightweight covering, as are also various lightweight aggregate concretes such as pumice concrete or concrete incorporating wood particles (e.g. Durisol). (For connection between concrete slabs and steel beams without composite action, see p. 256.) A fall for rainwater run-off from a concrete roof can also be obtained more economically by suitably sloping the roof beams than by means of a screed. The latter involves extra expense, increases the weight of the roof and, if the roof is constructed of prefabricated or precast components, introduces unnecessary moisture into the building.

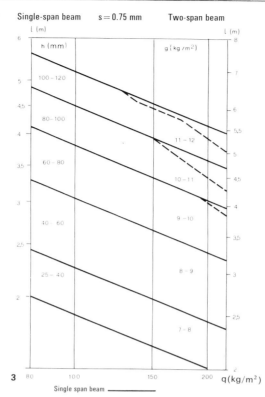
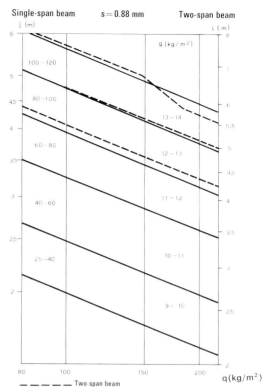

3

l = span of the roofing (m)
q = total load on the roofing, including the dead weight (kg/m²)
h = depth of the troughing (mm)
s = thickness of sheet (mm)
g = weight of sheet (kg/m²)

Example: Roof load with snow and weight of sheet, q = 150 kg/m², span l = 4.0 m.

s = 0.75 mm h = 100–120 mm g = 11–12 kg/m² single-span beam
s = 0.88 mm h = 80–100 mm g = 12–13 kg/m² single-span beam
s = 0.75 mm h = 60–80 mm g = 9–10 kg/m² two-span beam
s = 0.88 mm h = 60–80 mm g = 11–12 kg/m² two-span beam

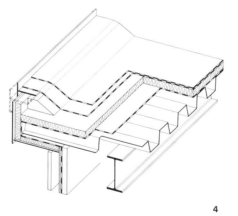

4 Trough decking projecting beyond the external wall of the building. Heat insulation and vapour barrier extend continuously around the perimeter. The edge upstand to the roof is formed by a wedge-shaped strip of wood or rigid foamed plastic.

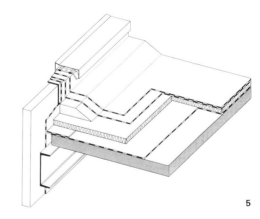

5 Concrete roof slab supported on steel beams. The external wall continues above the edge of the slab. Insulating and water-proofing layers are raised at the edge and covered by a profiled coping strip.

For references, see p. 276.

Roofs

Flat roof details

Domed rooflights

There is no problem fitting domed rooflights to roofs comprising in-situ concrete or large precast concrete slabs. However, narrow lightweight concrete roofing units must, at rooflights, be supported by trimmers.

1 + 2 Rooflights built into roof troughing require trimmers which are connected to the roof beams. The lights are supported by a timber frame or special sections. Under certain conditions, the closure sheet (e) which is supported by the longitudinal trimmer (a) can take over the function of the transverse trimmer (b).

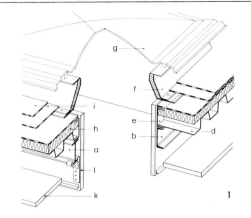

a trimmer
b secondary trimmer
c roof beam
d trough decking
e framing section
f rooflight frame
g rooflight
h heat insulation
i roof covering
k ceiling
l fascia

Rainwater discharge

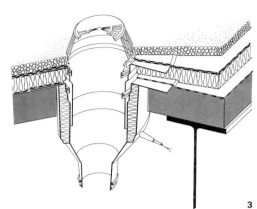

3 The outlets should, if possible, be so arranged on the roof that the downpipes are placed beside columns (see p. 241) or can suitably be accommodated in shafts. If there is sufficient space between the roof structure and the suspended ceiling, horizontal runs of rainwater pipe may be installed. The downpipes should have approximately 0.7 to 0.9 cm^2 of cross-sectional area per m^2 of roof catchment area. Inside the building the pipes should be insulated to avoid condensation forming on them. It is also possible to install artificial heating for use at rainwater outlets in cold weather.

Rails for external maintenance trolleys

Depending on the type of trolley used for raising and lowering the external maintenance cradle, rails attached to the external walls and/or the roof are needed. Rails fixed to the roof are subjected to either compressive or compressive and tensile forces. They are carried on various kinds of supports installed at specific spacings. Rail supports which are loaded only in compression can be laid on the roof covering, if the latter is strong enough to carry the loads. Rubber-tyred mobile trolleys do not require rails, but a load-spreading layer laid over the thermal insulation. Components loaded in tension must be secured by anchors penetrating the roof covering and attached to the structural parts of the roof. Horizontal wind forces on the rail-mounted trolley must also be resisted. At points where anchors penetrate the roofing, the openings must be carefully sealed to prevent the escape of heat through the insulation and vapour through the vapour barrier. It should also be borne in mind that the rail mountings are subjected to alternating loads.

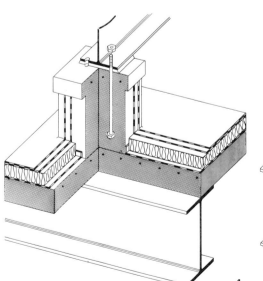

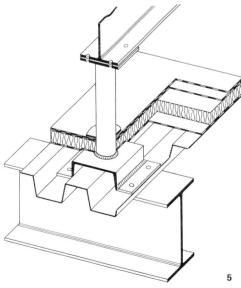

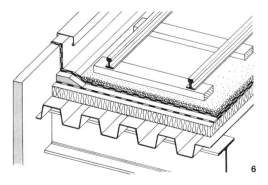

4 The rail is fixed to a concrete pedestal standing on the concrete roof slab. The insulation and waterproofing layers are raised around the pedestal.

5 On a roof comprising steel troughing, the rail may be fixed to a short tubular column which is provided with a welded collar at the junction with the insulation and waterproofing.

6 If the trolley has adequate counterweighting, the rails need not be anchored down to the roof structure, but may simply be laid on sleepers resting on the gravel covering of the roof.

Pitched roofs

In general, pitched roofs are employed only on relatively low buildings, such as single-storey industrial sheds and low-rise buildings up to about four storeys in height. Roofs over open-sided structures such as multi-storey car parks serve only to keep out the weather and do not require any thermal insulation.
Reference: *Geneigte Dächer* (*Pitched roofs*), Wiesbaden, 1974.

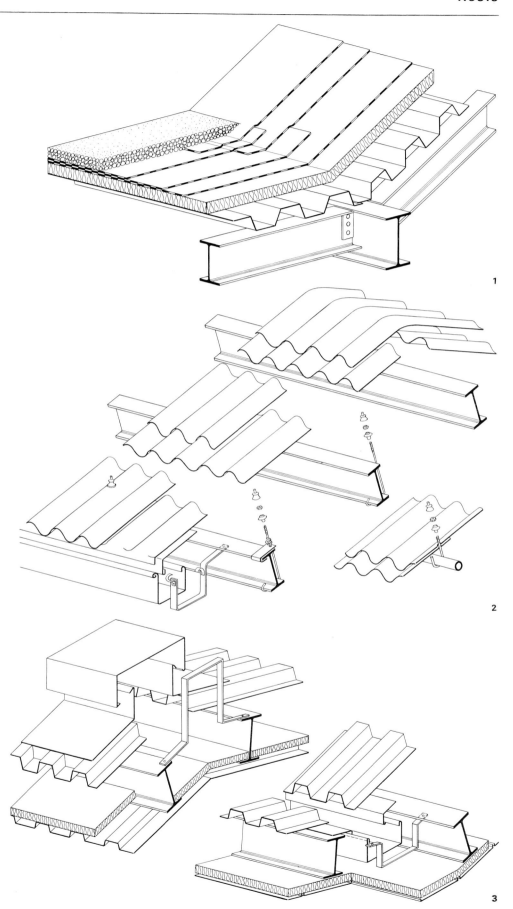

1 Pitched roof, externally insulated, as used for northlight roofing on multi-storey buildings. The drawing shows the transition from flat to pitched roofing; the latter is of similar construction, but without the layer of gravel. To reduce heat absorption from sunshine it is advisable to use light-coloured plastic coverings on the sloping surfaces.

2 Pitched roof with corrugated asbestos cement cladding, internally insulated. Thermal insulation fitted directly under the cladding or below the purlins. This is a highly developed form of roof construction, with standardised fixing and sealing devices. Maximum purlin spacing 1.45 m for 57 mm depth of corrugation, weight of sheet 13.25 kg/m^2, minimum roof pitch 7°; steep roofs can be covered in this way. These are typical figures. For further particulars see: Neufert, *Well-Eternit Handbuch* (*Corrugated asbestos cement manual*).

3 Pitched roof with steel or aluminium troughing internally insulated. Thermal insulation usually fitted below the purlins. Special units for ridge, eaves, gables, etc., are available. For approximate determination of load capacity, see diagram p. 301.
Steel troughing up to 15.00 m in length. If no joints are required, the units may be laid to very low angles of pitch. The units are galvanised and are available with plastic coating on both sides, in various colours. Plastic coating is advisable as extra protection.

External walls

Features of external walls for steel-framed buildings

The following characteristics peculiar to steel-framed buildings affect the construction and often also the external appearance of the façade:

- In steel-framed buildings the loads are transmitted by the columns, no loadbearing external wall being necessary. Hence, lightweight walls are generally more suitable than heavy ones. The columns may be wholly or partly incorporated into these walls.
- The location of the columns in relation to the external wall must therefore be considered, as they may, of course, be behind, in, or in front of the wall.
- As external columns can support the façades, mullions can often be dispensed with.
- The external wall can be secured quite simply to the structural frame by bolting or welding.
- Wind-bracings comprising diagonal members, whether installed inside or outside the façade, considerably affect the design of the external wall.
- Due attention should be paid to relative movements between the structural frame and the external wall. The deformations of a steel frame due to temperature variations, vertical live load and wind load — and, in tall buildings, the oscillations due to wind — are usually greater than those of a concrete frame.
- Because of accurate fabrication, dimensional tolerances in a steel frame are smaller than in structures built of other materials.
- The external wall must fulfil the requirements of structural fire protection. It must suitably protect the columns and beams of the frame.
- Erection of the external wall can often be combined with erection of the structural steelwork.
- As a rule, prefabricated units are used for constructing the external wall. These may be single components or may comprise two or more layers.

Functions of the external wall

In addition to the obvious functions of forming a physical and visual barrier (for example, to ensure privacy), the external wall has others to perform, many of which can adequately be performed by a single-layer wall system. The functions of the external wall are:

1. Protection against moisture

- In its simplest form the wall merely keeps out driving rain and snow.
- A fully developed external wall should keep out all moisture, including any tending to penetrate the wall under the action of wind.
- An important function is water vapour control. In winter the air in the interior of the building is warmer and more humid than the external air and is liable to give rise to condensation of moisture in the outer layers of the cladding. If the moisture which collects there cannot subsequently, in warmer weather, escape to the outside by diffusion, more and more will accumulate. This may go on for long periods, years perhaps. Eventually it may cause serious damage as a result of saturation of the wall, roof and floor components. For this reason, many building regulations require a continuous vapour barrier enclosing the entire building. The transition from roof to external wall may be awkward in this respect. Just as in the roof, in the wall the vapour barrier should be placed on the inside of the thermal insulation, in order to eliminate the risk of condensation.
- In rooms with moisture-laden air, where occasional condensation on the external walls is unavoidable, dripping can be prevented by a special absorbent layer which can store the condensate, gradually releasing it during subsequent periods of lower humidity in the room.

2. Protection against air movements

- In its simplest form, a wall merely gives protection from wind.
- A fully developed wall completely excludes wind.
- In some instances, the wall may also be required to prevent movement of air from inside the building to the outside.

3. Heat insulation and control

- Protection against direct sunshine by shading both the windows and closed wall surfaces. Such protection may be provided by projecting ledges, other horizontal and also vertical features, facings applied to closed surfaces, movable devices such as louvres or awnings, etc.
- Protection against direct sunshine by reflection: closed wall surfaces can be given a smooth light-coloured external finish; special solar insulating glass can be used for glazed areas.
- The main function of any fully developed external wall is to provide insulation against the passage of heat, in the sense that the amount of heat that goes through the wall is limited to a certain rate per unit time, characterised by the coefficient k, expressed in $m^2 \cdot °C \cdot h/kcal$.
- As a result of heat storage in relatively thick and heavy external walls a certain amount of natural equalisation is obtained. In summer, the heat due to sunshine is stored in the wall in the daytime, and is given off externally at night; in winter, the wall stores up heat from the interior of the building at times when the heating system is in operation, this heat subsequently being given off and serving to keep the interior warm during the intervals when the heating is switched off, more particularly at night.

Lightweight external walls have only a very limited heat storage capacity. On the other hand, they respond more rapidly and effectively to control by heating or cooling or by shading. Heat storage is therefore of minor significance in the walls of buildings comprising large glazed surfaces.

4. Sound insulation

- Insulation against external noise largely arising from vehicular traffic and aircraft.
- Sound insulation between storeys and between rooms in the same storey. The inner sheet or leaf of lightweight cladding often tends to transmit sound past sound-insulated partitions or floors.
- Sound absorption in rooms which have to fulfil specially stringent acoustic requirements.

5. Fire protection

- Protection against the spread of fire from storey to storey. The fire-resistance requirements applicable to the external wall are usually laid down in official regulations, as is also the specified minimum distances between windows, etc., with a view to restricting the spread of fire.
- The external wall provides the lateral closure of the space between the a floor and the ceiling below if this contains structural components which have no fire protection of their own. If the wall does not perform this function, some other means of closing the cavity at the perimeter of the building is necessary.
- Prevention of the spread of fire from a floor cavity to the storey above, more particularly if the cavity contains cables which can catch fire or air ducts which, in a fire, may become filled with hot gases.

6. Structural function

- The external wall must resist and transmit the forces exerted on railings and parapets and at window spandrels.
- Wind forces (suction and pressure) have to be resisted and transmitted to the loadbearing structure.
- Transmission of its own weight. Exceptionally: the performance of a loadbearing function in transmitting the floor and/or roof loads.
- Security against burglary.

Combination of requisite properties

Not all the properties enumerated above have to be present in an external wall.
The least exacting requirements are those which have to be fulfilled by the walls of open or semi-open buildings or parts of buildings, such as balconies, shelters, open multi-storey car parks, for example:

- parapets as a safeguard against accidents;
- screens as protection against entry by unauthorised persons;
- visual screening against glare, etc;
- protection against wind-driven snow or rain;
- protection against wind;
- protection against solar heat.

Fully enclosed but unheated buildings, such as warehouses, impose additional requirements on the external wall, for example:

- impermeability to rain, etc;
- impermeability to wind pressure;
- protection against fire.

Completely protected buildings involve further requirements, including always:

- heat-insulating capacity.

External wall construction

External walls

Principles

External walls may be classified according to various criteria. Considering their constructional features and their relationship to the loadbearing structure, the following basic types can be distinguished. There are no clear-cut categories, however, and transitions are gradual.

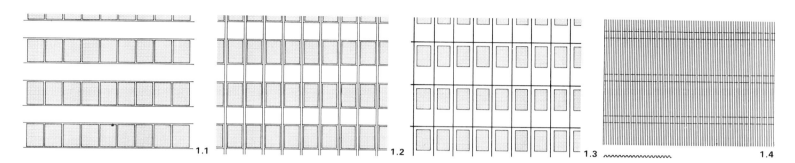

1.1 Band façades. The elevational treatment of the wall is characterised by pronounced horizontal features. The loadbearing members of the cladding are the continuous bands at spandrel level. These alternate with non-loadbearing bands, usually comprising the glazing.

1.2 Mullion façades. The mullions are the loadbearing elements of the wall cladding and their spacing usually corresponds to the width of the individual windows. The structural columns of the building may also serve as mullions. The mullions may be only one-storey high or may be continuous over several storeys.

1.3 Panel façades. The characteristic feature of the façade is the storey-height cladding panels, each corresponding to the width of a window.

1.4 Continuous cladding. Such façades generally exhibit no distinct subdivision into storeys and individual window features. The cladding units are of large size and may correspond to several storeys in height.

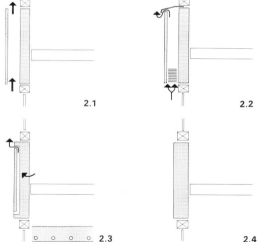

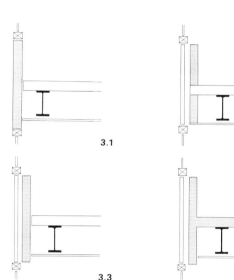

Venting

Depending on the degree of ventilation of the external wall structure, the following possibilities may be distinguished:

2.1 An outer leaf or panel installed in front of a cavity which is completely open at top and bottom. The outer leaf shades the actual wall and gives protection from driving rain. Sunshine on the wall will produce a chimney effect.

2.2 In this example, the cavity is also in free communication with the external air at top and bottom, but is not completely open. Waterproofing against penetration of rain is usually applied to the inner leaf.

2.3 Here the cavity is closed, but not hermetically sealed. It is vented a little in that it is in communication with the external air or with the interior of the building, so that equalisation of any differences in air pressure is effected.

2.4 A completely sealed wall. Any internal cavities are airtight enclosures.

Fire resistance

Regulations requiring fire-resisting spandrels and ceiling structures may result in the following solutions:

3.1 The external wall is of fire-resisting construction over the height of the spandrel, the floor structure and, possibly, the lintel of the storey below.

3.2 In this example, the external wall comprises a cladding which is not fire-resisting. Spandrel and floor cavity closures providing the necessary fire protection are constructed independently within the wall. See also p. 279.

3.3 Here again the wall cladding is not fire-resisting, and a separate spandrel, cavity-closing and lintel unit is installed inside.

3.4 Alternatively, the spandrel wall and the skirt may be cast monolithically with the concrete floor slab.

External walls
Materials and components for external walls

Brick and concrete infillings and claddings
Brickwork

Ordinary or lightweight bricks or sand-lime bricks, with or without external rendering, are placed within the frame. For high walls the masonry is additionally supported by vertical and horizontal steel members. Such infillings may be a half (**1.1**) or a whole brick (**1.2**) thick.

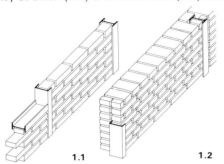

1.1   1.2

Concrete panels and slabs

These units, usually of precast concrete, are provided with a heat-insulating layer consisting of mineral fibre mat, chipboard, expanded polystyrene, etc.

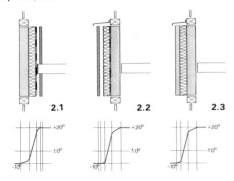

2.1   2.2   2.3

2.1 The heat-insulating layer on the inside of the panel enables the wall to store up heat on exposure to sunshine, but the dew-point is in either the concrete or in the insulation.
2.2 If the heat-insulating layer is applied on the outside of the panel, the latter can absorb heat from inside the building and act as a kind of storage heater that helps to keep the interior warm at night when the heating is switched off. The dew-point remains within the insulation.
2.3 The concrete slab is a composite three-layer unit, comprising an outer and an inner leaf with a layer of insulating material sandwiched between; the inner leaf performs the main supporting function and is thicker than the outer. The two are interconnected by stainless steel ties extending through the insulation. To obviate stresses in the panels due to temperature differences, rigid connections are provided at the centre of each panel and elastic connections at the edges.

Lightweight concrete panels and slabs

Because of their low weight and moderate cost, these are particularly suitable for steel-framed buildings.

- Aerated concrete is cast in large blocks which are then cut up into panels or slabs. No fittings, sockets, etc., can be concreted in, but the material is nailable. Used as single-layer cladding units. Bulk density 700 to 900 kg/m³.

- Lightweight concrete, made with expanded clay or shale or with pumice aggregate, with bulk density ranging from 1000 to 1600 kg/m³; or with expanded polystyrene when densities of 600 to 900 kg/m³ are possible. They are used as single-layer cladding units which can be cast to any desired shape and which can be provided with embedded fittings, etc.
- Multi-layer lightweight concrete cladding units. A typical panel of this type comprises a dense inner concrete leaf, a thick intermediate layer of insulating lightweight concrete made with wood particles, and an outer finishing layer of concrete (for example, Durisol panels).

Sections for mullions
Wood

Wood is often used in cases where the cladding is set back from the edges of the floors and is thus protected to some extent from the weather.

Steel

Hot-rolled, extruded or cold-formed steel sections are available. They may be one-piece or two-piece elements, the latter with an interlay of plastic for insulation. The materials used are plain steel, galvanised, painted, stove-enamelled or with plastic-coated finish, as well as weathering steel and stainless steel.

Aluminium

Extruded aluminium sections are available in a wide range of types and shapes, including quite complex ones, with integral ribs and other features for jointing and for insertion of sealing gaskets. Also decorative shapes. Plain finish or anodised, stove-enamelled or painted. As aluminium sections tend to be more expensive than those in steel, it may be advantageous to use a combination of a steel backing section with an aluminium covering section, with a suitable separating interlay to prevent galvanic action between the two metals.

Sandwich panels

A distinction may be drawn between self-supporting wall or spandrel panels and panels which are fitted between supporting mullions or columns. Depending on their physical function in relation to the interior atmosphere, open and closed panels are to be distinguished. Another distinction is between panels with mechanically bonded layers and those with adhesive-bonded layers. As a rule, a cladding panel comprises the outer skin, the insulating core, the vapour barrier and the inner skin.

Outer and inner skin materials.

Materials which may be shaped comprise:

- steel sheet: flat, cold-formed or drawn, often with stiffening ribs: ordinary steel with various finishes (galvanised, painted, stove-enamelled, plastic-coated), weathering steel or stainless steel;
- aluminium sheet: flat, cold-formed or drawn, profiled for stiffening, with various finishes (anodised, painted, stove-enamelled, plastic-coated) or plain;
- plastics.

Flat sheet materials:

- wood in the form of plywood or hardboard;
- asbestos cement: rough grey or white sheets, smooth sheets, such as Glasal;
- gypsum plasterboard (weather-resisting hard gypsum);

- fire-resisting asbestos-based panels (such as Promabest, Isoternit): only for the inner skin or for the outer skin of the inner leaf of an internally insulated wall.

Materials for the insulating core:

- fibrous materials, for example, mineral fibre products, usually loose;
- foamed plastics, which may be moulded, such as Styropor or Moltoprene;
- foamed glass in the form of tiles set in bitumen to produce vapourtight joints;
- foam introduced between the outer and inner skins, adhering to both, such as polyurethane;
- paper honeycomb, with or without foamed plastic filling.

Materials for vapour barrier:

- metal foil, for example, aluminium or copper;
- plastic foil.

Methods of sealing the panels

Steel or aluminium sheets:

3.1 Inner and outer skins with flanges fitting together in the manner of a box and sealed with adhesive strips.
3.2 Similar to preceding example, but with flanges joined by welding. Drawback: such edges form an easy heat-escape path.
3.3 Outer skin shaped by drawing, inner skin is a flat sheet, clipped on to a gasket of plastic or synthetic rubber.

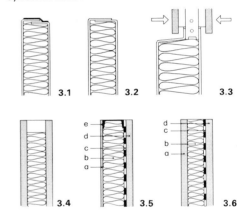

3.1   3.2   3.3

3.4   3.5   3.6

Panels with flat sheets

3.4 Flat sheets glued to edge sealing strips.
3.5 Seal formed by metal or plastic section to which the sheets are glued, composition:

  a flat metal or plastic sheet
  b insulation
  c vapour barrier, such as foil
  d inner skin, for example, plasterboard
  e channel edging of metal or plastic

3.6 Open sandwich panel with rigid insulating core, composition:

  a outer skin, for example, asbestos cement
  b insulation
  c vapour barrier
  d inner skin, for example, asbestos cement

Continuous cladding

- asbestos cement sheets: flat, corrugated or specially profiled;
- steel sheets: large corrugated or trough-section panels; galvanised, with or without additional plastic coating;
- aluminium sheets: trough sections and many other profile shapes; plain, anodised or plastic-coated;
- plastic panels.

Deformations

External walls

In designing the external wall, the following possible movements should be considered:
- for the connections between the wall and the structure: the relative movements between the wall and the rest of the building;
- for the joints in the wall cladding: the movements of the cladding units in relation to one another.

Deformation of the building

As far as external walls are concerned, it is not the absolute movements of the building which are important, but the relative movements of the cladding attachments with respect to one another. These comprise vertical and horizontal movements due to linear changes, as well as torsional movements of the floors of multi-storey buildings and the displacements of the floors relatively to one another in consequence of lateral deflection of the structure due to wind load.

Vertical movements

These arise from variations in the distance between successive floors, such variations being due to differences in deflection caused by live load and due to changes in the length of the columns caused by live load, wind and temperature.

1 Floor deflections may be significant. Under building regulations they are limited to a certain permissible fraction of the span.

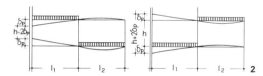

2 In the case of a cantilever it should be borne in mind that, whereas the loaded cantilever deflects downwards, the unloaded one may deflect upwards in consequence of load on the adjacent span, so that at the external wall the two opposing effects may be cumulative. The end sections of the cantilevers undergo angular rotation.

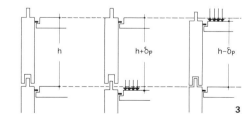

3 These diagrams illustrate the effect of floor deflection on the external wall.

Variations in length of columns

The changes in length that unclad external columns are liable to undergo vary between +3 and −6 mm, while such changes for internal columns range from +2 to −4 mm: these are very small changes, even when additive in extreme combinations.

Horizontal movements

Changes in the length of floors are of no practical significance to the external wall. Assuming a maximum temperature variation of ±10°C in the interior of the building, the corresponding linear variation for a 7.20 m wide panel is ±0.7 mm. Even if the edge face of the floor slab is externally exposed and thus subject to greater temperature ranges, this does not cause significant deformation, as the inner zones of the floor restrain it. The variations in length of the roof may be substantially larger (see p. 224 **8, 9**).

Attachment of external wall to floors or columns

Attachment to the columns is preferable in cases where they are closely spaced. If wall units are fixed to the edges of the floors, the joints between the units should be designed for the state of deformation that the floors are likely to have at the time of erection of the wall. It can generally be assumed that the floors will in practice never have full design live load acting on them (p. 242).
In buildings with floors cantilevering out a substantial distance from the outer columns it may be advantageous to suspend the whole external wall cladding from the top floor, with sliding attachments for securing the cladding to the other floors (see p. 204 **1**).

Deformation of the external wall

The outer layers of the wall are under the direct influence of the external climatic conditions, while the inner layers are subject to the conditions in the interior of the building. Thus the outer layers undergo major variations in temperature (−30 to +80°C), whereas the variations affecting the inner layers are quite small (+15 to +25°C). These conditions affect the external wall components in different ways.

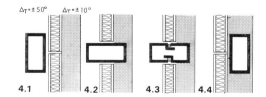

Mullions, transoms and framing sections

4.1 Members located outside the heat insulation undergo large linear variations because they are fully exposed to the external temperature conditions.
4.2 Members which are set within the wall and are in contact with the external atmosphere should be insulated at the inner face, otherwise they provide paths for the escape of heat. In addition, large differences in temperature between the inner and outer faces would cause the members to curve due to unequal expansion.
4.3 This latter problem can be avoided by using members which are composed of separate halves provided with a thin insulating insert at their joint. The halves can undergo thermal movements independent of each other.
4.4 Members which are located behind the heat-insulating layer are subject mainly to the relatively constant interior temperature conditions of the building and therefore undergo only slight thermal movements.
In addition to deformation due to variations in temperature, the mullions and other members are subject to deformation due to wind load.

Sandwich panels

5.1 Such panels undergo a certain amount of deformation, depending on their form of construction, as a result of differences in temperature at the outer and inner skins.
5.2 The outer skin is secured by individual ties (sometimes the only fastening being at the centre of the panel, to permit freedom of movement), the shear forces being transmitted by the insulating core. The outer skin undergoes deformation independent of the inner skin. The latter performs a loadbearing function, at least in so far as the panel itself is concerned, and is attached to the structural frame of the building, either directly or through mullions.

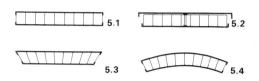

5.3 If the panel has a non-rigid core, which cannot transmit shear, the outer skin deforms almost independently of the inner skin, so that the panel develops a slight curvature.
5.4 If the panel is of one material throughout (such as aerated concrete) or if the outer and inner skins are rigidly interconnected (for example, by edge sealing strips), temperature differences cause curvature or warping. Furthermore, the panel as a whole undergoes a change in length dependent upon variations in the average temperature at mid-thickness.
Wind pressure is also a factor causing curvature of cladding panels.

Glazing

Sheets of glass should be freely supported so as to ensure that they will not be subjected to restraints and stresses due to movements of the cladding or the structure generally. Manufacturers of special glass, such as insulating glass, issue specific instructions for fitting it.

Superposition of deformations

In order to estimate the true relative movements of the various parts with respect to one another it is necessary to superimpose the deformations of the cladding panels and their supporting members and the displacements of the points of attachment due to the movements of the building as a whole. The connections and joints should be designed to cope with these conditions and the widths to be adopted for the joints should be determined accordingly.
Metal cladding may cause objectionable noise when the panels rub against one another when undergoing movement. This can be obviated by means of plastic interlays.
On all-welded metal-clad façades (as, for example, on the Civic Center in Chicago) the temperature differences do not induce movements, but produce restraint stresses within the cladding.

External walls

Attachment of the external wall to the loadbearing structure

Support conditions

At the points of attachment the external wall transmits the vertical loads due to its own weight and of any objects fixed to it (sun blinds, emergency balconies, radiators) and the horizontal loads arising from wind pressure and suction on the loadbearing system of the building. Depending on their structural function, the following forms of construction may be distinguished:

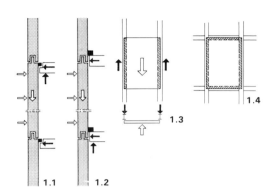

1 Storey-height external wall components are supported in the manner of a beam spanning from floor to floor. Components extending over two or more storeys function as continuous beams. They may be suspended from the floor above (**1.1**) or be supported on the floor below (**1.2**). The panels may be supported solely by the columns or mullions (**1.3**) or they may receive support from the floors as well, so that all four edges of each panel are supported (**1.4**).

Attachment of external wall to floor slab

The floor structure usually comprises a concrete slab, cast in situ or assembled from precast units, or sheet steel troughing filled with concrete. Devices for fixing the components of the wall to the concrete include anchor rails, dowels, bolts in threaded sockets embedded in the concrete, etc.

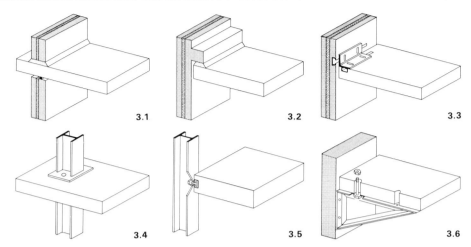

3.1 Concrete spandrel panel on a bed of mortar and supported by the floor slab.
3.2 Concrete cladding panel comprising an outer and an inner leaf, the latter provided with a supporting nib or ledge, also on a mortar bed.
3.3 Concrete cladding panel secured by means of steel angle provided with slotted holes to allow for tolerances.
3.4 Mullion secured to a floor by means of dowels.
3.5 Mullion secured to concrete floor slab by means of an embedded channel section anchor rail and T-head bolts.
3.6 Rigid connection of a cladding panel by means of a steel bracket secured by bolts screwed into threaded sockets embedded in the concrete floor slab. The sockets have to be accurately positioned, this method being suitable for precast concrete floors.

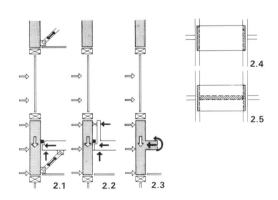

2 Loadbearing units of a horizontal band façade are each secured to one floor only. For stability, the band requires horizontal support below (**2.1**) or above the floor (**2.2**) or may be rigidly joined to the floor, forming a monolithic entity with it (**2.3**). In each bay between adjacent columns a panel or band unit may thus be supported at two edges by columns (**2.4**) or may additionally be supported by the floor (**2.5**). External wall claddings and other components of the wall may be secured direct to the loadbearing structure of the building, but usually appropriate fixing devices are interposed to provide compensation for dimensional inaccuracies.

Attachment of external wall to floor beams

Steel floor beams offer ideal possibilities for attachment by means of bolted or welded cleats.

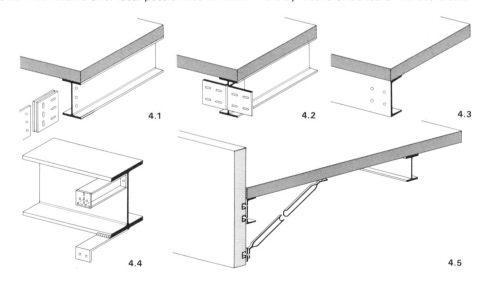

4.1 In this case the floor beam is perpendicular to the external wall. A gusset of the wall framework can be connected to the beam with a pair of splice plates, possibly with the interposition of plastic for insulation against heat loss due to metal-to-metal contact. The floor beam has to be accurately positioned, as only a very limited amount of horizontal adjustment of the cladding or its supporting framework within the plane of the wall is possible.
4.2 Slotted angles bolted to the web of the beam.
4.3 A channel section forming an edge beam extending parallel to the external wall provides ample scope for forming bolted connections.
4.4 If the floor beam is parallel to the wall but set some distance back from it, it is necessary to provide brackets for fixing the cladding.
4.5 Cladding panel bolted to the edge beam of the floor and additionally supported by struts.

Attachment of the external wall to the loadbearing structure

External walls

Attachment of external wall to steel columns

Steel columns installed behind the external wall offer numerous simple possibilities for attachment. The connecting elements must penetrate any fire encasement of the columns, or the external wall itself must perform the function of giving fire protection to the columns. Fire-resistant cladding panels must be secured by means of fixings which are also fire-resistant.

5.1 The simplest arrangement consists in bolting the cladding direct to the column.
5.2 Connections formed outside the protective casing.
5.3 Single bracket welded to a column set back from the wall.
5.4 Double bracket for heavier cladding.
5.5 Infill panels fitted in angle sections welded to the column flanges. Outer flange of the column remains exposed.
5.6 Outer flange of the column is provided with a welded bracket with a horizontal plate and a hole for securing the cladding by means of suspension rods or bolts. The horizontal plate is movable, secured by stops against slipping off its seat during erection, so that there is scope for adjustment in any direction. On completion of the erection of the cladding, the plate is fixed by welding.
5.7 Similar arrangement, but duplicated and attached to a larger bracket.
5.8 Connecting element bolted to the column allows separate fixing of the outer and the inner leaf of a cladding system with vented cavity.

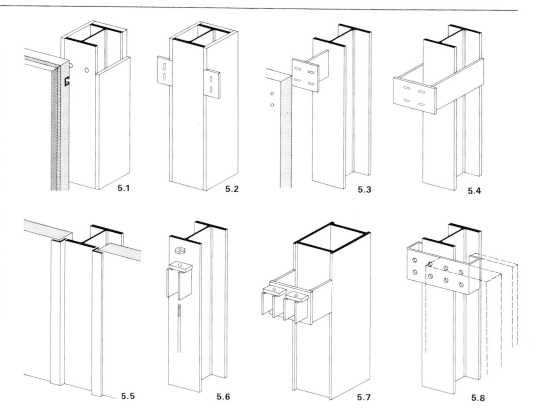

Compensation for tolerances

The joints in the external wall cladding must compensate for tolerances and for the differential deformation referred to on p. 307. Cladding units are manufactured to relatively close tolerances, metal units being more nearly precision-made in this respect than concrete ones. The tolerances on the dimensions of the supporting structure are wider. If the tolerances of the structure and those of the cladding are not interadjusted, large dimensional discrepancies are liable to occur, necessitating the use of interposed elements providing a large range of latitude to compensate for these differences.

6 The point of attachment of the cladding to the supporting structure may theoretically be displaced in three directions and be rotated about three axes in relation to the correct position and angular direction.

7 These diagrams illustrate the six degrees of freedom needed:

7.1 horizontal movement parallel to the plane of the wall (δ_x);
7.2 movement perpendicular to the plane of the wall (δ_y);
7.3 vertical movement parallel to the plane of the wall (δ_z);
7.4 rotation about the horizontal x-axis (α_x);
7.5 rotation about the horizontal y-axis (α_y);
7.6 rotation about the vertical z-axis (α_z).

8 Mullion attached by means of a system of angle fixings provided with slots to give adjustability in the three displacement and the three rotational directions.
9 Example of a supporting system for a mullion or cladding unit providing adjustability for displacement in three directions. The point bearing of the bolt additionally ensures rotational adjustability in all directions.

For further examples of cladding attachments with scope for adjustability and compensation, see pp. 310–315.

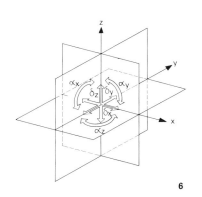
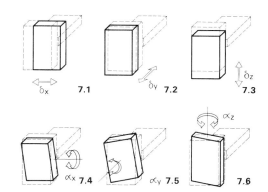
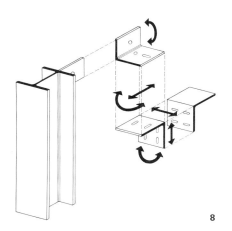
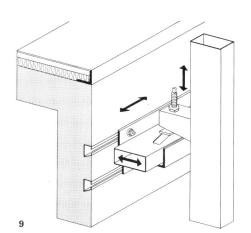

External walls
Façades in horizontal bands

The cladding comprises loadbearing and non-loadbearing bands, the former usually being closed, that is, without openings, and forming the window spandrels and/or the floor edge claddings. The glazing occupies the non-loadbearing bands in the façade. Some commonly employed forms of construction:

1 The spandrel bands, a, comprise the actual spandrel, b, the cladding, c, to the floor structure and the portion forming the lintel, d, for the storey below. The non-loadbearing bands installed between these spandrel bands may consist of either windows or closed infill panels.

2 The loadbearing band, e, comprises the cladding, c, to the edge of the floor structure, together with an upstand, f, and the lintel, d. The spaces between the loadbearing bands are filled either with a system of mullions and windows with spandrel panels or with storey-height closed infill panels.

3 Internally insulated cladding. The insulation comprises, for example, Durisol slabs spanning a distance of 7.50 m, secured to the columns by means of brackets and also supported against the floor slab. The windows are attached to anchoring rails. The external facing consists of anodised aluminium panels. (SFB Building, Berlin; architects: Tepez, Zander.)

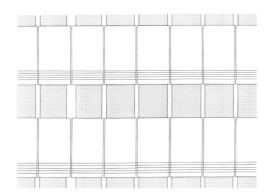

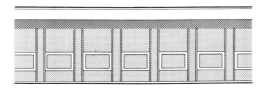

4 Lightweight concrete cladding units, comprising spandrels and lintels. The units are suspended from the structural steel frame by means of bolted connections permitting lateral and rotational tolerances in all directions. (Design for university buildings at Bielefeld, Germany.)

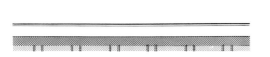

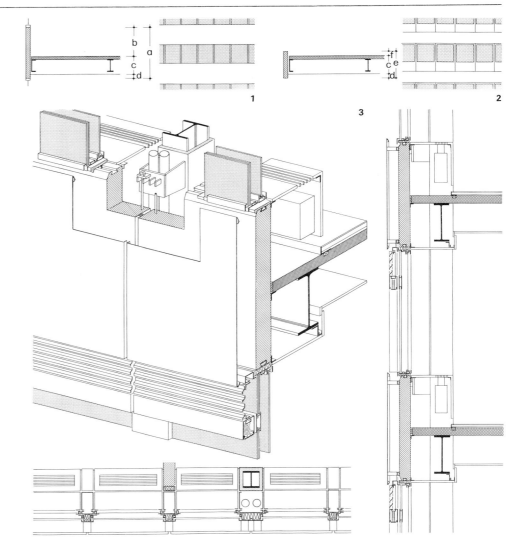

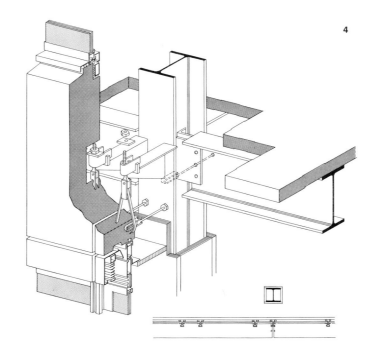

Façades in horizontal bands

5 Façade comprising large insulated window areas. The narrow loadbearing bands, covering the floor edges, are bolted to brackets which are welded to the tubular columns or to the castellated floor beams. The bands have aluminium sun-blind casings fitted externally. The infilling between the bands consists of fixed storey-height glazing for the upper storeys of the building and sliding windows for the ground floor.

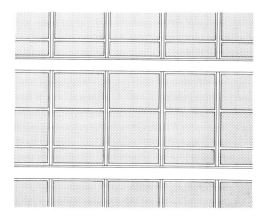

6 Steel cladding comprising loadbearing floor bands attached to the beams forming part of the standardised structural steel system (see p. 211 **12**). Mullions, each consisting of three component elements, are installed between these bands, the spaces between the mullions being filled in with closed panels, fixed glazing or movable windows. The floor bands are dimensioned to conform with the structural planning grid of 60 cm, while the mullions are designed for incorporation into the internal finishing grid of 70 cm and 113 cm in any desired sequence. The outer skin of the cladding is of weathering steel. The floor band is made fire-resistant by thermal insulation consisting of mineral-fibre mats. The non-loadbearing infill panels have a foamed polyurethane filling in their cavities.
Joints: sealing is effected by Neoprene gaskets around the infill panels or glazing units. The internal cover strip on each vertical joint corresponds to the position of a partition and provides the means of connecting the latter. (New buildings for Free University of Berlin; architects: Candilis, Josic, Woods, Schiedhelm; façade design based on an idea of Jean Prouvé.)

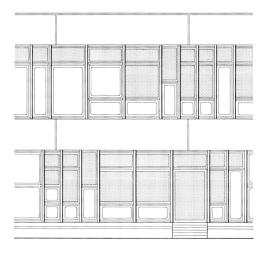

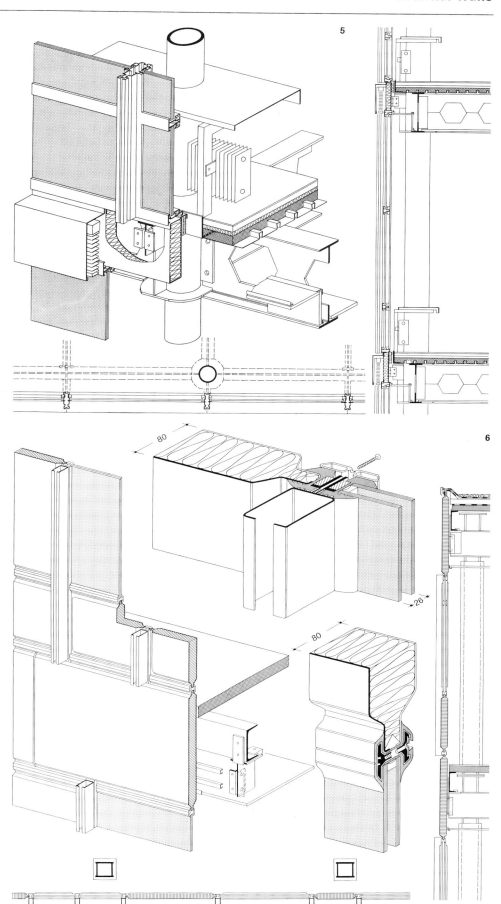

External walls

Mullion façades

In the main, two types may be distinguished:

1.1 The externally exposed mullions are self-supporting, or

1.2 the vertical features visible on the façade are sheathings or facings to the actual mullions which support them and which are located within or behind the façade.

If the outside structural columns of the building are also designed to perform the function of mullions — whereby savings in the cost of the façade and in erection time can be effected — they may be installed behind, within or in front of the façade, as indicated in **2.1**, **2.2**, **2.3** (see also p. 199).

In terms of structural design there is no significant difference between the following three arrangements:

3 cladding installed as curtain walling in front of the edges of the floors; or

4 cladding inserted between the floors, so that the floor structure is exposed as an architectural feature; or

5 recessed cladding set back some distance from the edges of the floors, so that the latter can form balconies that can be utilised, for example, as emergency escape routes.

6 Internally insulated cladding. Inserted between the external columns of the building are cold-formed sheet-metal frames supporting the inner leaf, which comprises a profiled aluminium panel with a bitumen coat on the outside and with a 30 mm backing of mineral-fibre insulating mat, a 60 mm cavity and a 55 mm layer of plasterboard. The outer leaf of the cladding consists of asbestos cement panels. (School building system developed by Hamburger Stahlbau.)

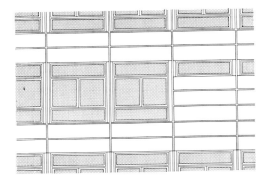

7 In this example, the dominant features are the white-painted outside columns which are partially clad with vermiculite sheets for fire protection at the rear. Window units with double glazing and spandrels with translucent glass are fitted behind the columns. (Faculty of Veterinary Medicine, Free University of Berlin; architects: Luckhardt, Wandelt.)

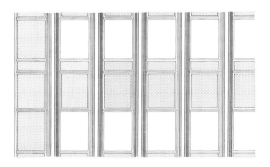

312

Mullion façades

External walls

8 Mullions consisting of cold-formed sections with Silan-Monoblock fire protection on the inside.
The spandrel panels are faced externally with Platal sheet metal units separated by a cavity from the fire-resisting layer which is backed by foamed plastic insulation and an internal covering of plastic-coated steel sheet. (Office building at Gevelsberg; architects: Jaenecke and Samuelson.)

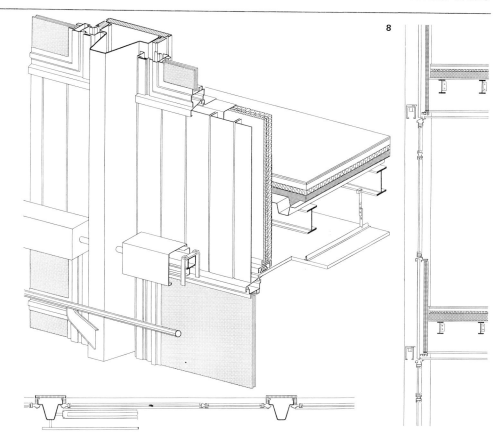

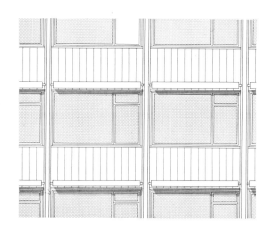

9 Aluminium mullions and transoms and 8 mm aluminium facings clipped together. Hot-dip galvanised fasteners and stainless steel fittings. Thermopane fixed glazing with Neoprene gaskets. Metal-to-metal joints sealed with butyl rubber strips.
The mullions are at 3.33 m centres with a storey-height of 3.65 m, pin-jointed to brackets with connections allowing three-way adjustment. Mullions are spliced at lintel level. Deformations are absorbed at the horizontal splices and in the vertical coupling sections of the mullions. Fire protection is provided by masonry backing to the spandrel panels and by reinforced concrete perimeter beams. The backing also serves as thermal insulation. (Deckel office building, Munich; architect: Henn.)

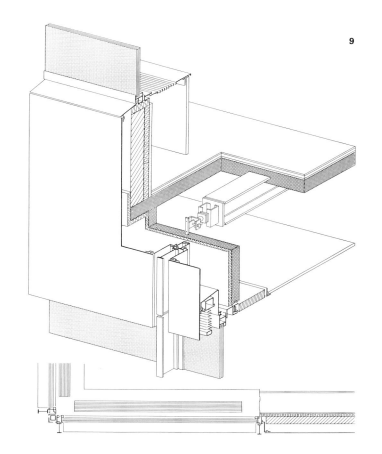

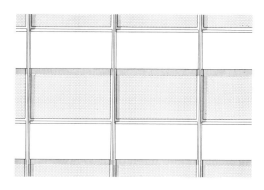

External walls

Mullion façades

10 In this example the cladding, in the form of a curtain wall, has a cavity with a certain amount of ventilation which allows vapour entering it by diffusion from the interior of the building to escape to the external atmosphere without detriment to the heat-insulating performance of the panels (type: Gartner). The mullions, consisting of IPE 100 rolled steel sections, are spaced at 2.50 m centres and are storey-height members separated by expansion joints. They are provided with welded connecting strips comprising threaded holes and bonded-on Neoprene gaskets.

The panels measure 2500 × 750 mm, with an overall thickness of 53 mm, consisting of 1.5 mm thick aluminium sheets bonded flat to a core of Aircomb, k = 1.08. The outer sheet of each unit has a white enamelled finish. Horizontal aluminium transoms add detail to the façade. (Kaufhof AG, Cologne; architect: Spielgen.)

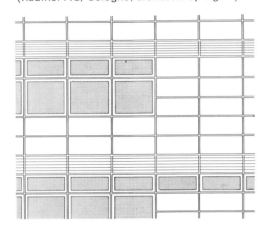

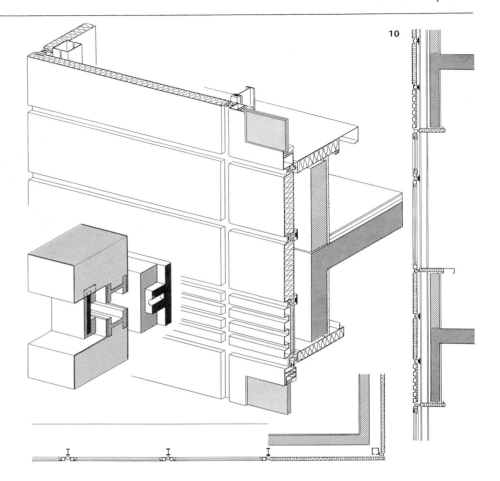

11 Suspended mullions, transoms of rectangular hollow section, hot-dip galvanised. Cladding panels comprise: 4 mm aluminium sheet, cavity, 3.6 mm asbestos cement sheet, 35 mm polyurethane foam, 0.4 mm aluminium foil, 4 mm asbestos cement sheet. (Electronic data processing building for Rollei Factory, Brunswick; architect: Henn.)

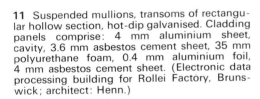

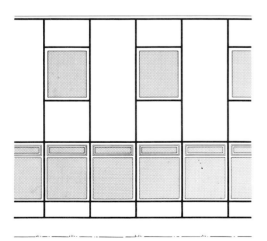

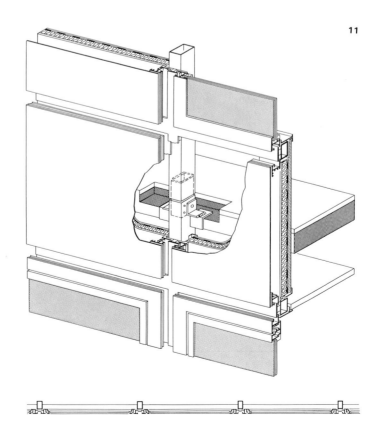

Mullion façades

External walls

12 Cor-Ten weathering steel cladding with welded joints. Concrete filling between steel column and cladding provides fire protection. Spandrel sheets have studs for anchorage to the concrete in which the lattice edge girder is encased. Windows have bronze-tinted glass which increases in thickness higher up in the building in order to resist the greater wind pressures acting there. (Civic Center, Chicago; architects: Murphy, Skidmore, Owings and Merrill.)

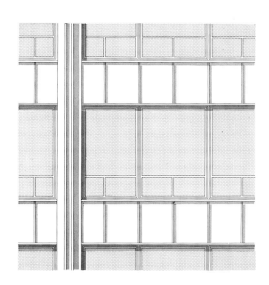

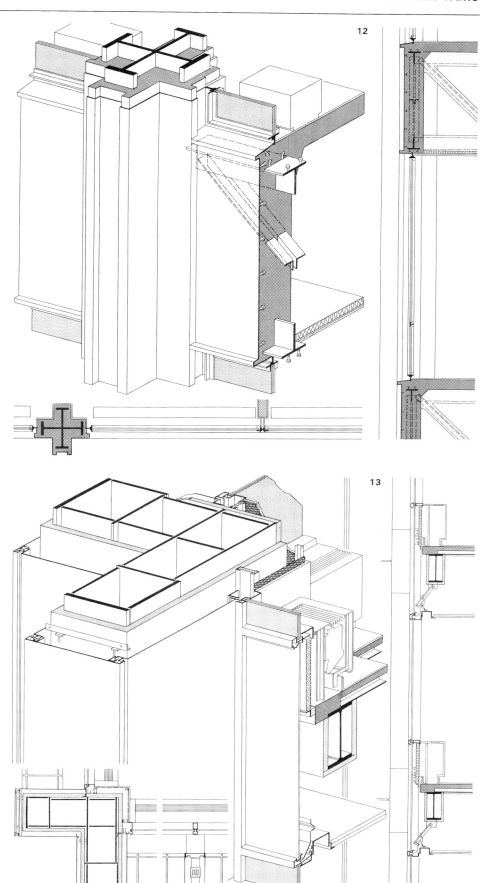

13 Corner column of a 26-storey building. The steel columns and diagonals have a fire-resistant casing and are sheathed in dark bronze-tinted aluminium. Thermal insulation is provided only in front of the air-conditioning units. Some of the windows are glazed with safety glass intended to be broken in order to serve as smoke escape vents in the event of a fire. See also p. 216 **9**. (Alcoa Building, San Francisco; architects: Skidmore, Owings and Merrill.)

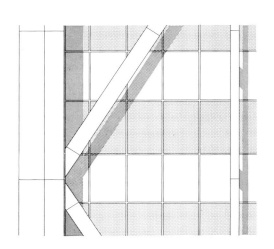

External walls — Panel elements • Cladding

These storey-height panels, corresponding to the width of a window, which are suspended from the floor structure, are provided with stiffeners to enable them to transmit the wind forces. The panels used on steel-framed buildings are nearly always of metal. Concrete panels of such size are not often used on a steel structure because of their great weight. The panels are usually supplied ready for erection, including the windows with their glazing.

1 Cast aluminium panels backed with 30 mm foamed polyurethane and vapour barrier are fitted between the floor edge beams, which are externally exposed. The cladding panels measure 280 × 118 cm. See also p. 144. (Siemens administrative building in Paris; architect: Zehrfuss.)

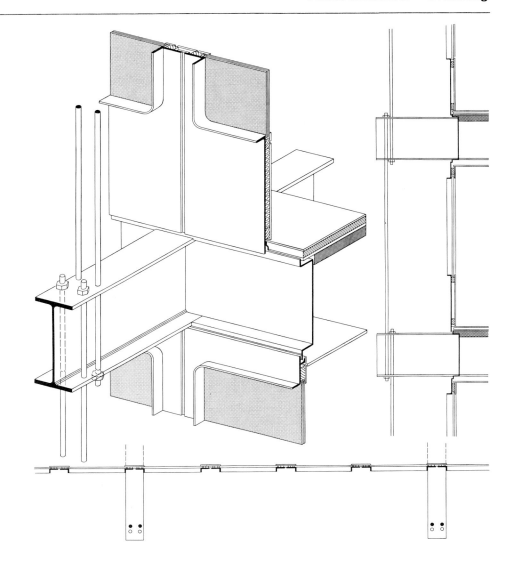

CLADDING

Arrangements for cladding large areas have been brought to a fine art. The outer skin normally consists of profiled sheets, the most common being coated steel troughing and corrugated asbestos-cement sheets. The heat-insulating layer and inner sheet are either connected to the external sheet in an integral unit (composite panel) or are secured independently to the supporting framework. This type of cladding is used mainly for shed-type industrial buildings and for this reason only two examples will be considered here.

1 The cladding units consist of two sheet-metal panels with rigid foam insulation sandwiched between. The panels, which are 12 to 15 m in length, are bolted to the structural steel frame and are covered with a plastic strip at the top (Hoesch Isowand system).

2 Corrugated asbestos cement panels as outer sheathing, with insulation and inner panels installed separately. The method of fixing is similar to that for roof cladding: see p. 303 **2**.

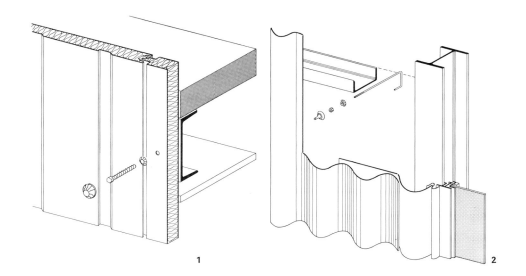

Means of escape, galleries • Rails for external maintenance equipment — External walls

Means of escape, galleries

The design of open balconies or galleries for multi-storey buildings depends on the purpose for which they are intended:

- regularly used horizontal circulation facilities for access to and from certain parts of each storey;
- escape routes for use in emergencies, more particularly connecting rooms for human occupation to staircases;
- service walkways for window cleaning, etc;
- solar control features, with access only in exceptional circumstances;
- features serving as baffles to check the spread of fire from storey to storey.

The requirements embodied in the building regulations depend on the functions such balconies have to perform: degree of fire resistance, incombustibility or barrier against smoke.

From the structural point of view, the following possibilities are available:

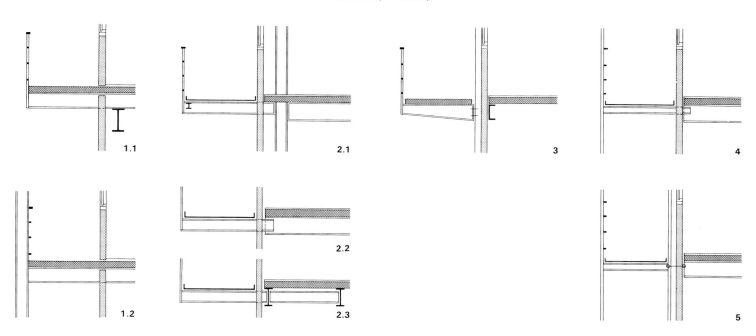

1 The external wall of the building is set back; the floor structure (slab and beams) continues to the edge of the balcony. This form of construction gives rise to heat insulation problems.
1.1 The floor slab and beams are cantilevered.
1.2 The external columns are right outside the building, at the edge of the balcony.
2 The balcony is supported by a cantilever fixed to the main steel frame and penetrating the cladding, so that in this case, too, there are insulation problems, though less serious than in the preceding example.
2.1 The cantilever is connected to the column. If the external columns are widely spaced, strong longitudinal beams must be provided along the outer edge of the balcony.
2.2 If the floor beams are perpendicular to the wall, the balcony cantilevers can be conveniently and easily connected to them.
2.3 Connection to floor beams parallel to the external wall.
3 The cantilever is connected to a mullion. Heat insulation problems are avoided, but the mullion is, of course, subjected to additional bending moment.
4 The outer edge of the balcony is supported by suspension rods attached to a cantilever beam at roof level. The inner edge is supported by mullions or external columns.
5 In this case, both the outer and inner edges of the balcony are supported by suspension rods, that is, the balcony is structurally independent of the external wall, being connected to the latter merely by ties or bracings at each floor.

Balcony floors

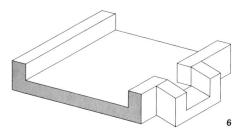 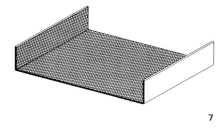 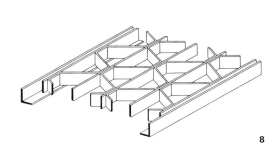

6 Precast concrete slabs, sometimes provided with edge upstands and water outlets, fire-resistant.
7 Steel plates with bent-up edge upstands and provided with non-slip patterned surface or plastic coating containing non-slip abrasive particles.
8 Gratings: self-supporting or framed in rolled steel sections. These gratings are supported on cantilevers and longitudinal beams consisting of light steel sections and may be used in cases where fire-resisting properties are not required.

Railings

Balconies should be provided with lightweight steel railings. Heavy concrete parapets are not suitable.

Rails for external maintenance equipment

For the cleaning and maintenance of the external surfaces, tall buildings are generally equipped with cradles or other appliances which are moved and operated from the roof. Most of these appliances require rails or other devices extending up the face of the building. With mullion façades, the mullions are often designed to perform this function; band and panel façades usually require special rails for the purpose. The forces that such rails have to resist are relatively small, so that their attachment to the façade presents no special problems.

Internal walls

Requirements applicable to internal walls

Depending on the duties they have to perform, internal walls have to fulfil various functions determined partly by conditions of use, partly by building regulations and partly by structural requirements. In the choice of a suitable wall, the question of cost obviously plays an important part.

1. Structural function

A general distinction can be drawn between loadbearing and non-loadbearing internal walls.

Loadbearing internal walls

Framed buildings normally have no walls intended for the specific purpose of transmitting vertical loads, but they frequently have walls which contribute to structural rigidity, that is, serving to transmit horizontal loads (pp. 219, 267). These are called shear walls and usually also receive a certain amount of vertical load, if only for counterbalancing the tensile forces that would otherwise be caused by bending moments. Such walls in most cases also perform a fire-resisting function in subdividing the building into fire compartments.

Non-loadbearing internal walls

Functionally these are partitions, merely separating room from room. They should possess the necessary strength and stability to support their own weight and the loads exerted by objects or appliances attached to them, as well as any horizontal forces due to such attachments and to live load, including impact effects. Such walls should be so designed and attached to the loadbearing structure of the building that they are not subjected to additional loads which they are not intended to take.

2. Visual screen

Usually the internal wall must be opaque, but sometimes transparency is required. The material, texture and colour of the wall are obviously important factors within the context of interior decoration. The cost, method of construction and, in many cases, the degree of prefabrication of internal walls depend on such factors.

3. Sound insulation

After fire protection, sound insulation is the most important function of the internal wall, especially in providing insulation against airborne sound transmission from room to room. The degree of insulation provided by a single-layer wall depends on its weight; in the case of a multi-layer wall it depends also on the thickness and nature of the material of the outer leaves, the nature of the insulating material sandwiched between, and on the manner of their interconnection. There are no reliable methods of calculating the degree of airborne sound reduction achieved by a wall. Such information must be obtained by means of tests.
Steel sections incorporated in partitions may affect their acoustic properties (see p. 321 **7**, p. 322 **3**, **4**).
Doors and windows of course play an important part in sound insulation and should in this respect be equal in quality to the wall or partition itself.
Even the acoustically perfect wall is largely useless if it is bypassed by sound transmission through a floor, ceiling or external wall.

Fulfilment of high acoustic requirements tends to add considerably to the cost of a wall. For full information on sound insulation in buildings it is necessary to consult the specialised literature.
Besides insulation, the sound-absorbing properties of a wall may be an important acoustic consideration, depending on the purpose for which a particular room is used.

4. Services

Pipes and wiring for sanitary and electrical installations are often fitted within internal walls. In the case of prefabricated partitions this necessitates special arrangements, for example, units specifically designed for bathrooms, kitchens, etc. Demountable partitions should preferably not contain any services, except perhaps electric wiring to switches or sockets.
Shelves and apparatus are often fixed to partitions, as in laboratories. In such cases, the partitions must be suitably designed and constructed for these extraneous functions.

5. Fire protection

Depending on the function it will be expected to perform in the event of an outbreak of fire, the internal wall may have to be of fire-resisting construction to separate the building into fire compartments and to protect the escape routes. Most building regulations impose the following requirements:

• Adequate fire resistance (in Germany generally 90 min) and extra strength for compartment walls (p. 320) and as enclosures to staircases (p. 293), described as 'protected shafts' in fire regulations.
• Walls separating one dwelling from another or separating rooms with a particular fire hazard from other rooms should provide similar resistance.
• Walls along escape routes, such as corridors, should have adequate fire-retarding capacity (30 min fire resistance).
• Partitions and other internal walls should not contain materials that will give off toxic or aggressive gases in the event of a fire. For instance, polyurethane evolves prussic acid and is therefore unsuitable as an insulating material for internal walls. PVC releases hydrochloric acid as a gas which has a corrosive action and may attack steel components.
• Double-leaf partitions can often be made to provide fire protection for structural steel components, if the leaves are of suitably fire-resistant materials. Thus, floor beams may be protected in this way, if their spacing corresponds to that of the partitions, the beams being incorporated within the latter. No suspended ceiling is needed, the partitions being extended to the underside of the floor slab.

Often, external as well as internal columns can with advantage be incorporated into partitions. Vertical lattice bracings within a structural frame are always concealed by sheathing which forms a double-leaf partition; the wall thickness is quite small if the bracing comprises only tension members consisting of flat or round bars.

6. Alterations

To extend the service life of a building it not infrequently becomes necessary to alter its purpose and function, necessitating changes in its internal layout. According to the degree of variability that they allow, the following distinctions in internal walls are possible:

• Walls which cannot be altered, for structural reasons: loadbearing walls which transmit vertical loads and serve as bracing;
for fire protection reasons: walls which enclose fire compartments or protect emergency escape routes;
for technical reasons: solid concrete walls, for example.
• Walls which can be altered only by demolition: this category includes partition walls of masonry, plaster or solid plasterboard construction.
If alterations are unlikely, but nevertheless cannot be ruled out during the course of the service life of the building, it is usually more economical to choose a type of construction which is inexpensive and will have to be demolished in the event of having to change the internal layout than to install expensive demountable partitions.
• Walls whose materials can to a great extent be recovered and re-used elsewhere if the wall has to be dismantled. Partitions comprising nailed or bolted timber studding or steel frames belong to this category.
• Demountable partitions, usually consisting of closed wall systems assembled from framed or panel-type units. These tend to be much more expensive than the other kinds of partition. There are very many types of demountable partitioning, varying widely in the amount of labour involved in taking them down and re-erecting them in a different position. Some systems require the services of specialist technicians, others can be handled by unskilled labour.
• Movable partitions for regular and fairly frequent opening and closing: sliding or folding partitions for temporary subdivision of rooms.

7. Prefabrication

For the components of the following types of partition it is possible to apply varying degrees of prefabrication:

• For concrete walls intended to serve as fire-resisting walls or staircase enclosures it may be convenient to use precast units which are erected with the structural steelwork.
• Large panels, often of storey-height, may be used for gypsum plaster walls.
• The units of timber framed partitions may be cut to the required size on the actual job (for example, gypsum plasterboard nailed to timber studding) or supplied fully prefabricated to the required size. The latter alternative is always adopted in the case of panel-type partition units or of prefabricated steel or aluminium frames for partitions. These frames are faced each side with plasterboard or some other suitable material.
• Demountable and movable partitions embody the most advanced degree of prefabrication.

Partition wall units are generally prefabricated to 120 cm width, the partitioning being based on fixed modular dimensions, generally 10, 30 and 60 cm. The partitions are always erected with the aid of top and bottom fixing rails or runners. Some systems additionally employ floor-to-ceiling posts to support the units.
For further information on modular partitions, including jointing, see p. 174.

318

Tolerances

Internal walls

Deviations from the theoretically correct dimensions arise as a result of unavoidable minor inaccuracies of fabrication (within the allowable tolerances) and as a result of deformations affecting the components. The dimensional tolerances acceptable for the structural framework of a building are larger than those for the finishing components subsequently fitted into it. Internal walls and partitions which are constructed or cut to size on the actual job can of course be adjusted to any dimensional variation of the supporting system. Prefabricated partition units should incorporate some method of compensating for tolerances.

The effect of such tolerances and deformations on fire-resisting walls is discussed on p. 320 and on shear walls on p. 267.

Effect of floor deflection

In the design of non-loadbearing partitions it is necessary to take proper account of the deformation of the building, more particularly the deflection of floors under vertical loads. The consequences of this are twofold:

1.1 To ensure that the partition will in no circumstances be subjected to loads which it is not intended to receive, the beam over a partition must have sufficient clearance in which to deflect without touching the partition. The space that must therefore be provided between the partition and the beam or floor over it may be filled with an elastic material.

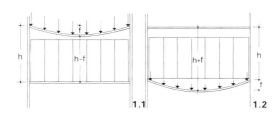

1.2 If the beam on which the partition rests deflects under load, the support conditions for the partition are modified. As a rule, the partition consists of incompressible panel-type units, but it can be made to adapt itself to the imposed deflection by the insertion of an elastic material under the bottom edge of the partition. In addition, the vertical joints should allow elastic movement between the panels.

The above-mentioned principles are often ignored, and this is a major reason why partitions become dislocated or damaged during the service life of the building.

Magnitude of the floor deflection

How large are the deformations that can reasonably be expected to occur in a multi-storey steel-framed building? Considerations similar to those pertaining to external walls are applicable also to internal walls (p. 307).

With full utilisation of the permissible stresses in the constructional material, steel beams often have lower stiffness than comparable concrete beams. Precautions to reduce the deflection of the steelwork, for example, by using a lower-strength steel (grade St 37 in lieu of St 52) and by not fully utilising the permissible stresses, are indeed effective, but add to the cost of the structure. Economic considerations will have to decide. It is often cheaper to introduce flexibility into the partitions than to increase the stiffness of the floor.

The mid-span deflection f is decisive in determining the amount of clearance to be provided between the floor and the top of a partition in the storey below.

The difference in deflection Δf between two adjacent units of a partition (see **3**) is largest near the ends of the span (Δf_1) and is equal to the tangent of the angular rotation α multiplied by the width b of the unit: $\Delta f_1 = b \tan \alpha$.

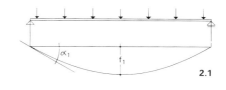

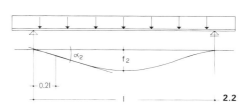

The largest value of α is obtained, both for a simply-supported beam, **2.1**, and for a continuous or fixed-end beam, **2.2**, as a function of the maximum deflection f at mid-span, namely:

$$\tan \alpha = \frac{16}{5} \frac{f}{l} = 3.2 \frac{f}{l}$$

This maximum angle α occurs at the bearings of a simply-supported beam, at about the 0.1 span points of a continuous beam, and at the 0.2 span points of a fixed-end beam.

On the assumption that only one-third of the imposed load is in practice a truly live load (p. 242) and that the standard width of a partition unit is 120 cm, it is possible to derive the following values in mm for Δf_1 and f for some frequently occurring spans:

$\frac{f}{l}$	$\frac{f}{1/3 \cdot l}$	1/3 Δf_1	l=6.0 1/3 f	7.2 1/3 f	8.4 1/3 f	9.6 1/3 f	12.0 1/3 f
1/200	1/600	6.4	10.0	12.0	14.0	16.0	20.0
1/300	1/900	4.3	6.7	8.0	9.3	10.7	13.4
1/500	1/1500	2.6	4.0	4.8	5.4	6.4	8.0
1/800	1/2400	1.6	2.5	3.0	3.5	4.0	5.0
1/1000	1/3000	1.3	2.0	2.4	2.8	3.2	4.0

Changes in width of joint

If the vertical joints between the partition units are sufficiently flexible the deformation pattern shown in **4** will develop under vertical load. The change in width of the joints between the ten panels, each of width b=120 cm (with an overall span of 12.00 m) and height h=3.00 m, with $f/l = \frac{1}{3} \times (1/500)$ and $\tan \alpha = 3.2$ (f/l) = (3.2/1500), is thus:

$$\delta = h \tan \alpha = \frac{3000 \times 3.2}{1500} = 6.4 \text{ mm}$$

at each end.

For the whole length of the wall: $2\delta = 12.8$ mm. Distributed over nine joints, this involves a compressive shortening of approximately $12.8/9 = 1.4$ mm per joint.

It should be borne in mind that, for partitions which are not installed until the internal finishing and fitting-up of the building has been completed, the first one-third of the imposed load is in fact produced by the weight of these partitions themselves and by the fittings and finishes in the rooms. On the other hand, where partitions are installed during erection of the structural frame of the building, there is also permanent deformation due to the deflection caused by all the permanent loads which are applied after installation of the partitions, such as floor finishings, ceilings and services.

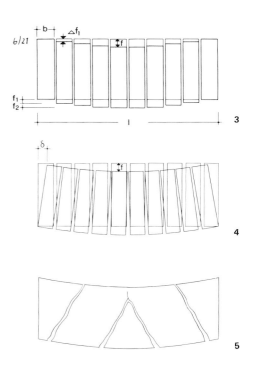

Rigid partitions

From the above considerations it follows that rigid internal walls and partitions, say, in brickwork, will crack more or less as shown in **5** if they have fairly long spans. If such cracks, which not only spoil the appearance but also impair the sound-insulation properties, are to be avoided, the wall should be subdivided into units connected by joints permitting a certain amount of movement.

Internal walls

The object of compartment walls is to prevent the spread of fire within the building (internal compartment walls) or to other buildings (external compartment walls). To fulfil these requirements such walls should:

- possess adequate fire resistance (90 min);
- preserve their stability under fire conditions.

This last-mentioned property, more particularly envisaging the stability of the wall under the action of unpredictable loads, is investigated by subjecting the components to horizontal impact force and eccentric load in a fire resistance test. For walls serving to separate fire compartments or performing a similar major fire-resisting function, the following minimum thickness requirements are laid down in the Federal Republic of Germany:

25 cm for lightweight concrete (density not exceeding 1800 kg/m^3)
24 cm for masonry
20 cm for normal-weight concrete (density exceeding 1800 kg/m^3)
14 cm for normal-weight concrete if the wall is laterally restrained.

Under German building regulations, fire-resisting walls in a building should be spaced between 50 and 60 m apart.

At re-entrant corners the possibility of the spread of fire over a distance of 5 m must be prevented by locating the compartment wall away from the corner, **1.1**; alternatively, part of the external wall should be constructed as a compartment wall, **1.2**.

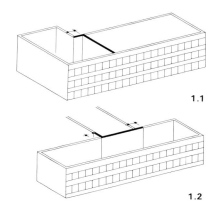

External compartment walls

This aspect really belongs to the subject of external walls, but frequently an external wall has to be made fire-resisting because, as a result of subsequent extension to the building or the construction of neighbouring premises, it eventually becomes an internal wall. A truly external compartment wall is required in circumstances where there is a small space between two adjacent buildings. In Germany an external compartment wall is required if this space is less than 5 m. In addition to possessing the requisite fire resistance, such walls must of course have the usual properties of external walls, such as watertightness, heat-insulating capacity, etc.

Internal compartment walls

Such walls are often located at expansion joints in buildings. Services are allowed to penetrate an internal compartment wall if they are made of incombustible materials and suitable precautions are taken to prevent spread

Compartment walls

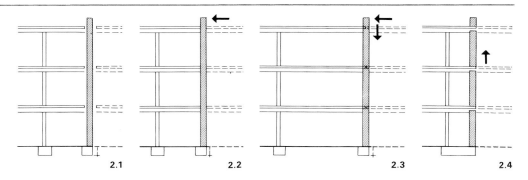

of fire through them, for example, fire-check flaps in air ducts. Electric cables call for particular caution because if they catch fire, the fire is liable to spread along them. Door openings should be provided with self-closing fire-resisting doors.

From the structural point of view the following possibilities must be considered:

2.1 The compartment wall possesses independent structural stability.
2.2 The compartment wall derives lateral support from one or more parts of the building, so that it is stiffened transversely to its plane, but may at the same time serve as a shear wall transmitting horizontal loads acting in its own plane.
2.3 The compartment wall transmits vertical loads from the floors, but derives lateral support from the building. It may also serve as a shear wall.
2.4 The compartment wall is supported by the floor structure and must not be subjected to vertical loads or to bending moments within its own plane. The wall is therefore allowed only to stand on each floor, while at the top in each storey the wall must be provided with a movable support serving only to restrain the wall laterally, but not forming a rigid connection. Instead, there must be a gap between the wall and the soffit of the floor above, this gap being filled with a compressible fire-resisting material, such as mineral fibre or asbestos. The filling must also act as a smoke barrier.

Free-standing plane fire-resisting walls often constitute a complicating factor in connection with steel erection. Walls such as those in **2.1**, **2.2** and **2.3**, and **3.1**, **3.2** and **3.3**, can be constructed in in-situ concrete up to a height of two or three storeys before erection of the steel. For high-rise buildings the construction of the fire-resisting walls must keep ahead of erection unless these walls are to be installed later. If the former alternative is adopted, it necessitates careful planning of the operations to ensure that the steelwork and concrete contractors' activities do not interfere with one another.

Fire-resisting walls as shown in **2.2**, **2.3** and **2.4**, and **3.2**, **3.3** and **3.4**, can be assembled from precast concrete components simultaneously with steelwork erection. Fire-resisting walls supported by the floors of the building (**2.4**, **3.4**) can often be constructed more advantageously in brickwork.

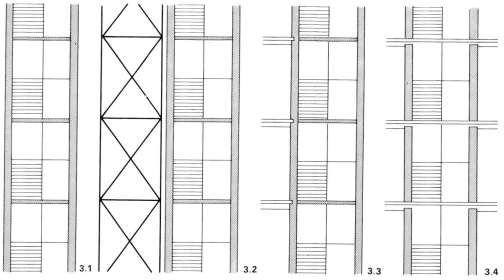

Staircases

The walls of staircase enclosures in high-rise buildings usually have to fulfil the same requirements as those applicable to walls separating fire compartments. Structurally, the following possibilities call for consideration:

3.1 the walls are stable in themselves;
3.2 they derive lateral support from the building structure;
3.3 they participate in the transmission of structural loads and/or stiffen the building;
3.4 they are supported on each floor of the building.

For further details, see p. 294.

Types of partition, connections — Internal walls

Types of partition

1 Solid partition walls which are usually constructed with the framework of the building but which are capable of demolition:

- brickwork of all kinds, one brick, a half brick or a quarter brick thick, with or without plastering, or double-leaf construction, or with suitable facing;
- plaster walls constructed by applying coats of plaster to woven wire fabric or expanded metal lathing;
- walls constructed from gypsum plaster slabs or panels, arranged in a bonded pattern or as storey-height units, comprising joints bonded with neat plaster or with mortar, of single-leaf or double-leaf construction, with plastered finish or with facings.

2 Demountable frame-and-panel walls which are installed at the time of finishing and fitting-up the building; they include:

- studding comprising timber or aluminium or steel light-gauge sections.
- panels nailed or screwed to the framework comprising asbestos cement, chipboard, plasterboard, etc; special fire-resisting forms of construction are available, also sandwich construction with insulating core, faced with hardboard and finished with plastic sheet, for example;
- insulating mats, for example, in mineral wool.

3 Freely demountable partition systems which can be installed at any time during the finishing operations or later. These comprise frame-and-panel systems or post-and-panel systems, erected with the aid of rails or runners at floor and ceiling levels:

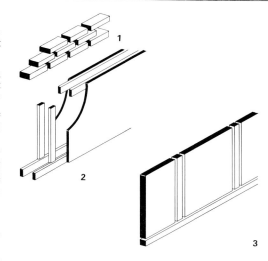

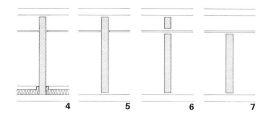

- partitions composed of timber units;
- partitions composed of sheet-steel units;
- aluminium framing with various types of cladding.

Partition connections

A partition is installed between two floors, or between floor and ceiling, and must be so connected at top and bottom that any vertical and horizontal forces to which the partition is subjected are resisted and that the requisite sound insulation is ensured.

4 The partition is installed between the structural floors above and below. It penetrates the suspended ceiling and the floating floor surfacing. This arrangement provides very good sound insulation, but is very troublesome when it comes to taking down the partition and installing it in another position. Such partitions are more suitable for separating walls between adjacent dwellings or walls of rooms or halls having to fulfil stringent sound-insulation requirements.

5 The partition stands on the surfacing of the floor below and penetrates the ceiling to reach the structural floor above. Services in the space above the ceiling may have to pass through the partition.

6 The partition stands on the surfacing of the floor below, or on the floor finish or covering, and terminates below the ceiling. A separate closure must be installed in the space above the ceiling to ensure sound insulation. The partition is demountable, but the separate closure must usually be left in position if it is penetrated by services and a new one will have to be installed over the partition when the latter is re-erected.

7 The partition is similar to that in the preceding example, but now the space above the ceiling is left open. The ceiling itself should be sound-insulating and possess enough strength to provide lateral support for the partition at the junction.

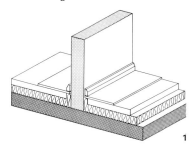

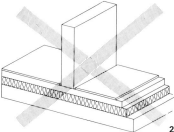

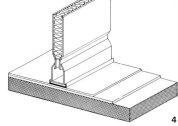

Bottom connection of partitions

1 In the case of a floating floor system the insulating layer is raised at the junction with the partition, so that the actual floor surfacing does not touch the latter, thus ensuring that there is no direct transmission of sound from one to the other.

2 From the acoustic point of view it is not a good solution to stand the partition on the floating floor surfacing, as this will transmit sound waves under the partition into the next room. This drawback can be mitigated by interrupting the surfacing with a narrow slot on each side of the partition.

3 Demountable partitions may stand on the finished floor surface or — if it is sufficiently even — on the structural floor itself.

4 If partitions are likely to be taken down and re-erected frequently, it is most convenient to stand them on the floor covering.

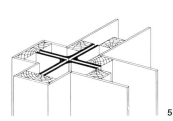

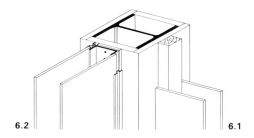

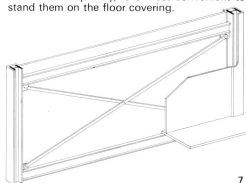

Side connection

5 Side connection of partitions to a cruciform steel column. In this case the partition consists of plasterboard on a timber frame. Suitably constructed, it can serve also as fire protection to the column.

6 Partitions — timber-framed (**6.1**) and steel-framed (**6.2**) — connected to an encased universal column. The column may constitute an acoustic bridge, unless a layer of compressible material is incorporated in the column casing.

7 Fire protection of vertical lattice bracing by incorporating it in a double-leaf partition. If the diagonals consist of round or flat bars, the overall thickness of the partition can be kept small. For connections, see p. 265.

Internal walls — Top connection of partitions

The arrangements for securing the partition at the top should provide sound insulation equivalent to that of the partition itself. This can most simply be accomplished by connecting the partition to the structural floor above, as in **1**, **2**, **3**, **4** and **7**. The connection becomes complicated if the partition has to be penetrated by services, **1**, or beams, **5**, **6**, **7**, or ribs of the floor structure, **8**, **9**. In such cases, it should be investigated whether a more economical solution can be obtained with a sound-insulating suspended ceiling which is continued over the partition, **11**. Construction of a closure within the floor cavity, **10**, is practicable only if there are not too many services in the cavity. In buildings with complex and sophisticated services the only solution that leaves full freedom of layout for the services is that shown in **11**.

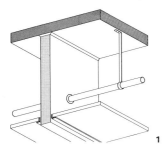

1 Solid partition joined to the structural floor, with interposition of a resilient strip in the joint. Suspended ceiling is interrupted at the partition. Good sound insulation.

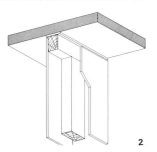

2 Timber-framed partition connected to the underside of a concrete floor slab with interposition of a resilient strip. Good sound insulation, but if services have to pass through the wall, special care must be taken to preserve such insulation.

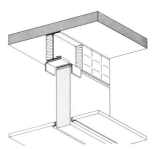

3 Connection of partition to floor beam with movable fixing and resilient seal. The steel beam is encased in rock-wool mats for sound insulation.

4 Floor beam is enclosed within the partition, whose two leaves provide fire protection. No suspended ceiling. To ensure good sound insulation, the leaves are not in direct contact with the steel beam: resilient material is interposed.

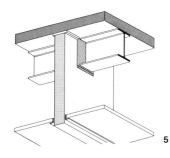

5 Solid partition extends to the underside of the floor and is penetrated by beams, with resilient material inserted for sound insulation and also to allow some flexural movement of the beam.

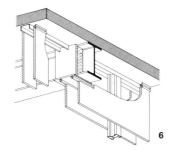

6 Steel beam passing through steel-framed partition. The beam has mineral-wool filling between its flanges at the junction and has a resilient pad inserted under it.

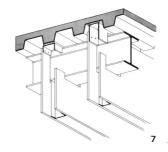

7 Steel troughing with the partition placed parallel to the troughs. Connection with resilient sealing strip. The layout grid of the partitioning must of course correspond to the spacing of the troughs. Penetration of beam through partition as in preceding example.

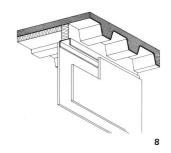

8 If the partition runs transverse to the troughing, special profiled closing strips must be used at the top to provide a horizontal junction for connecting the partition.

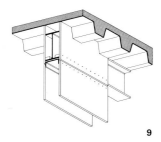

9 In this example the partition, likewise running transverse to the troughing, is connected to the underside of a floor beam, which is laterally enclosed in side panels cut out to match the profiling. These panels are fixed to the beam to allow relative movement between beam and partition.

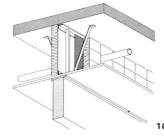

10 Partition terminates under the suspended ceiling. The space above the ceiling is separately partitioned with mineral-wool quilting applied to wire mesh stretched across a light framework. Alternatively, mineral-wool mats may hang loosely down in the cavity. Struts attached to the underside of the structural floor provide stability for the partition.

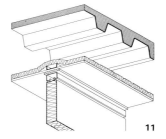

11 Partition terminates under the suspended ceiling and is stabilised by it. Sound insulation is by a mineral-fibre mat over the ceiling. This need provide only half the insulating capacity of the partition, as the sound has to pass twice through it on going from one room to the other. This arrangement provides complete freedom for the layout of services in the floor cavity above the ceiling and the partitions can be installed in any position without having to worry about sound insulation.

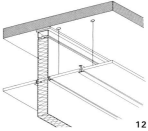

12 The modular suspended ceiling in this example comprises a grid of positions at which the partitions, extending up to the structural floor, can be installed without involving damage to the ceiling. There is only limited scope for services in the floor cavity. This system is suitable for offices without air-conditioning or ventilation, the electrical wiring being installed in the floor surfacing rather than in the cavity under the floor.

Non-demountable partitions

Internal walls

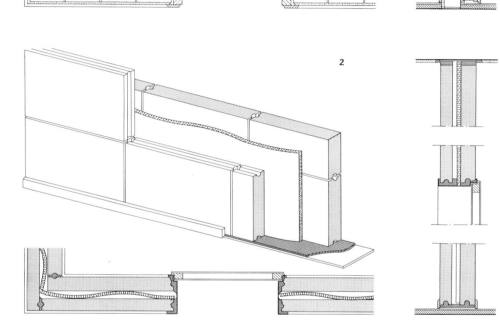

1 Double-leaf brick partition wall with plasterboard facing backed by a sound-insulating material (mineral-fibre mats or wood-wool acoustic boards) fixed to a timber frame. The brick wall cannot accommodate itself to flexural movements of the floor above and, therefore, adequate space, sealed with a resilient material, should be left at the top. To remove the partition, it will have to be demolished.

2 Double-leaf partition composed of prefabricated solid gypsum planks or slabs with a filling of insulating material in between. It should be continued up to the structural floor above. The individual slabs have bonded joints (formed with neat gypsum plaster, for example) finished flush with the external faces, which can then be papered or painted. Top and bottom connections should be resiliently sealed, with mineral-fibre material or foam rubber. In this case, too, removal of the partition will involve its demolition.

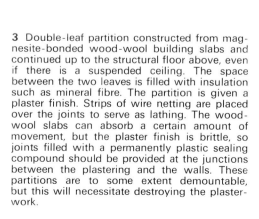

3 Double-leaf partition constructed from magnesite-bonded wood-wool building slabs and continued up to the structural floor above, even if there is a suspended ceiling. The space between the two leaves is filled with insulation such as mineral fibre. The partition is given a plaster finish. Strips of wire netting are placed over the joints to serve as lathing. The wood-wool slabs can absorb a certain amount of movement, but the plaster finish is brittle, so joints filled with a permanently plastic sealing compound should be provided at the junctions between the plastering and the walls. These partitions are to some extent demountable, but this will necessitate destroying the plasterwork.

Internal walls — Demountable partitions

4 Timber frame with insulation material and two leaves of nailed gypsum plasterboard, it being possible to attach the partition to the false ceiling. The insulation comprises 20 to 30 mm thick rock-wool quilting fixed to one side. The plasterboard is 12.5 mm thick. Resilient sealing strips are provided at the top and bottom junctions to absorb movement. These partitions are demountable, but not movable in the sense of being capable of re-erection in a different position.

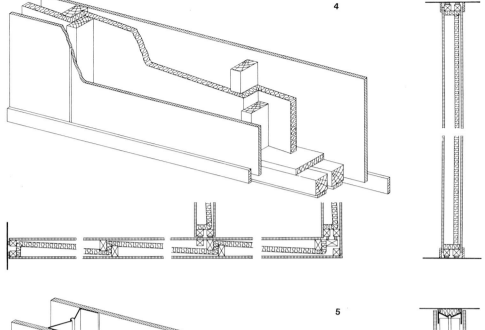

5 Sheathing of gypsum plasterboard fixed to a steel frame consisting of cold-formed sections. The latter comprise posts which are clipped into M-section cold-formed guide runners at top and bottom. Posts and runners possess sufficient inherent resilience to absorb a certain amount of movement. If the joints are left open or are merely closed with cover strips, this type of partitioning is fully demountable and capable of re-erection.

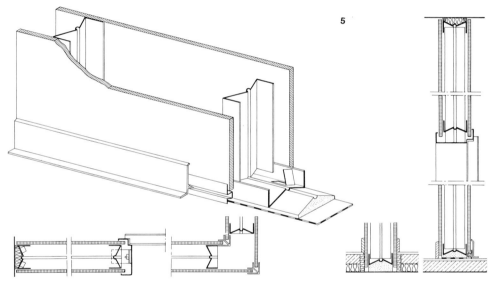

6 Cold-formed steel frame sheathed with gypsum plasterboard. The two-part (split-type) posts are held in cold-formed channel-section runners at top and bottom and are spaced at 500 or 625 mm centres. The space between the two leaves contains 40 or 60 mm thick mineral fibre insulation. The faces of the partition may be provided with a plastic laminate finish. Movements are absorbed by compressible seals made of bitumen-saturated fabric. In addition, the lipped channel sections at the joints of the sheathing panels allow some movement in relation to the supporting framework. If plastic-coated panels are employed, this partitioning is fully demountable and capable of re-erection.

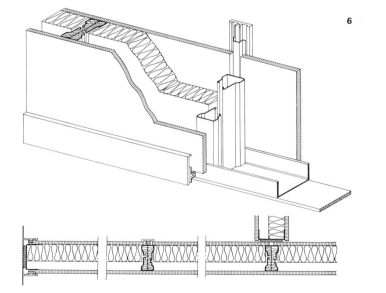

Removable partitions Internal walls

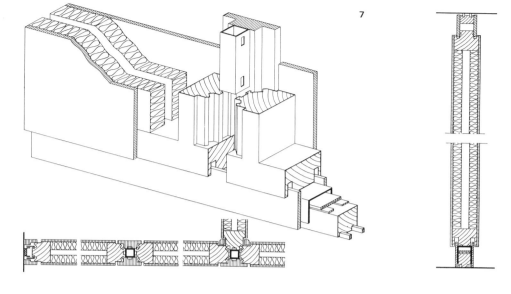

7 Double-leaf units consisting of chipboard panels fixed to a timber frame. The partition can be erected on the finished floor surface and it can also be connected to a suspended ceiling. The sheathing may alternatively consist of glued-on plywood or hardboard with 30 mm thick rock-wool mats bonded to the inner face. The units can be quickly clipped into slots in the square hollow section posts which allow some sliding movement and which, together with rubber sealing elements, allow the partition as a whole to absorb movement. The joints are masked with cover strips flush with the face of the partition. With pre-finished units, painted or plastic-coated, such partitions are fully removable.

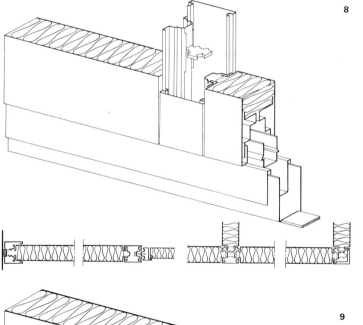

8 Double-leaf units consisting of sheet-steel sandwich panels which are fitted and locked together without posts. The partition can be installed on the finished floor surface and can be connected to a suspended ceiling. The insulating material in the panels is loose mineral wool held in position by stiffening ribs on the inner faces of the steel sheet. The units are assembled with the aid of stud fixings. The joints are provided with clip-on cover strips flush with the face of the partition. The method of connection to floor and ceiling allows up to 6 cm movement. This partitioning can be readily dismantled and used elsewhere.

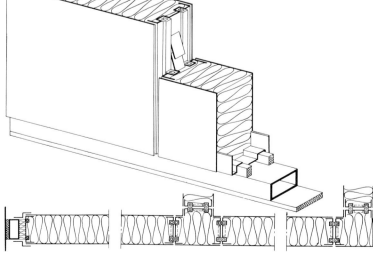 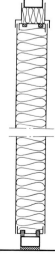

9 Double-leaf sheet-steel sandwich panel units fixed between hollow top and bottom runners. This partition can also be installed on the finished floor surface, while connection to a suspended ceiling is possible. The storey-height units comprise steel sheet (with damping to suppress drumming noise) applied to a frame, with mineral-wool filling in the intervening cavity. Units are interlocked by means of steel tongued clip fixings. Movement is absorbed by expanded rubber sections and by sliding action in the connections. This type of partition can also be easily removed.

Internal walls

Removable partitions

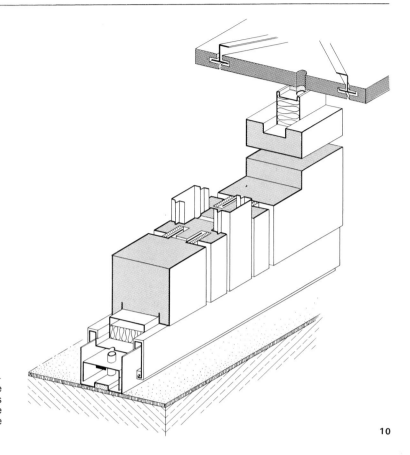

10 This partitioning system comprises cold-formed steel posts and stove-enamelled steel-sheathed units with a core of felted mineral fibre. The panels have modular dimensions of 70 and 130 cm. The partition stands on the floor covering and terminates at the ceiling. Deflections of the floor structure are absorbed in the bottom runner. These partitions are easy to dismantle.

Fire Protection

Effects of fire	
Destruction, heat, fumes and gases	327
Preventive measures against fire	
Stability of buildings	328
Escape routes	329
Fire compartments	329
Firefighting equipment	329
German fire regulations	330
Stability of structures under fire conditions	
Fire as a structural load	330
Fire load	331
Categories of fire resistance for buildings	332
Fire resistance rating of structural components	332
Choice of fire resistance rating	333
Fire behaviour and fire protection of steel components	
Behaviour of unprotected steel components in a fire	334
Protection of steel components against fire	335
Twelve rules for fire safety in steel buildings	335
References	336

EFFECTS OF FIRE

Along with gravity, wind, earthquakes, variations in temperature, sound, moisture, chemical and electrical action, the effects of fire must, from the structural designer's point of view, be regarded as one of the phenomena to which a building may be exposed at some time during its service life. The effects of fire should therefore be investigated along with those of the other influences. Buildings and their components should be so designed that they can withstand these effects throughout the duration of the fire or for a certain specified period. Fire damage is caused by the combustion of combustible materials, which are:

1. a permanent part of the structure:
- loadbearing members;
- components performing a space-enclosing function;
- components forming part of the services; or

2. not attached to the structure:
- furnishings;
- stored materials;
- consumables, including fuel.

The effects of a fire are:
- destruction of combustible materials;
- generation of heat;
- production of fumes and gases.

Destruction by fire

Loadbearing components made of combustible materials (for example, timber) undergo progressive reduction of cross-sectional area on exposure to fire, which may eventually result in structural failure.

Partial destruction may cause components to lose their unity, and danger will arise from:
- the collapse of certain components;
- spread of fire to adjacent rooms and premises;
- burning pieces falling from the building.

Heat

The heat generated during a fire is transmitted by radiation or by convection (hot gases of combustion) to structural components which, even if they are incombustible themselves, may suffer damage in consequence. The heat causes:
- expansion;
- reduction in strength.

Thermal expansion causes lengthening or, in the event of unequal heating on two sides, curvature of the components affected. This in turn may cause displacements and/or restraint forces which are liable to damage loadbearing members, possibly resulting in structural failure. Such effects must therefore be investigated at the design stage.
Restraint forces arise more particularly in rigid frame structures in consequence of the expansion of the cross beams. This may push the columns off their bearings and thus cause collapse of the framework. Expansion of floor slabs may endanger other components, which is something that must be taken into account in the design of expansion joints (p. 223).
Some materials, such as metals and concrete, undergo a loss of strength at elevated temperatures. For the behaviour of steel in this respect, see pp. 334, 335.

Fumes and gases

The gases developed in a fire in a building are dangerous because:
- they reduce or blot out visibility;
- they are liable to cause asphyxiation;
- they are poisonous and thus endanger human life;
- they chemically attack certain materials.

Reduced visibility makes escape from the burning building more difficult and hampers rescue and firefighting operations. Lack of oxygen in the air may cause asphyxiation.
Toxic constituents of the gases include carbon monoxide and the poisonous gases which may be formed, such as prussic acid, which evolves in the combustion of certain types of plastic.
Chemical action may arise, for example, from hydrochloric acid, which is formed when PVC burns and which is condensed and precipitated on various parts of the structure. The acid can penetrate into concrete and cause corrosion of reinforcement. See reference 2, p. 336. On the other hand, exposed steel surfaces are not so much in danger from hydrochloric acid, as the acid can subsequently be removed from them fairly easily, without having caused any significant damage.
In the Federal Republic of Germany the 'Recommendations for the use of combustible materials in building construction' give rules for the proper use of such materials as coverings and insulating layers.

Fire protection

Preventive measures • Stability of buildings

Fire protection in its widest sense comprises all measures and precautions aimed at the prevention of fire and at combating it in the event of an outbreak. Passive fire protection is concerned more particularly with prevention. It is subject to statutory regulations. Thus, the building regulations in most countries stipulate that buildings should be so constructed and maintained that:

• the outbreak and the spread of fire are as much as possible prevented, and that, in the event of a fire;
• effective firefighting operations and the rescue of human beings and animals are possible.

Structural fire protection
Causes of fire

Outbreaks are prevented as far as possible by the application of regulations controlling potential fire hazards (heating installations, electrical equipment). The intensity of a fire can be reduced by imposing restrictions on the quantity of combustible materials used in the finishes and furnishings of a building, since fire severity is proportional to the combustible contents ('fire load') of the building.
The fire load constituted by finishings and fittings forming an integral part of the building can be controlled by specifying the use of particular materials and excluding others.
The fire load constituted by furnishings, stored materials, fuel, etc., can be estimated on the basis of statistical data and can be controlled by the enforcement of safety regulations (pp. 331, 332).

Fire containment

A distinction is to be drawn between measures to prevent the spread of fire from one building to another and those to prevent or retard the spread of fire within a building.
Spread of fire to adjacent premises can be prevented by spacing the buildings not less than a certain distance apart or by separating them by means of external compartment walls.
Within a building the spread of fire can be checked by compartmentation, that is, the building is subdivided into compartments with fire-resisting walls and floors. Automatic fire-extinguishing equipment can also help to confine the fire within a particular part of the building.

Access for fire brigade

To allow fire engines to reach the premises, it is necessary to provide adequate access ways of a certain width and height and also of sufficient loadbearing capacity. In the interior of the building there should, as far as possible, be safe means of access for firemen to get at the centre of the outbreak. Hydrants of specified size and capacity should be provided to supply water for firefighting.

Rescue operations

Safe escape routes are a first requirement for the saving of human life in the event of an outbreak of fire. They are essential and there is no substitute for them. Escape routes should be as free as possible from gases and fumes. Alarm systems are necessary in buildings occupied by large numbers of people.

Structural and operational fire protection

Structural fire protection may be subdivided into passive and active defence against the spread and development of fire in buildings. Passive precautions comprise:

• measures to safeguard the stability of buildings under fire conditions;
• subdivision of the building, both horizontally and vertically, into compartments for fire containment;
• provision of escape routes.

Active precautions comprise:

• fire alarms, smoke and fire detectors;
• automatic fire-extinguishing equipment such as sprinklers.

The scope of preventive measures

The objects of structural fire protection are:

• the protection of human life;
• the prevention of material damage to the building and to adjacent property.

Protection of persons (primary fire protection)

The protection and saving of human life must be the primary concern of any system of fire protection. The extent and scope of the precautions to be taken cannot be measured by an economic yardstick, but must be judged in terms of the potential hazard to life. In circumstances where there is little or no danger to people in the event of a fire, the safeguards to protect people can be correspondingly reduced.

Protection of property (secondary fire protection)

With regard to the prevention of material damage, priority must be given to those measures which help to prevent the spread and growth of a fire and which prevent the collapse of buildings.
Material damage to the building itself and to adjacent property is measurable in financial terms. The risk involved will depend, inter alia, upon the structural fire protection facilities and arrangements, which can be assessed and expressed in terms of insurance premium payable. In this way, there is established an economic relationship between the potential loss or damage (or the insurance premium) and the cost of the fire protection measures applied. In the general interests of the national economy, the sum of the risk involved (or the premium payable to cover the risk) and the money spent on fire protection should be a minimum.

Period of effectiveness of fire protection

Measures for structural fire protection need be effective only for a limited length of time, for example:

• escape routes should remain passable in safety until all the people in the building have had time to get out;
• compartment walls and important loadbearing components should remain functional for the duration of the fire;
• other loadbearing components should be able to perform their function during rescue and firefighting operations.

'Total protection' can therefore never be the aim of fire precautions. Protection against fire is necessary only to a limited extent and/or for a limited length of time. If these limits are set too high, they will be wasteful, resulting in expenditure in excess of what is warranted by the needs of the situation. Most building regulations allow considerable scope for the designer's individual judgment in deciding what protective measures are needed in a given set of circumstances.

Criteria for the extent and scope of fire protection

The requirements imposed by structural fire protection, and therefore the scope of the measures to be applied, depend on many factors. As a general principle: where there is nothing that can burn, there is nothing to protect; where there are no occupants, there are no lives to save. Factors which affect such issues are, for example:

• the number of people likely to be present in a building at any given time: stringent requirements for assembly halls or department stores, much less stringent ones for industrial shed-type buildings or multi-storey car parks.
• the mobility of the people in the building, for example, stringent requirements for hospitals, children's or old people's homes, etc., as against less stringent ones for schools (p. 180) or sports halls;
• the fire load of the building, that is, the amount of its combustible contents: stringent requirements for residential buildings and department stores, less stringent ones for schools and multi-storey car parks.
• the area and layout of the building in plan;
• the height of the building: special requirements are applicable to high-rise buildings, the top floors of which are above the height that can be reached by standard fire escape ladders (above 22 m in the Federal Republic of Germany), whereas the requirements applicable to single-storey or two-storey buildings are far simpler;
• the time that is likely to expire before arrival of the fire brigade; thus industrial premises which have their own works fire brigade or premises such as theatres or exhibition halls, where trained firemen are on duty at all relevant times, need not comply with such stringent requirements.

STABILITY OF BUILDINGS

A building must preserve its stability under fire conditions in order to make rescue and firefighting operations possible and to limit the material damage that occurs.

High-rise buildings

The loadbearing structure of a multi-storey building should survive the burn-out of the fire compartment or compartments affected by an outbreak.

Low-rise buildings

For rescue purposes it is generally sufficient for a low building to preserve its stability for only a relatively short period (say 30 min). If the

Escape routes • Fire compartments • Firefighting equipment

fire has not been put out within this length of time, the finishes and fittings of the building will have suffered so much damage that is seems unnecessary to incur any special expense in striving to preserve the structural framework intact.

Various requirements

To preserve the structural stability of a building does not necessarily mean that the structural components must always be of fire-resisting construction. Too often, this is accepted without question, although the interrelation of fire load and the severity of the conditions to which the components are exposed is now sufficiently well understood.
Also, the various components of a building should be considered in the light of differing criteria. The requirements should vary according to the importance of the respective components in terms of maintaining the stability of the structure. This principle is generally valid. Thus, it would be wrong, for example, to extend the requirement for fire-resisting construction to comprise each and every part of a building, including floors and roofs. For details concerning the design of structural components to withstand fire, see p. 331.

Cost of protecting structural steelwork

Protection of the loadbearing structure, more particularly the steelwork, constitutes only a relatively small proportion of the total expenditure in respect of structural fire protection. Thus steel buildings do not in this respect differ significantly from buildings constructed from other materials. Direct protection of the steel, in so far as it is not provided by other components which have to be installed anyway, involves only a tiny fraction of the total cost of construction.

ESCAPE ROUTES

Protected escape routes are a first requirement for saving the occupants of a building on fire and for the safety of the rescuers and firefighting personnel. The death toll in a fire is hardly ever due to premature collapse of structural components, but often to inadequate or impassable escape routes. It is also true that people trapped in a burning building are killed not so much by the flames as by asphyxiation by smoke or toxic fumes. It is not always possible to evacuate a high-rise building completely in the event of fire. It is therefore necessary to provide safe zones in the building and to make such zones able to be found quickly in an emergency.

Fire alarms

To enable the occupants of a building to reach a safe refuge in time, they should receive the earliest possible warning of the outbreak of fire. Fairly large buildings in which considerable numbers of people may be present at any particular time should therefore be equipped with fire alarms and/or a loudspeaker system.

Marking of escape routes

Escape routes should be properly marked and signposted. Such markings should be installed low down, since smoke rises and will first impair visibility in the upper parts of corridors.

Corridors

Corridors serving as escape routes must remain serviceable for a certain definite period, that is, their walls should have a specified fire resistance. They should be kept free from obstruction; in particular, no combustible materials should be stored in them. Combustible materials should be used in the construction of corridors only after their flame-spreading and smoke-developing properties have been carefully checked.

Compartment walls

If compartment walls intersect escape routes, the latter should, at such walls, be provided with self-closing doors which open in the direction of escape and have a certain specified fire resistance.

Stairs

On all floors of a building it should be possible to reach a staircase within a certain distance from any point on that floor. The German safety regulations require that this distance shall not exceed 35 m and that the staircase shall be in an enclosure whose walls have a certain specified strength and are provided with appropriate doors. This applies to so-called essential staircases. External stairs, accessible from arcades or escape balconies, are also acceptable in this sense.

Evacuation of smoke

Smoke evacuation from escape routes is very important. Staircase enclosures should have vents at the top, preferably of a type which opens automatically in the presence of smoke.

Emergency exits

It is essential to provide a sufficient number of emergency exits of adequate size which can readily be opened in the event of danger.
The following examples, based on the German fire regulations for buildings, illustrate the principles governing the location of essential staircases:

- internal staircases are permitted in buildings up to three storeys in height;
- in buildings exceeding three storeys the staircases should be located alongside an external wall and have windows;
- a high-rise building should have at least two staircases, one of which should be located alongside an external wall;
- only one of these two staircases need go down to ground floor level if the bottom exit of the other gives access to a part of the building not more than 22 m in height;
- in a high-rise building only one staircase need be provided if it is designed and constructed as a safety staircase accessible only along an external open gallery;
- buildings may be provided with external emergency stairs accessible along external open galleries of fire-resisting construction.

FIRE COMPARTMENTS

The fire compartments within a building are enclosed by internal compartment walls and floors. Two or more storeys may be combined into one compartment.

Compartment walls

For requirements see p. 333, for construction see p. 320. Doors in compartment walls should have equal fire resistance and be self-closing.

Floors

For requirements see p. 333, for construction see p. 279 et seq.

External walls

Another route for the spread of fire from storey to storey is via the external walls. When the window panes are shattered, the flames can shoot out and set fire to the storey above. This danger can be reduced or eliminated by suitably increasing the distance that the flames will have to travel. Requirements for the fire resistance of external walls are given on p. 333, for construction see pp. 279, 305.

Protected shafts

Vertical circulation shafts — staircase and lift enclosures, service shafts — are usually treated as independent fire compartments, for the danger of fire propagation through such shafts, especially service shafts, is very serious. Ducted services passing through fire-resisting walls from one compartment to another also deserve particular attention. Air-conditioning and ventilation ducts should be provided with automatically closing shutters or flaps; exhaust air ducts which may, in the event of a fire, serve also for smoke extraction should themselves be of airtight and fireproof construction. The danger of fire developing in electric cables and its propagation along the cables into other compartments has already been pointed out.

FIREFIGHTING EQUIPMENT

Automatic equipment can prevent the spread of a fire. Such active fire protection may in certain circumstances justify some relaxation in passive precautions:

- by allowing larger fire compartments to be constructed under the safety regulations;
- by allowing the requirements for the fire resistance of structural components to be lowered.

Increasing the size of the compartments may be an important concession in multi-storey buildings such as department stores, buildings containing large assembly halls, office buildings containing open-plan offices, etc.
Fixed firefighting apparatus, usually fed with water, is of the two following general types:

- spray systems which are turned on by hand;
- spray systems which come into operation automatically (sprinklers).

Fire protection — Stability of structures under fire conditions • Fire as a structural load

Sprinkler systems

Sprinklers, which are usually linked to alarm systems, are the most effective type of fixed firefighting apparatus (see reference 7, p. 336). As they come into action at a relatively low temperature (generally 70°C), they detect and combat the fire at its very outset. For example, a typical sprinkler head will respond to an outbreak in about 3 to 5 min when fitted at a ceiling height of 5 m; for a height of 10 m the response time is between 5 and 7 min. Only those sprinklers which are located over or near the actual centre of the outbreak of fire will come into operation, so that water damage to the premises is only local. Each sprinkler head can monitor a floor area of about 12 m². Statistical data show that in most outbreaks of fire not more than five sprinkler heads are actuated, more rarely up to 25.

A sprinkler installation is not very expensive and its cost is soon repaid by the cuts in insurance premiums that it secures for the building owners.

Sprinkler equipment

1 A sprinkler system comprises: a number of sprinkler heads fed with water through a network of pipes, the alarm system and the water supply system.

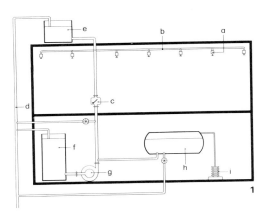

Diagram of a sprinkler system
- a sprinkler head
- b network of pipes
- c fire alarm device
- d main supply pipe
- e high-level water tank
- f low-level water tank
- g centrifugal pump
- h pressurised water tank
- i air compressor

Wet and dry systems

In the wet system, which is the more common, the pipes are kept constantly full of water right up to the sprinkler heads. In the dry system, which is used in unheated buildings in which there is a danger of freezing in winter, the feed pipes are kept filled with air under pressure. When a sprinkler head opens in response to a rise in temperature, this pressure drops, and water is admitted to the feed pipes. The response time is somewhat longer than in the wet system.

Sprinkler heads

A sprinkler head comprises a deflector plate and some kind of seal or plug which responds to a rise in temperature by allowing the water to flow. There are three main types:

2 The type comprising a frangible glass bulb which prevents the water from emerging. The bulb contains a liquid which expands when the temperature rises, thus shattering the bulb and releasing the water.

3 In this type a pair of levers hold a plug in position and are themselves held by a fusible link which releases them at a specified temperature.

4 In the third type the plug is held in position by a salt crystal which melts on reaching a certain temperature, thus releasing the plug.

These devices are produced in a range designed to respond at various temperatures from 70° to 200°C, each temperature rating being indicated by a distinctive colour.

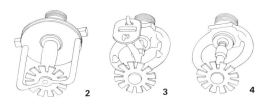

Network of pipes

The network of feed pipes is subdivided into sections, each supplying up to 500 (exceptionally 1000) sprinkler heads.

Fire warning system

The flow of water actuates a warning device in each section of the pipe network. This device may in turn set off a fire alarm or transmit a signal to the fire brigade.

Water supply

A sprinkler head requires between 60 and 100 litres of water per minute. A sprinkler installation should, if possible, have two independent water supplies, one of which functions automatically, while the other should be 'inexhaustible'. Normally, water at a rate of 3 m³ per minute and per section of the pipe network should be available for a period of one hour, always being delivered at sufficient pressure. The following sources of supply may be employed:

- the public water mains (or a comparable source);
- high-level storage tanks;
- open low-level tanks with centrifugal pumps;
- pressurised water tanks.

High-level or pressurised tanks of 180 m³ capacity are usually employed as the first source of supply, backed up by the public mains as the second. The latter can be rated as an 'inexhaustible' supply only if it can draw upon a reservoir of not less than 500 m³ capacity.

GERMAN FIRE REGULATIONS

Various sets of regulations enforcing structural safety have to be complied with in the Federal Republic of Germany:

- Building regulations of the individual Länder based on a set of model building regulations.
- Regulations relating to the construction and operation of business premises, department stores, assembly halls, garages, etc., also based on corresponding model regulations.
- Implementing decrees for the enforcement of the building regulations of the Länder.
- German Standard DIN 4102 *Fire behaviour of constructional materials and structural components*:

Sheet 1: Definitions, requirements and testing of materials;
Sheet 2: Definitions, requirements and testing of components;
Sheet 3: Definitions, requirements and testing of special components;
Sheet 4: Co-ordination of definitions;
Sheet 5: Explanatory comments on sheets 1 to 4.

- German Standard DIN 18 230 *Structural fire protection in industrial buildings*.
- Recommendations for the use of combustible materials in building construction.

STABILITY OF STRUCTURES UNDER FIRE CONDITIONS

The fire regulations of most countries contain more or less strictly enforceable rules for the design and construction of structures and their component parts with regard to their behaviour under fire conditions. These rules are chiefly based on experience, because it has hitherto not been considered possible to calculate the structural effects of fire on the lines of a structural analysis. In borderline cases the existing regulations often leave much to the designer's own judgment.

More recent research has, however, gone a long way towards making it feasible to analyse the structural effects of a fire. Thus, fire has come to be treated in the same way as an imposed load. Appropriate methods are embodied already in the regulations introduced in the Netherlands, Sweden and Switzerland. In Germany a procedure for estimating the structural effects of fire in industrial buildings is embodied in DIN 18 230 mentioned above.

FIRE AS A STRUCTURAL LOAD

In order to calculate the effects of fire on the loadbearing members of a structure and to determine their dimensions and details, the following investigations are necessary:

- evaluation of the fire load;
- determination of the behaviour of the fire;
- determination of the transfer of heat in the components;
- calculation of the temperature of the components;
- the design of the components.

Fire load

The fire load can be calculated from the weight of the combustible materials present within a given space and from the heat liberated by their combustion. The portion of fire load due to permanent finishings and fittings of the building can be determined at the design stage, whereas the portion due to the movable contents of the building can be predicted only on the basis of statistical data relating to particular types of building.

Fire load

Fire protection

Rate of fire growth in buildings

This is characterised by the time–temperature curve within the compartment or room in the building considered. The Swedish regulations give a method for calculating this curve from the magnitude of the fire load and the ratio between the size and height of window openings and the area of all the enclosing surfaces of the space considered.

Alternatively, the actual fire may, through the determination of an equivalent fire duration, be converted into the standard fire. The standard fire and the equivalent fire conforming to the standard curve will produce the same maximum temperature effect in a structural component.

Temperature of components

From the temperature of the gases which envelop a structural component during a fire it is possible to calculate the amount of heat transmitted to the component. The temperature is calculated for each component individually, taking due account of the insulation provided by any encasement or cladding.

Design of components

If this encasement or cladding prevents the temperature in the structural component from rising above the critical value T_1 at which permissible loading can still just be supported, then fire as an imposed load need not be considered in the design of that component itself. But if calculation shows the temperature in the loadbearing cross-section to rise above T_1, a design based on reduced permissible stresses may be adopted for steelwork, see p. 334.

Fire as a structural load

With this approach, fire is thus treated as a load amenable to structural analysis, that is, components of a building can be designed, by taking account of the actual fire hazard and of the amount of heat likely to be absorbed in a fire, by just the same procedures as those applied in designing for gravity loads, wind loads or temperature differences.

Method indicated in German DIN 18 230

In this Standard the procedure is simplified by calculating a category of fire resistance from data for the building, more particularly its shape and its fire load. Instead of a detailed calculation, as allowed by the Swedish regulations, it is possible to determine the necessary category of fire resistance for a component as a function of the protection required for the building in question. Structural components are assigned to various categories on the basis of fire tests performed with the international standard time–temperature curve. Of course, with this method it is also possible to deduce the requisite fire resistance of individual components by theoretical calculation, depending on the insulating capacity of the encasement or cladding and on the cross-sectional proportions.

FIRE LOAD

The fire load is the quantity of heat ΣQ_i produced in the fire compartment by the total combustion of all the combustible material. Thus

$$Q_i = G_i \cdot H_u \quad (\text{Mcal})$$

where Q_i = quantity of heat generated by each material (Mcal)
G_i = quantity of material (kg)
H_u = calorific value of the material at normal moisture content (Mcal/kg).

This value is often expressed in kg/m² of equivalent timber, the calorific value of timber being 4 Mcal/kg. The calorific values per kg and per dm³ of some important materials are given in Table 1.
The fire load per unit area in a compartment of a building is thus:

$$q = \Sigma Q_i / A \quad (\text{Mcal/m}^2)$$

where A denotes the floor area of the compartment (m²).
When, in a part of the compartment, the fire load exceeds the average fire load for the compartment by a certain amount, DIN 18 230 stipulates that the resistance of the part in question should be calculated on the basis of the higher load.

Table 1: Calorific value of some building and other materials

Material	Density kg/dm³	Calorific value Mcal/kg	Mcal/dm³
Hardwood, paper	0.800	4.0	3.2
Softwood	0.600	4.5	2.7
Wood-wool slab, DIN 1101	0.500	1.5	0.75
Polyvinyl chloride		5.0	
Phenolic resin		6.0	
Polyurethane		5.5	
Petrol	0.8	10.5	8.4
Fuel oil	1.0	10.5	10.5
Bitumen felt	1.0	4.0	4.0
Rubber	1.2	8.0	9.6
Linoleum	1.3	5.0	6.5

Fire growth rate

The behaviour of any particular material is characterised by the time–temperature curve developed in the compartment in which combustion takes place. This curve in turn can be related to the international standard time–temperature curve, on which the category of fire resistance of structural components can be based. This is done by multiplying the fire load by a rating factor m (see below), the actual relationship applied having been determined on the basis of fire resistance tests. The time–temperature curves for certain materials are presented in **1**; the standard fire curve is shown for comparison (see also p. 332, **1**). The effect of the fire load in tests performed with one and the same material is demonstrated in **2**. The examples show that the temperatures of natural fires mostly lie well below the standard curve, although higher temperatures may be registered during a short phase at the beginning of the fire.

The rating factor m envisaged in DIN 18 230 takes account of the surface area of the combustible material per unit volume thereof, expressed in the material thickness, d. Because of the larger surface area a number of thin slices of a combustible material will burn more fiercely than an equal quantity of that material in one undivided piece.

Rating factor m — Table 2

Combustible material	Least dimension (cm)	Factor m
Wood-wool, loose paper	d < 1	1.3
Boards, furniture	1 < d < 4	1.1
Planks, square-sawn timber	4 < d < 10	0.9
Beams	10 < d < 20	0.7
Logs, paper bales	d > 20	0.3
Petrol		1.3
Fuel oil		1.2

The concept 'equivalent fire load' is introduced, defined as:

$$Q'_i = Q_i \cdot m_i = G_i \cdot H_u \cdot m_i \quad (\text{Mcal})$$

This is determined for each individual material in the fire compartment under consideration.

The equivalent fire load per unit area is:

$$q' = \Sigma Q'_i / A \quad (\text{Mcal/m}^2)$$

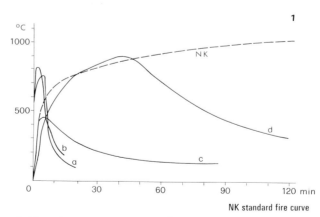

1 Time–temperature curves for various materials (see reference 3, p. 336):

a fuel oil in open tank   30 Mcal/m²
b hardboard as cladding   30 Mcal/m²
c paper   200 Mcal/m²
d wood in 45 × 45 mm battens   200 Mcal/m²

NK standard fire curve

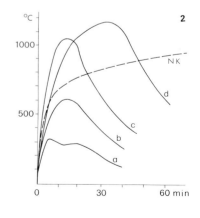

2 Effect of fire load on the time–temperature behaviour of burning wood, in the form of 45 × 45 mm battens (see reference 3, p. 336):

a 7.5 kg/m² = 30 Mcal/m²
b 15 kg/m² = 60 Mcal/m²
c 30 kg/m² = 120 Mcal/m²
d 60 kg/m² = 240 Mcal/m²

Fire protection — Categories of fire resistance • Fire resistance rating of structural components

The rating factor is expressed by:
$$m = \sum Q'_i / \sum Q_i$$
and thus: $q' = q \cdot m$ (see ref. 4, p. 336). Values of q' have been determined for various buildings, depending on their purpose and mode of use.

Comparison

The actual fire load which, according to Law (ref. 3, p. 336), occurs in buildings such as schools, offices, hospitals, etc., with wooden furniture is of the order of $q = 80 - 100$ Mcal/m². This is a figure based on numerous investigations of actual buildings in these categories. For comparison, in Table 3 the fire load is calculated for a typical classroom in a school according to the method of DIN 18 230:

Fire load in a classroom — Table 3

Material	G_i (kg/m²)	H_u (Mcal/kg)	q (Mcal/m²)	m	q'_i (Mcal/m²)
0.5 cm linoleum	6.5	5.0	32.5	1.0	32.5
Furniture (wood)	10.0	4.5	45.0	1.1	49.5
Paper	2.0	4.0	8.0	1.3	10.4
Other materials					5.0
			85.5		97.4

$$m = \frac{q'}{q} = \frac{97.4}{85.5} = 1.13$$

The European Convention for Constructional Steelwork carried out statistical surveys of the combustible contents of 480 offices, etc., in administrative buildings and arrived at an average value of $q = 92$ Mcal/m². 90% of these rooms had a fire load of less than 164 Mcal/m² and 95% had less than 188 Mcal/m².

CATEGORIES OF FIRE RESISTANCE FOR BUILDINGS

In DIN 18 230 five categories are quoted, as shown in Table 5. The fire load and the category are linked by the equation

$$q_r = q' \cdot f \quad (\text{Mcal/m}^2)$$

where q_r is the calculated fire load. The 'building factor' f, introduced for simplification, incorporates a number of factors which take account of the area of the fire compartment and the number of storeys and which is specifically applicable to industrial buildings. Nonetheless, values essential for multi-storey buildings are quoted in Table 4.

The values are lower for situations where adequate openings for the escape of heat are available, single-glazed windows constituting such openings. The category of fire resistance CF for each compartment is established on the basis of the calculated fire load q_r, as indicated in Table 5.

Building factor f — Table 4

Area of fire Compartment A (m²)	Number of storeys					
	1	2	3	4	5	≥6
< 1 600	1.0	1.2	1.3	1.4	1.5	1.6
1 600 – 3 000	1.2	1.44	1.56	1.68	1.80	1.92
3 000 – 5 000	1.4	1.68	1.82	1.96	2.10	2.24
5 000 – 7 000	1.6	1.92	2.08	2.24	2.40	2.56
7 000 – 10 000	1.8	2.16	2.34	2.52	2.70	2.88
> 10 000	2.0	2.40	2.60	2.80	3.00	3.20

Category of fire resistance CF — Table 5

Calculated fire load q_r (Mcal/m²)	Category of fire resistance CF
< 75	I
75–150	II
150–300	III
300–450	IV
450–600	V
> 600	V*

* only possible when special precautions are taken

FIRE RESISTANCE RATING OF STRUCTURAL COMPONENTS

Except in cases where an ultimate-load calculation based on the actual expected time–temperature curve is permitted, structural components are classified according to the minimum period of fire resistance for which they can continue to perform their loadbearing function in the fire test. The standard fire curve determines the temperature as a function of time in the testing furnace (**1**). The fire resistance is thus subdivided into a number of ratings each of which is 30 min longer than the preceding one. The standard curve is internationally established (ISO), and the fire resistance rating is also widely adopted throughout the world. In the Federal Republic of Germany the ratings indicated in Table 7 are adopted.

The materials of which the structural components consist are themselves classified according to combustibility in DIN 4102, as indicated in Table 6:

Combustibility of materials — Table 6

Class of material	Property of material
A	
A_1	incombustible
A_2	
B	combustible
B_1	difficult to ignite
B_2	average inflammability
B_3	easy to ignite

The principal requirements specified for individual components in these Standards are:

Loadbearing components

Loadbearing components should be tested while carrying the permissible working load. If this is not practicable, then the temperature of steel columns should not be allowed to exceed an average of 400°C, while individual values should not exceed 500°C. Columns assigned to fire resistance rating F 90 and upwards should be able to withstand the sudden cooling action of water sprayed on them in the hot condition without fracture or spalling of the encasement so as to expose bare steel. Also, from rating F90 upwards, structurally important constituents of components should consist of materials of class A; and for rating F 180 the components should consist entirely of class A materials: see Table 6.

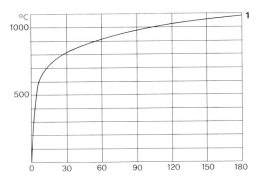

Standard fire curve for fire resistance rating

Non-loadbearing components

In fire tests, non-loadbearing components must not collapse under their own weight.

Space-enclosing components

These should prevent the passage of fire. No inflammable gases should develop on the face remote from the fire, and the average temperature on this side of the wall must not exceed 140°C, while individual values must not exceed 180°C. From fire resistance rating F 90 upwards those parts which, on destruction, would allow the fire to proceed unchecked (and walls not more than 50 mm thick: entirely) should consist of class A materials, while space-enclosing components designed for F 180 should be constructed entirely of such incombustible materials.

Space-enclosing ceilings from rating F 90 upwards should comprise a continuous dense layer, at least 50 mm thick, consisting of class A materials. Partitions should be able to withstand an impact test of 2 kg·m after undergoing the fire test.

Fire resistance rating — Table 7

Period of fire resistance (min)	Fire resistance rating	valid for	Designation	Fire resistance rating	valid for	Fire resistance rating	valid for	Designation
≥ 30	F 30	load-bearing components	fire-retarding fh	W30		T 30		fire-retarding barriers
≥ 60	F 60	non-load-bearing components	fire-resistant fb	W60	External walls	T 60	Fire barriers	fire-resistant barriers
≥ 90	F 90			W90		T 90		
≥120	F120	compartment walls	highly fire-resistant			T120		highly fire-resistant barriers
≥180	F180	space-enclosing components				T180		

Choice of fire resistance rating

Fire protection

Compartment walls

These should be constructed as space-enclosing walls of fire resistance rating F 90, should consist of class A materials throughout, should contain no cavities, and should preserve their stability even under eccentric load and 300 kg·m impact test, besides conforming to certain minimum thickness requirements, depending on the material in question.

External walls

External walls should check the propagation of fire from storey to storey outside the building. The fire resistance applicable to these walls extends also to the fixings, fastenings and cladding joints and they must not collapse under their own weight in a fire nor lose their cohesion. External walls are tested on the inner face with the full standard fire curve, whereas the outer face need only withstand a temperature of 650°C (which must not be reached until at least ten minutes have elapsed). On exposure to fire on the outside of the building, the wall cladding must not produce any inflammable gases. With fire inside the building, the average temperature on the outer face of the wall must not exceed 140°C, with individual values not above 180°C.

Fire barriers

Fire barriers, which in the present context comprise doors and similar closures in walls and ceilings with fire resistance ratings from F 30 to F 180, as well as in compartment walls, should fulfil specific requirements. Other regulations cover such features as liftshaft doors, glazing, roof coverings and air ducts.

General principle

An important general principle is that suspended ceilings, encasement and cladding which are attached to structural components to improve their fire resistance should form a constructional whole with them. They must not be tested separately, but always in combination with the component they are intended to protect. For example, a suspended ceiling by itself is not fire-resistant: the structural floor together with the ceiling and the side closures of the floor cavity form a fire-resistant whole. The same applies to the cladding around a column: cladding and column together constitute the fire-resistant assembly.

CHOICE OF FIRE RESISTANCE RATING

In Germany, the fire resistance rating to which any particular type of structural component should belong is laid down in the building regulations. This assignment to a particular rating is based on various criteria, for example, the purpose for which the building is used and the number of storeys it comprises, but without reference to fire load (see left-hand side of Table 8). This omission must be regarded as a serious drawback in that it often results in excessively severe demands and unnecessarily high expenditure in respect of structural fire protection.

On the right-hand side in Table 8, the requirements for fire resistance ratings (not to be confused with the number of storeys in the building) as laid down in DIN 18 230 are compared, in so far as comparison is possible. This table is the closing link in procedure whereby, starting from the fire load, followed by the determination of the category of fire resistance, the fire resistance rating of each type of structural component is determined by a purely analytical process of calculation which greatly reduces the latitude allowed to subjective judgment by the designer.

Critical temperature of steel

Steel is an incombustible material. Its mechanical properties, such as yield point, tensile strength and modulus of elasticity, are, however, a function of the temperature. The yield point is often, but incorrectly, regarded as the criterion for the behaviour of steel under conditions of fire. In fact, since fire is a catastrophic loading case, it is justifiable to take account of the reserves of strength that the metal can offer above its yield point and up to the final collapse of the structural component. The instant of failure is characterised by the temperature then attained: the so-called critical temperature. This is not a constant for steel in general but is substantially dependent on the grade of steel employed in the component concerned. For full utilisation of the permissible stress, the critical temperature for steel grade St 37 is 560°C, while that for St 52 is 580°C (see p. 334, **2**). In an actual fire the structure is generally not subject to its full design load (live load, wind, etc.), so that in practice the temperature at which failure ensues is higher, for example, at 650°C for St 37 and at 670°C for St 52, assuming the actual stress in the steel to be about half the allowable stress.

In a statically indeterminate system designed on the basis of elastic theory the available strength reserve in this respect has an effect similar to that of a reduction of the stresses below the allowable limits. Thus a continuous beam over a large number of supports has the same failure temperature as a simply-supported beam in which the maximum stress is only three-quarters of the allowable stress. The values stated here are applicable to encased steel members. The failure temperatures measured at the surface of unprotected steel members, exposed directly to the heat, are at least 50°C higher because of the higher rate of heating in such circumstances.

Effect of the structural system

Attainment of the failure temperature at a particular section does not necessarily always result in collapse of the structural component in question. Statically indeterminate structures designed by the elastic method still have reserves of strength. When the failure temperature is first reached at a particular section of a continuous beam, a plastic hinge develops there, but the beam still retains its load capacity. Under fire conditions a continuous steel beam sags in its several spans, with the result that the beam is stressed mainly in tension instead of in bending. Provided that the overall coherence of the system is maintained — by appropriate detailing of the connections — it offers further reserves of strength, that is a higher value may be adopted for the critical temperature. If these relationships are duly taken into consideration, it is possible to effect savings in the cost of fire protection by reducing the stringency of the requirements to be fulfilled, this being reflected, for example, in the use of smaller thicknesses of protective material.

Fire protection requirement   Table 8

Function	Component	Regional building regulations North Rhine–Westphalia				Din 18230 Category of fire resistance				
		≤2	3...5 storeys	>5	Multi-storey buildings	I	II	III	IV	V
Load-bearing components	Columns, walls, main beams	F 30	F 90	F 90	F 90		F 30	F 60	F 90	F 120
	Floors	F 30 trT=A	F 30 trT=A	F 90	F 90		F 30	F 30	F 60	F 90
	Roof structures							F 30	F 30	F 30
Components enclosing fire compartments	Walls		F 90 Br	F 90 Br	F 90 Br	F 60	F 90	F 90	F 90 Br	F 120 Br
	Floors			F 90	F 90	F 30	F 30	F 60	F 90	F 120
	Closure of openings			T 90	T 90	T 30	T 30	T 90 (2×T 30)	T 90 (2×T 30)	T 90 (2×T 30)
Space-enclosing components	Party walls	F 90	F 90	F 90	F 90					
	Corridor walls		F 30	F 30	F 30					
	Staircase enclosure walls		F 90 d=Br	F 90 d=Br	F 90 d=Br					
	Staircases		trT=A	F 90 trT=A	F 90 trT=A					
	External walls		W 30 or =A	W 30 or =A	W 30 or =A					

Key to symbols:

Br	Strength of a compartment wall
d=Br	Thickness as for a compartment wall of the same material
=A	of incombustible material
trT=A	Incombustible loadbearing component

Fire protection

Behaviour of unprotected steel components in a fire

Effect of cross-sectional shape

The length of time that elapses before a structural component reaches its failure temperature depends on the rate of heat absorption by the component. Generally speaking, to reach a certain temperature a component with a large cross-section must absorb more heat than one with a smaller cross-section. The rate of absorption is higher, that is heat is transferred more rapidly to the structural component concerned, if its surface area is large in relation to its mass. Thus, the rate of heat absorption and therefore the rate of temperature rise is high for a slender open section, whereas it is lower for a closed tubular section or box-section because in such components the heat has direct access to one side of the steel only. The criterion is the ratio U/F, where U is the perimeter of the section in cm and F is its area in cm^2.

In **1** the set of curves marked d = 0 represents the fire resistance period, in minutes, of the unprotected steel section. It appears that a structural component with cross-sectional dimensions characterised by U/F = 0.33 and a failure temperature of 550°C has a fire resistance of 30 min even when unprotected, therefore, it is assignable to fire resistance class F 30. This degree of fire resistance is attained, for example, by a 60.5 mm thick flat bar, since U = 2 cm and F = 6.05 cm^2 per cm width, and therefore U/F = 2/(6.05) = 0.33; it is also attained by a box-section with a wall thickness of 30.3 mm, in which case U/F = 1/(3.03) = 0.33. Equivalent to the box-section in this respect is an open I-section in which the spaces between the flanges are filled with concrete: treated in this way, the following I-sections attain 30 min fire resistance: I 600, IPBl 1000, IPB 650, IPBv 240.

Used as a continuous beam or alternatively as a simply-supported beam with 75% utilisation of allowable stress, a fire resistance of 30 min is attained by an unprotected IPBv 300 section. Under the same structural conditions, but with concrete filling between the flanges, this same fire resistance period is attained by: I 360, IPB 600, IPBl 450, IPB 320, IPBv 100 (all sections with greater depth have more than 30 min resistance).

The curves in **1** indicate the encasement thickness to be provided when a component with a certain U/F ratio has to attain a certain fire resistance t_F. These relationships are based on experimental work and calculations carried out at international level and are intended for incorporation into national regulations.

Heat dissipation

The period of fire resistance of a structural component is extended if some of the heat transferred to it is in turn transmitted to other parts. Thus a hollow steel column has a higher resistance when filled with concrete than when left empty, even if the concrete is not considered to contribute to the structural strength. A similar improvement in fire resistance has been observed in composite beams whose top flange is cooled by transfer of heat to the concrete in which it is embedded. In tests, unprotected composite beams have been found to attain 30 min resistance, while continuous composite beams have attained even more.

Deformations

For judging the behaviour of steel structures the following considerations are important. Structural steels reach their yield point and thus begin to undergo plastic deformation long before failure occurs. As a rule, impending failure or collapse is preceded by increasingly large deformations, such as deflections of beams. This fundamental pattern of behaviour is evident also under fire conditions, except that the stresses at which the loadbearing capacity becomes exhausted are lower than at normal temperatures. Therefore a steel structure, generally speaking, does not collapse suddenly when attacked by fire. There are unmistakable warning signs, namely, large deformations.

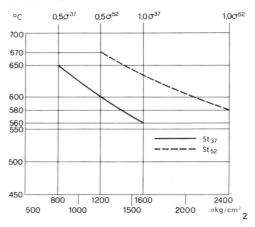

2. Critical temperatures of St 37 and St 52 grade steel

Re-use of steelwork components

When a steel beam or column cools after a fire, it regains its original strength. Deformed structures regain substantial loadbearing capacity, which is an important advantage in terms of safety during demolition operations.
Steel members which have remained straight can certainly be re-used. Bent or buckled ones can readily be replaced. An American speaker at a conference once expressed this in the following words: 'If it's bent, cut it out; if it's straight, paint it'.
Fire damage to steel buildings is easily detectable; there is no reason to fear that there may be hidden damage lurking below the surface.

METHODS OF FIRE PROTECTION FOR STEEL COMPONENTS

Structural steelwork, like structural components consisting of other materials, must possess the required fire resistance. If this cannot be obtained with unprotected steel sections, protection must be supplied in the form of casings, intumescent materials, water-filling or screening.

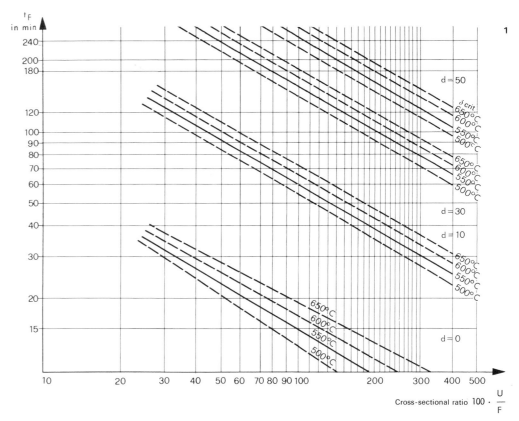

Relationship between period of fire resistance, cross-sectional ratio and thickness of encasement assumed to be Vermitecta cladding.

IPB 300
F = 149 cm^2
U = 90 cm
$\frac{U}{F}$ = 0.6/cm

IPB 300
F = 149 cm^2
U = 173 cm
$\frac{U}{F}$ = 1.16/cm

Protection of steel components against fire • Twelve rules for fire safety in steel buildings — Fire protection

Casings

This general designation comprises a wide range of materials and methods for the cladding of steelwork components as protection against the action of fire.

A commonly used method of solid encasement is to embed the steel columns or beams in in-situ concrete poured into temporary formwork. Alternatively, various types of bricks or blocks may be used to surround the steel. Lighter forms of construction, whereby a box-like enclosure is built round the column or beam, may consist of fire-resistant building boards or slabs or may be formed by plaster applied in situ to expanded metal wrapped around the steelwork. The plaster may be lime, cement or gypsum plaster or may incorporate vermiculite, pearlite or asbestos; it is generally applied in several coats.

Casings consisting of slabs or boards may be made of such materials as vermiculite, pearlite, asbestos, gypsum or asbestos cement. Sometimes specially shaped interlocking precast lightweight concrete, gypsum plaster or asbestos cement blocks are used. The slabs or boards are attached to the steelwork by adhesive bonding and/or by nailing or screwing them to one another. The steelwork requires a protective treatment against corrosion. A suitable priming coat will usually suffice for the purpose (p. 358).

Casings formed direct on the steelwork by spraying usually consist of a cement-based plaster containing asbestos or vermiculite and applied in several layers. Sprayed asbestos (or limpet asbestos) thus consists of a mixture of finely divided asbestos fibres and portland cement which takes the profile of steel components to which it is applied, so that it can be used on any structure regardless of shape or size. On sections with very deep webs it may be necessary to embed wire mesh in the casing. As a rule, the casing is sprayed onto the unpainted steel surface from which rust and mill-scale have been removed. A light film of rust, such as may form in the interval between blast-cleaning at the fabricating works and the application of the casing to the erected steelwork, is harmless. A disadvantage of the method is that it is messy; it may cause dirtying of metal or concrete façade claddings, and for this reason the sprayed casing is usually applied to the steelwork before the cladding panels are installed. Freshly sprayed surfaces should be protected from rain; the roof should therefore have been erected before spraying is carried out. In cold weather it may be necessary to take precautions to protect the freshly applied casing from frost. The casing provides corrosion protection.

Besides being cast in-situ around the steelwork, concrete as a casing material may alternatively be applied by spraying (guniting process), usually requiring the embedding of wire mesh in the casing. Solid concrete casings cast in situ are usually reinforced with mesh or with small-section reinforcing bars to restrain shrinkage cracking and to ensure strength of the encasement.

Intumescent materials

Intumescent materials are applied as coatings (paint or mastic) or as sheets. These materials develop their protective action only in response to exposure to heat, which causes them to swell and thus form a protective layer of appropriate thickness. With this protection a fire resistance of 30 min can be obtained in most cases, with 60 min or more for heavier or thick-walled sections. These coatings are sensitive to moisture, and their use should therefore be confined to the interiors of buildings. Types suitable for external use are being developed, and the resistance of these materials against ageing is also under investigation. Complete corrosion protection of the steel, comprising a priming coat and finish coat, is necessary under the intumescent coating.

Water-filling

The filling of hollow steel sections with water is the most effective protection against fire, ensuring that the steelwork will remain structurally functional throughout the duration of the fire. The water is circulated through the structure, the heated water being cooled for recirculation or replaced by fresh water from the mains. This principle has been applied to columns, and a system of this kind is described on p. 240. The principle could possibly be extended to horizontal members.

Screening

A wall or ceiling of fire-resistant material can be designed to enclose otherwise unprotected steelwork within it, for instance, in the cavity between two leaves of a wall. This is usually the most economical method of fire protection, since the space-enclosing components are required for other reasons anyway and can at little or no extra cost be so constructed as to form, with the steelwork they protect, a system possessing the desired degree of fire resistance. For examples, see pp. 293, 321.

TWELVE RULES FOR FIRE SAFETY IN STEEL BUILDINGS

Requirements

Rule 1

The requirements as to structural fire protection should be appropriate to the fire hazard presented by the components concerned.

Rule 2

With each type of building and each purpose for which a building is used are associated particular requirements; these should be uniformly and consistently formulated.

Rule 3

Fire protection measures are not necessary in circumstances where structural components are not at risk or where protection is intentionally dispensed with.

Protection of human life and property

Rule 4

Protection of life and protection of property should be considered separately from each other.

Rule 5

Priority should be given to the protection of life and of third-party property; the requirements imposed by building authorities should be confined to these aspects.

Rule 6

The protection to be provided for the building owner's own property depends on economic considerations and only in exceptional cases on considerations of public interest.

Risk and insurance premiums

Rule 7

The risk of loss of the contents of a building is, generally speaking, not affected by the fire resistance of the structural components; hence fire insurance premiums should be calculated separately for the building and for its contents respectively.

Rule 8

Structural and operational measures which enhance the fire safety of a building, and the fact that steel structural components are re-usable, should be reasons for reducing the premiums.

Steelwork under fire conditions

Rule 9

Steel does not burn; steel components will continue to perform their loadbearing function so long as they do not reach the critical temperature (from 500° to 750°C, depending on structural conditions). If above-critical temperatures are not expected to occur, no fire protection measures are necessary.

Rule 10

Above the critical temperature the loadbearing capacity may be destroyed; increasingly large deformations give clear warning of approaching failure or collapse.

Rule 11

By means of suitable protective measures steel structural components can be given any required degree of fire resistance.

Rule 12

Fire damage to steel structures is readily detectable and can be speedily put right.

Examples of structural fire-protective measures

Protection of columns: pp. 238–241.
Protection of structural floors: pp. 279–281.
Suspended ceilings: pp. 281–284.
Enclosures for staircases: p. 293.
External walls: p. 305.
Fire-resisting compartment walls: p. 320.
Partitions: pp. 321–326.

REFERENCES

1 Schubert: Feuerlösch- und Wärmeabzugsanlagen (Firefighting and heat-extraction equipment), *Bauwelt* 1972, p. 1356.

2 Locher and Sprung: Einwirkung von salzsäurehaltigen PVC-Brandgasen auf Beton (Effect on concrete of PVC combustion gases containing hydrochloric acid), *Beton* No. 70, p. 63 ff.

3 Law: *Fire loads, Planning and design of tall buildings*, Technical Committee 8.

4 Schneider: *Bewertung des unterschiedlichen Brandverhaltens von Stoffen bei natürlichen Bränden* (Assessment of different behaviour of materials in natural fires). Zentralblatt für Industriebau 72, p. 230 ff.

5 Ehm: *Structural behaviour, Planning and design of tall buildings*, Technical Committee 8.

6 Bongard and Portmann: *Brandschutz im Stahlbau* (Fire protection in steel construction), Deutscher Stahlbau-Verband, Cologne, p. 25.

7 *Automatischer Brandschutz* (Automatic fire protection), Verband der Sachversicherer e.V., Cologne, 1970.

8 Bongard: *Wann kann auf den Brandschutz von Stahlbauteilen ganz oder teilweise verzichtet werden?* (When can fire protection of steelwork be wholly or partly dispensed with?), Unpublished draft.

Preliminaries	
Types of contract	338
Division of responsibilities for the scheme	339
Bills of quantities	340
Planning execution of scheme	340
Co-ordination of general planning Coding	342
Payment scheme	343
Fabrication	
Organisation of fabrication	344
Processes of fabrication of structural steelwork	345
Layout of a fabricating shop	346
Erection	
Planning of erection	347
Methods of erection	348
Cranes and derricks	350
Site Organisation, scaffolding, safety precautions	352

Execution of the Scheme

PRELIMINARIES

The preliminary activities that must always precede the actual construction work include:

duties to be performed by the building owner;
work by local authorities;
technical planning;
commercial and legal arrangements;
duties to be performed by the contractor;
job planning.

Duties of the building owner or his representative

Decision to build
Provision of the site for the building
Provision of finance
Specification of the purposes and functions of the building
Awarding of the contract for the work

Matters to be dealt with by the local authorities

Town planning aspects
Architectural aspects
Legal authorisations
Authorisation by the building inspection authority

Technical planning and design activities

• Architects' services in the following stages:

Preliminary design
Design
Formalities in securing a building licence
Construction documents

• Consulting engineers' services in the sectors of:

Structural engineering
Heating and ventilating
Electrical engineering
Specialist engineering services (kitchens, laboratories, lifts, etc.)
in general, these services are required in the stages of preliminary design:
Design
Construction documents (for example, formwork and reinforcement drawings, workshop drawings, detail design)

• Consultants' services in the sectors of:

Soil mechanics
Physical problems (sound insulation, thermal insulation, protection against dampness)
Fire protection

Commercial and legal steps to be taken by the building owner and the contractor in connection with the planning of the scheme

Cost estimating
Calculation of quantities
Costing the job
Drawing up bills of quantities and specifications
Inviting tenders and awarding the contract

Contractor's duties

• In the preliminary stages:

Planning of the execution of the scheme

• During the course of construction:

Co-ordination of constructional operations
Supervision of work
Measuring-up and settlement of accounts

General planning

The contractor will be able to meet his contractual obligations in terms of cost and completion time only by careful preparation and planning of the scheme and by subsequent close adherence to the plans and schedules drawn up.

• A master plan should be drawn up, listing and timing all the subsequent design and planning operations required. This master planning should begin at a very early stage in the scheme and should include the procedures associated with securing the site and the finance for the job, as well as technical planning, the operations involving public authorities and the commercial and legal operations associated with tendering and the subsequent award of the contract (pp. 339, 340).
• This should be supplemented by cost planning aimed at minimising the cost of the job by optimum control of those operations which for economy depend on speed of execution (p. 343).
• Planning the actual execution of the scheme (pp. 340–342).

The following discussion of the subject is confined to those procedures which are typical of steel-framed buildings.

Preliminaries

Types of contract

1. Contract based on bill of quantities

The preparatory work by the designer is so far advanced that the tenderer is able to appreciate all the circumstances affecting cost. All the materials to be supplied and installed and all the operations involved in constructing the building are listed as items in a bill of quantities prepared by the client for pricing by the contractor. The quantitative risk falls upon the building owner. The tender is based upon the quantities. A lump sum contract is only possible if the tenderer has had the opportunity to check the quantities stated.
An example of a bill of quantities for structural steelwork, floors and fire protection is given on p. 340.

2. Contract for the supply of components based on specification only

The components to be supplied are fully specified, but their structural design is left to the bidder, who is also responsible for determining the quantities. In cases where there may be subsequent additions or modifications to the scope of the contract, an itemised schedule of prices may be included. As an alternative to the bill of quantities contract, this type of contract is suitable in cases where the tenderer is allowed to put up alternative design proposals of his own, for example, using his own proprietary components or systems.

3. Contract for building based on functional specification

The building to be constructed is described by specifying the number, size, potential use and equipment of all the rooms, together with all other requirements and functions. The contractor's tender may include the architectural and structural design work. This type of contract is especially appropriate in cases where the contractor wishes to use his own construction systems, as he can so adjust his design to their technical possibilities that an optimum overall solution is obtained.

Scope of the work undertaken by the steelwork contractor *Not included in carcase work

	Steelwork	Floors, stairs	Fire protection	Remaining carcase work	Roof, external wall*	Space-enclosing internal fitting-out*	Plumbing, heating, air-conditioning	Electrical services	Lifts, escalators, etc.
		Carcase work				Fitting-out		Installation of services	
1	▬▬								
2	▬▬▬▬								
3	▬▬▬▬▬▬								
4	▬▬▬▬▬▬▬▬▬▬▬▬								
5	▬▬▬▬▬▬▬▬▬▬▬▬▬▬▬▬▬▬								

Steel frame

Basically, the steelwork contractor will fabricate, supply and erect the components for the structural steel frame of the building.

Steel frame, floors, stairs, fire protection

In addition, it is now common practice in Germany (although not elsewhere) to entrust to the steelwork contractor various other constructional operations closely bound up with the steelwork, more particularly:

• *Floors and stairs:*

(a) Composite floors (with in-situ or precast concrete floor slabs), in order to guarantee the soundness of the structure, since the slab and the steel beams form a structural whole (pp. 256–259).
(b) Prefabricated floors, as the design (pp. 271–280) and laying (p. 341) of the floor units have to be carefully co-ordinated with the erection of the steel frame.
Often, the erection of the floor units is carried out with the same lifting equipment. The floor may also be utilised as erection scaffolding (p. 351).

(c) Prefabricated stairs which are supported by the structural steel frame (p. 294), the reason here also being that stairs and steelwork are erected at the same time. In addition, the stairs speed construction as it is then not necessary to use ladders (p. 351).

• Fire protection (p. 335):

(a) Sprayed asbestos or similar encasement, in order to guarantee the soundness of the work, as such casings also have to provide corrosion protection for the steel (p. 358).
(b) Box casings, as these have to be designed in combination with the steelwork.

• Foundations and basements, provided that these are relatively simple or of limited extent. Otherwise, this work is usually entrusted to specialist civil engineering firms.
• Compartment walls and vertical circulation cores, as the construction of these features is so closely related to the arrangements for the erection of the steelwork. This method corresponds with that practised in reinforced concrete construction, and allows a comparison to be made between the two forms of construction.

Carcase work

The award of the whole of the carcase work to the steelwork contractor facilitates the co-ordination of planning and construction. It is also possible to add the installation of the internal components.

PACKAGE DEALS

1. Contracts placed with individual contractors

The building owner awards contracts for various parts of the job directly to individual contractors. He, or more usually his representatives, must co-ordinate and supervise these contractors' activities. In case of dispute, it falls upon him to apportion blame.

2. Contracts for certain groups of trades or for the whole job placed with main contractors

The main contractor who undertakes the complete structural work or indeed the construction of the finished building is usually the contractor who puts up the structural frame and thus has the major share of the job. For a steel-framed building this is usually the steelwork

Legal conditions for contractors

Extent of contract	Structural performance	Contract management	Guarantee and liability for sub-contractors' performance	Liability for cost of sub-contractors' performance
Individual contracts	Contractor	Resident engineer		
Groups of trades	Main contractor	Main contractor		
or	General contractor	General contractor	General contractor	
Total undertaking		General contractor	General contractor	General contractor

Division of responsibilities for the scheme

Preliminaries

contractor. The major attraction of this procedure to the client is the rationalising effect that can thus be achieved, which usually so reduces costs that it more than offsets the main contractor's extra charges over and above the prices charged by his sub-contractors.

The main contractor undertakes full responsibility for the job, including the guarantees and the obligation to complete the work within the contractually agreed time. He appoints the agent or contract manager in charge of site operations. Supervision of these operations, however, always remains the responsibility of the client or his representatives.

In Germany, the following legal forms are normally adopted:

- Principal contractor: The principal contractor executes that proportion of the work which he has specifically undertaken as his own share. He makes arrangements with sub-contractors on behalf and for the account of the client and does not himself enter into contractual relations with them. The client may, however, delegate to the principal contractor certain rights with respect to the sub-contractors.
- General contractor: The general contractor has rights and obligations like the principal contractor, but he makes his arrangements with sub-contractors on his own behalf and additionally bears the full price risk for all supplies and services to be provided.
- General organising contractor: The arrangements are similar to those concerning a general contractor, but in this case the contractor confines his services to organising and co-ordinating the work of sub-contractors without himself directly engaging in construction work.

3. Joint ventures • Consortia

In tendering for large projects a number of contractors may join forces in joint ventures or for the execution of the work they may form a working association which may, as far as the client and sub-contractors are concerned, act in any of the ways envisaged under point 2 above.

Joint ventures

A joint venture is a legally independent corporate body. It is usually composed of two or more contractors who jointly carry out a particular type of work and who jointly enter into contractual arrangements with the client, towards whom they are jointly and severally liable. The client deals with the body as a single entity in all matters, including payment.

Consortium

A consortium is an association of contractors which differs from a joint venture in that it does not enjoy independent legal status. The members of the consortium are likewise jointly and severally liable to the client, but with regard to payment the client deals with the members individually. This form of association is frequently adopted for prefabricated construction jobs for which the fabricating operations are carried out in the various member firms' workshops, while the work to be undertaken jointly by the members comprises only the – in terms of cost – relatively minor operation of erecting the prefabricated components. Two or more steelwork firms often enter into a consortium. Also, this form of association is commonly adopted when a steelwork firm and a general civil engineering firm work together on the same job as co-operating but independent partners.

Holding consortium

On a very large project it may occur that the civil engineering firms engaged on it join forces in the form of a joint venture while the steelwork contractors form a consortium. The two groups together form a so-called 'holding consortium'.

DIVISION OF RESPONSIBILITIES FOR THE SCHEME

The technical design and planning operations are carried out by the client or by his appointed representatives, more particularly architects and consulting engineers, or in certain cases by the contractor. The contractor may be entrusted with the design work in circumstances where special techniques or methods of construction have been developed by a construction firm and are to be used for the job. The relationships and division of responsibilities between the various participants are illustrated in the accompanying diagram.

Under a bill of quantities contract, the steelwork contractor is usually entrusted with the detailing of the structure and with the preparation of the workshop drawings (model b). Less frequently, this is done by structural consultants (model a).

If the contract is based on specification only, the steelwork contractor also undertakes the general design and structural calculations (model c), usually comprising the steel frame, the floors and fire protection. This type of contract is quite common with steelwork, as most firms have their own design offices which are experienced and specialise in this kind of work. Competitive tendering between different steel construction and proprietary flooring systems is thus made possible.

In the case of a contract based on a functional specification (model d) the steelwork contractor undertakes the entire design of the building together with the structural design and other technical matters. In this way complete building systems can enter into competition with one another.

Relationships governing the design and execution of structural work

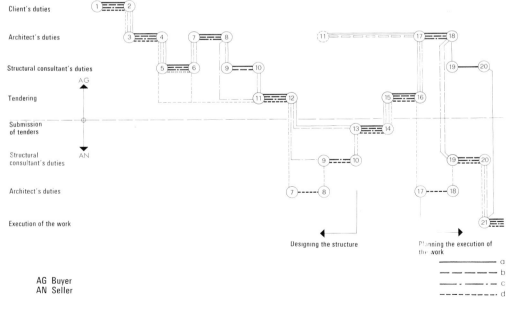

AG Buyer
AN Seller

1 Decision to build
1–2 Formulation of service requirements
3–4 Preliminary architectural design
5–6 Preliminary structural design
7–8 Architectural design
9–10 Structural design
11–12 Preparation of tender documents
13–14 Preparation of tender
14–15 Submission of tenders
15–16 Inspection of tenders
16 Awarding of contract
17–18 Construction documents
19–20 Construction documents for structural work
21 Start of construction work

339

Preliminaries

Bills of quantities • Planning execution of the scheme

In preparing the bill of quantities it should be stipulated whether the structural steelwork is to be paid for on the basis of actual weight or calculated weight. In the Federal Republic of Germany the latter procedure may be applied in accordance with the Standard DIN 18 335, in which case it should be specified what extra weights are to be allowed for structural connections. In order to simplify this often very laborious procedure for determining the quantities for payment, it is, in structural steelwork, advisable to keep the number of pricing items in the bill down to the minimum, confining these to items which comprise significant differences in the price-determining factors. In the following specimen bill of quantities separate items have been included for design, for surface protection of the steel and for erection. Alternatively, these could be lumped in with the unit prices for the 'supply' items, as is in fact usually done in bills for small jobs. The items listed in this specimen contain alternatives; in certain cases an item could be subdivided.

Specimen bill of quantities for a steel structure with floors and fire protection

1. Structural steelwork
1.1 Design
   Item 1.1.01 Structural design   lump sum....
   Item 1.1.02 Workshop drawings   lump sum....
   Note: The two items may be combined or be included in the unit prices of the 'supply' items listed below.
1.2 Supply
   Item 1.2.01 ....t columns as rolled/welded/tubular sections, grade St 37, including fittings: fabricate and deliver to site,   per t....
   Item 1.2.11 ....t main beams as rolled/welded sections/castellated beams/lattice girders, St 37, otherwise as item 1.2.01,   per t....
   Item 1.2.21 ....t secondary beams as rolled/welded sections, castellated beams/lattice girders/composite beams including shear studs, St 37, otherwise as item 1.2.01,   per t....
   Item 1.2.31 ....t vertical/horizontal bracings of rolled/welded sections, St 37, otherwise as item 1.2.01,   per t....
   Item 1.2.41 ....t portal-frame columns/beams, welded, St 37, otherwise as item 1.2.01,   per t....
   Item 1.2.51 ....t other steelwork for stairs/parts embedded in concrete/fixings for external wall claddings, etc., St 37, otherwise as item 1.2.01,   per t....
   Item 1.2.61 ....No. apertures in beam webs untrimmed/trimmed circular, dia. $\leq$ 100, 100–200, 200–300, $\geq$ 300 mm, rectangular 200 × 300, 300 × 400, 300 × 500 mm   each....
   Item 1.2.71 ....t steelwork of items 1.2.01–1.2.61 in material grade St 52, WT 37, WT 52   extra per t....
   Item 1.2.81 ....t steelwork in items 1.2.01–1.2.61 surface preparation to standard grade 2/3 and priming with one/two priming coats zinc chromate paint, thickness 40/80 μ per t....
1.3 Erection
   Item 1.3.01 Site layout and facilities comprising, for example, huts for offices and other purposes, fence around site, job notice board, access roads, water and electricity supply, lifting appliances, other construction machinery and tools: set up, keep available during the contractual construction period, and dismantle,   lump sum....
   Item 1.3.02 Keeping the site facilities of item 1.3.01 available for extension of the construction period, for which the client shall accept no liability,   per month....
   Item 1.3.03 ....t steelwork of items 1.2.01–1.2.07 to be erected,   per t or lump sum....
   Note: Item 1.3.01 may be included in item 1.3.03. Items 1.3.01 and 1.3.03 may be included in the unit prices for the 'supply' items.

2. Floor slab
   Item 2.01 ....m² in-situ concrete flooring ....cm thick, concrete class...., surface rough/smoothed, in fire-resistant construction, reinforced with ...., with/without composite action: to be constructed, including formwork,   per m²....
   Item 2.11 ....m² precast reinforced concrete slabs without/with composite action, width ..../..../...., length ..../..../.... cm thickness, concrete class ...., surface as-formed/smooth for direct application of floor covering, in fire-resistant construction, reinforced with fabric mesh: construct and lay, including in-situ filling of joints between slabs and at columns,   per m²....
   Item 2.21 ....m² steel sheet troughing, comprising galvanised troughed profile ...., sheet thickness ....mm, without/with action as horizontal diaphragm/composite action, with reinforcement/as permanent formwork: supply and lay, including concrete topping class ...., ... cm thickness over top of steel sheet, surface rough/smoothed/flat for direct application of floor covering,   per m²....
   Item 2.31 ....No. apertures in the floor slab, circular/rectangular size....,   each....
   Item 2.41 ....No. fittings: supply/fix,   each....

3. Fire protection
   Item 3.01 ....m² fire protection to the rolled/box-section/tubular columns, encased with sprayed plaster/concrete class....following the profile of the section, boxed-in/round casing consisting of ...., fire rating F 30/F 90,   per m²....
   Item 3.11 ....m² fire protection to secondary beams/main beams/bracing members encased with .... sprayed plaster/boxed-in casing with ...., fire rating F 30/F 90,   per m²....
   Item 3.21 ....m² fire protection to soffit of steel sheet troughing, consisting of .... sprayed plaster/ .... fire-resistant panel, fire rating F 30/F 90,   per m²....
   Note: Fire protection may be included in an item and be priced as a lump sum.

PLANNING EXECUTION OF SCHEME

Keeping within specified time limits, which frequently constitute essential features of building contracts, is possible only by the suitable planning of all the operations involved in the execution of the construction job. All activities from the arrival of materials to the completion of the building must be comprised in the planning. Planning for structural steelwork is fundamentally no different from that for other methods of construction, but it shares with all prefabricated techniques the need for greater precision and more strictly enforced time schedules than are needed for in-situ building methods. This is because, with prefabrication, work carried out in physically separate and distant factories and workshops has to be co-ordinated and because the structural components are often fabricated long before they are actually due to be erected on site.

For relatively small construction jobs the time schedule may be embodied in a simple bar chart. For larger projects it is necessary to use critical path analysis.

In countries with a fairly cold climate the seasons play an important part in connection with job planning. Periods of frost or snowfall may necessitate the interruption of concreting operations, make other types of work more difficult, or may necessitate cumbersome and expensive protective measures to enable work to continue.

Erection of steelwork is virtually independent of weather conditions, as is also the laying of metal flooring or installation of precast concrete units (except the operation of filling the joints with in-situ concrete or mortar) for floors, stairs, walls, etc. For this reason, there is generally no winter interruption in steelwork. The construction time can therefore be shortened by several months in comparison with cold-weather-sensitive construction methods, and this should of course be reflected in the time schedule for the job as a whole.

Planning execution of scheme

Preliminaries

Multi-storey steel-framed buildings

1 Typical time schedule chart for multi-storey steel-framed building. The foundations and the concreting operations for the basement have been completed before the onset of winter. The steel frame and other prefabricated parts of the building are erected in the winter months. By the beginning of the following winter, the roof and external walls are weatherproof and the building is heated, so that internal finishing can be carried out in the second winter.

Steelwork

2 Time schedule for the supply and erection of a steel structure. Critical times:

- Ordering the material: after this it will no longer be possible to alter the structural sections.
- Start of fabrication: two or three weeks before this date it will no longer be possible to alter the structural design without disruption of production, extra costs and delays in completion of the building.

With structural steelwork — as with all methods using prefabricated components — a certain preliminary period prior to the start of erection is needed.

The preliminary periods are:

- From the time of placing the order with the steelwork contractor to the ordering of the rolled steel (period needed for preparing the structural design to the point where the required steel sections have been determined) $\frac{1}{2}$ to 2 months
- For delivery of the rolled steel (during this period the workshop drawings are produced) $1\frac{1}{2}$ to $2\frac{1}{2}$ months
- Period from start of fabrication in the shops to delivery of the first components $\frac{1}{2}$ to 2 months
- Total preliminary period $2\frac{1}{2}$ to $6\frac{1}{2}$ months
- Average preliminary period   approx. 4 months

During this period, the excavation and foundations are carried out.
Where a large project is subdivided into a number of sections, the critical times for each section must be determined.
The period of $1\frac{1}{2}$ to $2\frac{1}{2}$ months needed for obtaining delivery of the rolled steel is due to the fact that the requisite sections are generally rolled to order. In view of the many different sections, the different lengths required and the various possible grades of steel, only a limited range of rolled material can be held in stock. The delivery periods of the steel mills are bound up with the production process. A mill rolls a certain steel section for a number of days and then the rolls on the mill are changed to produce a different section. Thus, any particular section is rolled only at intervals ranging from one to two months. For large sections and for special grades of steel, it is moreover necessary to cast separate ingots to serve as the basic material for rolling each length of beam or column ordered.
Rolled sections in current demand, in steel grade St 37-1, are generally available from stock in lengths of 12 to 15 m. For fairly small and light steel structures, it is thus possible to obtain very quick delivery if the designer confines himself to these popular sections.

Floors

In the construction of floors in a steel-framed building, there will be, depending on the method of flooring employed, a preliminary period before floor construction can start and a complementary period after completion of the erection of steelwork.

3 Floors constructed from precast concrete slabs, with or without composite action with the steelwork, require a preliminary period of about four months, that is, about as long as for the steelwork itself, while the operations of filling the joints with in-situ concrete or mortar require a follow-up period of about one month.
4 In-situ concrete floors require only a short preliminary period for design and for obtaining the reinforcement, but there is sometimes a long follow-up period, since the concreting of the floors tends, unless there is particularly good site organisation, to lag behind the erection of steelwork.
5 Flooring constructed with steel sheet troughing involves a preliminary period of two to three months. The units are then laid together with the structural steelwork. The in-situ concrete filling involves a follow-up period of about one month after the end of steelwork erection. The fire protection can, with some time-overlap, be installed during steelwork erection and requires a follow-up period of about one month.

Preliminaries

Co-ordination of general planning

In order to conform to the time schedule for the execution of the work it is essential to have the technical documents, more particularly the working drawings, always available in good time. It is therefore necessary to establish a master plan for the preparation of the drawings. Just as for the actual construction work, this necessitates estimating the time required for the designing, planning and drawing-office work. This involves a number of different specialist skills, all of which must be correctly co-ordinated and tied in with the master plan. It is important to produce at an early stage a clearcut breakdown of the various sectors of design work involved.

Example of flow of information, showing how the various specialist sectors are interlinked

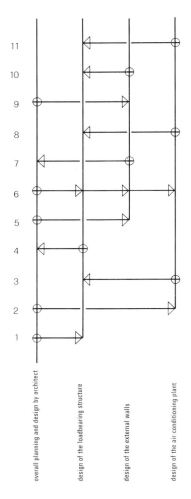

1 Basic design drawing to structural designer
2 Basic design drawing to air-conditioning engineer
3 Air-conditioning engineer states how much floor cavity space he requires
4 Structural designer proposes floor construction depth
5 Basic design drawing to cladding designer
6 Decision on floor construction depth, storey height to all specialist sectors
7 Proposal for cladding design
8 Openings in ground floor
9 Approval of cladding
10 Cladding fixing details
11 Openings in first, second and third floors

342

For some sectors of a building project it is possible, once the basic dimensions have been established, to go ahead and prepare designs and working drawings independently. On the other hand, the design of the loadbearing structural system is bound up with nearly all the other crafts and specialist skills participating in the project and therefore requires particularly careful co-ordination and timing. The construction drawings produced affect in one way or another the work of many people and many technological disciplines: structural designers, detailers, sanitary engineers, heating and air-conditioning engineers, lift design engineers, cladding designers, checkers, etc.

For the prefabricated components of a building, including more particularly the steelwork components, it is important to produce detail drawings at an early stage, to allow identical or similar components to be produced economically in substantial quantity. Such mass-produced components may be required at widely differing times during the execution of the job. In order to have them available in store for correct despatch at the appropriate time, it is desirable to produce ordering schedules (p. 347) suited to the particular method of storekeeping employed and also to apply a system of coding to all the components (as described below).

Depending on the size and complexity of the project, the number of persons participating in its execution, and the available time, the degree of master-planning, the design and drawing-office work vary in intensity, but in any case the exact times should be specified when each drawing is to be made available to each of the engineers and others who will require it in connection with their respective shares in the work. Very careful time scheduling is necessary for this.

CODING

Master-planning of the various design operations should include the coding of the individual components and their drawings in a well-conceived coding system ultimately comprising all the crafts and specialist skills involved in the project. For the coding of the drawings the following categories of drawings are to be distinguished:

1. The architect's layout drawings to 1:500, 1:200, 1:100 and 1:50 scales, giving all the important dimensions of the building and indicating the interconnection and interaction of all its parts.
2. The layout drawings of the various specialist sectors also contain the principal dimensions and indicate the types and locations of the components employed. They serve for co-ordination with all other specialist sectors and are required for the erection of the components shown.
3. The detail drawings of the components contain all detailed information and are used for fabrication and manufacturing purposes. In addition to precise dimensions and tolerances, they contain information on the materials used and instructions for the fabricating shop. They are accompanied by parts lists giving precise particulars of materials and quantities.

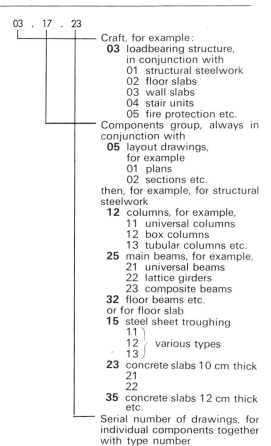

For the loadbearing system of a steel-framed building the following simple coding system for drawings and components has proved very suitable:

If there are not enough subdivisions, each of the three number groups can be given a third or a fourth position. In this way even a very large construction project can be conveniently and unambiguously coded. The formation of series of individual components is facilitated in that, for all those components with identical characteristics, the first two number groups are identical, regardless of the position which the component occupies in the structure.

This coding is, however, not sufficient for erection purposes and therefore for despatch and possibly for storage of the components in the fabricating yard. For large projects, it is therefore necessary to introduce a second coding system which gives information on the position occupied by the component in the structure (erection coding) and which ties up with the coding system for the drawings. From the second group of figures a number is adopted to characterise the components group, for example, 01.13.23 means a tubular steel column of type 23.

Coding

Preliminaries

The erection coding might be presented as follows:

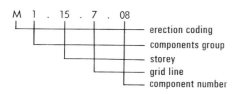

- erection coding
- components group
- storey
- grid line
- component number

The tubular column of type 23 is therefore installed in the 15th storey on grid line 7 as the 8th column.
Both coding systems are applied to the layout drawings and to the components themselves. Examples relating to the layout drawings for main and secondary beams are given in **1** and for floor slabs in **2** for a steel-framed building.
The two systems are interlinked by components lists which give information on the position of any particular type of component and also what type is required at any particular location in the structure.
The coding system should be capable of incorporation into a data processing programme. However, care should be taken to ensure that the numbering system does not become unwieldy because of too many interconnections.

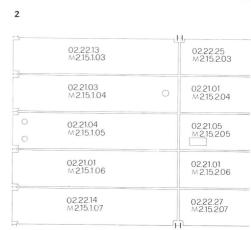

PAYMENT SCHEME

The timing of the payment received by the contractor from the client affects the price of the job. It involves an interest gain for one party and an interest loss to the other; the amount concerned may well be a substantial proportion of the total construction cost. It is therefore necessary to stipulate the dates for payment in the building contract. In most cases these are linked to the contractor's performance in terms of progress. The conditions of payment vary between jobs which involve prefabrication and those which do not.

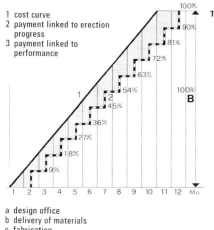

1 cost curve
2 payment linked to erection progress
3 payment linked to performance

a design office
b delivery of materials
c fabrication
d erection

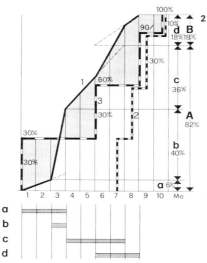

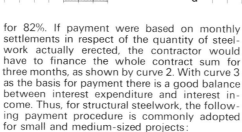

A costs off site
B costs on site

1 For buildings not involving prefabrication it is normal practice to make monthly interim payments, according to the progress of the job, with 10% deducted as guarantee. The diagram shows the relationships, in the form of summation curves, for costs (1) and payments (2). The shaded area between the two curves represents the proportion of the costs for which the contractor has to make provision.

2 If there is prefabrication, a substantial proportion of the cost is incurred, not on the actual building site, but in fabricating shops or other places of manufacture. These off-site activities also require interim payments. The diagram represents a payment scheme for the steelwork of a medium-sized building (about 600 tons of steel): in this case 18% of the construction cost is absorbed by work on the site, while the work in the contractor's fabricating shop accounts for 82%. If payment were based on monthly settlements in respect of the quantity of steelwork actually erected, the contractor would have to finance the whole contract sum for three months, as shown by curve 2. With curve 3 as the basis for payment there is a good balance between interest expenditure and interest income. Thus, for structural steelwork, the following payment procedure is commonly adopted for small and medium-sized projects:

30% on awarding the contract;
30% during fabrication or, in the case of small structures, at the commencement of erection;
30% during erection in instalments or, in the case of small structures, on completion of erection.

3 These relationships are even more clearly demonstrated in the case of a project involving a long prefabrication and a short erection period (as is commonly encountered in prefabricated building construction). The time schedule and payment scheme for a large steel-framed building shows, for example, that the preliminary work off the site takes eleven months, followed by only three months' site work. The material is delivered in four instalments, so that the cost curve presents a stepped shape. If curve 2 were adopted, the contractor would have to finance the job almost entirely. For this reason, it is usual, on major projects, to agree that payment be effected in fixed interim amounts distributed over the entire duration of the contract period, as shown by curve 3.

Fabrication

Organisation of fabrication

The time schedules for the design work and for the execution of the construction work, respectively, are linked by the planned timing of the fabrication operations for the steelwork and the manufacture of the precast concrete components. The control of these production operations is an internal matter for the steelwork contractor and his sub-contractors. However, it involves the designer of the building in that he must ensure that the necessary drawings are available to the contractor in good time for the latter to adhere to his production programme. It is therefore important that the designer should have an adequate understanding of production control in both the steelwork and concrete industries.

The fabrication time required obviously depends on the capacity of the fabrication shops. A production unit (PU) can fabricate a certain quantity per unit time (for example, an eight-hour working shift). A production unit may be one work-place in a workshop, a drilling machine, a transfer machine or a concrete precasting mould. The daily output can be stepped up by increasing the number of production units or by increasing the number of working shifts per unit.

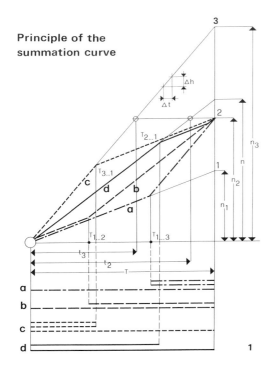

Principle of the summation curve

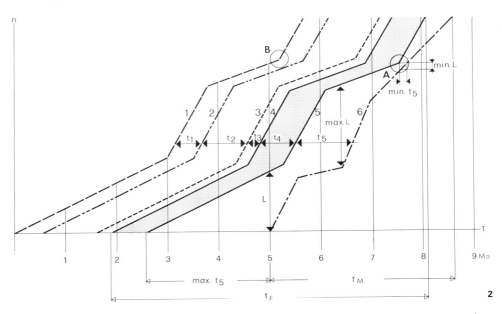

PU	Production unit/shift
Δt	Time interval, for example, 1 hour, 1 shift, 1 day, 1 week
Δh	Number of components produced in time interval Δt
n_1	Number of components that can be produced with 1 PU in time T
$n_2 = 2n_1$	Number of components that can be produced with 2 PU in time T
$n_3 = 3n_1$	Number of components that can be produced with 3 PU in time T
n	Number of components needed in time T
t_3	Requisite production time for n with 3 PU
t_2	Requisite production time for n with 2 PU
$T_{1..2}$	Latest time for change-over from 1 to 2 PU
$T_{1..3}$	Latest time for change-over from 1 to 3 PU
$T_{3..1}$	Earliest time for change-over from 3 to 1 PU
$T_{2..1}$	Earliest time for change-over from 2 to 1 PU

1 to 6 Summation curves

1 Points of time at which the design must have been decided
2 Ordering the materials
3 Drawings received by the production planning department, together with arrival of necessary materials
4 Start of fabrication
5 End of fabrication, components put into storage
6 Erection, components taken from storage

t_1	Time needed for structural calculations
t_2	Designing and detailing, also delivery time for materials
t_3	Preparation by production planning department
t_4	Fabrication of components
max. t_5	Longest storage period
min. t_5	Shortest storage period
max. L	Largest quantity in storage
min. L	Smallest quantity in storage
t_F	Total fabrication time
t_M	Total erection time

1 The summation curve provides a clear and convenient method of controlling the necessary capacity. With one PU it is possible to produce n_1 components in the available time T. Actually, n components are required. If two or three PU are used, the time required is $t_2 < T$ or $t_3 < T$, respectively. If it is intended to use one PU during the whole production period and to add a second and possibly a third PU during a portion of that period, it is necessary to know the earliest or the latest possible points of time for effecting the change-over. These times can be read from the diagram.

2 The fabrication operations are governed by summation curve 6 for delivery of components for erection (= latest completion time for fabrication) and summation curve 3 which determines the arrival of the drawings in the production planning department and also the arrival of the materials for fabrication (= earliest start of fabrication). The period t_3 in between is available for preparatory work in the production planning department, while t_4 is available for fabricating the components. With this method of representation it is possible to detect bottlenecks at an early stage — such as that at A, due to the late decision on design at B (curve 1) and resulting temporarily in less fabrication.

The horizontal distances between the curves represent the lengths of time occupied by the activities concerned. The vertical distance from curve 5 (end of fabrication = transfer to storage) indicates the quantity L put into storage. It appears that considerable storage space is likely to be required in the fifth and the seventh month.

The diagram may be extended by the addition of curves for structural calculations (1) and for detailing and material ordering (2), as well as other curves, if desired.

Process of fabrication of structural steelwork

A steel fabricating works comprises the following departments:

- yard for storage of rolled steel sections, plates, etc;
- surface preparation of the steelwork;
- fabrication of components;
- assembly and connection of the members into units for erection;
- storage of the finished products.

Stockyard

Only relatively small quantities of popular steel sections and plates are normally held in stock. Most of the material required for a given job is ordered directly from the steel mills or from stockholders. Overhead travelling cranes, often provided with lifting magnets, and various other mobile units are used for off-loading and picking up the material.

Surface preparation

Mill-scale and rust are removed by blast-cleaning, generally before fabrication of individual members, by means of automatic equipment, the abrasive employed usually being steel shot or grit. The abrasive is either projected by compressed air or is thrown by high-speed impeller wheels. For various standards of rust removal that may be specified, see p. 358. Immediately after blast-cleaning, the steel should be given a thin coat of blast primer (which permits welding), and subsequently, after fabrication, one or two priming coats of paint.

Fabrication of individual members

This comprises the cutting and drilling of the material. The actual operations involved are different for sections, flat products and bars. (For definitions, see p. 355.)
Members consisting of rolled steel sections are usually cut to the desired length by sawing, **1**. Modern steel fabricating shops are equipped with semi-automatic or fully-automatic sawing machines. Inclined cuts, **2** and **3**, are also sometimes sawn, but are more often executed by flame-cutting. Special cuts are always formed by flame-cutting, as in **4, 5** and **6**; so are holes in webs, etc., **7**. It should be noted that the flame cuts should not lie within the fillets, **8**. Corners of rectangular holes and re-entrant cuts should be drilled in advance in order to obviate cracking due to notch effects that are otherwise liable to occur here.
Smaller circular holes in large rolled steel members are always drilled in their marked positions. In modern fabricating shops the drilling operations are fully automated and performed on transfer machines.
Members consisting of smaller sections are cut to length by sawing, flame-cutting or shearing. Further cutting to shape may also be done in some cases by shearing, otherwise by flame-cutting. Holes are drilled or, in thin-walled sections, punched.
Flat products can be cropped to length in a guillotine or can be flame-cut.
A multiple-head straight-line slitting machine, operating with oxygen-gas burners, can produce a number of parallel cuts simultaneously, **9**, or can, with cam-guided control, produce curved cuts. For weld preparation, a double bevel cut and root face cut can be made simultaneously with a three-burner head, **10**. Small parts may be cut from steel plate either manually or with semi-automatic or fully automatic flame-cutting machines controlled by templates or drawings or by electronic co-ordinate control.
After cutting and other preparatory operations, the members are transported singly or stacked on pallets – depending on the individual weight and size of the members – to the assembly shop.

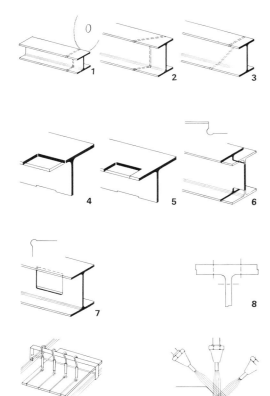

Assembly

The parts to be assembled are secured by bolting or tack welding. For components produced in quantity, the amount of labour can be reduced by means of assembly jigs, whereby the accuracy of fabrication is moreover increased. Shop connections are mostly welded, site connections are usually bolted, for speed and convenience (p. 356).

Welding

Welding is carried out manually or by mechanical methods. Plate girders and box girders are usually fabricated on semi-automatic welding lines which produce two or four welds simultaneously. The heat applied during welding may cause distortion of the steel members, in which case they must be straightened. Some components have to be fabricated to a specified curvature, for example, cambered girders. Straightening is done in the cold state on a straightening press or with the aid of heating. Shrinkage effects can often be conveniently utilised for straightening.

Special machines

The equipment of many fabricating shops also comprises such machines as planers, milling machines, folding presses, forming machines for sheet steel, etc.

Mechanical handling

The internal handling and transport facilities within the fabricating shop have a considerable effect on the economy of its operations. Modern works are equipped with roller tables and transfer skids for moving heavy components. Small components are stacked on pallets which are handled by overhead travelling cranes or stacker trucks; large ones are lifted and moved by cranes, often equipped with lifting beams with magnets attached to avoid the need for hooks and slings. Rail-mounted trucks are used for transport between one building and another.

Storage

The finished components are kept in a storage yard until despatch to the construction site. A large yard requires carefully planned systematic operating procedures for efficient storage and retrieval of components.

Fabrication

Layout of an ideal fully mechanised steel fabricating shop

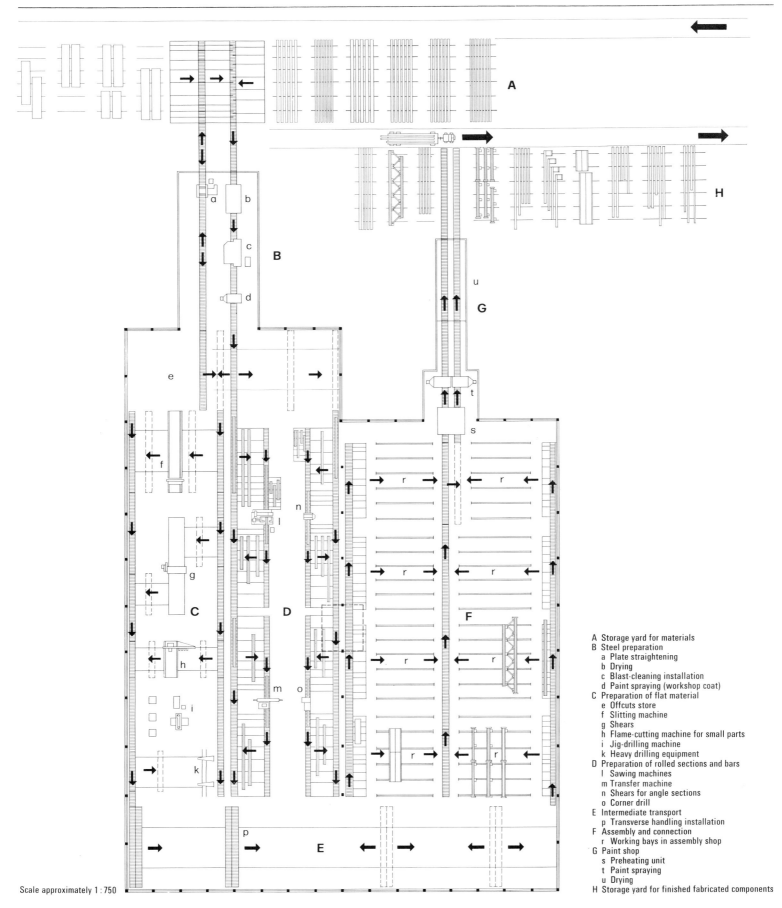

A Storage yard for materials
B Steel preparation
  a Plate straightening
  b Drying
  c Blast-cleaning installation
  d Paint spraying (workshop coat)
C Preparation of flat material
  e Offcuts store
  f Slitting machine
  g Shears
  h Flame-cutting machine for small parts
  i Jig-drilling machine
  k Heavy drilling equipment
D Preparation of rolled sections and bars
  l Sawing machines
  m Transfer machine
  n Shears for angle sections
  o Corner drill
E Intermediate transport
  p Transverse handling installation
F Assembly and connection
  r Working bays in assembly shop
G Paint shop
  s Preheating unit
  t Paint spraying
  u Drying
H Storage yard for finished fabricated components

Scale approximately 1 : 750

Planning of erection

Erection

In planning the execution of a construction project for a multi-storey steel-framed building it is important to understand the erection processes and their timing in order to dovetail them into the other building operations. The very speed of steelwork erection makes careful planning necessary. The effects of seasonal and weather conditions are mentioned on p. 340.

Steelwork erection may be combined with the erection of other prefabricated components (such as precast concrete floor slabs, steel floor troughing, cladding, staircase and wall components), for which the same lifting appliances are used. In addition, these appliances can be used for lifting various items of internal fittings and finishings (partitions, floor-covering materials, pipes, ducts, etc.) to the required floors.

Steelwork erection is preceded by the construction of the foundations and of those structural concrete features serving to stiffen the building, more particularly cores. If the construction time available is short, it may be advisable to use precast concrete as far as possible, for example, for basement columns and floors, and to use steel bracing and other stiffening systems in lieu of concrete cores.

To ensure smooth and speedy erection it is essential that adequate all-weather access roads be made available for transport vehicles, as well as tracks and hardstandings for mobile cranes (p. 351).

Erection operations:

- off-loading of structural components delivered to the site;
- lifting of components to the position in which they have to be installed;
- securing of components with temporary connections;
- plumbing and levelling the structure;
- establishment of permanent structural connections between the components by bolting or welding or with in-situ concrete or mortar between precast concrete components;
- erection and dismantling of scaffolding and props for giving temporary support to components.

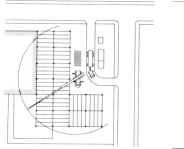
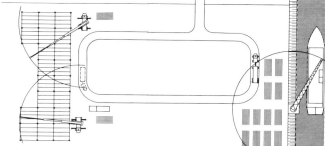

1 Erection without intermediate storage of components on site

Steel frames can be erected by lifting the components directly from the delivery vehicle. On cramped sites, say, in city centres, this may be essential. Good organisation of deliveries of components in accordance with an accurately planned time schedule is essential for this. To compensate for possible delays in delivery, for example, due to traffic conditions on the roads, a small stockpile of steelwork components is kept within reach of the erection crane. Large steel units and precast concrete slabs are erected directly from the vehicle.

2 Erection with intermediate storage

The setting-up of an intermediate storage yard on the construction site takes up much space, besides requiring a special crane for off-loading the components from the vehicles and subsequently reclaiming them from the store. Also, vehicles for on-site transport to the actual place of erection are needed. Despite the extra cost involved, this procedure is adopted where deliveries of steel components to the site cannot be accurately timed or where large quantities are delivered at one and the same time, for example, by barge.

Erection time

The speed of erection depends on the number of cranes employed on the job and on the number of lifting cycles that these cranes perform per working shift. The nature and accessibility of the site determines how many cranes can effectively be used.

The number of lifting cycles attainable per shift depends on the lifting height, the accessibility of the parts of the structure concerned, the relative ease or awkwardness of slinging the loads, and the weather conditions.

In a shift of between 8 and 10 hours a crane can perform between 10 and 20 lifting cycles under difficult conditions; under ordinary conditions this number is generally between 20 and 40. On average, with single-shift operation and one crane, between 200 and 300 tonnes of steelwork, including the associated flooring units, can be erected per month.

The time taken up by one lifting cycle comprises the time taken for slinging the component, that is, attaching it to the lifting hook of the crane, the actual lifting and slewing time, the time taken to set down the load and detach it from the hook, and the time for lowering the hook for the next load.

In circumstances where lifting times tend to be long, a number of smaller components may be handled simultaneously — bundled or palleted — and placed temporarily on an upper floor within easy reach of the actual point of erection. The crane then handles the components one by one to erect them. Also, individual components may be assembled into larger units at ground level with the aid of special lifting appliances, such units then being lifted into position by the erection cranes. For examples, see pp. 348–351.

Erection planning

Once the sequence of erection has been developed (pp. 348–349), the job can be split up into a number of erection operations. The working positions, radii and heights of the cranes are determined for each operation and the number of lifting cycles required is estimated. With structural steelwork it is thus possible to achieve precise timing of all the operations, to establish accurate time schedules for the quantities of materials required at specific stages of the job and to predetermine the exact sequence of erecting the various components.

Also, schedules for the despatch of components from the fabricators' storage yards can be prepared in good time, and furthermore — if the load capacity of the transport vehicles to be employed is known — it is possible to draw up haulage schemes allowing the fabrication and storage of the components to be carefully programmed and controlled (p. 344). The erection coding of the components, as described on p. 343, should preferably be established at an early stage (as soon as the layout drawings are available), even before the components have been fabricated and have been assigned their type numbers. With such detailed planning of fabrication, storage and erection, it is possible to achieve the shortest possible erection times and thus reduce costs.

3 Aids for adjusting columns

a Axes on steel plates
b Slab base
c Nicks on the slab base
d Chisel marks on the end plates
e Punch mark for the plumb line
f Packing plates for adjusting height

Setting out

To be able to comply exactly with the relevant erection tolerances, it is necessary to be able to make precise measurements, both before and during erection, based on a system of triangulation and levelling.

Care taken during the erection of the columns will facilitate the levelling of the floors which should be checked again before the final bolting of the connections. In high-rise buildings optical instruments can be employed with greater accuracy than plumb lines. When the deflection of composite floors is prevented by propping during the concreting of the slabs, the height of the props must be checked by levelling (p. 257).

Erection | Methods of erection

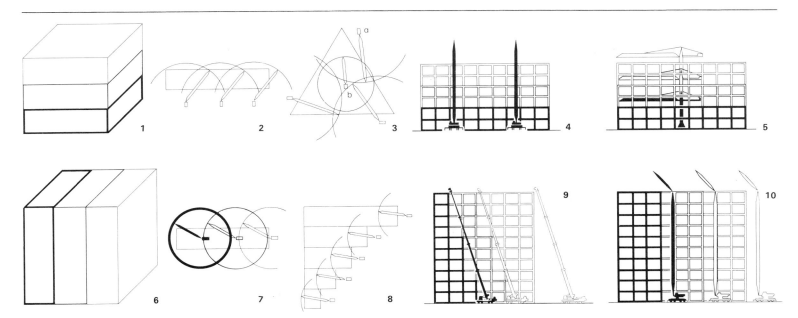

A steel-framed building may be erected storey by storey, **1**, or bay by bay, **6**. Frequently, a combination of the two methods provides the best solution.

Storey-by-storey erection

With this method, each storey is completely erected, preferably including the floors if these are of prefabricated construction, before work on the next storey begins. This method is always used for tower blocks. In the case of slab blocks, **2**, or buildings covering a relatively large area, **3**, the crane must perform time-consuming lateral movements for the erection of each storey or, alternatively, several cranes must be used in order to bring every part of the area of the building within reach of a crane. Truck-mounted mobile cranes and tower cranes which travel on tracks have to stand outside the building, **2**, **3a**, **4**. Fixed tower cranes can stand inside the building, **3b**, **5**, and rise with it. Storey-by-storey erection has the advantage that the internal finishing operations such as the laying of floor surfacings, services, etc., can start in all parts of a storey as soon as its erection has been completed. Fittings which are liable to suffer from exposure to moisture cannot be installed until the floor over the storey concerned has been laid and rendered waterproof (and the ingress of rain through staircase and liftshafts has also been prevented).

Bay-by-bay erection

A bay or a number of bays is erected entirely, from ground floor to roof, by means of the crane standing in one working position, **6**. The crane may stand beside the building, **2**, or on the centre-line, **7**, and, after the erection of each bay, it is moved along the appropriate distance for the erection of the next bay. The procedure commonly adopted for the erection of buildings with a large area on plan is indicated in **8**. A stepped variant of bay-by-bay erection may be adopted when truck-mounted cranes are used, **9**. On the other hand, a crane with a vertical mast can erect a complete bay from one position, **10**. The roof can be completed over each bay, so that finishing and fitting-up work can begin in floors below.

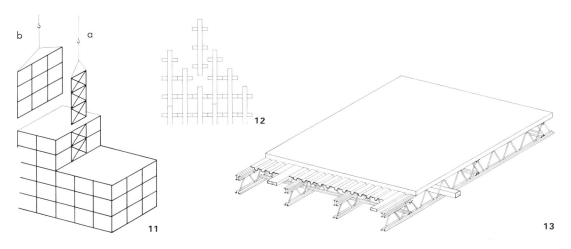
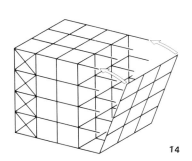

Pre-assembly

Pre-assembly of components into larger units can speed up the actual work of erection. If the resulting units are of transportable size, the pre-assembly operations can be carried out in the fabricating shop; otherwise they can be done on site in a special workshop or in the open, within reach of the erection crane. The following parts of a multi-storey frame can be suitably pre-assembled:

- vertical bracings, **11a**;
- assemblies of columns and main beams, **11b**;
- external columns with beam stubs attached, as exemplified by the façade assembly units for the Standard Oil Building in Chicago, **12** (see also p. 212, **17**);
- floor bays, as exemplified by the assembly comprising lattice girders and troughed decking used in the World Trade Center, New York, **13**;
- beam-and-column assemblies corresponding to the full height and/or width of the building, pre-assembled on the ground and swung up to the vertical position, **14**; this method has been used more particularly in housing construction in France. See also pp. 80–81.

Methods of erection

Erection

Erection of suspended buildings

In the case of a suspended building, erection starts with the internal core. Next, the top cantilever girders are installed. The floors of the building are then hoisted, starting at the top. It is necessary to take into account that, as the dead load increases with each additional storey erected, the core will undergo compressive shortening, while the suspension members will stretch. There is thus a certain amount of differential movement between the inner and the outer end of each floor beam extending from the core to the perimeter.

15 The beams and floor slabs or troughing units are lifted and fastened either individually or as pre-assemblies.

16 The whole floor is completely assembled on the ground and lifted as a single unit.

Erection of tower blocks

The cranes are equipped with high-speed winches to reduce the lifting times. Extendable hoists for the transport of men and small items of material should closely follow the upward progress of erection. It is important to install the permanent stairs along with the framework and floors, as they make for speedier and safer communication between floors than is possible with ladders.

17 Tower crane, stationary or mobile, standing beside the building.

18 Tower crane installed within the building, usually in a liftshaft.

19 Cranes attached to the concrete core, constructed in advance with sliding formwork.

20 Climbing cranes attached to the structural steelwork.

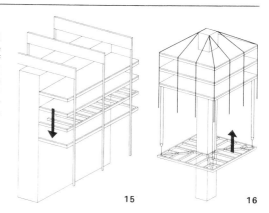

15   16

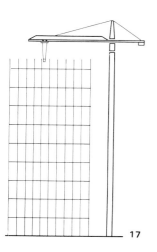 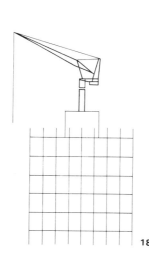 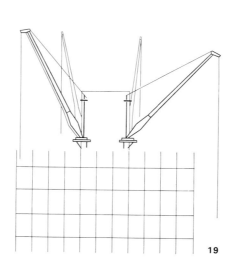 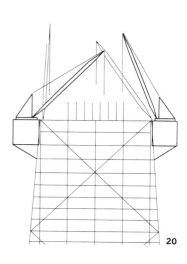

17   18   19   20

Lift-slab process

The upper slabs of the building are constructed at ground level within the plan area of the building itself and are lifted up along the columns (which pass through suitable holes formed in the floor structure). The lifting appliances are mounted on top of the columns and are hydraulically controlled to synchronise with one another. As a rule, jacks are used, the lifting operation thus being performed in a succession of relatively short strokes permitting accurate control to ensure that the floor remains horizontal. The jacks exert forces of around 70 tonnes each. During the lifting of the floors, the structure must be carefully braced to ensure its lateral stability. See p. 190 **1**.

21 Lifting jack extended to its full length of travel: the load hangs by the suspension rods from the upper transverse member; lower gripping device is released.

22 Jack is closed: load now hangs from the lower transverse member; upper gripping device is released.

23 Floor lifted into position is secured to the column by means of inclined struts.

Reference: Nussbaumer/Lindorfer, Lift-Slab-verfahren (Lift-slab process). *Der Bauingenieur*, 46/19, p. 122.
Hubdeckenverfahren Hochtief AG, 3/1972, p. 37.

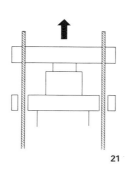 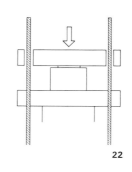 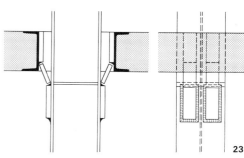

21   22   23

Cranes and derricks — Erection

The lifting appliances for steelwork erection raise the loads vertically and also move them horizontally. They may be installed in one or more fixed positions or may be mobile on a rail track or a pneumatic-tyred truck. They may be adjustable for height or be extendable to keep pace with the increasing height of the building, or may be designed to climb as erection proceeds. The last-mentioned type of appliance is known as a climbing crane; it is attached to some suitable part of the structure and is raised with it. The length of time for which a crane is needed on a job depends on its period of use and on the time needed for transporting it to and from the site, setting it up and dismantling it. The time absorbed by these subsidiary operations varies with the weight of the crane; they must be in some reasonable proportion to the actual working time if economical use of the crane is to be achieved.

Erection cranes should be capable of rapid and precise movements, to keep the lifting cycles as short as possible and to facilitate erection of the components. The magnitude of the load that a crane can lift varies with the radius of operation, being determined by the permissible moment ($=$ load $\times$ radius). With most cranes, erection can be carried out at wind forces up to 5 or 6.

Climbing cranes

1 Climbing crane secured to a column of the structural frame. The crane can raise itself by alternately releasing and clamping the attachments. The columns are provided with special attachment points for the crane.

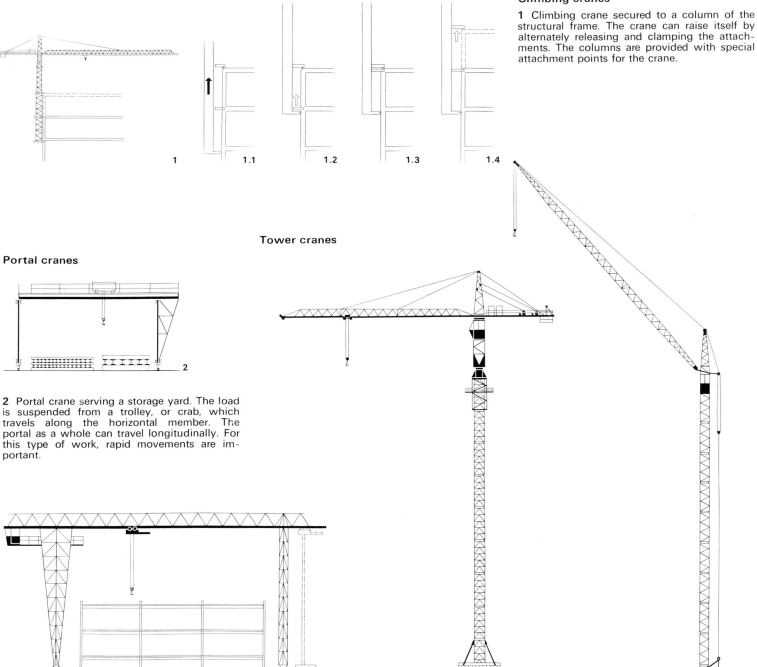

Portal cranes

2 Portal crane serving a storage yard. The load is suspended from a trolley, or crab, which travels along the horizontal member. The portal as a whole can travel longitudinally. For this type of work, rapid movements are important.

Tower cranes

3 Goliath crane for the erection of slab blocks. Erection can be done storey by storey or bay by bay.

4 Tower crane with horizontal jib with travelling crab. The height of the crane can be increased by the addition of further units to extend the mast.

5 Tower crane with luffing jib for steelwork erection. This particular type has a load capacity of 4.3 t, a load moment of 80/100 tm, and a jib length of 40 m. Stationary or tail-mounted mobile versions.

Cranes and derricks — Erection

Derricks

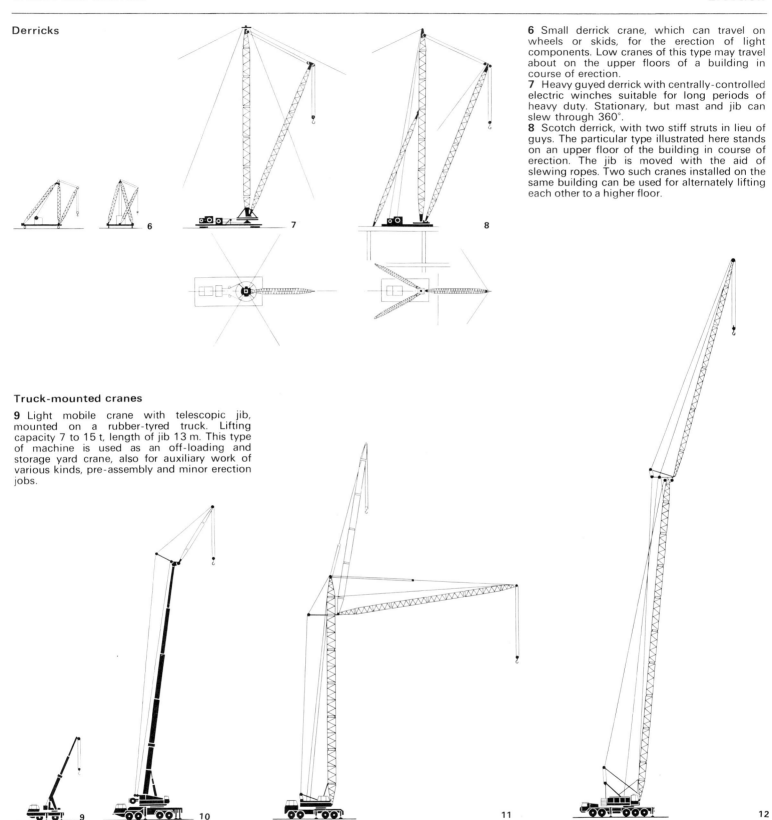

6 Small derrick crane, which can travel on wheels or skids, for the erection of light components. Low cranes of this type may travel about on the upper floors of a building in course of erection.

7 Heavy guyed derrick with centrally-controlled electric winches suitable for long periods of heavy duty. Stationary, but mast and jib can slew through 360°.

8 Scotch derrick, with two stiff struts in lieu of guys. The particular type illustrated here stands on an upper floor of the building in course of erection. The jib is moved with the aid of slewing ropes. Two such cranes installed on the same building can be used for alternately lifting each other to a higher floor.

Truck-mounted cranes

9 Light mobile crane with telescopic jib, mounted on a rubber-tyred truck. Lifting capacity 7 to 15 t, length of jib 13 m. This type of machine is used as an off-loading and storage yard crane, also for auxiliary work of various kinds, pre-assembly and minor erection jobs.

10 Medium-heavy mobile crane with telescopic jib or lattice jib, 40 m long, 40 to 80 t lifting capacity. Suitable as an erection crane for buildings up to about 15 storeys high.

11 Mobile crane with lattice tower and jib for the erection of tall buildings. See also p. 348, **9**, **10**.

12 Heavy mobile crane, 100 to 500 t capacity, jib up to 160 m long, for heavy loads (even at large radii of operation) in the erection of high-rise buildings.

Erection — Site organisation • Scaffolding • Safety precautions

Site organisation

The erection of steelwork and the space-enclosing components erected along with it requires relatively little space on the construction site. The site arrangements comprise:

Fencing

The fence enclosing the site, and the contractors' sign, should conform with local requirements and regulations.

Access roads

The roads giving access from the public highway to the off-loading bay on the site should be paved and have a width of between 3 and 4 m, so that they can be used by heavy transport vehicles under all weather conditions. Reliable roads are an important prerequisite for steelwork erection. They should be properly signposted so that drivers unfamiliar with the site can find their way on them (even when the roads are muddy or covered with snow). Turning areas should be provided. Any pipelines or other services already laid on the site should be suitably bridged where they are crossed by roads.

Cranes

Cranes move about on paved surfaces (truck-mounted mobile cranes) or on rails (tower cranes). If a crane has to stand on a floor over a basement, it should be checked whether the floor is strong enough or needs to be propped. Quite often it is cheaper to design the floor to carry the crane load than to install props. In other cases it may be better to stand the crane in the basement and to construct the floor over the basement in the same manner as the upper floors of the building, that is, supported on steel beams, which are erected by the crane.

Storage yard

An intermediate storage yard for the temporary holding of steelwork prior to erection should be located close to the actual place of erection. The amount of storage space required, including the access roads, is between 3 and 4 m^2/t when components are laid individually on the ground, and between 1 and 2 m^2/t for stacked components. Paved roads will be needed for transporting the steelwork from the yard to the place of erection.

Sheds and huts

For jobs of short duration the site offices, canteen facilities, lavatories and washing facilities may be accommodated in caravan trailers. On sites where work is to continue for a fairly long period, demountable huts or even permanent buildings may have to be installed. Steelwork erection does not require a large labour force, so that these sheds and huts do not take up much space. Per crane and per shift there will be between 10 and 12 men, plus one to three persons for supervisory duties. Some small sheds for tools and for bolts and nuts should be provided near the men's huts.

Electricity, water

Erection cranes and extensive welding operations absorb heavy currents. If self-propelled mobile cranes are used for erection, it is usually sufficient to lay on an electricity supply for lighting and manual power tools only. Water is needed only for washing and sanitary use.

Scaffolding

Scaffolding constitutes an important part of the site facilities. A distinction may be drawn between scaffolding that performs mainly a supporting function, scaffolding to serve as working platforms for the steelwork erectors, and safety scaffolding for the protection of people from falling objects.

Scaffolding for temporary support

Scaffolding which performs a supporting function is used to relieve certain structural components of load until they are ready to perform their loadbearing duty unaided. The steel or timber propping sometimes inserted under composite beams until the concrete slab has hardened comes within this class of temporary support. Adjustable props may also be used for the correction of levels, etc.

Workmen's scaffolding

This type of scaffolding provides access for the erectors to the joints where the structural connections have to be formed. It comprises:

- walkways laid over the beams in the interior of the building;
- ladders and temporary stairs;
- hoists for light loads and persons.

If the floors in a steel-framed building are constructed from precast concrete slabs or troughed steel decking and are erected at the same time as the steelwork, walkways across the floor beams are not necessary, accessibility is much improved and erection as a whole is speeded up. The same considerations apply to staircases which are installed along with the steelwork.
In steel-framed buildings, tubular scaffolding is not used for steelwork erection, but for fixing the wall cladding. All scaffolding should be reliable and be provided with handrails.

Safety scaffolding

- This scaffolding is provided to protect workmen undertaking the following trades working below those parts of the structure where erection operations are in progress. It is unnecessary in a building of which the floors are installed at the time of steelwork erection.
- Safety scaffolding may also be needed as protection for passers-by on public highways adjacent to a construction site.

Protection from winter weather conditions

Such measures may be necessary on steel-framed buildings when concreting operations have to be carried out, such as the construction of in-situ concrete floor slabs, filling of joints between precast slabs, filling of steel decking, floor surfacings, sprayed casings for fire protection of steelwork. Since these operations do not affect the steelwork erection itself, however, they can be postponed until warmer weather arrives. Otherwise the various storeys of the building will have to be closed in and artificially heated. To enclose a steel framework in a temporary cladding is not very expensive, as the frame itself forms the support, so that only the protective sheeting has to be provided.

Safety precautions

Steelwork erection often has to be carried out at great heights and in places difficult of access. Objects may fall and cause accidents. For these reasons there are strict safety regulations to be complied with, issued by the building inspection and/or other authorities such as waterways and electricity boards. The principal general precautions in connection with steelwork are:

Safeguarding of access roads, scaffolding, etc.

Proper safeguarding of the access ways to the construction site and of the arrangements for men to gain access to the various parts of the structure itself requires careful advance planning and constant maintenance. Ladderways and scaffolding should be strongly and safely constructed in accordance with regulations and be so secured that no parts thereof can shift during use. Ladderways should be provided with handrails. All accessible staircases and floors should likewise have railings. If erection of the wall cladding follows closely upon steelwork erection, it is sufficient, during the short interim periods, to mark the edges of floors with strings to which small flags are attached. Openings in floors should be covered. As the access ways within the building are to be used also by workmen for the subsequent finishing and fitting-up operations and will often have to be altered in the course thereof, a special gang appointed to supervise and maintain the access ways is necessary on a large site.

Safeguarding of loads

The presence of persons directly under suspended loads and under erection operations is dangerous and should be prohibited. Construction operations which precede or follow the steelwork erection should be so co-ordinated as to leave adequate safety distances. Otherwise safety scaffolding must be provided.
There are special regulations for the safe slinging of loads to be handled by cranes and for the testing of ropes, chains, hooks and other lifting tackle.

Fire protection

Sparks from welding operations present a fire hazard. Suitable extinguishing equipment should be available in readiness.

Personal protection

Personal protective equipment on construction sites includes:

- helmets,
- safety boots with steel toe-caps,
- safety belts for steelwork erectors,
- goggles to be worn during grinding operations,
- goggles to be worn by welders.

Emergency measures

A copy of the appropriate accident prevention regulations should be available for inspection on the construction site. Equipment for use in emergencies, such as stretchers, first-aid kits, etc., should likewise be available and be distinguishable by clearly visible markings and inscriptions. Notice-boards giving the addresses and telephone numbers of local doctors, hospitals, first-aid posts, fire brigade and police station should be displayed on every construction site.

Steel

Steel
- General structural steels 353
- Special steels 354
- Structural steel products 355
- Methods of connection 356

Corrosion protection
- Surface preparation, priming coats, finishing coats 358
- Design, maintenance, costs 359

STEEL

Steels are classified, according to their chemical composition, as mild, low-alloy and high-alloy steels. All steels, including unalloyed steels, contain admixtures of elements other than iron which are present in accurately specified quantities and are major factors in determining the properties of the steels concerned. In particular, the carbon content is important; it is this that distinguishes steel from pig iron.

Manufacture of steel

The relationship between the chemical composition of steel, the technology of its manufacture and its working properties have been the subject of very considerable research over a long period. The amounts of alloying additions can be controlled within close limits, so that the steelworks can guarantee a high degree of uniformity in the composition and properties of the steel they supply.

Structural steels

For building purposes the 'general structural steels' complying with German Standard DIN 17 100 are of greatest importance. The analogous British Standard is BS 4360: *Weldable structural steels*.

The 'general structural steels' comprise the St 37 mild steel (designated as Fe 37 in the Euronorm Standards) and the St 52 low-alloy steel (Euronorm Fe 52). The corresponding British steels are respectively of grades 43 and 50. In addition, weathering steels, stainless steels and high-tensile steels are used in building construction (pp. 354, 355).

Physical properties

The steels used in structural engineering all have almost identical physical properties:

specific gravity $\gamma = 7.85$ kg/dm^3
coefficient of thermal expansion $\alpha_t = 12 \times 10^{-6}/°C$
modulus of elasticity $E = 2.1 \times 10^6$ kg/cm^2

Rolled steel products

Structural steel is supplied in the form of hot-rolled products comprising sectional steel (structural sections, bars) and flat products (flat bars, flats, strip and plates): see p. 355. Some products are subjected to further processing in the cold state, as in the manufacture of cold-rolled strip and certain kinds of wire. In structural engineering, castings and forged steel are also used for components such as bearings.

General structural steels

The users of structural steels should possess adequate knowledge of their mechanical and technological properties.

Mechanical properties

The stress–strain diagram provides the most important information on the mechanical behaviour of steel. It is obtained by plotting the stress against the strain in the tensile test.

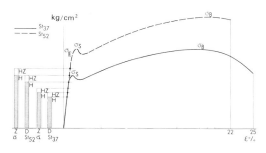

D Compressive forces H Main loads
Z Tensile forces HZ Main and additional loads

- Elastic range: Inspection of the stress–strain diagram of a structural steel shows that up to a certain limit – the elastic limit σ_E – the strains ε are very small and proportional to the stress σ. In this range of stresses the modulus of elasticity $E = \sigma/\varepsilon$ is constant. Up to the elastic limit the deformations are elastic, that is, the strains disappear when the load is removed: the test specimen thus returns to its original length. When the stress is increased above the elastic limit, a change in behaviour occurs: the specimen now undergoes a certain amount of permanent (plastic) strain in addition to elastic strain. At the yield point σ_s the stress–strain curve displays a marked increase in (permanent) strain while there is no increase in stress. Not all steels have a pronounced yield point, however. High-tensile steel has a stress-strain diagram which is characterised by a more or less gradual curvature beyond the initial straight portion that corresponds to elastic behaviour. For such steels a conventional yield point is defined, usually the so-called 0.2% proof stress, this being the stress at which the test specimen, on removal of the tensile load, is left with a permanent strain of 0.2%.

- Plastic range: If the specimen is loaded to stresses beyond the yield point, it undergoes substantial plastic deformation. On unloading it is found, despite its permanent increase in length, to have regained the strength and elasticity it originally had before loading. Frequently repeated loading and unloading of a steel specimen will raise its yield point without increasing the ultimate stress. Loading the specimen far beyond the yield point is accompanied by considerable increase in length with only a slight increase in stress, until finally the specimen fractures on, or shortly after, attaining the stress σ_B, which is known as the tensile strength.

- Strain at failure: The strain ε_B measured at fracture of the tensile test specimen is very large, often exceeding 20% of the original gauge length (see diagram). It is usually referred to simply as the 'elongation' and is an important quality of the steel. Steel with a high elongation value is a safer material in being less brittle than steel with a low elongation value. Elongation is a measure of the ductility of a steel: thus a ductile steel does not fracture suddenly, but gives warning of potential structural failure by its large deformations. In general, it is also an important property of steel that overloading or the occurrence of restraint stresses (for example, due to temperature variations). If the yield point is exceeded, it merely causes some permanent deformation but does not impair the strength of the steel.

- Allowable stresses: In general, the yield point, that is, the stress at which plastic strain commences, is taken as the limiting value of the stress, in relation to which there should be an adequate margin of safety. This applies more particularly to elastic design, based on the assumption of elastic behaviour of the steel. The desired margin is embodied in the factor of safety laid down in the design code or other regulations to which the steelwork designer has to conform.

In Germany a distinction is drawn between so-called main loads and additional loads. The factor of safety for the allowable stress due to the main loads, that is, the ratio of the yield point to this allowable stress, is 1.50. For the permissible stress due to the main loads acting together with the additional loads, the factor of safety is 1.33.

- Plastic design: Certain reserves of structural strength still exist after the yield point of steel has been passed. These are utilised in the plastic

353

Steel — General structural steels · Special steels

design method, which in many cases provides an alternative to the conventional elastic design method. The criterion of ultimate behaviour in this case is the collapse of the structure as a result of the yield point being exceeded at certain sections, which are then considered to behave as 'plastic hinges': for example, in a continuous beam these would be the sections at the mid-span points and over the supports. The factor of safety (the so-called 'load factor') against collapse varies with the type of load involved.

Mechanical properties

The properties which characterise the quality of the steel are determined by means of tests.
The notch impact value and transition temperature characterise the weldability of steel. The weldability, however, depends not only on the properties and the grade of the steel, but also on the welding process and the consumables employed, as well as on the design and detailing of the structural component concerned.

Nomenclature of structural steels (Table 1)

The principal structural steels envisaged in DIN 17 100 (Euronorm 25) are St 37 (Fe 37) and St 52 (Fe 52). The distinctive numbers represent the minimum tensile strength of the steels in kg/mm^2. In the German classification presented in this table the steel grade designation is followed by a numeral indicating the quality group (1–2–3) and is preceded by a symbol indicating whether it is rimming steel (U), killed steel (R), or specially killed steel (RR); furthermore a symbol indicating the as-delivered condition may be added: untreated (U), normalised (N).
These quality designations are of importance for judging the suitability of steels for welded construction and for use in dynamically loaded structural components. The responsibility for the choice of the grade and type of steel to use rests with structural designer or with the steelwork fabricator's welding engineer.

Allowable stresses

The allowable stresses in the design of steel structures in Germany are indicated in Table 1 and also in the diagram on p. 353.

Table 1

Steel grade DIN 17 100	Method of casting	As-delivered condition	Steel grade Euronorm 25	Chemical composition Ladle analysis				Guaranteed mechanical values			Allowable stresses in steel structures DIN 1050				
				C %	P %	S %	N %	σ_B kg/cm²	σ_S kg/cm²	σ_B %	H HZ	σ_D compression kg/cm²	σ_Z tension kg/cm²	shear kg/cm²	bearing kg/cm²
U St 37–1	U	U, N	Fe 37–A	0.20	0.070	0.050	—								
R St 37–1	R	U, N	Fe 37–A												
U St 37–2	U	U, N	Fe 37–B 3 U	0.18	0.050	0.050	0.007	3700 — 4500	2400	25	H HZ	1400 1600	1600 1800	900 1050	2800 3200
R St 37–2	R	U, N	Fe 37–B 3 NU	0.17											
St 37–3	RR	U N	Fe 37–C 3 Fe 37–D 3	0.17	0.045	0.045	0.009								
St 52–3	RR	U N	Fe 52–C 3 Fe 52–D 3	0.20	0.045	0.045	0.009	5200 — 6200	3600	22	H HZ	2100 2400	2400 2700	1350 1550	4200 4800

For exceptions and further particulars the standards in question should be consulted.

SPECIAL STEELS

Weathering, stainless and high-tensile steels

Weathering steels

Unless they are protected with paint or some other medium, conventional structural steels, which contain small quantities of manganese and silicon, corrode in the presence of moisture and form rust, which is attended by an increase in volume. When, however, small quantities of chromium, copper and nickel or vanadium are added to the elements already mentioned, products are obtained which, on exposure to the weather, become covered with a tight oxide skin, or patina, which largely protects the steel from further corrosive attack. Such steels are known as weathering steels because they weather to a variety of brown and purple shades.
Depending on the climate to which the weathering steels are exposed, the process of weathering takes from about one to three years. Alternate wetting and drying promotes the oxidation but to obtain the best appearance the steelwork should be blast-cleaned at some stage during fabrication. With a few notable exceptions, such as the Ford Foundation Building in New York, weathering steels are not normally used inside buildings. Although, with experience, greater liberties are being taken, weathering steels are not being used in the bare state under very aggressive atmospheric conditions, as in the immediate vicinity of the sea or in certain industrial atmospheres. Nevertheless, in Japan especially, where weathering steels have been painted in such corrosive conditions, it has been demonstrated that the steel requires repainting at far less frequent intervals.

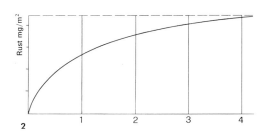

2

Diagram **2** shows a curve giving the average rate at which oxidation may be expected to take place in the early life of a structure. Nevertheless, in those countries where it rains frequently the initial stages of weathering may be prolonged as the rain tends to wash the rust from such exposed components as external columns. Unless special care is taken in the detailing of cladding and other adjacent components, they may be so stained that their appearance is affected. The alternative is to choose materials, such as bricks, coloured in various shades of brown, blue, purple or grey on which it is difficult to detect stains. In addition, pavings around buildings can be protected from staining by the interposition of a strip of gravel between the exterior of the building and the paved surface. Similarly, concrete surfaces may be given a silicone coat from which it is relatively easy to remove rust stains.
Weathering steels are used for loadbearing structural components as well as for non-loadbearing cladding. They can be fabricated like conventional structural steels of the same grade. Both electrodes and high-strength friction-grip bolts are available which are compatible with and will weather to the same colour as the weathering steel components.
In Germany, weathering steels bear the designation WT prefixed to the grade of steel: WT St 37–2, WT St 37–3 and WT St 52–3. British weathering steels comply with BS 4360 and are prefixed with the designation WR. As in most other steel-producing countries, they are available in Grade 50 only. The corresponding American standard is ASTM A.588.
Because additional alloying elements are included in weathering steels, they are more expensive than the conventional structural steels

Structural steel products

having the same mechanical properties. The extra cost varies with the type of rolled steel product involved, but generally speaking it is roughly the same as the paint system which would be required if conventional steels were used. As, in suitable environments, weathering steels require little or no maintenance, they are economical from the start.

References:
- Lorenz, Wetterfester Stahl im Stahlhochbau (Weathering steel in steel structures), *Deutsche Bauzeitung*, 6/1972.
- *Richtlinien für die Lieferung, Verarbeitung und Anwendung wetterfester Baustähle* (Regulations for the supply, fabrication and application of weathering steels), Stahlbau-Verlags GmbH, Cologne.
- Merkblatt 434 *Wetterfester Baustahl* (Brochure 434 *Weathering steel*), Beratungsstelle für Stahlverwendung (Steel Information Centre), Düsseldorf.

Stainless steels

The stainless steels used in structural engineering are alloy steels containing at least 12% chromium and usually 1% silicon and 1% manganese. Very high-quality stainless steels additionally contain other alloying constituents, particularly molybdenum and nickel. Stainless steels are resistant to chemically aggressive agents. They must comply with German Standard DIN 17 440.

Because of their relatively high cost, stainless steels are used only in small quantities in building construction, for example, as fixing devices for wall claddings in places which are not subsequently accessible for inspection and maintenance, and as light-gauge panels and cold-formed sections for claddings, windows, doors, handrails, etc.

High-tensile structural steels (fine-grained steels)

For special purposes, low-alloy high-tensile steels with a guaranteed yield point of up to 70 kg/mm^2 and characterised by a fine grain-structure are available. Because of their low carbon content they possess good weldability. There are many proprietary brands of such steels, in various grades, differing in the alloying elements they contain.

Thus, in Germany the grades StE 47 and StE 70 are used in steel buildings. The letter E denotes that the figure following it refers to the yield point (kg/mm^2).

The high yield point of such steel is fully utilised only if the maximum stress in the structural component is a tensile stress or — but only in members of low slenderness ratio — a compressive stress. In all cases where the stability of the component is the criterion, the allowable stress is governed, not by the yield point, but by the modulus of elasticity, which is no higher in high-tensile steels than in general structural steels. It would therefore be uneconomical to use an expensive high-tensile steel for struts and other compression members in a steel structure.

Rational solutions can be obtained by the judicious use of different types of steel for members in different parts and/or performing different functions in a steel-framed building. For example, in a high-rise building the columns of the bottom storeys are often made of high-tensile steel, while the upper storeys have columns made of general structural steels.

STRUCTURAL STEEL PRODUCTS

Hot-rolled sections

The series of I-sections of 80 mm depth and upwards produced in Germany are listed in Table 2. The IPE and IPB series conform to the Euronorm Standard.

IPE sections

The I-series is being steadily superseded by the IPE series, in which the beams, while possessing equal load capacity, weigh less and are moreover easier to connect structurally because of the parallel flanges. This series is especially suitable for use as beams and in composite construction. The derived IPEo and IPEv series are not standardised at national level, but conform to the manufacturers' own standards.

Channel sections

The present sections with tapered flanges are due to be replaced by parallel-flanged channels in the near future.

IPB sections

In multi-storey structures, the three IPB series are employed mainly as columns and beams.

Design charts

for IPB sections as columns: p. 229;
for IPB and IPE sections as beams: pp. 245–247;
for IPE sections as composite beams: p. 259;
for IPB and IPE sections as castellated beams: p. 251.

Rolled steel sections — Table 2

With tapered flanges		With parallel flanges					
I		IPE	IPEo *	IPEv *	IPBl HEA	IPB HEB	IPBv HEM
80	80						
100	100	100	100	100	100	100	100
120	120	120	120	120	120	120	120
140	140	140	140	140	140	140	140
160	160	160	160	160	160	160	160
180	180	180	180	180	180	180	180
200	200	200	200	200	200	200	200
220	220	220	220	220	220	220	220
240	240	240	240	240	240	240	240
260	260				260	260	260
		270	270	270			
280	280				280	280	280
300	300	300	300	300	300	300	300
320	320				320	320	320
		330	330	330			
340					340	340	340
	350						
360		360	360	360	360	360	360
380	380						
400	**400**	400	400	400	400	400	400
		450	450	450	450	450	450
		500	500	500	500	500	500
		550	550	550	550	550	550
600		600	600	600	600	600	600
					650	650	650
					700	700	700
					800	800	800
					900	900	900
					1000	1000	1000

* Non-standard sections

Smaller rolled sections

The following are more particularly of importance in building construction:

- equal angle sections with legs 20 to 200 mm in length;
- unequal angle sections with legs in lengths from 30/20 to 200/100 mm;
- standard and broad-flanged T-sections, I-sections and channels less than 80 mm in depth;
- round, square and flat bars.

Universal plate

This product, which ranges in width from 150 to 1250 mm and in thickness from 5 to 65 mm, is rolled in a universal plate mill, which has vertical as well as horizontal rolls.

Plate and sheet

These are produced in rolling mills which have horizontal rolls only. In Germany the following classification is adopted:

- sheet: below 3 mm thickness;
- light plate: from 3 to 5 mm thickness;
- heavy plate: more than 5 mm thickness.

In Britain a conventional classification is to describe material not over $\frac{1}{8}$ inch thick as 'sheet' and to apply the term 'plate' to material more than $\frac{1}{8}$ inch thick.

Plate girders, box girders and box-section columns are fabricated from plate by welding (see design chart, p. 230).

Strip

Strip is a thin rolled product which is coiled up after the last rolling pass.

Steel — Structural steel products

Hollow sections

Hollow sections, more particularly circular, are produced seamless by a hot rolling process or are formed with a straight or spiral welded seam. Circular hollow sections are rolled to order and can be supplied in any desired size up to 622 mm dia. For welded sections there is virtually no upper limit to the diameter. In multi-storey buildings, hollow sections are now used as columns (see design chart, p. 231). Square or rectangular hollow sections, with a range of wall thicknesses, are very suitable for exposed structural components of all kinds. Table 3 gives some typical sizes.

References:

Merkblatt 387 *Rechteck-Hohlprofile* (Brochure 387 *Rectangular hollow sections*) and Merkblatt 224 *Runde Hohlprofile für den Stahlbau* (Brochure 224 *Circular hollow sections for structural steelwork*), Beratungsstelle für Stahlverwendung (Steel Information Centre), Düsseldorf.

Special sections

These comprise the special sections for steel window frames, produced by hot rolling or by extrusion. Extruded sections are used more particularly in those cases where the required cross-sectional shape could otherwise be produced only by machining processes involving the removal of substantial amounts of metal or by welding or bolting together two or more component sections. Hollow sections can also be produced by extrusion.
Improved surface texture and/or closer tolerances are obtainable by cold drawing of hot-rolled or extruded sections.

Reference:

Merkblatt 300 *Das Profilstahlrohr im Fenster- und Türenbau* (Brochure 300 *The tubular steel section in window and door construction*), Beratungsstelle für Stahlverwendung (Steel Information Centre), Düsseldorf.

Cold-formed sections

Light-gauge sections can be produced from sheet or strip by cold-forming. Depending on the forming method employed, a distinction is to be drawn between cold-rolled sections, which are generally produced from strip on rolling machines, and folded sections, produced from sheet or strip on a folding press.

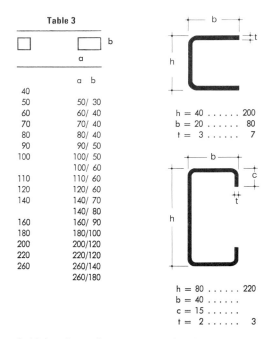

Table 3

a	b
40	
50	50/ 30
60	60/ 40
70	70/ 40
80	80/ 40
90	90/ 50
100	100/ 50
	100/ 60
110	110/ 60
120	120/ 60
140	140/ 70
	140/ 80
160	160/ 90
180	180/100
200	200/120
220	220/120
260	260/140
	260/180

$h = 40 \ldots\ldots 200$
$b = 20 \ldots\ldots 80$
$t = 3 \ldots\ldots 7$

$h = 80 \ldots\ldots 220$
$b = 40 \ldots\ldots$
$c = 15 \ldots\ldots$
$t = 2 \ldots\ldots 3$

Cold-forming allows greater freedom in the shaping of sections than is possible with hot-rolling, so that a high degree of adaptation of a section to its structural function can be obtained, thus providing optimum utilisation and low weight. The increased strength due to work-hardening associated with cold-forming is sometimes deliberately used as a means of increasing the loadbearing capacity of the members produced.
All steels possessing adequate cold-forming capacity can be used, including stainless steels and weathering steels. The properties of stainless steels (attractive appearance, durability, ease of maintenance) are utilised for light-gauge components performing no appreciable loadbearing function, such as cladding panels, door and window trimmings, etc.
Cold-formed structural steel sections are seldom employed in multi-storey buildings, as the commonly available sections are not strong enough for use as loadbearing components in such structures. On the other hand, cold-formed plain or lipped channel sections (as illustrated), which are produced in a range of depths up to 200 mm and 220 mm respectively, are quite suitable for lightweight roof framing. An example of a cold-formed lattice girder is the so-called X-girder, which has proved very suitable for roofs: see p. 253.

Reference:

Kaltprofile (*Cold-formed sections*), Beratungsstelle für Stahlverwendung (Steel Information Centre), Düsseldorf.

Trough sections

Many manufacturers produce trough-section units in a variety of non-standardised cross-sectional shapes. The lighter grades of such sections are used as wall and roof cladding, while the heavier ones are used as decking for the construction of floors in multi-storey buildings. The units are supplied in lengths up to 18 m, in widths of approximately 1 m, and in thicknesses ranging from 0.56 to 2.00 mm. They are galvanised and may additionally be plastic-coated.

- For structural details of
wall cladding, see pp. 312, 316;
roof cladding, see pp. 300–303;
floor decking, see pp. 274–276.
- Design charts: p. 203 (roof cladding), p. 275 (floor decking).

METHODS OF CONNECTION

The methods of making connections between structural steel components may be subdivided into those which can subsequently be released without destroying the fastener and those which cannot be so released. Also a distinction may be drawn between welded connections and those formed by dowel-type elements, more particularly rivets and bolts.

Rivets

A riveted connection cannot be released without destroying the rivets. Hot riveting as a means of forming structural connections in steelwork is now largely obsolete. Blind rivets, which can be inserted from one side, are sometimes used for connecting steel sheet (p. 274).

Bolts

Bolted connections can subsequently be released without destruction of the bolts. As black bolts and turned and fitted bolts act in bearing and shear, their design must be based on the cross-sectional area of the core of the screw thread.

Black bolts

In Europe, these bolts, which are not manufactured to a high degree of precision, are usually inserted into holes 1 mm greater in diameter than the nominal diameter of the bolt. In a connection comprising a number of bolts, not all of them will be bearing against the sides of their holes at the start of loading. The connection develops its full structural action only after a certain amount of slip has occurred. For this reason the permissible load capacity per bolt is lower than for turned and fitted bolts.

Close-tolerance turned and fitted bolts

Turned bolts are machined to truly circular shape of exact size. The shank fits tightly in the hole (in DIN 7968 tolerance h11 is stipulated for the shank and H11 for the hole), and for this reason these bolts are allowed to carry more load than black bolts of the same nominal diameter. Turned bolts are supplied as bright bolts or galvanised or cadmium-plated.

Methods of connection

Corrosion protection

High-strength friction-grip bolts

Friction-grip bolts are torqued to such a shank tension that they hold the connected components purely by friction. As the bolt is inserted into a hole with a clearance of about 2 mm, it is assumed that the bolt does not act in shear or bearing. The contact surfaces must be free of loose mill-scale, thick rust and paint and should preferably be blast-cleaned. The bolts may be tightened with power tools or with manual torque wrenches which cut out when the specified torque has been applied.

Welding

The term is applicable to any process whereby the parts to be connected are joined by fusion, with or without the use of consumable material. The following methods are used in steelwork:

Electric arc welding

This is the commonest method, the pool of weld metal being formed by an electric arc, while additional material is supplied by the melting of a filler rod, such as an electrode. The molten pool of metal must be protected from contact with the air. Depending on how this protection is provided and on the mode of feeding the filler rod or wire, various techniques are available:

- Manual welding with coated electrodes. The coating melts and forms a protective slag over the pool.
- Manual welding with automatic feed of the bare filler wire and protection of the pool by an inert shielding gas, usually CO_2 (shielded arc welding).
- Machine welding with automatically fed bare filler wire, protection being provided by the melting of flux powder on the surface of the molten metal (submerged-arc welding).

The last-mentioned method can be used only for horizontal welds. The two manual methods can additionally be used for vertical and overhead welds. In the fabrication of steelwork, components may, if necessary, be manipulated into the most suitable position for applying the welding technique adopted for the job.

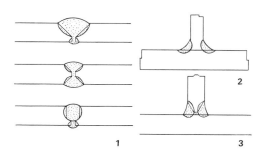

Three typical butt welds are illustrated in **1**; fillet welds are shown in **2** and **3**.
Loadbearing welds should be executed only by skilled welders under the supervision of expert welding technicians or engineers. Welds for dynamically loaded components are subject to special conditions. Important welds should be inspected by X-ray or ultrasonic methods.

Flash butt welding

This technique is used for joining components of small cross-section, the electric arc being formed between the components themselves. No filler metal is added. The method is suitable for the splicing of reinforcement and the attachment of studs as:

- dowels (for example, for fastening external wall panels);
- anchorage in concrete;
- shear connectors for composite construction.

Flash butt welding may be executed by hand or automatically on an electronically controlled machine.
Stud shear connectors are usually welded to the steel beams in the workshops, rather than on the construction site. It should be noted that this method of welding involves high amperage. Under suitable conditions, the studs can be welded to the beams through profiled steel decking on the site.

Spot welding

This method is often used for the welding of thin sheet, for example, in the fabrication of cladding panels. The welds are produced by local fusion of a small area of the metal.

CORROSION PROTECTION

Steels, except the stainless and the weathering types, usually have to be given some form of corrosion protection. The methods of initially applying and subsequently maintaining the protective systems commonly employed in multi-storey steel-framed buildings are outlined here. For fuller information, the relevant literature must be consulted.

Chemical process

Corrosion depends particularly on the relative humidity and on the amount of aggressive substances in the atmosphere. As may be seen in the accompanying diagram, appreciable corrosion occurs only when the humidity exceeds about 70%. It is caused by the reaction of atmospheric oxygen with iron at the surface of the steelwork. Under conditions of low atmospheric humidity, say, in a dry subtropical climate or in the interior of a building in a temperate climate, this reaction can take place only to a very limited extent.
In humid surroundings the occurrence of corrosive attack by oxygen can be prevented by passive or by active protection methods.

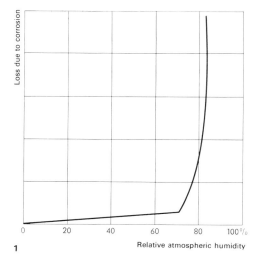

1

Passive protection is provided by a coating which serves to prevent oxygen from penetrating to the surface of the steel. Even quite small cracks or pores in the coating may, however, provide an entry for oxygen and thus initiate rust formation, which can then proceed rapidly, as rusting is associated with an increase in volume which disrupts the protective coating. Used merely by itself, passive protection is therefore not an adequate method.

Active protection is provided by a coating containing metals which can combine with the oxygen and thus render it harmless to the steel. This function is performed, for instance, by the lead in red lead or zinc in zinc chromate (passivating coatings) or by zinc or cadmium in metal coatings. However, such coatings alone are not an adequate protection either, as the oxidation process to which they are subjected will eventually consume the protective metals.

A protective system that will continue to be effective for a long time must therefore consist of an active coat covered by a protective coat. The former may consist of one or more coats of priming paint or of a metal primer. The covering may consist of a finishing coat of paint.

Corrosion protection — Surface preparation • Priming coats • Finishing coats

Under dry atmospheric conditions, other measures that prevent access of oxygen to the steel may also be effective as corrosion protection, for example, cladding.

The embedding of the steel in concrete can also be regarded as a form of passive protection. However, chemical reactions are involved here and it is generally considered that concrete provides more than merely passive resistance to corrosive attack.

Correct detailing of steelwork can do much to reduce the corrosion hazard: see p. 359.

Surface preparation

The priming coat of paint or metal can adhere permanently to the steel and protect it efficiently only if the steel surface has been thoroughly cleaned to remove rust, mill-scale and other contaminants, such as oil. Not only the loose but also the tightly adhering mill-scale should be removed, as corrosion products involving an increase in volume may form under it, so that the mill-scale and the protective coatings are dislodged from the steel.

In German practice, there are three grades of surface preparation:

grade 1: free from all loosely adhering matter;
grade 2: metallically clean surface with traces of rust and old paint in the pores of the metal;
grade 3: metallically bright surface.

As a rule, blast-cleaning is applied, that is, the steel surface to be prepared for its protective coating is subjected to a jet of abrasive particles delivered by compressed air or by a mechanical impeller. In modern fabricating shops the surface preparation is usually applied before fabrication. It is important to use the correct abrasive and blasting conditions to ensure that the roughness of the cleaned steel surface is suited to the thickness of the coatings to be applied, so that the peaks will be effectively covered. The depth of roughness should preferably not exceed $70\,\mu$.

Cleaning the steel by manual wire-brushing, scraping, hammering, etc., is inadequate for high-class corrosion protection and is no longer used except for minor repair jobs.

Priming coats. Metal coatings

The prerequisite for an efficient priming coat is the thorough surface preparation of the steel and its early protection after such preparation. The most extensively used primers are:

with passivating pigments: red lead and zinc chromate;
with metallic pigments: zinc dust.

They are almost invariably applied by airless spraying. Because of the lead poisoning hazard, the spraying of red lead calls for special precautions, and in subsequent welding operations there is moreover a health hazard to the welders. For these reasons, zinc chromate is increasingly being used as a primer in lieu of red lead. Zinc dust as a priming pigment is more expensive, but is less susceptible to mechanical damage and has better weather resistance. It is therefore suitable for steelwork involving a long erection period before the finishing coats are applied.

Priming coats

For steelwork exposed to the open air the priming coat system should have a thickness of $80\,\mu$ and be applied in one or two coats. Adequate thickness of coating more particularly at arrises and corners should be ensured. In general, a thickness of $40\,\mu$ will suffice for steelwork which is in the interior of buildings, in rooms with normal humidity conditions.

When the steel is blast-cleaned prior to fabrication, it is usually provided at once with a thin preliminary coat of paint, 10 to $15\,\mu$ thick, which allows welding and which is reckoned — in so far as it is intact — to form part of the required priming. For the safety reasons already mentioned, zinc chromate is the only suitable type of paint for this purpose.

After erection of the steelwork, any damage to the priming coats suffered during transport or handling should be carefully made good. Furthermore, the bolts should be painted, as they are usually supplied without any corrosion protection of their own. For use in aggressive conditions, however, metal-plated bolts are obtainable.

Metal coatings

The most commonly employed protective coating metal is zinc, applied by hot-dip galvanising or by spraying.

• Hot-dip galvanising can be applied even to long components, as galvanising baths up to 20 m in length are available. It is a very effective method of protection, especially for external steel features which are relatively inaccessible or awkward from the viewpoint of maintenance, such as walkways, railings, etc. Galvanising may give adequate protection without necessitating any further covering treatment, though cladding units may additionally be plastic-coated. The thickness of the layer of zinc deposited by hot-dip galvanising is usually between 80 and $100\,\mu$. Components in which rolling or welding stresses are locked may undergo distortion as a result of heating in the galvanising bath and will then have to be straightened afterwards.

Trough-section sheet-steel units are normally produced from strip which has been coated on a continuous galvanising line. Cold-formed sections may be produced from galvanised sheet or strip.

• Sprayed zinc coatings may be applied in the works or on site to completed steel structures of any shape and size. The thickness of the layer of zinc is generally between 80 and $150\,\mu$. The steel surface must be adequately roughened before spraying and this treatment should be immediately followed by the application of pore-filling finishing coats of paint. Blast-cleaning on erected steelwork is practicable only under appropriate local conditions.

Finishing coats, cladding, encasement

Finishing coats

To prevent aggressive substances from gaining access to the priming coats and thus causing their early deterioration, certain requirements must be fulfilled by the finishing coats. Thus, their binding medium should be chemically compatible with that of the priming coats. As a rule, one or two finishing coats, each 30 to $50\,\mu$ thick, are applied. These coats contain the colouring pigments. Painting should be done only in dry weather and at temperatures between $+5°$ and $+50°C$.

Structural components which will, after erection, no longer be accessible for painting should receive their finishing coats in the fabricating shops: for example, the top flanges of roof beams to which troughed steel decking units will be fixed, or the outer flanges of columns which will have cladding panels attached to them.

The total thickness of the paintwork (priming plus finishing coats) should be as follows:

• for unclad steelwork exposed in interior spaces with ordinary humidity conditions: $130\,\mu$
• in industrial atmospheres or in major cities: $160\,\mu$
• in marine atmospheres: $200\,\mu$
• in particularly aggressive atmospheres: $220\,\mu$

Cladding

Less stringent requirements for corrosion protection are applicable to steel components which are in the interior of buildings and are protected by cladding from contact with moist air:

• For components between the two leaves of a partition of in preformed hollow fire encasement not directly adhering to the steel, it will generally suffice to apply a priming coat of $40\,\mu$ thickness, without a finishing coat.
• The same applies to steelwork in enclosed floor cavities of relatively low depth which contain few services and which are completed and closed in at an early stage. If these conditions are not satisfied, it is necessary to give the steel a finishing coat of $40\,\mu$ in addition to the priming coat of $40\,\mu$ thickness.
• Trough-section units for floors and for external claddings are normally supplied galvanised, which is sufficient corrosion protection for normal conditions. However, overhanging eaves should in addition be given a finishing coat of $40\,\mu$ thickness.

Encasement

Fire protection encasement which follows the profile of the steelwork and is in direct contact with it usually takes the form of concrete or sprayed asbestos.

• To ensure good adhesion, sprayed protection should be applied to a blast-cleaned steel surface. A light film that may form during the interval between fabrication and spraying is regarded as harmless. The casing, in addition to providing fire protection, also protects the steel from corrosion.
• A 4 cm thick cover of normal-weight concrete, with an embedded wire mesh cage, ensures reliable corrosion protection of the steel. In circumstances where steel components are not completely embedded in concrete, it is necessary to pay due attention to the transition zones — especially if the components are exposed to the external atmosphere — for it is here that corrosion is most likely to start. The same applies to the boundary zones between external columns and paved surfaces outside a building (see p. 233, **6**).

Interior spaces with high humidity levels

Steelwork in rooms or other spaces with high levels of humidity inside buildings should be protected in the same way as components in contact with the external atmosphere. With proper precautions there is no objection to using steel in such surroundings. In industry, more particularly in steelmaking and the chemical industry, steel structures are extensively used even in very aggressive atmospheres.

Steelwork in swimming baths containing water with a high chlorine content can be given the same protective treatment as steel structures in marine atmospheres. A higher standard of corrosion protection is also generally required

Detailing the steelwork

In the open air, those parts of steelwork are especially liable to corrode which become covered with dirt or dust, as these deposits absorb and hold moisture, thus forming points of attack. This hazard can be considerably reduced by careful detailing, more particularly with a view to avoiding pockets or other features where moisture is likely to collect. Simple shapes with plain smooth surfaces are preferable to complex ones.

All steelwork in contact with the open air or in humid internal spaces should be so designed as to be properly accessible for the initial application of paintwork and for subsequent maintenance.

Welded hollow sections which are completely sealed require no internal corrosion protection.

Maintenance and cost of corrosion protection paintwork

Maintenance

Modern protective coatings have a long life. Experience gained by the German Federal Railways, whose corrosion protection regulations are based on comprehensive tests, demonstrates that the service life of a good protective system under average climatic conditions is between 10 and 15 years. In a favourable climate with dry and clean air it will last longer than this, whereas its life will be shorter in aggressive atmospheres, such as in industrial regions or in the vicinity of the sea.

At the end of this period, however, it normally does not become necessary to renew the entire protective system, only the finishing coats. This need to repaint structural steelwork at certain intervals of time, though of course costing money, is not without a positive aspect in that a new coat of paint improves the appearance of the structure.

It is advisable to inspect the paint on steelwork at regular intervals — once a year, for example — and to put right any minor damage at once. The results of the inspection should be recorded. With conscientious and expert maintenance the annual cost can be very substantially reduced — by something like 50% — in relation to the cost of completely stripping and repainting the steel after complete deterioration of the old protective paint system. In the interior of buildings it is, generally speaking, not necessary to renew corrosion protection paintwork, so that no maintenance costs arise from this source. In rooms with high levels of humidity or with aggressive fumes, however, structural steelwork should be given the same attention as if it were exposed to the external atmosphere.

Cost of corrosion protection

For multi-storey steel-framed buildings the cost of protective measures against corrosion, as an approximate percentage of the total cost of the structural steelwork, is as follows:

surface preparation, grade 2		1.8%
1 priming coat	40μ	2.4%
2 priming coats	80μ	4.6%
1 finishing coat	40μ	2.5%
2 finishing coats	80μ	4.9%

The cost of the steelwork is about 10 to 15% of the overall construction cost. Bearing in mind that most of the steel components are in the interior and require only a priming coat and that many components may not have to be painted at all because they are encased for fire protection, the cost of corrosion preventive treatment of an average multi-storey steel-framed building is somewhere between about 0.1 and 1.2% of the total cost of the building.

References

- Bierner: *Handbuch für den Rostschutzanstrich* (*Handbook of corrosion protection paintwork*), Curt R. Vincentz-Verlag, Hanover.
- *Schutzanstrich von Stahlkonstruktionen in der Industrie; Richtlinien und technische Vorschriften des VDEh, der AGI und des DNA* (*Protective painting of steel structures in industry; regulations and technical code of practice of the VDEh, AGI and DNA*), Verlag Stahleisen mbH, Düsseldorf; Curt R. Vincentz Verlag, Hanover, 1971.
- German Standard DIN 55 928: *Schutzanstrich von Stahlbauwerken Richtlinien* (*Protective painting of steel structures, regulations*), edition 6.59.
- *Technische Vorschriften für den Rostschutz von Stahlbauwerken* (*RoSt*) (*Technical code of practice for the protection of steel structures from corrosion*), German Federal Railways, DV 807, 1963.
- *Vorläufige Richtlinien für die Auswahl von Fertigungsanstrichen FA bei der Walzstahlkonservierung im Stahlbau* (*Provisional regulations for the selection of paint coatings for the preservation of rolled steel in structural steelwork*), DASt, edition 6.68.
- AGI-Arbeitsblatt K 20, Blatt 1: *Schutz von Stahlkonstruktionen; Oberflächenbehandlung, Anstriche, Metallbezüge* (AGI working memorandum K 20, sheet 1: *Protection of steel structures; surface preparation, painting, metal coatings*).
- Van Oeteren: *Konstruktion und Korrosionsschutz* (*Design and corrosion protection*), Curt R. Vincentz Verlag, Hanover, 1967.
- Offprints 593, 534, 588, 587 of the Fachabteilung Feuerverzinken im Fachverband Stahlblechverarbeitung (Hot-dip Galvanising Division of the Steel Plate and Sheet Fabricating Trade Association), Hagen.
- Merkblätter (Brochures) 101, 179, 269, 325, 329, 359, 400, issued by the Beratungsstelle für Stahlverwendung (Steel Information Centre), Düsseldorf.

Photographs taken by the photographers mentioned below were provided by the contributors to Part Two, 'Examples of Multi-Storey Steel-Framed Buildings' (page numbers given: a =above; b =below; m =middle; l =left; r =right):
Allegri, U., Brescia (73 a) · Baer, M., Berkeley (156, 157) · Bauters, S. P. R. L., Brussels (68 al) · Belva, A. (164 al) · Bérenger, P., Bagnolet (77 br) · Bezzola, L., Bätterkinden (89 a) · Biaugeaud, J., Arcueil (134, 135) · Bryan & Shear Ltd, Glasgow (71) · Cartoni, A., Rome (82 b) · Dermond, B. A., Zürich (61 a) · Donat, J., London (62 a) · Dufresne, G. (121) · Ehrmann, G., Paris (76) · Falke, J. (154) · Fotocinecolor, Rome (142) · Foto-Orgel-Köhne, Berlin (124, 125) · Freeman, J. R., London (147 m) · Guillat, G., Paris (144, 145) · Gunder, W., Binningen (74) · Hanisch, M., Essen (105 a) · Hassenberg, O., Hanover (111) · Hedrich-Blessing, Chicago (92, 93) · Heidersberger, H., Wolfsburg (94, 140) · Industrial Foto, Rome (109) · Kindermann, Berlin (128) · Kramer, C., Rotterdam (148 ar) · Kutter, E., jr, Luxembourg (126, 127) · Lavallette, G. M. (65 br, 103, 153 a) · Lepri, P., Piombino (63 ar, m) · Mako, M., Paris (123 br +l) · Martin, R., Brussels (158 l) · Maurer, J., Bruges (90 b, 91 ar) · Meier-Menzel, H.-J., Murnau (96 a, 97 b, 114, 115 a) · Middendorf, E. T., Berlin (118) · Moosbrugger, B., Zürich (91 al, bl, br) · Moreau, Jean-J. (112) · Nicolini, T. (138, 139b) · Poteau, J., Lille (100, 101) · Priivits, H., Stockholm (116) · Publicam, Hilversum (104, 148 al, 149 ar) · Publifoto, Milan (143 m) · Werkfoto Rüterbau (119 a, 137) · Savio, O. (82 a) · Scherb, K., Vienna (86) · SEPT, Saint-Cloud (123 a) · Sergquels & Dietens, Brussels (155 b, 158 ar) · SFB-Bauplanung (129 a, bl, br) · Siebold, M., Avusy (59) · Snoek, H., London (83, 146) · Stab-Foto, Stockholm (117) · Stoller, E., New York (99, 160, 161) · Studios EJ, West Wickham, Kent (147 a, b) · Sundahl, S., Saltsjö-Duvnäs (60) · Svensson, K. G., Falun (72 b) · Uetz, J. P., Seeberg (115 b) · Wagner, W., Vienna (87 a, m, r) · Wandelle, P., Paris (95, 122, 152) · Wolgensinger, M., Zürich (132, 133 a).